Daily Values for Food Labels

The Daily Values are standard values, developed by the Food and Drug Administration (FDA) for use on food labels. Daily Values for protein, vitamins, and minerals reflect average allowances based on the RDA. Daily Values for nutrients and food components, such as fat and fiber, that do not have an established RDA but do have important relationships with health are based on recommended calculation factors as noted.

Nutrient	Amount
Protein [a]	50 g
Thiamin	1.5 mg
Riboflavin	1.7 mg
Niacin	20 mg NE
Biotin	300 μg
Pantothenic acid	10 mg
Vitamin B_6	2 mg
Folate	400 μg
Vitamin B_{12}	6 μg
Vitamin C	60 mg
Vitamin A [b]	5000 IU
Vitamin D [b]	400 IU
Vitamin E [b]	30 IU
Vitamin K	80 μg
Calcium	1000 mg
Iron	18 mg
Zinc	15 mg
Iodine	150 μg
Copper	2 mg
Chromium	120 μg
Selenium	70 μg
Molybdenum	75 μg
Manganese	2 mg
Chloride	3400 mg
Magnesium	400 mg
Phosphorus	1000 mg

[a] The Daily Values for protein vary for different groups of people: pregnant women, 60 g; nursing mothers, 65 g; infants under 1 year, 14 g; children 1 to 4 years, 16 g.
[b] The Daily Values for fat-soluble vitamins are expressed in International units (IU), an old system of measurement. The current RDA and tables of food composition use a more accurate system of measurement. Equivalent values are as follows: for vitamin A, 875 μg RE; for vitamin D, 10 μg; for vitamin E, 9 mg α-TE.

Food Component	Amount	Calculation
Fat	65 g	30% of kcalories
Saturated fat	20 g	10% of kcalories
Cholesterol	300 mg	Same regardless of kcalories
Carbohydrate (total)	300 g	60% of kcalories
Fiber	25 g	11.5 g per 1000 kcalories
Protein	50 g	10% of kcalories
Sodium	2400 mg	Same regardless of kcalories
Potassium	3500 mg	Same regardless of kcalories

Note: Daily Values were established for adults and children over 4 years old. The values for energy-yielding nutrients are based on 2000 kcalories a day.

Glossary of Nutrient Measures

kcal: kcalories; a unit by which energy is measured.

g: grams; a unit of weight equivalent to about 0.03 ounces.

mg: milligrams; one-thousandth of a gram.

μg: micrograms; one-millionth of a gram.

mg NE: milligrams niacin equivalents; a measure of niacin activity.

mg α-TE: milligrams alpha-tocopherol equivalents; a measure of vitamin E activity.

μg RE: micrograms retinol equivalents; a measure of vitamin A activity.

IU: international units; an old measure of vitamin activity determined by biological methods (as opposed to new measures that are determined by direct chemical analysis).

t h o m s o n . c o m

changing the way the world learns[SM]

To get extra value from this book for no additional cost, go to:

http://www.thomson.com/wadsworth.html

thomson.com is the World Wide Web site for Wadsworth/ITP and is your direct source to dozens of on-line resources. *thomson.com* helps you find out about supplements, experiment with demonstration software, search for a job, and send e-mail to many of our authors. You can even preview new publications and exciting new technologies.

thomson.com *It's where you'll find us in the future.*

ABOUT THE AUTHORS

Sharon Rady Rolfes, M.S., R.D., received her B.S. in psychology and criminology in 1974 and her M.S. in nutrition and food science in 1982 at the Florida State University. She is a founding member of Nutrition and Health Associates, an information resource center that maintains an ongoing bibliographic data base tracking research in over a thousand nutrition-related topics. Her other publications include the textbooks *Understanding Nutrition, Understanding Normal and Clinical Nutrition*, and *Nutrition for Health and Health Care* and a multimedia CD-ROM called *Nutrition Interactive*. In addition to writing, she also lectures at universities and at professional conferences and serves as a consultant for various educational projects. She maintains a professional membership in the American Dietetic Association.

Linda Kelly DeBruyne, M.S., R.D., received her B.S. in 1980 and her M.S. in 1982 in nutrition and food science at the Florida State University. She is also a founding member of Nutrition and Health Associates, where her specialty areas are life cycle nutrition and fitness. Her other publications include *Nutrition and Diet Therapy, Nutrition for Health and Health Care, Making Life Choices, The Fitness Triad: Motivation, Training, and Nutrition*, and the *Nutrition Interactive* CD-ROM. As a consultant for a group of Tallahassee pediatricians, she teaches infant and child nutrition classes to parents. She is a member of the American Dietetic Association.

Eleanor Noss Whitney, Ph.D., received her B.A. in biology from Radcliffe College in 1960 and her Ph.D. in biology from Washington University, St. Louis, in 1970. Formerly on the faculty at the Florida State University, and a dietitian registered with the American Dietetic Association, she now devotes full time to research, writing, and consulting in nutrition, health, and environmental issues. Her earlier publications include articles in *Science, Genetics*, and other journals. Her textbooks include *Nutrition Concepts and Controversies, Understanding Nutrition, Understanding Normal and Clinical Nutrition, Nutrition and Diet Therapy*, and *Essential Life Choices* for college students and *Making Life Choices* for high-school students. Among her present interests are energy conservation, solar energy uses, alternatively fueled vehicles, and ecosystem restoration.

SECOND EDITION

Life Span Nutrition

CONCEPTION THROUGH LIFE

Sharon Rady Rolfes

Linda Kelly DeBruyne

Eleanor Noss Whitney

West / Wadsworth

I(T)P® An International Thomson Publishing Company

Belmont, CA ■ Albany, NY ■ Bonn ■ Boston ■ Cincinnati ■ Detroit
Johannesburg ■ London ■ Madrid ■ Melbourne ■ Mexico City
New York ■ Paris ■ Singapore ■ Tokyo ■ Toronto ■ Washington

Publisher: Peter Marshall
Associate Development Editor: Laura Perkinson
Marketing Manager: Becky Tollerson
Project Editor: Sandra Craig
Print Buyer: Barbara Britton
Permissions Editor: Veronica Oliva
Production: Greg Hubit Bookworks
Text and Cover Design: Devenish Design

Cover Photographs: All photos from PhotoEdit: (top) David
Young-Wolff; (row 2, left) Myrleen Ferguson Cate, (right)
Tom McCarthy; (row 3, left and center) David Young-Wolff,
(right) Tom McCarthy; (bottom) Tom McCarthy
Photo Research: PhotoEdit
Illustrations: Lotus Art
Composition and Prepress: G&S Typesetters, Inc.
Cover Printer: Phoenix Color
Printer and Binder: Quebecor/Fairfield

*This book is printed on
acid-free recycled paper.*

Printed in the United States of America
 3   4   5   6   7   8   9   10

For more information, contact Wadsworth Publishing Company, 10 Davis Drive, Belmont, CA 94002,
or electronically at http://www.thomson.com/wadsworth.html

International Thomson Publishing Europe
Berkshire House 168–173
High Holborn
London, WC1V 7AA, England

International Thomson Editores
Campos Eliseos 385, Piso 7
Col. Polanco
11560 México D.F. México

Thomas Nelson Australia
102 Dodds Street
South Melbourne 3205
Victoria, Australia

International Thomson Publishing Asia
221 Henderson Road
#05-10 Henderson Building
Singapore 0315

Nelson Canada
1120 Birchmount Road
Scarborough, Ontario
Canada M1K 5G4

International Thomson Publishing Japan
Hirakawacho Kyowa Building, 3F
2-2-1 Hirakawacho
Chiyoda-ku, Tokyo 102, Japan

International Thomson Publishing GmbH
Königswinterer Strasse 418
53227 Bonn, Germany

International Thomson Publishing Southern Africa
Building 18, Constantia Park
240 Old Pretoria Road
Halfway House, 1685 South Africa

Library of Congress Cataloging-in-Publication Data
Rolfes, Sharon Rady.
     Life span nutrition : conception through life / Sharon Rady
Rolfes. Linda Kelly DeBruyne ; Eleanor Noss Whitney. — 2nd ed.
          p.   cm.
     Includes bibliographical references and index.
     ISBN 0-534-53834-7
     1. Nutrition. 2. Children—Nutrition. 3. Aged—Nutrition.
I. DeBruyne, Linda K.   II. Whitney, Eleanor Noss.   III. Title.
QP141.R64   1998
613.2—dc21   97-36769

DEDICATIONS

Nourishment means more than providing food; it also means giving love. Anyone joining my family at one of our gatherings would see an abundance of both. I dedicate this book to all the members of my family—including those who have departed and those who have yet to arrive.—*Sharon*

To the memory of my grandmother, Helen Garfunkel, whose commitment to her faith and to her family was steadfast, and whose firm but loving ways taught me to respect and appreciate life.—*Linda*

To Max and Muffin, Emily and Rebecca, from their admiring Grandma. —*Ellie*

Contents in Brief

Contents

Preface

In writing this book, we have gathered information from the thousands of scientific research articles that have contributed to the understanding of how nutrition influences people throughout their lives. In relaying this wealth of scientific research, we have tried to maintain a writing style that is both educational and enjoyable to read. Because our readers will be interested in life span nutrition for both professional and personal reasons, we have addressed this information from two perspectives. Those who plan to serve others as dietitians, nutritionists, educators, researchers, and caregivers of special age groups should find within these pages the knowledge they need to design programs that will meet their clients' needs. Those who want to learn how to provide the best possible nutrition for their own families will find that information as well.

The chapters introduce the stages of the life span sequentially. First, Chapter 1 presents background information that is applicable to all stages of the life span—the nutrients, recommended dietary intakes, diet-planning principles, and nutrition assessment. Then discussion of the life span begins with a "Prelude to Pregnancy," which introduces the concept that attention to good nutrition before conception supports a healthy pregnancy. Chapter 2 examines the normal course of a healthy pregnancy and how nutrition supports this crucial time of fetal development. Chapter 3 continues the discussion of pregnancy with a look at factors (such as maternal age, malnutriton, and substance abuse) that can increase the risk of a poor outcome, methods of assessing a pregnant woman's risks and nutrition status, and assistance programs for those in need. Chapter 4 starts with the information new mothers need to decide whether to breastfeed or formula-feed their newborns, and concludes with a discussion of lactation and the nutrient needs of a breastfeeding mother. Chapter 5 examines the unique needs of infancy and the ways infants change during the all-important first year of life. Chapter 6 then follows the course of childhood to puberty, and Chapter 7 continues the story through adolescence. Chapter 8 examines early adulthood with an emphasis on weight control; Chapter 9 discusses middle adulthood with a focus on health promotion and disease prevention; and Chapter 10 presents older adulthood with a spotlight on aging and the special nutrition needs of the later years.

Each chapter is organized in roughly the same way, so that the reader can know what to expect. For every stage of life, the chapter presents the characteristics of normal growth and development, the energy and nutrient needs and common nutrient deficiencies, nutrition assessment, nutrition-related concerns, and food choices and fitness to support health.

Interspersed with the chapters are Focal Points, features that highlight the latest research on special topics of interest to each stage of the life span. No Focal Point follows the introductory chapter, so they begin with Focal Point 2A, which describes inheritance and conception, and Focal Point 2B, which presents medical disorders that require special nutrition intervention during pregnancy. Focal Point 3 describes the devastating effects of alcohol consumption during pregnancy. Focal Point 4 provides a "how to" lesson in breastfeeding and formula-feeding infants. The three Focal Points that follow Chapter 5 examine the nutrient needs of high-risk infants, the relationships between iron deficiency and infant behavior, and the dietary support required to treat inborn errors of metabolism, respectively. Chapter 6 also has three Focal Points, which cover lead toxicity, the nutrient needs of sick children, and the early development of chronic diseases, respectively. The two Focal Points for Chapter 7 feature eating disorders and the menstrual cycle. Focal Point 8 takes a look at the few connections between nutrition and contraception. Focal Point 9 examines the role alcohol plays in interfering with good nutrition and health. And finally, Focal Point 10 offers valuable information on the interactions between drugs and nutrients.

This book provides additional special features as well. Definitions of terms appear in the margins adjacent to the text where they are first mentioned. Also presented wherever appropriate are the Healthy People 2000 nutrition-related priorities (Appendix B presents them in full). Healthy People 2000 is a report developed by the U.S. Department of Health and Human Services that establishes national objectives in health promotion and disease prevention for the year 2000.

The appendixes are valuable references for a number of purposes. Appendix A provides tables, charts, and forms used in nutrition assessment. Appendix B presents the Recommended Dietary Allowances (RDA), the U.S. RDA and Daily Values, the nutrition-related priorities of Healthy People 2000, and Nutrition Recommendations from the World Health Organization (WHO). Appendix C covers information for Canadians: the Recommended Nutrient Intakes (RNI) and instructions on reading food labels. Appendix D presents a table of infant formula composition, and Appendix E features the composition of vitamin-mineral supplements for pregnant women, infants, and children. Appendix F lists nutrition resources, including several Internet addresses.

Speaking of the Internet, a comment on its use in the study of nutrition (or any subject, for that matter) may be appropriate. The Internet offers endless opportunities to obtain high-quality information. Unfortunately, it also delivers an abundance of incomplete, misleading, or inaccurate information. Simply put: Anyone can publish anything. To discriminate a credible and helpful source from all the others, users must adopt standards similar to those used for printed publications. They must consider whether the author's educational degrees, credentials, and affiliations qualify him or her to speak authoritatively on nutrition. They also need to determine whether the information is based on valid scientific research and note whether credible references are provided. To explore several credible nutrition websites on the Internet, start with a visit to our website at:

http://www.wadsworth.com/nutrition

A few notes on style: we have used the term *parent* to refer to many kinds of caregivers, knowing that in many situations it is not literally a biological parent who may be responsible for a child's nutrition. We have addressed portions of the text directly to this person, even though not all readers are presently in that role in life. In our view, these elements of style make reading more pleasant than the generic passive voice, and we hope readers will not take them amiss.

We have learned and benefited from writing this book, and have found that the information in it serves us well in our lives, both as professionals and as parents and family members. We hope it serves you equally well.

Sharon Rady Rolfes
Linda Kelly DeBruyne

ACKNOWLEDGMENTS

We must begin our thank-you list with Gary Woodruff, for his encouragement to write this book in the first place, and with Ellie Whitney, for her many contributions to its first edition. Our publisher Peter Marshall, marketing manager Becky Tollerson, project editor Sandra Craig, developmental editors Laura Perkinson and Jane Bass, and production editor Greg Hubit deserve a round of applause for their coordination of review, production, and marketing. We are grateful to Linda Patton for her efficient library research work, Sally Mayo for her patient and competent word processing, and Sabrina McGriff for her assistance with a variety of office tasks. To the many others involved in designing, indexing, typesetting, and marketing, we tip our hats in appreciation.

We give thanks daily for our children, Lyle, Zak, Tyler, and Marni, who have given us many experiences that brought the information in this book to life. We appreciate the love and enthusiastic support of "the Toms," and of our partner Fran Sizer in helping us to complete this project.

Finally, we thank our reviewers for enhancing the quality and accuracy of this text:

Betty Alford
Texas Women's University

Leta P. Aljadir
University of Delaware

Vivian M. Bruce
Retired Professor
University of Manitoba

Anne M. Dattilo
University of Georgia

Debra Hollingsworth
McNeese State University

Bernice H. Kopel
Oklahoma State University

Mary J. Oakland
Iowa State University of Science and Technology

Tejaswini Rao
SUNY—Buffalo State College

Jackie Runyan
Iowa State University

Diana Spillman
Miami University

Anne VanBeber
Texas Christian University

Suzy Weems
Stephen F. Austin University

Carol Windham
Utah State University

Kathleen Yadrick
University of Southern Mississippi

Nutrition Overview

Nutrition Basics

The Nutrients

Recommended Dietary Intakes

Principles of Meal Planning

Dietary Guidelines

Food Labels

Special Concerns for Specific Stages of Life

Nutrition Assessment Basics

Historical Information

Anthropometric Data

Physical Examinations

Biochemical Analyses

The Team Approach

diet: the foods and beverages a person eats and drinks.

nutrients: substances obtained from food and used in the body to provide energy and structural materials and to regulate growth, maintenance, and repair of the body's tissues; nutrients may also reduce the risks of some chronic diseases.

energy: the capacity to do work. The energy in food is chemical energy. The body can convert this chemical energy to mechanical, electrical, or heat energy.

foods: products derived from plants or animals that can be taken into the body to yield nutrients for the maintenance of life and the growth and repair of tissues.

When people shop for foods, they are buying nutrients.

E ach person enters this world with a unique genetic map that determines the primary ways that person's physical and mental characteristics will develop throughout life. You must accept many of these characteristics without option for change, but can influence others within genetically defined limits. One of several ways to ensure the optimal growth, maintenance, and health of the body is through proper nutrition. Ideally, the diet should supply sufficient amounts of all the nutrients to meet the needs incurred by the physiological demands of growth, reproduction, lactation, disease, and aging.

All people—pregnant and lactating women, infants, children, adolescents, and adults—need the same nutrients, but the amounts they need vary depending on their stage of life. This chapter reviews the nutrition basics that apply to everyone and introduces some of the factors that influence nutrition in each phase of life. It also provides an overview of nutrition assessment, to prepare for the sections in each chapter that give special consideration to assessment during specific times of life.

Nutrition Basics

People at every stage of the life span need energy and the full array of nutrients to support the body's many activities. Standards have been established quantifying how much of each nutrient people need at various ages. Selecting foods appropriate to each stage of development that will deliver nutrients without excess energy requires a person to use certain diet-planning principles. This section reviews the nutrients, recommended intakes, and dietary principles.

The Nutrients

Both foods and the human body are composed of nutrients and other materials. The six classes of nutrients include carbohydrates, fats, proteins, vitamins, minerals, and water. In the body, three of these nutrients (carbohydrates, fats, and proteins) can be used to provide energy. In contrast, vitamins, minerals, and water do not yield energy.

The Energy-Yielding Nutrients. The body uses the energy-yielding nutrients to fuel all its metabolic and physical activities. When the body metabolizes these nutrients, the bonds between their atoms break. The carbon and hydrogen atoms combine with oxygen, yielding carbon dioxide and water for excretion. As the bonds between the carbon and hydrogen atoms break, they release energy. Some of this energy is given off as heat, but some is used to send electrical impulses through the nerves, to synthesize body compounds (including fat), and to fuel muscular activity.

If the body does not use these nutrients to fuel metabolic and physical activities, it rearranges them into storage compounds such as glycogen and fat, to be drawn on between meals and overnight when fresh supplies run low. If more energy is consumed than is expended, the result is weight gain. Figure 1-1 shows the ways the body uses carbohydrate, fat, and protein when food is abundant, as well as when intake is limited.

The amount of energy a food provides depends on how much carbohydrate, fat, and protein it contains and can be measured in calories—tiny units of energy so small that a single apple provides tens of thousands of them. To ease calculations, energy is expressed in 1000-calorie units known as kilocalories (shortened to kcalories, but commonly called "calories"). This book uses the term *kcalorie* and its abbreviation *kcal* throughout, as do other scientific books and journals. When completely metabolized in the body, a gram of carbohydrate yields about 4 kcalories of energy; a gram of protein also yields 4 kcalories; and a gram of fat yields 9 kcalories.

The six classes of nutrients:
- *Carbohydrate.*
- *Fat.*
- *Protein.*
- *Vitamins.*
- *Minerals.*
- *Water.*

energy-yielding nutrients: the nutrients that yield energy for the body's use:
- Carbohydrates.
- Fats.
- Proteins.

Figure 1-1   Feasting and Fasting

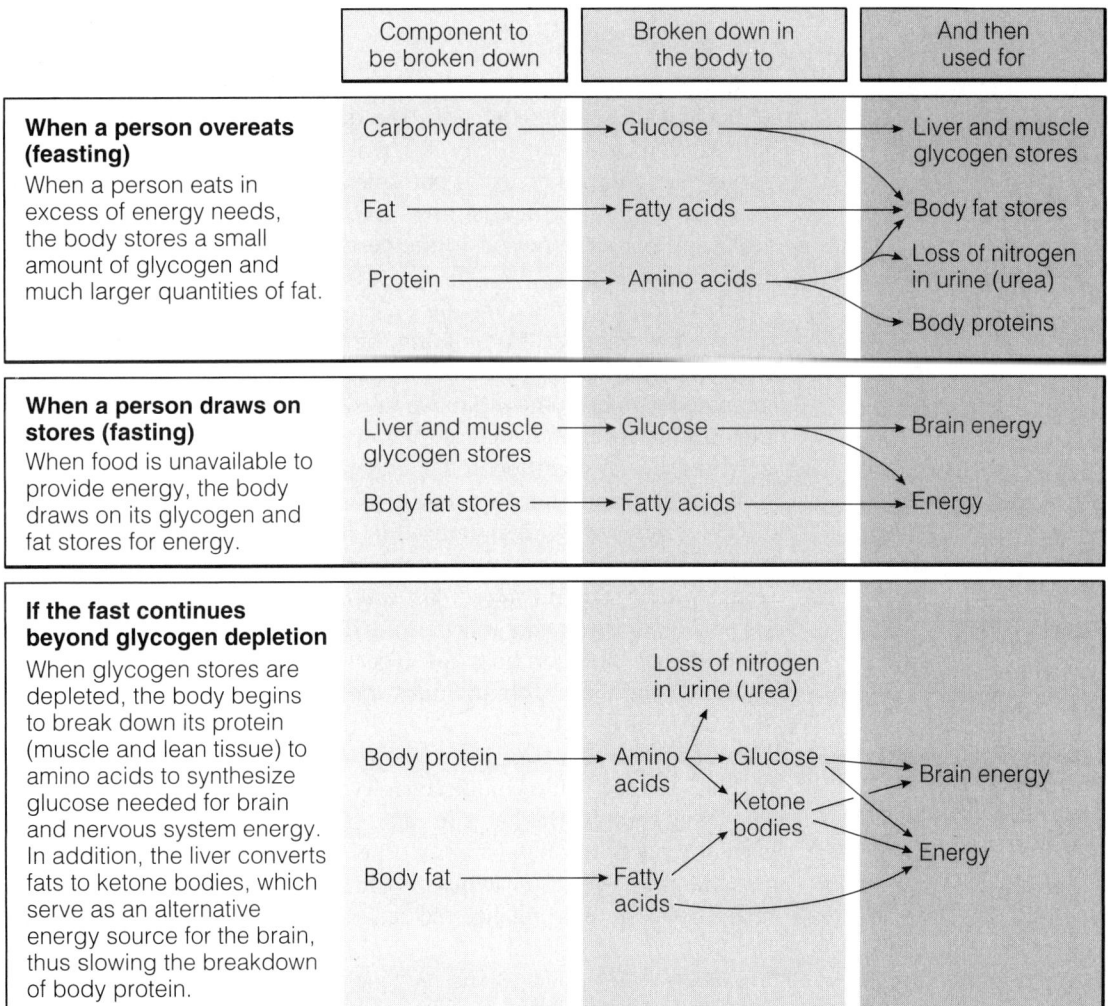

	Component to be broken down	Broken down in the body to	And then used for
When a person overeats (feasting) When a person eats in excess of energy needs, the body stores a small amount of glycogen and much larger quantities of fat.	Carbohydrate → Glucose →	Fat → Fatty acids →	Protein → Amino acids → Liver and muscle glycogen stores / Body fat stores / Loss of nitrogen in urine (urea) / Body proteins
When a person draws on stores (fasting) When food is unavailable to provide energy, the body draws on its glycogen and fat stores for energy.	Liver and muscle glycogen stores → Glucose → / Body fat stores → Fatty acids →		Brain energy / Energy
If the fast continues beyond glycogen depletion When glycogen stores are depleted, the body begins to break down its protein (muscle and lean tissue) to amino acids to synthesize glucose needed for brain and nervous system energy. In addition, the liver converts fats to ketone bodies, which serve as an alternative energy source for the brain, thus slowing the breakdown of body protein.	Body protein → Amino acids ← Glucose / Ketone bodies / Body fat → Fatty acids		Loss of nitrogen in urine (urea) / Brain energy / Energy

kcalories: units by which energy is measured. One kcalorie is the amount of heat necessary to raise the temperature of 1 kg of water 1°C.

1 g carbohydrate = 4 kcal.
1 g protein = 4 kcal.
1 g fat = 9 kcal.

To calculate the energy available from a food, multiply the number of grams of carbohydrate, protein, and fat by 4, 4, and 9, respectively. Then add the results together. For example, 1 slice of bread with 1 tablespoon of peanut butter on it contains 16 grams carbohydrate, 7 grams protein, and 9 grams fat:

$$16 \text{ g carbohydrate} \times 4 \text{ kcal/g} = 064 \text{ kcal.}$$
$$7 \text{ g protein} \times 4 \text{ kcal/g} = 028 \text{ kcal.}$$
$$9 \text{ g fat} \times 9 \text{ kcal/g} = 081 \text{ kcal.}$$
$$\text{Total} = 173 \text{ kcal.}$$

From this information, you can calculate the percentage of kcalories each of the energy nutrients contributes to the total. To determine the percentage of kcalories from fat, for example, divide the 81 fat kcalories by the total 173 kcalories:

$$81 \div 173 = 0.468 \text{ (rounded to 0.47).}$$

Then multiply by 100 to get the percentage:

$$0.47 \times 100 = 47\%.$$

Health recommendations that urge people to limit fat intake to 30 percent of kcalories refers to the day's total energy intake, not to individual foods. Still, if the proportion of fat in each food choice throughout a day exceeds 30 percent of kcalories, then the day's total surely will too. Knowing that this snack provides 47 percent of its kcalories from fat alerts a person to the need to make lower-fat selections at other times that day.

In addition to providing energy, carbohydrate, fat, and protein provide the raw materials for building the body's tissues and regulating its many activities. In fact, protein's role as a fuel source is relatively minor compared with both the other two nutrients and its other roles. Proteins are found in structures such as the muscles and skin, and help regulate activities such as digestion and energy production.

1 g alcohol = 7 kcal.

Alcohol is not a nutrient, but like carbohydrates, fats, and proteins, it yields energy when metabolized in the body (7 kcalories per gram). When consumed in excess of energy needs, alcohol, too, can be converted to body fat and stored. But when alcohol contributes a substantial portion of the energy in a person's diet, the harm it does far exceeds the problems of excess body fat. (Focal Points 3 and 9 describe the effects of alcohol on health and nutrition.)

vitamins: essential nutrients required in small amounts by the body for health.

The Vitamins. There are 13 different vitamins, each with its own special roles to play.* One vitamin enables the eyes to see in dim light, another helps protect the lungs from air pollution, and still another helps make the sex hormones—among other things. When you cut yourself, one vitamin helps stop the bleeding and another helps repair the skin. Vitamins busily help replace old red blood cells and the lining of the

*The water-soluble vitamins are vitamin C and the eight B vitamins: thiamin, riboflavin, niacin, vitamins B_6 and B_{12}, folate, biotin, and pantothenic acid. The fat-soluble vitamins are vitamins A, D, E, and K.

digestive tract. Almost every action in the body requires the assistance of vitamins.

The Minerals. Some 16 minerals are known to be essential in human nutrition.* Others are still being studied to determine whether they play significant roles in the human body. Still other minerals are important because they are not essential nutrients, but environmental contaminants that displace the nutrient minerals from their workplaces in the body, disrupting body functions. This problem is exemplified by the discussion of lead toxicity in Focal Point 6A.

minerals: small, inorganic atoms or molecules; some minerals are essential nutrients required in small amounts.

The roles of the vitamins and minerals in nutrition are diverse and numerous enough to fill whole books. Table 1-1 summarizes some basic information about them.

Water. Water constitutes about 55 to 60 percent of an adult's body weight and a higher percentage of a child's. The obvious dietary sources of water are water itself and other beverages; in addition, nearly all foods contain water. Water is also generated during metabolism (recall that when the energy-yielding nutrients break down, their carbon and hydrogen atoms combine with oxygen to yield carbon dioxide and water). The water derived from these three sources normally balances perfectly with daily water losses.

Without exaggeration, nutrients provide the physical and metabolic basis for nearly all that we are and all that we do. Now, exactly how much of each nutrient does the body need?

Recommended Dietary Intakes

For more than fifty years, nutrition experts have produced a set of nutrient and energy standards known as the Recommended Dietary Allowances (RDA). These standards, based on careful examination and interpretation of scientific evidence, are revised periodically. A major revision of the RDA is currently under way, and the first in a series of reports to replace the RDA was published in 1997. The revised recommendations are called Dietary Reference Intakes (DRI) and reflect the collaborative efforts of both the United States and Canada. Among the changes are revisions in the age groupings, shown in Table 1-2 on p. 10.[1] Until 1997, the RDA were the only standards available and they will continue to serve health professionals until DRI can be established for all nutrients. For this reason, this book uses both the 1989 RDA and the 1997 DRI for selected nutrients where appropriate. This section introduces the RDA in a general way, and the chapters to come discuss specific nutrient recommendations for each stage of the life span.

The 1989 Recommended Dietary Allowances (RDA) and the 1997 Dietary Reference Intakes (DRI) for selected nutrients are presented on the inside front cover.

The 1990 Canadian standards, Recommended Nutrient Intakes (RNI), are presented in Appendix B.

---

*The major minerals are found in the human body in amounts larger than 5 grams and include calcium, phosphorus, potassium, sodium, chloride, magnesium, and sulfur. The trace minerals are found in the human body in amounts less than 5 grams and include iron, iodine, zinc, chromium, selenium, fluoride, molybdenum, copper, and manganese.

Table 1-1   The Vitamins and Minerals—A Summary

Fat-Soluble Vitamins	Chief Functions in the Body	Significant Sources
Vitamin A (retinol, retinal, retinoic acid); precursor is provitamin A carotenoids such as beta-carotene	Vision; maintenance of cornea, epithelial cells, mucous membranes, skin; bone and tooth growth; reproduction; hormone synthesis and regulation; immunity; cancer protection	Retinol: fortified milk, cheese, cream, butter, fortified margarine, eggs, liver Beta-carotene: spinach and other dark leafy greens, broccoli, deep orange fruits (apricots, cantaloupe) and vegetables (squash, carrots, sweet potatoes, pumpkin)
Vitamin D (calciferol, cholecalciferol, 1,25-dihydroxy vitamin D); precursor is the body's own cholesterol	Mineralization of bones (raises blood calcium and phosphorus by increasing absorption from digestive tract, withdrawing calcium from bones, stimulating retention by kidneys)	Synthesized in the body with the help of sunlight; fortified milk, fortified margarine, egg yolks, liver, fatty fish
Vitamin E (alpha-tocopherol, tocopherol, tocotrienol)	Antioxidant, stabilization of cell membranes, regulation of oxidation reactions, protection of polyunsaturated fatty acids (PUFA) and vitamin A	Polyunsaturated plant oils (margarine, salad dressings, shortenings), leafy green vegetables, wheat germ, whole-grain products, liver, egg yolks, nuts, seeds
Vitamin K (menadione, menaquinone, phylloquinone, naphthoquinone)	Synthesis of blood-clotting proteins and a blood protein that regulates blood calcium	Bacterial synthesis[a] in the digestive tract; liver, leafy green vegetables, cabbage-type vegetables, milk

Water-Soluble Vitamins	Chief Functions in the Body	Significant Sources
Thiamin (vitamin B_1)	Part of TPP (thiamin pyrophosphate), a coenzyme used in energy metabolism; supports normal appetite and nervous system function	Occurs in all nutritious foods in moderate amounts; pork, ham, bacon, liver, whole-grain or enriched breads and cereals, legumes, nuts
Riboflavin (vitamin B_2)	Part of FMN (flavin mononucleotide) and FAD (flavin adenine dinucleotide), coenzymes used in energy metabolism; supports normal vision and skin health	Milk, yogurt, cottage cheese, meat, leafy green vegetables, whole-grain or enriched breads and cereals
Niacin (nicotinic acid, nicotinamide, niacinamide, vitamin B_3); precursor is dietary tryptophan	Part of NAD (nicotinamide adenine dinucleotide) and NADP (its phosphate form), coenzymes used in energy metabolism; supports health of skin, nervous system, and digestive system	Milk, eggs, meat, poultry, fish, whole-grain and enriched breads and cereals, nuts, and all protein-containing foods

[a]Vitamin K needs cannot be met from bacterial synthesis alone; however, it is a potentially important source in the jejunum and ileum, where absorption efficiency ranges from 40 to 70%.

(continued)

Table 1-1 The Vitamins and Minerals—A Summary (*continued*)

Water-Soluble Vitamins	Chief Functions in the Body	Significant Sources
Vitamin B_6 (pyridoxine, pyridoxal, pyridoxamine)	Part of PLP (pyridoxal phosphate) and PMP (pyridoxamine phosphate), coenzymes used in amino acid and fatty acid metabolism; helps to convert tryptophan to niacin; helps to make red blood cells	Leafy green vegetables, meats, fish, poultry, shellfish, legumes, fruits, whole grains
Folate (folacin, folic acid, pteroylglutamic acid)	Part of THF (tetrahydrofolate) and DHF (dihydrofolate), coenzymes used in new cell synthesis	Leafy green vegetables, legumes, seeds, liver
Vitamin B_{12} (cobalamin and related forms)	Part of methylcobalamin and deoxyadenocobalamin, coenzymes used in new cell synthesis; helps to maintain nerve cells; reforms folate coenzyme	Animal products (meat, fish, poultry, shellfish, milk, cheese, eggs)
Pantothenic acid	Part of Coenzyme A, used in energy metabolism	Widespread in foods
Biotin	Part of a coenzyme used in energy metabolism, fat synthesis, amino acid metabolism, and glycogen synthesis	Widespread in foods
Vitamin C (ascorbic acid)	Collagen synthesis (strengthens blood vessel walls, forms scar tissue, provides matrix for bone growth), antioxidant, thyroxin synthesis, amino acid metabolism, strengthens resistance to infection, helps in absorption of iron	Citrus fruits, cabbage-type vegetables, dark green vegetables, cantaloupe, strawberries, peppers, lettuce, tomatoes, potatoes, papayas, mangos

Major Minerals	Chief Functions in the Body	Significant Sources
Calcium	The principal mineral of bones and teeth; also involved in muscle (including heart muscle) contraction and relaxation, nerve functioning, blood clotting, blood pressure, and immune defenses	Milk and milk products, small fish (with bones), tofu (bean curd), greens (broccoli, chard), legumes
Phosphorus	A principal mineral of bones and teeth; part of every cell; important in genetic material, part of phospholipids, used in energy transfer and in buffer systems that maintain acid–base balance	All animal tissues (meat, fish, poultry, eggs, milk)

(continued)

Table 1-1 The Vitamins and Minerals—A Summary (*continued*)

Major Minerals	Chief Functions in the Body	Significant Sources
Magnesium	Involved in bone mineralization, the building of protein, enzyme action, muscular contraction, transmission of nerve impulses, and maintenance of teeth	Nuts, legumes, whole grains, dark green vegetables, seafood, chocolate, cocoa
Sodium	An electrolyte that maintains normal fluid balance and electrolyte balance; assists in nerve impulse transmission and muscle contraction	Table salt, soy sauce; moderate amounts in meats, milks, breads, and vegetables; large amounts in many processed foods
Chloride	An electrolyte that maintains normal fluid balance and acid–base balance; part of the hydrochloric acid found in the stomach, necessary for proper digestion	Table salt, soy sauce; moderate amounts in meats, milks, breads, and vegetables; large amounts in many processed foods
Potassium	An electrolyte that maintains normal fluid balance and electrolyte balance; facilitates many reactions, including the making of protein; supports cell integrity; assists in the transmission of nerve impulses and the contraction of muscles, including the heart	All whole foods: meats, milks, fruits, vegetables, grains, legumes
Sulfur	As part of proteins, stabilizes their shape by forming sulfur–sulfur bridges; part of the vitamins biotin and thiamin and the hormone insulin; involved in the body's detoxification process (combines with toxic substances to form harmless compounds)	All protein-containing foods (meats, fish, eggs, milk, legumes, nuts)

Trace Minerals	Chief Functions in the Body	Significant Sources
Iodine	A compound of the thyroid hormones that help to regulate growth, development, and metabolic rate	Iodized salt, seafood, bread, dairy products, plants grown in iodine-rich soil, and animals fed those plants
Iron	Part of the protein hemoglobin, which carries oxygen from place to place in the body; part of the protein myoglobin in muscles, which makes oxygen available for muscle contraction; necessary for the utilization of energy as part of the cells' metabolic machinery	Red meats, fish, poultry, shellfish, eggs, legumes, dried fruits

(*continued*)

Table 1-1   The Vitamins and Minerals—A Summary (*continued*)

Trace Minerals	Chief Functions in the Body	Significant Sources
Zinc	Part of many enzymes; associated with the hormone insulin; involved in making genetic material and proteins, immune reactions, transport of vitamin A, taste perception, wound healing, the making of sperm	Protein-containing foods (meats, fish, poultry, whole grains, vegetables)
Copper	Necessary for the absorption and use of iron in the formation of hemoglobin; part of several enzymes and a factor that helps to form the protective covering of nerves	Meats, drinking water
Fluoride	An element involved in the formation of bones and teeth, helps make teeth resistant to decay	Drinking water (if naturally fluoride containing or fluoridated), tea, seafood
Selenium	Part of an enzyme system that works with vitamin E to protect body compounds from oxidation	Seafood, meat, grains
Chromium	Associated with insulin and required for the release of energy from glucose	Meats, unrefined foods, fats, vegetable oils
Cobalt	Part of vitamin B_{12} and therefore involved in nerve cell function and blood formation	Foods of animal origin (meats, milk and milk products)
Molybdenum	Facilitator, with enzymes, of many cell processes	Legumes, cereals, organ meats
Manganese	Facilitator, with enzymes, of many cell processes	Widespread in foods

The RDA are the amounts of selected nutrients considered adequate to meet the known nutrient needs of practically all healthy people. They are not minimum requirements, nor are they necessarily optimal intakes. They are safe and adequate recommendations that include a generous margin of safety. People with medical problems may have special nutrient needs that require supplemented or restricted intakes.

Diet planners use the RDA as guidelines in planning and evaluating diets for groups of people, such as schoolchildren. The RDA can be used to estimate the risks of deficiencies for an individual only if his or her intakes are averaged over a sufficient length of time (at least three days, preferably more).

With careful planning, the RDA for energy and each nutrient can be met from a variety of foods. This goal is difficult and unnecessary to achieve on a daily basis, however. The RDA apply to *average* daily

Table 1-2 Age Groupings for Dietary Recommendations

United States RDA (1989)	Canada RNI (1991)	DRI (1997)
0–6 months	0–4 months	0–6 months
7–12 months	5–12 months	7–12 months
1–3 years	1 year	1–3 years
	2–3 years	
4–6 years	4–6 years	4–8 years
7–10 years	7–9 years (M, F)	
11–14 years (M, F)	10–12 years (M, F)	9–13 years (M, F)
	13–15 years (M, F)	
15–18 years (M, F)	16–18 years (M, F)	14–18 years (M, F)
19–24 years (M, F)	19–24 years (M, F)	19–30 years (M, F)
25–50 years (M, F)	25–49 years (M, F)	31–50 years (M, F)
51+ years (M, F)	50–74 years (M, F)	51–70 years (M, F)
	75+ years (M, F)	71+ years (M, F)
Pregnant	Pregnancy 1st trimester 2nd trimester 3rd trimester	Pregnancy (may be subdivided as needed)
Lactating 1st 6 months 2nd 6 months	Lactating	Lactating (may be subdivided as needed)

Source: Committee on Dietary Reference Intakes, *Dietary Reference Intakes for Calcium, Phosphorus, Magnesium, Vitamin D, and Fluoride* (Washington, D.C.: National Academy Press, 1997), pp. 1-8–1-11.

intakes. The length of time over which a person's intake can deviate from the average without risk of deficiency or overdose varies for each nutrient, depending on the body's use and storage of the nutrient. For most nutrients (such as thiamin and vitamin C), deprivation would lead to rapid development of deficiency symptoms (within days or weeks); their RDA reflect average intakes over at least three days. For others (such as vitamin A and vitamin B_{12}), deficiencies would develop much slower (within months or years); their RDA reflect average intakes over several months.

The RDA are generous, and although they do not necessarily cover every individual for every nutrient, they probably should not be exceeded by much. Some nutrients can be toxic at intakes not far above the RDA, and people's tolerances for high doses of nutrients vary.

In contrast to the RDA for nutrients, the RDA for energy are not generous, but are set at the mean of the population's estimated requirements. For energy, too much is as bad for health as too little because excess energy may lead to obesity, and not enough energy may cause undernutrition.

Protein is the only energy-yielding nutrient for which an RDA has been established. The Committee on Dietary Allowances assumes that the protein is consumed along with enough carbohydrate and fat to provide sufficient energy and that other nutrients in the diet are adequate. An adequate intake of carbohydrates and fats spares proteins from being used for energy and allows them to perform their unique roles.

Principles of Meal Planning

Knowing the individual nutrients and their recommended intakes enables planners to select foods to create diets that supply all the needed nutrients in the appropriate amounts for good health. When planning meals, keep the following principles in mind:

■ Adequacy (obtaining from foods all nutrients in sufficient quantities).

■ Balance (ensuring that all food groups are represented in recommended proportions).

■ kCalorie control (managing energy intake to meet the body's needs without being excessive or insufficient).

■ Moderation (providing enough, but not too much, of a food or beverage by following recommended serving sizes and number of servings).

■ Variety (eating a wide selection of different foods within each food group).

The single principle that helps most to accomplish all these objectives is:

■ Nutrient density (selecting foods that deliver abundant nutrients relative to the amount of food energy).

To plan a diet that achieves all the dietary principles just outlined, a planner needs not only knowledge but tools. One of the most commonly used tools is the USDA's Food Guide Pyramid, which builds a diet from five clusters of foods that are similar in origin and nutrient content. Figure 1-2 lists the most notable nutrients of each group, the foods within each group categorized by nutrient density, the number of servings recommended, and the serving sizes.

Dietary Guidelines

The Food Guide Pyramid helps consumers plan a diet that fulfills two criteria: It provides enough of the essentials, and it avoids the excesses. The planner evaluating a diet thus has two questions to ask. One is, How does the diet compare with the RDA (or the Canadian RNI), which makes specific recommendations for food energy, protein, vitamins, and minerals? The second question is, How well does the diet control intakes of fat, saturated fat, cholesterol, salt, sugar, and alcohol? The current RDA were not designed to address this second set of concerns; another set of recommendations, the USDA's *Dietary Guidelines*, provide

To ensure an adequate and balanced diet, eat a variety of foods daily, choosing different foods from each group.

nutrient density: a measure of the nutrients a food provides relative to the energy it provides. The more nutrients and the fewer kcalories, the higher the nutrient density.

Five food groups:
■ *Breads, cereals, and other grain products.*
■ *Vegetables.*
■ *Fruits.*
■ *Meat, poultry, fish, and alternates.*
■ *Milk, cheese, and yogurt.*

Figure 1-2 The Food Guide Pyramid

Breads, Cereals, and Other Grain Products

These foods are notable for their contributions of complex carbohydrates, riboflavin, thiamin, niacin, iron, protein, magnesium, and fiber.

6 to 11 servings per day.

Serving = 1 slice bread; 1/2 c cooked cereal, rice, or pasta; 1 oz ready-to-eat cereal; 1/2 bun, bagel, or English muffin; 1 small roll, biscuit, or muffin; 3 to 4 small or 2 large crackers.

- ■ Whole grains (wheat, oats, barley, millet, rye, bulgur), enriched breads, rolls, tortillas, cereals, bagels, rice, pastas (macaroni, spaghetti), air-popped corn.
- ▨ Pancakes, muffins, cornbread, crackers, cookies, biscuits, presweetened cereals, granola, taco shells, waffles.
- ☐ Croissants, fried rice, doughnuts, pastries, cakes, pies.

Vegetables

These foods are notable for their contributions of vitamin A, vitamin C, folate, potassium, magnesium, and fiber, and for their lack of fat and cholesterol.

3 to 5 servings per day (use dark green, leafy vegetables and legumes several times a week).

Serving = 1/2 c cooked or raw vegetables; 1 c leafy raw vegetables; 1/2 c cooked legumes; 3/4 c vegetable juice.

- ■ Bean sprouts, broccoli, brussels sprouts, cabbage, carrots, cauliflower, corn, cucumbers, green beans, green peas, leafy greens (spinach, mustard, and collard greens), legumes, lettuce, mushrooms, potatoes, sweet potatoes, tomatoes, winter squash.
- ▨ Candied sweet potatoes.
- ☐ French fries, tempura vegetables, scalloped potatoes, potato salad.

Fruits

These foods are notable for their contributions of vitamin A, vitamin C, potassium, and fiber, and for their lack of sodium, fat, and cholesterol.

2 to 4 servings per day.

Serving = typical portion (such as 1 medium apple, banana, or orange, 1/2 grapefruit, 1 melon wedge); 3/4 c juice; 1/2 c berries; 1/2 c diced, cooked, or canned fruit; 1/4 c dried fruit.

- ■ Apricots, cantaloupe, grapefruit, oranges, orange juice, peaches, strawberries, apples, bananas, pears, unsweetened juices.
- ▨ Canned or frozen fruit (in syrup); sweetened juices.
- ☐ Dried fruit, coconut, avocados.

Meat, Poultry, Fish, and Alternates

These foods are notable for their contributions of protein, phosphorus, vitamin B_6, vitamin B_{12}, zinc, magnesium, iron, niacin, and thiamin.

2 to 3 servings per day.

Serving = 2 to 3 oz lean, cooked meat, poultry, or fish (total 5 to 7 oz per day); count 1 egg, 1/2 c cooked legumes, 4 oz tofu, or 2 tbs nuts, seeds, or peanut butter as 1 oz meat (or about 1/3 serving).

- ■ Poultry (light meat, no skin), fish, shellfish, legumes, egg whites.
- ▨ Lean meat (fat-trimmed beef, lamb, pork); poultry (dark meat, no skin); ham; refried beans; whole eggs, tofu, tempeh.
- ☐ Hot dogs, luncheon meats, ground beef, peanut butter, nuts, sausage, bacon, fried fish or poultry, duck.

Key:

- ■ Foods generally highest in nutrient density (good first choice).
- ▨ Foods moderate in nutrient density (reasonable second choice).
- ☐ Foods lowest in nutrient density (limit selections).

(continued)

Figure 1-2 The Food Guide Pyramid (*continued*)

Milk, Cheese, and Yogurt

These foods are notable for their contributions of calcium, riboflavin, protein, vitamin B_{12}, and, when fortified, vitamin D and vitamin A.

2 servings per day.

3 servings per day for teenagers and young adults, pregnant/lactating women, women past menopause.

4 servings per day for pregnant/lactating teenagers.

Serving = 1 c milk or yogurt; 2 oz processed cheese food; 1 1/2 oz cheese.

- ■ Nonfat and 1% low-fat milk (and nonfat products such as buttermilk, cottage cheese, cheese, yogurt); fortified soy milk.
- ▨ 2% reduced-fat milk (and reduced-fat products such as yogurt, cheese, cottage cheese); chocolate milk; sherbet; ice milk.
- ☐ Whole milk (and whole-milk products such as cheese, yogurt); custard; milk shakes; ice cream.

Fats, Sweets, and Alcoholic Beverages

These foods are notable for their contributions of sugar, fat, alcohol, and food energy. No servings are suggested, because these foods provide few nutrients. Note that some of the following items, for example, doughnuts, are high in both sugar and fat. Alcoholic beverages are not classed as foods; they contribute few nutrients, but do provide food energy and so are included in this miscellaneous group. Miscellaneous foods not high in kcalories, such as spices, herbs, coffee, tea, and diet soft drinks, can be used freely.

- ☐ Foods high in fat include margarine, salad dressing, oils, mayonnaise, sour cream, cream cheese, butter, gravy, sauces, potato chips, chocolate bars.
- ☐ Foods high in sugar include cakes, pies, cookies, doughnuts, sweet rolls, candy, soft drinks, fruit drinks, jelly, syrup, gelatin, desserts, sugar, and honey.
- ☐ Alcoholic beverages include wine, beer, and liquor.

Note: Serve children at least the lower number of servings from each group, but in smaller amounts (for example, 1/4 to 1/3 cup rice). Children should receive the equivalent of 2 cups of milk each day, but again in smaller quantities per serving (for example, 4 half-cup portions). Pregnant women may require additional servings of fruits, vegetables, meats, and breads to meet their higher needs for energy, vitamins, and minerals.

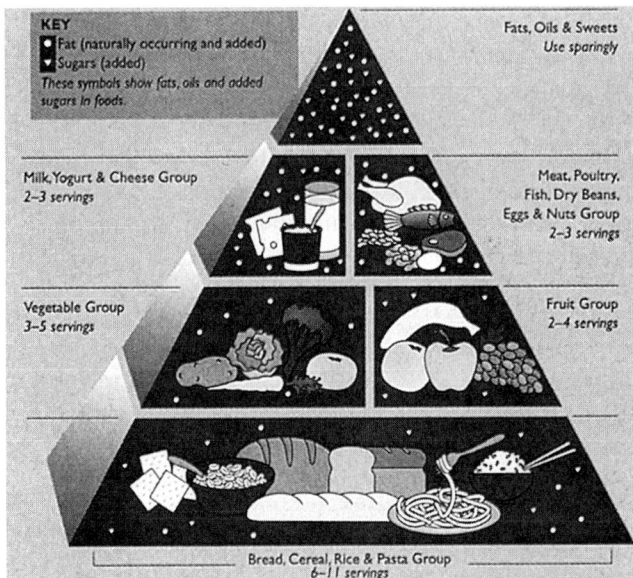

Food Guide Pyramid

A Guide to Daily Food Choices

The breadth of the base shows that grains (breads, cereals, rice, and pasta) deserve most emphasis in the diet. The tip is smallest: Use fats, oils, and sweets sparingly.

Table 1-3 The USDA's *Dietary Guidelines for Americans, 1995*

- Eat a variety of foods.
- Balance the food you eat with physical activity; maintain or improve your weight.
- Choose a diet with plenty of grain products, vegetables, and fruits.
- Choose a diet low in fat, saturated fat, and cholesterol.
- Choose a diet moderate in sugars.
- Choose a diet moderate in salt and sodium.
- If you drink alcoholic beverages, do so in moderation.

Note: These guidelines are designed for healthy people over two years old.

Source: U.S. Department of Agriculture and U.S. Department of Health and Human Services, *Dietary Guidelines for Americans* (U.S. Government Printing Office, 1995).

advice about food choices that will promote health and prevent disease (see Table 1-3).

The first two guidelines encourage people to eat a variety of foods to get the nutrients needed to support good health and to balance food intake with physical activity in order to maintain or improve body weight. The next two guidelines urge a shift in the balance of energy nutrients: They encourage people to increase their carbohydrate and reduce their fat intakes by choosing a diet that is abundant in grains, vegetables, and fruits and low in fat, saturated fat, and cholesterol. The last three guidelines recommend a diet moderate in sugars, salt and sodium, and alcoholic beverages for those who partake. Together, these seven guidelines point the way toward better health. Table 1-4 presents *Canada's Guidelines for Healthy Eating*, which offer similar suggestions.

✗ Healthy People 2000

The Healthy People 2000 report sets national objectives in health promotion and disease prevention for the year 2000. The 21 nutrition-related priorities are listed in Appendix B and appear in the text where their subjects are discussed.

✗ Healthy People 2000

Increase to at least 90% the proportion of restaurants and institutional foodservice operations that offer identifiable low-fat, low-kcalorie food choices, consistent with the *Dietary Guidelines for Americans*.

Table 1-4   *Canada's Guidelines for Healthy Eating*

- Enjoy a variety of foods.
- Emphasize cereals, breads, other grain products, vegetables, and fruits.
- Choose lower-fat dairy products, leaner meats, and foods prepared with little or no fat.
- Achieve and maintain a healthy body weight by enjoying regular physical activity and healthy eating.
- Limit salt, alcohol, and caffeine.

Source: These guidelines derive from *Action Towards Healthy Eating: The Report of the Communications/Implementation Committee and Nutrition Recommendations. . . . A Call for Action: Summary Report of the Scientific Review Committee and the Communications/Implementation Committee,* which are available from Branch Publications Unit, Health Services and Promotion Branch, Department of Health and Welfare, 5th Floor, Jeanne Mance Building, Ottawa, Ontario K1A 1B4.

Food Labels

How do you begin to design a diet that meets the *Dietary Guidelines*? Start with the foods you enjoy eating, and then try to make improvements, little by little. When shopping, think of the Food Pyramid and read food labels to help you choose nutrient-dense foods within each group.

✗   **Healthy People 2000**

Increase to at least 85% the proportion of people aged 18 and older who use food labels to make nutritious food selections.

Consumers read food labels to learn about nutrition and its possible connections with health.

Food labels appear on virtually all foods and provide much useful nutrition information. According to law, every food label must prominently display and express in ordinary words:

■ The common or usual name of the product.

■ The name and address of the manufacturer, packer, or distributor.

■ The net contents in terms of weight, measure, or count.

■ The ingredients in descending order of predominance by weight.

■ The serving size and number of servings per container.

■ The quantities of specified nutrients and food constituents.

The information in the first three items is useful, of course, but it is the last three items that tell consumers about the nutritional value of a product. Figure 1-3 highlights key information on a food label.

⫪ Healthy People 2000

Achieve useful and informative nutrition labeling for virtually all processed foods and at least 40% of fresh meats, poultry, fish, fruits, vegetables, baked goods, and ready-to-eat carry-away foods.

Special Concerns for Specific Stages of Life

This text focuses on the nutrient needs of people in each stage of life—pregnancy, lactation, infancy, childhood, adolescence, and throughout adulthood—and on the special nutrition attention each stage requires. As mentioned previously, people need the same nutrients throughout their lives, but the amounts vary. Personal factors affecting nutrient needs include age, sex, genetics, and lifestyle habits such as physical activity. A brief look at some nutrition concerns during specific times of life will set the stage for the chapters to come.

Healthy body weight, for example, is important in every stage of the life span, but for different reasons at different times. Women who enter pregnancy at the appropriate weight and who gain at a certain rate have the best chance of bearing full-term, healthy infants—infants whose health and survival outlooks are optimal. An infant's weight at birth is a predictor of the infant's future health status. A child's weight is of interest as a measure of growth, and therefore underweight may reflect poor nutrition. Overweight is also a matter of concern. Childhood obesity urgently needs preventive attention if the adult years are to find this health hazard manageable. Overweight in adults is a risk factor for many different diseases, as Chapter 9 explains.

Iron status is another example of the different emphases in various stages of life. During pregnancy, a woman must consume enough iron both to support her enlarged blood volume and to provide fetal stores to

Figure 1-3 An Example of a Food Label

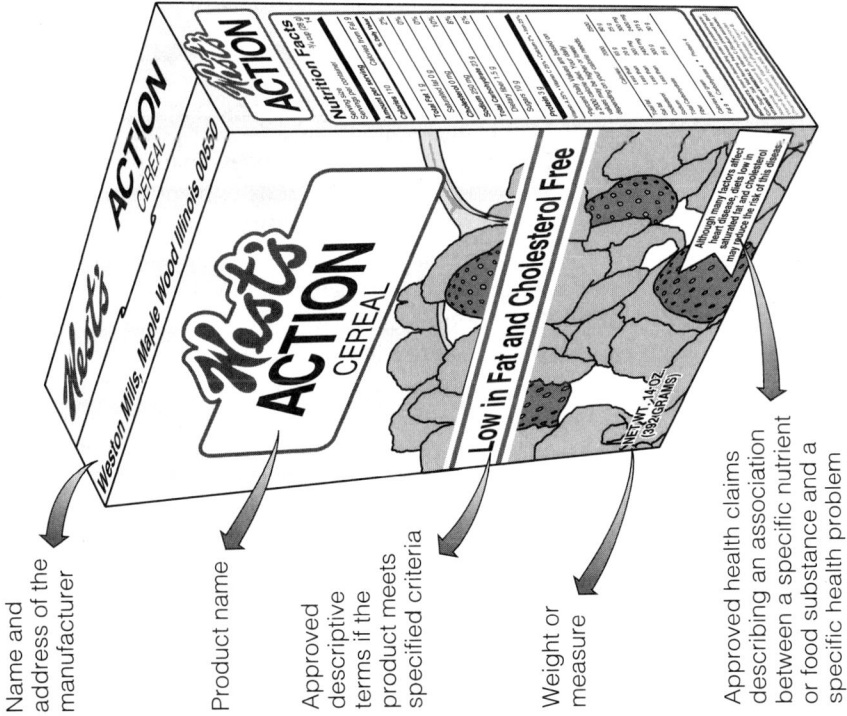

Name and address of the manufacturer

Product name

Approved descriptive terms if the product meets specified criteria

Weight or measure

Approved health claims describing an association between a specific nutrient or food substance and a specific health problem

West's
ACTION
CEREAL

Weston Mills, Maple Wood Illinois 00550

West's
ACTION
CEREAL

Low in Fat and Cholesterol Free

Although many factors affect saturated fat, and cholesterol may reduce the risk of this disease.

NET WT. 14 OZ. (392 GRAMS)

Standard serving sizes and number of servings per container facilitate comparison shopping

kCalorie information and quantities of nutrients per serving, in actual amounts

Quantities of nutrients as "% Daily Values" based on a 2000-kcalorie energy intake allow consumers to compare the product against recommendations

Daily Values for selected nutrients for a 2000- and a 2500-kcalorie diet (see inside front cover for additional information)

kCalorie per gram reminder

Ingredients listed in descending order of predominance by weight

West's
ACTION
CEREAL

Nutrition Facts

Serving size 3/4 cup (28 g)
Servings per container 14

Amount per serving	
Calories 110	Calories from fat 9
	% Daily Value*
Total Fat 1 g	2%
Saturated fat 0 g	0%
Cholesterol 0 mg	0%
Sodium 250 mg	10%
Total Carbohydrate 23 g	8%
Dietary fiber 1.5 g	6%
Sugars 10 g	
Protein 3 g	

Vitamin A 25% • Vitamin C 25% • Calcium 2% • Iron 25%

*Percent Daily Values are based on a 2000 calorie diet. Your daily values may be higher or lower depending on your calorie needs.

	Calories:	2000	2500
Total fat	Less than	65 g	80 g
Sat fat	Less than	20 g	25 g
Cholesterol	Less than	300 mg	300 mg
Sodium	Less than	2400 mg	2400 mg
Total Carbohydrate		300 g	375 g
Fiber		25 g	30 g

Calories per gram
Fat 9 • Carbohydrate 4 • Protein 4

INGREDIENTS, listed in descending order of predominance: Corn, Sugar, Salt, Malt flavoring, freshness preserved by BHT. **VITAMINS and MINERALS:** Vitamin C (Sodium ascorbate), Niacinamide, Iron, Vitamin B6 (Pyridoxine hydrochloride), Vitamin B2 (Riboflavin), Vitamin A (Palmitate), Vitamin B1 (Thiamin hydrochloride), Folic acid, and Vitamin D.

last for half a year following the infant's birth. In infancy and early childhood, iron deficiency anemia is common, so dietary iron continues to be a nutrition focus. At puberty when menstruation begins, iron needs increase by an estimated 50 percent for girls because of menstrual iron losses. At menopause, iron needs for women decline as losses diminish. For men, iron overload is twice as prevalent as is iron deficiency, which may have implications for the development of both heart disease and cancer.

Calcium needs also receive mention at every stage and are important for different reasons from one time to the next. Throughout early life, calcium contributes to the growth of bone. In adulthood and in pregnancy, adequate calcium intakes may help defend against hypertension. In later life, continued adequate calcium intake may help to minimize the inevitable bone loss of old age.

Just as nutrient needs are important for different reasons at different times of life, so too is the avoidance of excesses. Consider fat and cholesterol intakes, for example. For adults, limiting fat, saturated fat, and cholesterol helps reduce the risk of cardiovascular disease, especially for men. For infants and children, however, too little fat and cholesterol in the diet may impair normal growth and development.[2] But it may also be important for children to avoid excess intakes of fat and cholesterol; Chapter 6 discusses the controversy over children's fat intake recommendations. The chapters to come deal with the key nutrients and with the dietary recommendations that apply to each stage of life.

Nutrition Assessment Basics

To give all the details of nutrition assessment procedures would entail writing another textbook. Any student of nutrition, however, should know the basics of a proper nutrition assessment procedure, for two reasons.

First, anyone seeking to deliver competent health care must be able to evaluate nutrition status. Health care facilities must conduct nutrition assessments as a routine part of the initial workup on every client—pregnant women, infants, children, adolescents, and all others—to ensure that diets meet the energy and nutrient requirements of normal growth and health. Medical care must also employ nutrition assessment techniques, or refer all clients to specialists skilled in them, so that poor nutrition will not hinder responses to treatment and recovery from illnesses.

Second, anyone seeking to obtain competent health care needs to know what to expect in the way of nutrition care. Nutrition is such a popular subject that many unqualified people claim it as their province. Thousands of people with fake degrees in nutrition, more than in any other field, offer useless advice. The knowledgeable consumer needs to know enough to avoid incompetent care.

The same basic nutrition assessment principles apply to all stages of the life span, but the details of emphasis differ from one stage to the next. The basics are presented here; the particulars for each stage of life are in the chapters to come. To avoid clutter, many of the tables and charts routinely used have been collected into Appendix A at the end of this book; the letter *A* in references to "Table A-1," "Figure A-1," and "Form A-1" alerts the reader to turn to that appendix, if desired.

Nutrition assessment evaluates many factors that influence or reflect nutrition status, using many sources. To prepare a nutrition assessment, the assessor, usually a registered dietitian or a physician trained in clinical nutrition, uses:

Nurses and registered dietetic technicians (RDT) are the professionals who most often help dietitians complete nutrition assessments.

- ■ Historical information.
- ■ Anthropometric data.
- ■ Physical (clinical) examinations.
- ■ Biochemical analyses (laboratory tests).

Each of these methods involves collecting data in various ways and interpreting each finding in relation to the others in order to create a total picture.

The accurate gathering of this information and its careful interpretation form the basis for a meaningful evaluation. Gathering information is a time-consuming process, though, and time is often scarce in the health care setting. A strategic compromise is to screen clients by col-

Taken as a whole, the information gathered during a nutrition assessment helps define a person's nutrition status.

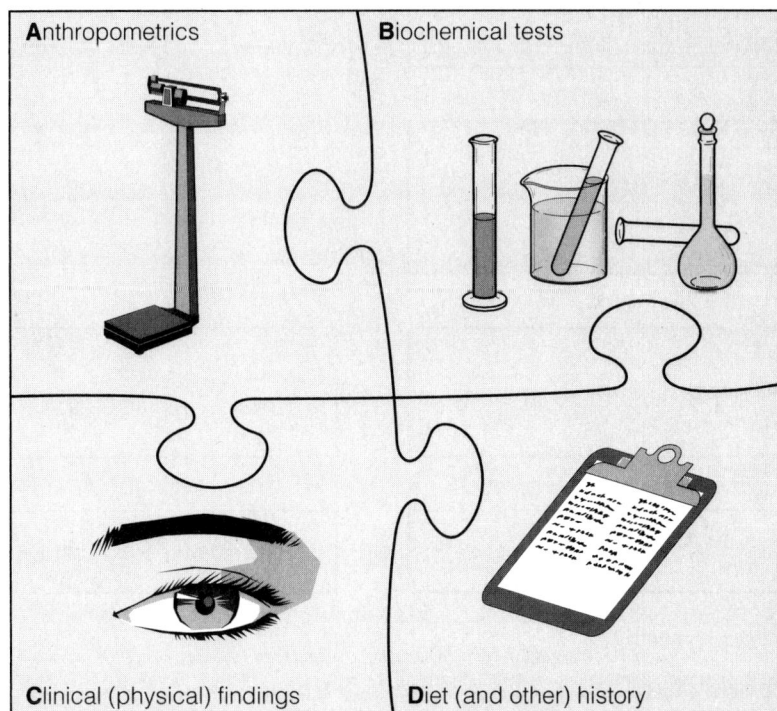

Anthropometrics

Biochemical tests

Clinical (physical) findings

Diet (and other) history

nutrition screening: the use of preliminary nutrition assessment techniques to identify people who are malnourished or who are at risk for malnutrition.

lecting preliminary data. Data such as height–weight and hematocrit are easy to obtain and can alert health care workers to potential problems. Nutrition screening identifies clients who will require additional nutrition assessment (see Figure 1-4).

The dietitian has the primary responsibility for assessing nutrition status and developing and implementing a client's nutrition care plan.

✶ Healthy People 2000

Increase to at least 75% the proportion of primary care providers who provide nutrition assessment and counseling and/or referral to qualified nutritionists or dietitians.

Figure 1-4 Nutrition Screening

A preliminary evaluation screens for possible nutrition disorders. If results of initial tests are abnormal, follow-up tests are conducted to further evaluate the disorder. If results of these tests indicate no nutrition disorder, a reevaluation is conducted at a later time. If results indicate abnormalities, steps are taken to correct them.

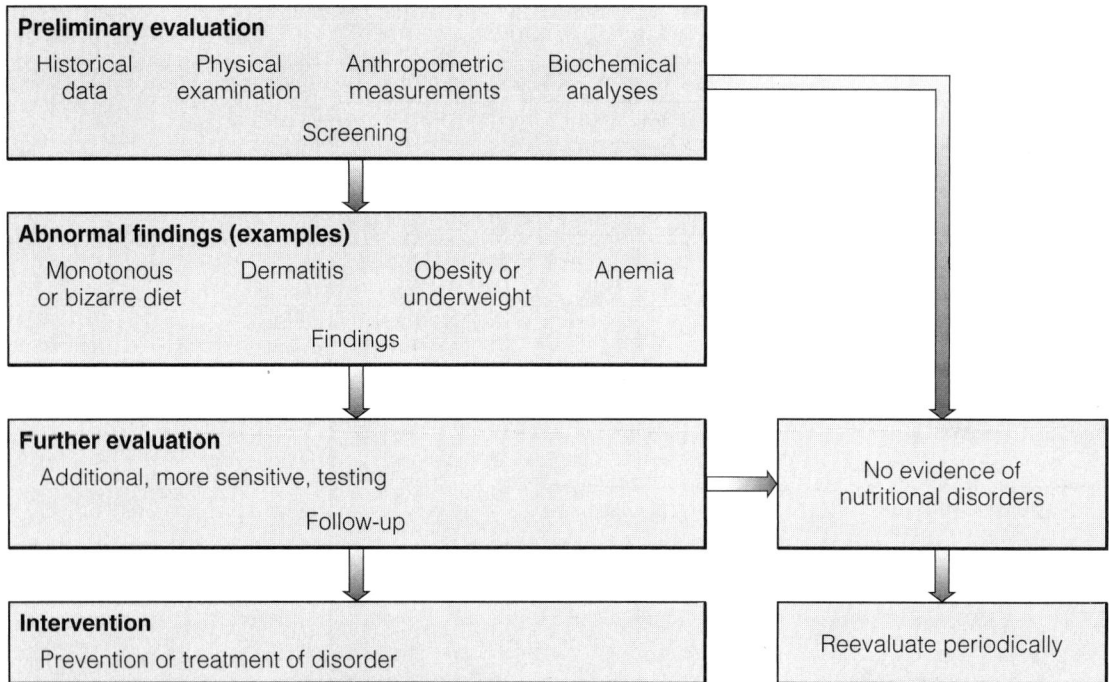

Preliminary evaluation

| Historical data | Physical examination | Anthropometric measurements | Biochemical analyses |

Screening

↓

Abnormal findings (examples)

| Monotonous or bizarre diet | Dermatitis | Obesity or underweight | Anemia |

Findings

↓

Further evaluation

Additional, more sensitive, testing

Follow-up

→ No evidence of nutritional disorders

↓

Intervention

Prevention or treatment of disorder

Reevaluate periodically

Source: Adapted from S. J. Fomon and coauthors, *Nutritional Disorders of Children: Prevention, Screening, and Followup*, DHEW (HSA) publication no. 76-5612 (Washington, D.C.: U.S. Government Printing Office, 1976), inside front cover.

The physician, nurse, dietetic technician, social worker, pharmacist, physical therapist, and occupational therapist also make valuable contributions. To the extent that all these people apply their nutrition knowledge, technical skills, and interpersonal skills, the care plan will be realistic and attainable.

Historical Information

Clues about present nutrition status become evident with a review of a person's historical records. A thorough history identifies risk factors associated with poor nutrition status and provides a sense of the whole person (see Table 1-5). An adept history taker uses the interview both to gather facts and to establish rapport with the client. A complete history can identify risk factors that will alert an assessor to nutrition problems (see Table 1-6). (Form A-1 in Appendix A shows the kinds of questions asked to collect such information.) The following paragraphs review the major areas of nutrition concern in a person's history: health, socioeconomic factors, drugs, and diet.

Health History. The assessor can obtain health histories from records completed by the attending physician, nurse, or other health care professional.* In addition, conversations with the person can uncover valuable information previously overlooked because no one thought to ask or because the client was not thinking clearly when asked. An accurate, complete health history can reveal conditions that increase the risks for malnutrition. Diseases and their treatments can have either immediate or long-term effects on nutrition status by interfering with ingestion, digestion, absorption, metabolism, or excretion of nutrients.

health history: the medical record.

*Traditionally, the health history has been called the medical history. The term *health history* now seems more appropriate, however, since the contents describe a client's health status. Current trends in the medical profession emphasize health promotion and disease prevention.

Table 1-5 Nutrition Assessment: Historical Data Routinely Used

Type of History	What It Identifies
Health history	Health factors that affect nutrition status
Socioeconomic history	Personal, financial, and environmental influences on food intake, nutrient needs, and diet therapy options
Drug history	Medications and nutrient supplements that affect nutrition status
Diet history	Nutrient intake excesses or deficiencies and reasons for imbalances

Socioeconomic History. Socioeconomic factors profoundly affect nutrition status. The ethnic background and educational level of both the client and the other members of the household influence food availability and food choices. Level of income also influences the diet. In general, the quality of the diet declines as income falls. At some point, the ability to purchase the foods required to meet nutrient needs is lost; an inadequate income puts an adequate diet out of reach. Agencies use

Table 1-6 Nutrition Assessment: Risk Factors for Poor Nutrition Status

Health History
- Acquired immune deficiency syndrome (AIDS)
- Alcoholism
- Anorexia (lack of appetite)
- Anorexia nervosa
- Bulimia nervosa
- Cancer
- Chewing or swallowing difficulties (including poorly fitted dentures, dental caries, and missing teeth)
- Chronic obstructive pulmonary disease
- Circulatory problems
- Constipation
- Crohn's disease
- Decubitus ulcers
- Dementia
- Diabetes mellitus

- Diarrhea
- Diseases of the GI tract
- Drug addiction
- Fever
- Heart disease
- Hormonal imbalance
- Hyperlipidemia
- Hypertension
- Infection
- Kidney disease
- Liver disease
- Lung disease
- Malabsorption
- Mental illness
- Mental retardation or deterioration
- Multiple pregnancies
- Nausea
- Neurological disorders

- Organ failure
- Overweight
- Pancreatic insufficiency
- Paralysis
- Physical disability
- Pneumonia
- Pregnancy
- Radiation therapy
- Recent major illness
- Recent major surgery
- Recent weight loss or gain
- Smoking
- Surgery of the GI tract
- Trauma
- Ulcerative colitis
- Ulcers
- Underweight
- Vomiting

Socioeconomic History
- Access to groceries
- Education
- Ethnic identity
- Income

- Kitchen facilities
- Number of people in household

- Occupation
- Religious affiliation
- Social activities

Drug History
- Amphetamines
- Analgesics
- Antacids
- Antibiotics
- Anticancer agents
- Anticonvulsant agents
- Antidepressant agents

- Antidiarrheal agents
- Antihyperlipemic agents
- Antihypertensive agents
- Antiulcer agents
- Catabolic steroids
- Diuretics
- Hormonal agents

- Immunosuppressive agents
- Laxatives
- Oral contraceptives
- Sulfonylurea agents
- Vitamin and other nutrient preparations

Diet History
- Deficient or excessive food intakes
- Frequently eating out
- Intravenous fluids (other than total parenteral nutrition) for 7 or more days

- Monotonous diet (lacking variety)
- No intake for 7 or more days

- Poor appetite
- Restricted or fad diets
- Unbalanced diet (omitting any food group)

poverty indexes to identify people at risk for poor nutrition and to qual-ify people for government food assistance programs.

Low income affects not only the power to purchase foods but also the ability to shop for, store, and cook them. A skilled assessor will note whether a person has transportation to a grocery store that sells a sufficient variety of low-cost foods, and whether the person has access to a refrigerator and stove.

Drug History. The many interactions of foods and drugs require that health care professionals pay special attention to any person who takes drugs routinely. Nutrients and drugs may interact in many ways:

- Drugs can alter food intake and the absorption, metabolism, and excretion of nutrients.
- Foods and nutrients can alter the absorption, metabolism, and ex-cretion of drugs.

Focal Point 10 summarizes the mechanisms by which these interactions occur and provides specific examples.

Adverse interactions are most likely in elderly people, especially those with serious illnesses that raise nutrient needs, because these people are most likely to be in poor nutrition status, to suffer chronic disorders, to use many drugs, and to take them over long periods of time. If a person is taking any drug, the assessor records the name of the drug; the dose, frequency, and duration of intake; the reason for taking the drug; and signs of any adverse effects (Form A-1). This information becomes one of the puzzle pieces that must be fitted together to obtain a picture of the person's nutrition status.

Diet History. A diet history provides a record of a person's eating habits and food intake and can identify possible nutrient imbalances. The accurate recording of such data requires skill. Food choices are an im-portant part of lifestyle and often reflect a person's philosophy. The as-sessor who asks nonjudgmental questions about eating habits and food intake encourages trust and enhances the likelihood of obtaining accu-rate information.

Assessors evaluate food intake using various tools such as the 24-hour recall, the usual intake record, the food record, and the food fre-quency checklist. (Sample forms used for these types of diet histories are Forms A-2, A-3, and A-4.) Such tools depend on an accurate account of portion sizes and food composition.[3] Food models or photos and mea-suring devices can help clients identify the types of foods and quantities consumed. The assessor also needs to know how the foods are prepared and perhaps when they are eaten.

The 24-hour recall provides data on one day only and is commonly used in nutrition surveys to obtain estimates of the typical food intakes for a population. An advantage of the 24-hour recall is that it is easy to obtain. It is also more likely to provide accurate data, at least about the past 24 hours, than estimates of average intakes over long periods. The usefulness of this tool is limited, however, in that it may not provide

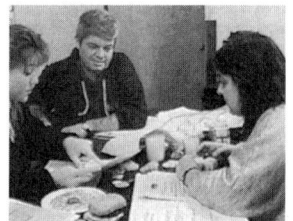

Clients provide food intake data most accurately when guided by a dietitian who uses plastic models of foods to illustrate portion sizes.

enough accurate information to allow for generalizations about an individual's usual food intake. The previous day's intake may not be typical, for example, or the person may be unable to report portion sizes accurately, or may conceal or forget information about foods eaten. This limitation is partially overcome when 24-hour recalls are collected on several nonconsecutive days.

To obtain data about a person's usual intake, an inquiry might begin with "What is the first thing you usually eat or drink during the day?" Similar questions follow until a typical daily intake pattern emerges. This method uses the same form as the 24-hour recall (Form A-2) and can be especially useful in verifying food intake when the past 24 hours have been atypical. It also helps the assessor verify food habits. For example, one person may always eat an afternoon snack; another may never eat breakfast. A person whose intake varies widely from day to day, however, may find it difficult to answer such general questions, and in that case, a food record over several days should be used to estimate nutrient intake.

To gather food intake data from a food record, the assessor instructs the person to write down all foods and beverages consumed, the time of day, the amounts consumed, and methods of preparation. Often the person must record other information as well, depending on the purpose of the food record. When the purpose of the record is to help a person change eating behaviors and lose weight, the record might also include information about the person's mood, the occasion (party, holiday, family meal), behaviors associated with eating food (watching TV, driving in the car on the way to work, sitting at the table with family), and physical activity. When the purpose of the food record is to establish blood glucose control, records include details of drug administration, physical activity, and the results of blood glucose monitoring. When the purpose of the record is to establish food tolerances (such as an allergy or the amount of lactose a person can handle), food records also include symptoms associated with eating (for example, cramps, diarrhea, nausea, or hives).

Food records work especially well with cooperative people but require considerable time and effort on their part. A prime advantage is that the recordkeeper assumes an active role and may for the first time become aware of personal food habits and assume responsibility for them. It also provides the assessor with an accurate picture of the person's lifestyle and factors that affect food intake. The major disadvantages stem from poor compliance in recording the data and conscious or unconscious changes in eating habits that may occur while the person is keeping the record.

Another diet history approach is to use a food frequency checklist to ascertain how often an individual eats a specific type of food. This information helps pinpoint food groups, and therefore nutrients, that may be excessive or deficient in the diet. That a person ate no vegetables yesterday may not seem particularly significant, but never eating vegetables is a warning of possible nutrient deficiencies. When used with the usual intake or 24-hour recall approach, the food frequency record enables the assessor to double-check the accuracy of the information obtained.

Diet Analysis. After collecting food intake data, the assessor estimates nutrient intakes, either informally by using food guides (such as the pyramid) or formally by using food composition tables. An informal evaluation is possible only if the assessor has enough prior experience with formal calculations to "see" nutrient amounts in reported food intakes without calculations. Even then, such an informal analysis is best followed by a spot check for key nutrients by actual calculation. The formal calculation can be performed either manually (by looking up each food in a table of food composition, recording its nutrients, and adding them up) or by using a computer diet analysis program. These intakes are compared with standards, such as the RDA, dietary guidelines, or Food Guide Pyramid, to determine how closely the diet meets the recommendations.

Diet histories can be superbly informative, but they also have limitations. For example, a computer diet analysis tends to imply greater accuracy than is possible to obtain from data as uncertain as those that provide the starting information. Nutrient contents of foods listed in tables of food composition or stored in computer databases are averages, and for some nutrients, incomplete. In addition, the available data on nutrient contents of foods do not reflect the amounts of nutrients a person actually absorbs. Iron is a case in point: Its availability from a given meal may vary from as high as 35 percent to less than 2 percent, depending on the person's iron status; the relative amounts of heme iron, nonheme iron, vitamin C, meat, fish, and poultry eaten at the meal; and the presence of inhibitors of iron absorption such as tea, coffee, and nuts.

Furthermore, reported portion sizes may not be correct. The person who reports eating "a serving" of greens may not distinguish between 1/2 cup and 1 whole cup; only trained individuals can accurately report serving sizes. Children tend to remember the serving sizes of foods they like as being larger than serving sizes of foods they dislike.

Thus any comparison of reported nutrient intakes with nutrient needs provides many opportunities for error. Most professionals learn to use shortcut systems to obtain rough estimates of nutrient intakes and then use calculation methods to pinpoint suspected nutrient deficiencies or imbalances.

When combined with other sources of information, a diet history confirms or eliminates the possibility of suspected nutrition problems. Consider the concerns during special stages of life mentioned earlier. Body weight was the first one. When a person is overweight or underweight, a diet history can detect an excessive or deficient food energy intake. Other areas of concern were iron and calcium status. A food frequency checklist may be especially valuable for assessing dietary iron and calcium, because the most significant food sources of each of these nutrients are found in a single food group: meat and meat alternates for iron, and milk and milk products for calcium. The absence or scarcity of

Meat, **f**ish, and **p**oultry contain a factor known as the **MPF factor** that enhances the absorption of nonheme iron present in the same foods or in other foods eaten at the same time. **Nonheme iron** is poorly absorbed, but commonly found in both plant-derived and animal-derived foods. In contrast, **heme iron** is well absorbed, but only found in foods derived from the flesh of animals.

these foods on a food frequency checklist will alert a clinician to obtain more information about iron and calcium intakes; to look at other factors that influence iron and calcium status; and to proceed with biochemical tests of iron status, as described in Appendix A. If the diet history reveals a high fat intake, an evaluation of other indicators of cardiovascular disease risk may be necessary and appropriate diet advice given as well.

Nutrient intakes in sufficient amounts do not guarantee adequate nutrient status. Likewise, insufficient intakes do not always indicate deficiencies, but instead alert the client to possible nutrition concerns. Because each person digests, absorbs, metabolizes, and excretes nutrients in a unique way, individual needs vary. Intakes of nutrients are only pieces of a puzzle that must be put together with other indicators of nutrition status to extract meaning.

Anthropometric Data

anthropometrics: relating to measurements of the physical characteristics of the body, such as height and weight (*anthropos* = human; *metric* = measuring).

Anthropometrics are physical measurements that reflect body composition and development (see Table 1-7). They serve three main purposes: first, to evaluate the progress of growth in pregnant women, infants, children, and adolescents; second, to detect undernutrition and overnutrition in all age groups; and third, to measure changes in body composition over time.

Health care professionals compare anthropometrics taken on an individual with population standards specific for sex and age and with previous measures of the individual. Measurements reveal how individuals compare both with others in a similar population and with themselves over time.

Taking measurements is easy and requires minimal equipment. Mastering the techniques requires proper instruction and practice to ensure reliability. Once the correct techniques are learned, variation is minimized.

Height and Weight. Height and weight are among the most common and useful assessment measurements. Length measurements for infants and children up to age three and height measurements for children over three are particularly valuable in assessing growth and therefore nutrition status. Chapters 5 and 6 describe preferred methods of measuring height in infants and children. For adults, height measurements alone are not critical but help to estimate desirable weight and to interpret other assessment data.

Unfortunately, many health care professionals merely ask clients how tall they are rather than measuring their height. Self-reported height is often inaccurate and should be used only as a last resort when measurement is impractical (in the case of an uncooperative client, an emergency hospital admission, or the like).

For measuring weight, beam balance and electronic scales are the most accurate types of scales. Bathroom scales are inaccurate and inappropriate in a professional setting. To make repeated measures useful,

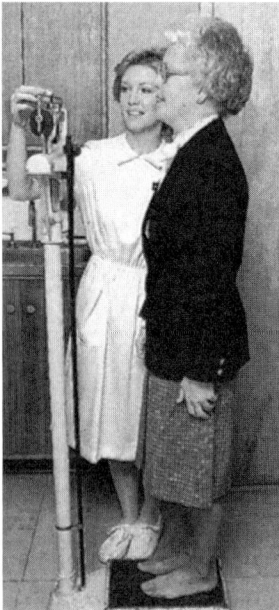

Beam balance scales provide accurate weight measurements for adults.

Table 1-7 Nutrition Assessment: Anthropometric Measurements Routinely Used

Type of Measurement	What It Reflects
Height-weight	Overnutrition and undernutrition; growth in pregnant women, infants, children, and adolescents
%DBW/IBW, %UBW,ᵃ recent weight change	Overnutrition and undernutrition
Head circumference	Brain growth and development in infants and children under two
Midarm circumference	Muscle mass and subcutaneous fat
Fatfold	Subcutaneous and total body fat
Midarm muscle circumference	Muscle mass (i.e., protein status)

ᵃ%DBW/IBW = percentage of desirable body weight or ideal body weight; %UBW = percentage of usual body weight.

standardized conditions are necessary. Each weighing should take place at the same time of day (preferably before breakfast), in the same amount of clothing (without shoes), after the person has voided, and on the same scale. As with all measurements, observed weight is immediately recorded in either kilograms or pounds.

Assessment of weight for height requires comparison with standards. The standards used for infants and children, the growth charts, are described in Chapters 5 and 6, respectively. For adults, health care professionals typically compare weights with weight-for-height tables, as shown in Chapter 8.

Body Mass Index. Another way health professionals assess body weight is to use a standard derived from the height and weight measures—the body mass index (BMI):

$$BMI = \frac{\text{weight (kg)}}{\text{height (m)}^2}$$

BMI values correlate with disease risks and indicate that both underweight and overweight incur health risks. The BMI ranges considered desirable, overweight, and obese are given for pregnant women, adolescents, and adults in their respective chapters.

Fatfold Measurements. Measures of body fatness are more meaningful than measures of mere weight, both for determining whether weight is appropriate and for assessing disease risks. A practical method for estimating body fatness is to use a fatfold caliper—a device that measures the thickness of a fold of fat on the back of the arm, below the shoulder blade, on the side of the waist, or other specific sites (see Figure A-2). About half of the fat in the body lies directly beneath the skin, so the thickness of this subcutaneous fat reflects total body fat.

body mass index (BMI): an index of a person's weight in relation to height, determined by dividing the weight (in kilograms) by the square of the height (in meters).

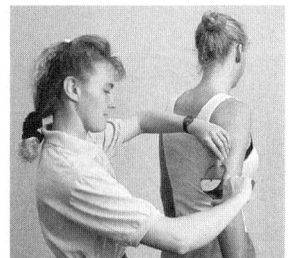

Obtaining an accurate fatfold measurement requires training in the use of a caliper that has been calibrated.

central obesity: excess fat around the trunk of the body; also called **upper-body fat**.

visceral fat: fat stored within the abdominal cavity in association with the internal organs, as opposed to fat stored directly under the skin (**subcutaneous fat**).

A major limitation of the fatfold test is that fat may be thicker under the skin in one area than another. This limitation can be overcome by taking fatfold measurements at three or more different places on the body (including both central- and lower-body sites) and comparing each measurement with standards for that site.

The distribution of fat on the body may be more critical than fatness alone. Visceral fat that is stored around the organs of the abdomen presents a greater risk to health than fat elsewhere on the body and increases the risk of premature death.[4] This distribution of fat is referred to as central obesity or upper-body fat and, independently of total body fat, is associated with increased risks of heart disease, stroke, diabetes, hypertension, and some types of cancer.[5]

Physical Examinations

One clinician astutely summarized the role of the physical examination in nutrition assessment this way: "To me, physical examination proves the saying 'A picture is worth a thousand words.'"[6] Indeed, health care professionals can often tell from looking at people whether they are overweight, underweight, lethargic, confused, or unable to feed themselves, to give a few examples.

With closer examination, a skilled assessor can use a physical examination to search for signs of nutrient deficiencies or toxicities. Every part of the body that can be inspected can offer such clues: the hair, eyes, skin, posture, tongue, fingernails, and others. Like other assessment methods, such an examination requires knowledge and skill, for many physical signs can reflect more than one nutrient deficiency or toxicity or even conditions not related to nutrition. For example, cracked lips may be caused by sunburn, windburn, dehydration, or any of several B vitamin deficiencies, to name just a few possible causes. For this reason, physical findings by themselves are especially unreliable for diagnosing a nutrition problem. Instead, their value is in revealing possible nutrient imbalances or in confirming data collected from other assessment measures.

A peek inside the mouth provides clues to a person's nutrition status.

Physical examinations can reveal a lot about a person's nutritional health. Table 1-8 lists general physical signs of malnutrition. (Table A-5 lists a few signs of specific vitamin and mineral imbalances.)

Biochemical Analyses

All the approaches to nutrition assessment discussed so far are external approaches. Biochemical analyses or laboratory tests help determine what is happening to the body internally. Most tests are based on analyses of blood and urine samples, which contain nutrients, enzymes, and metabolites (end products) that reflect nutrition status.

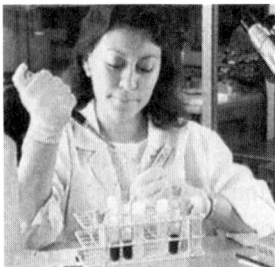

Blood and urine samples reflect the body's use of nutrients.

Interpreting biochemical data requires an understanding of the long metabolic sequences that lead to the production of the compounds seen in blood and urine. No single test can reveal nutrition status because many factors influence test results. The low blood concentration of a nutrient may reflect a primary deficiency of that nutrient, but it may

Table 1-8 Nutrition Assessment: General Physical Signs That May Indicate Malnutrition

	Healthy	Malnourished
Hair	Shiny, firm in the scalp	Dull, brittle, dry, loose; falls out
Eyes	Bright, clear pink membranes; adjust easily to darkness	Pale membranes; spots, redness; adjust slowly to darkness
Teeth and gums	No pain or cavities, gums firm, teeth bright	Missing, discolored, decayed teeth; gums bleed easily and are swollen and spongy
Face	Good complexion	Off-color, scaly, flaky, cracked skin
Glands	No lumps	Swollen at front of neck and cheeks
Tongue	Red, bumpy, rough	Sore, smooth, purplish, swollen
Skin	Smooth, firm, good color	Dry, rough, spotty; "sandpaper" feel or sores; lack of fat under skin
Nails	Firm, pink	Spoon-shaped, brittle, ridged
Behavior	Alert, attentive, cheerful	Irritable, apathetic, inattentive, hyperactive
Internal systems	Heart rate, heart rhythm, and blood pressure normal; normal digestive function; reflexes and psychological development normal	Heart rate, heart rhythm, or blood pressure abnormal; liver and spleen enlarged; abnormal digestion; mental irritability, confusion; burning, tingling of hands and feet; loss of balance and coordination
Muscles and bones	Good muscle tone and posture; long bones straight	"Wasted" appearance of muscles; swollen bumps on skull or ends of bones; small bumps on ribs; bowed legs or knock-knees

Note: The physical signs shown here are consistent with malnutrition but not diagnostic of it.

also be secondary to the deficiency of one or several other nutrients or to a disease. Taken together with other assessment data, however, laboratory test results help to make a total picture that becomes clear with careful interpretation. Biochemical analyses are especially useful in helping detect a subclinical deficiency by uncovering early signs of malnutrition before the clinical signs of a classic deficiency disease appear.

Laboratory tests most commonly performed to detect nutrition-related health status include measures of protein-energy status, blood lipid measures, blood glucose determinations, and measures of vitamin and mineral status. Protein-energy malnutrition (PEM) is uncommon in the United States, but is a major malnutrition problem in other parts of the world.

Blood Lipids. The determination of blood lipid concentrations provides a way of identifying people who are likely to develop cardiovascular disease. Table FP6C-1 in Focal Point 6C shows percentile classifications of blood cholesterol concentrations for children, and Table 9-2 in Chapter 9 shows initial classifications for adults and recommended follow-up.

primary deficiency: a nutrient deficiency that develops because dietary intake of the nutrient is inadequate.

secondary deficiency: a nutrient deficiency that develops because the body's handling of the nutrient is inadequate. For example, a disease may reduce absorption, accelerate use, hasten excretion, or destroy the nutrient; consequently, a deficiency will develop even when the nutrient is supplied in amounts sufficient for most healthy people.

Blood Glucose. Tests of blood glucose can identify people who may have diabetes mellitus. A fasting glucose concentration of greater than 140 milligrams per 100 milliliters blood on more than one occasion establishes a positive diagnosis. Another tool for diagnosing diabetes is the glucose tolerance test. Diabetes during pregnancy (discussed in Focal Point 2B) poses special concerns for both mother and infant.

Vitamin and Mineral Status. Laboratory tests used to assess vitamin and mineral status (see Table A-6) are particularly useful when combined with diet histories and physical findings. Vitamins and minerals present in the blood and urine sometimes reflect recent rather than long-term intakes, however, which can make detecting subclinical deficiencies difficult. Furthermore, because many nutrients interact, the amounts of other nutrients in the body can affect a lab value for a particular nutrient. It is also important to remember that conditions other than nutrients (such as dehydration or diseases) influence biochemical measures.

subclinical deficiency: a nutrient deficiency in the early stages, before the outward signs have appeared.

Iron Status. Because iron deficiency is an extremely common nutrient deficiency and affects people at all stages of life, its assessment is particularly important. To assess iron status accurately, clinicians need to understand that iron deficiency develops in stages.[7] This section provides a brief overview of how assessors detect these stages, and Appendix A provides the standards for assessing iron status. In the first stage of iron deficiency, iron stores diminish. Measures of serum ferritin reflect iron stores and are most valuable in assessing iron status (see Table A-9).

Stages of iron deficiency:
■ *Iron stores diminish.*
■ *Transport iron decreases.*
■ *Hemoglobin production falls.*

The second stage of iron deficiency is characterized by a decrease in the iron transported within the blood: Serum iron falls, and the iron-carrying protein transferrin *increases* (an adaptation that enhances iron absorption). Together, these two measures can determine the severity of the deficiency—the more transferrin and the less iron in the blood, the more advanced the deficiency is (see Tables A-10 and A-11).

serum ferritin: a measure of iron status that provides a noninvasive estimate of the body's iron stores.

The third stage of iron deficiency occurs when the supply of transport iron diminishes to the point that it limits hemoglobin production. (Hemoglobin is the iron-containing pigment of the red blood cells, whose function is to carry oxygen.) Now the hemoglobin precursor, erythrocyte protoporphyrin, begins to accumulate as hemoglobin and hematocrit values decline (see Table A-12).

transferrin: the iron carrier protein. **Mucosal transferrin** transfers iron in the mucosal cells of the intestine to **blood transferrin**, which carries iron in the blood to the rest of the body.

Hemoglobin and hematocrit tests are easy, quick, and inexpensive, so they are the tests most commonly used in evaluating iron status; their usefulness is limited, however, because they are late indicators of iron deficiency. Furthermore, other nutrient deficiencies and medical conditions can influence their values. Diagnosis of iron deficiency requires the use of age-specific standards for infants and children, and for pregnant women, standards specific to pregnancy (see Tables A-7 and A-8).

hemoglobin: the globular protein of the red blood cells that carries oxygen from the lungs to the cells throughout the body (*hemo* = blood; *globin* = globular protein).

erythrocyte protoporphyrin (PRO-toe-POR-fe-rin): a precursor to hemoglobin.

Calcium Status. The assessment of calcium status is difficult because a calcium deficiency is not evident in a blood sample; blood calcium re-

hematocrit: measurement of the volume of the red blood cells packed by centrifuge in a given volume of blood; the volume reflects red blood cell size and number.

mains normal no matter what the bone content may be. Even an X ray of the bones reveals a calcium deficit only when it is so advanced as to be virtually irreversible. Evaluation of calcium status therefore depends on dietary assessment to reveal low calcium intakes and medical history to identify risk factors for excessive adult bone loss.

The Team Approach

Whether conducting a nutrition assessment or any other health care activity, wise health care professionals work together, sharing their unique expertise and skills to benefit their clients' health. Health care teams vary from institution to institution, but most often, core members of the team include physicians, nurses, dietitians, and pharmacists; other team members vary according to the specific medical problem.

Figure 1-5 shows an example of a health care team that provides nutrition support, listing each member's duties and their overlapping responsibilities. By working together, the members can integrate their knowledge and provide the client with the benefits of their combined expertise.

To appreciate the team approach, remember the old adage "Two heads are better than one." In this case, several heads are better still. Clients benefit when they have many eyes noting problems and many brains searching for solutions. In short, teamwork works.

Considerable knowledge has been gained from research into the nutrient needs and from the nutrition assessment of people in various stages of life. This text describes that knowledge, applies it to help improve people's nutrition status and health, and explores questions yet to be answered.

The team approach helps ensure safe and effective nutrition support.

Figure 1-5 The Nutrition Support Team

Members of the nutrition support team share many of the responsibilities for client care, but each member also provides unique services.

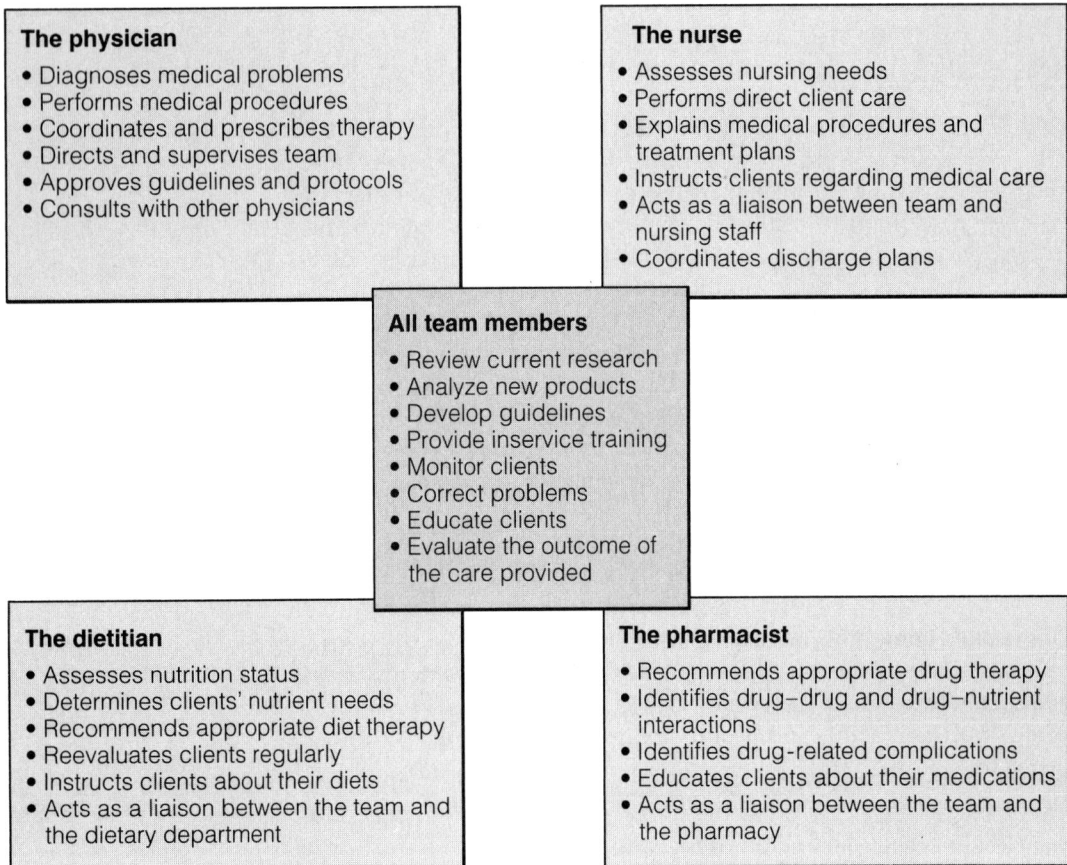

The physician
- Diagnoses medical problems
- Performs medical procedures
- Coordinates and prescribes therapy
- Directs and supervises team
- Approves guidelines and protocols
- Consults with other physicians

The nurse
- Assesses nursing needs
- Performs direct client care
- Explains medical procedures and treatment plans
- Instructs clients regarding medical care
- Acts as a liaison between team and nursing staff
- Coordinates discharge plans

All team members
- Review current research
- Analyze new products
- Develop guidelines
- Provide inservice training
- Monitor clients
- Correct problems
- Educate clients
- Evaluate the outcome of the care provided

The dietitian
- Assesses nutrition status
- Determines clients' nutrient needs
- Recommends appropriate diet therapy
- Reevaluates clients regularly
- Instructs clients about their diets
- Acts as a liaison between the team and the dietary department

The pharmacist
- Recommends appropriate drug therapy
- Identifies drug–drug and drug–nutrient interactions
- Identifies drug-related complications
- Educates clients about their medications
- Acts as a liaison between the team and the pharmacy

CHAPTER ONE NOTES

1. E. R. Monsen, New Dietary Reference Intakes proposed to replace the Recommended Dietary Allowances, *Journal of the American Dietetic Association* 96 (1996): 754–755.
2. M. T. Pugliese and coauthors, Parental health beliefs as a cause of non-organic failure to thrive, *Pediatrics* 80 (1987): 175–182.
3. L. R. Young and M. Nestle, Portion sizes in dietary assessment: Issues and policy implications, *Nutrition Reviews* 53 (1995): 149–158.
4. P. Björntorp, Regional adiposity, in *Obesity*, eds. P. Björntorp and B. N. Brodoff (Philadelphia: Lippincott, 1992), pp. 579–586.
5. M. Zamboni and coauthors, Obesity and regional body-fat distribution in men: Separate and joint relationships to glucose tolerance and plasma lipoproteins, *American Journal of Clinical Nutrition* 60 (1994): 682–687; E. M. Emery and coauthors, A review of the association between abdominal fat distribution, health outcome measures, and modifiable risk factors, *American Journal of Health Promotion* 7 (1993): 342–353; F. X. Pi-Sunyer, Health implications of obesity, *American Journal of Clinical Nutrition* 53 (1991): 1595S–1603S.
6. K. Hammond, Nutrition focused physical assessment, *Support Line*, August 1996, pp. 1–4.
7. V. Herbert, Everyone should be tested for iron disorders, *Journal of the American Dietetic Association* 92 (1992): 1502–1509.

Prelude to Pregnancy

A prelude to pregnancy must, by its nature, focus primarily on women—before and between pregnancies. Prepregnant women's needs differ from those of men, from those of pregnant and lactating women, and from those of women after the childbearing years. A man's nutrition may affect his fertility and possibly the genetic contributions he makes to his children; hence men's nutrition is discussed in later chapters where appropriate. It is through women that nutrition before conception exerts the greater influence, however, because a woman's body is the environment in which new human beings are grown.

A woman's body accomplishes remarkable feats. For approximately the first decade of life, her body's sole focus is on its own growth and maintenance. Then her body begins to release mature ova for reproduction, and after fertilization it provides for the development of a whole new human being. Even after the infant's birth, the woman's body offers nourishment for several more months. The health of the woman's body *before conception* influences her fertility, the health of the infants she may later conceive and bear, and her own health later in life.

The time before conception provides a unique opportunity for a woman to prepare herself physically, mentally, and emotionally for the many changes to come. To prepare her body as the most suitable environment for a fetus to develop in requires her to establish healthful habits.

Maintain a Healthy Body Weight. Any woman who hopes to become pregnant in the future is well advised to prepare by achieving and maintaining a healthy weight now. A woman's weight *before conception* influences her health throughout the pregnancy and the fetus's development. Both underweight and overweight women, and their newborns, face greater risks than others. By learning to eat nutrient-dense foods in the appropriate amounts before pregnancy, women can establish, in advance, eating habits that will support a healthy pregnancy later.

Choose a Diet Adequate in Nutrients. The diets that women of childbearing age typically choose rarely provide enough calcium, folate, or iron. Nutrient needs, especially for these three nutrients, intensify in pregnancy. To help maximize bone density and improve calcium status before the start of a pregnancy, a young woman should eat three to four

To prepare for a healthy pregnancy, young adults need to select foods that will provide needed nutrients and maintain a healthy body weight.

servings of calcium-rich foods daily. To consume enough calcium without exceeding her energy allowance, a woman must select fat-free and low-fat milk products and calcium-rich vegetables and legumes such as broccoli, turnip greens, and pinto beans.

Vegetables and legumes are also rich sources of folate, another nutrient of concern for women of childbearing age. An adequate folate intake during the month before conception and throughout the first trimester of pregnancy can help prevent one of the most common types of birth defects. All woman of childbearing age who are capable of becoming pregnant should eat plenty of fruits, vegetables, and legumes daily to ensure an adequate folate intake.

Women are also in a precarious state with respect to iron sufficiency during the procreative years. The combination of high losses (due to repeated loss of menstrual blood) and low intakes makes it a challenge to meet iron needs. All too often, women enter pregnancy with depleted iron stores. To *prevent* iron deficiency, women need to eat an iron-rich, absorption-enhancing diet. To *treat* iron deficiency, women may also need supplements. The absorption of iron taken as ferrous sulfate or as an iron chelate is better than from other kinds of supplements. Absorption also improves when the woman takes iron supplements between meals or at bedtime, on an empty stomach, and with liquids other than milk, tea, or coffee, which inhibit absorption. Whether a woman needs iron, or any other nutrient supplement, depends on many factors that only a thorough nutrition assessment can identify.

Except for iron, young women can normally get all the nutrients they need by eating a varied diet of whole foods. Unfortunately, many do not eat this way. Women who do not may benefit from multivitamin-mineral supplements that provide between 50 and 150 percent of the RDA for each nutrient.

Certainly, for the best possible health before pregnancy, women need to pay attention to their nutrient intakes. Another lifestyle factor that enhances health goes hand in hand with nutrition—fitness.

Be Physically Active. A healthy woman conditioned to physical activity before pregnancy can normally continue exercising throughout her pregnancy, but conception is not a good time to begin strenuous workouts. A woman who wants to be physically active *when* she is pregnant needs to become physically active *beforehand*.

Avoid Harmful Environmental Factors. In addition to nutrition, maternal health and fetal development also depend on genetics and other environmental factors. Because a woman's lifestyle habits and exposures before pregnancy continue to influence her body's chemistry when she is pregnant, the choices she makes can influence the course of a pregnancy she is not even planning at the time.

Environmental factors can affect fetal development in two ways—by altering the genes themselves and by altering their expression (discussed in Focal Point 2A). Genes provide potential—that is, the capacity to reach a certain developmental level. That potential can fail to be fully

Exercise prepares young adults for a variety of physical challenges—including pregnancy.

realized because of poor nutrition, chemical damage, radiation, lack of oxygen, or other conditions, which can all interfere with correct gene copying or prevent full or normal expression of genes. Congenital abnormalities, then, can result from adverse physical or chemical conditions during development of the ovum and spermatozoa or the fetus. The major nutrition-related conditions that have adverse consequences include malnutrition, obesity, and alcohol and other drug use, and are discussed in detail in Chapters 2 and 3.

During pregnancy, a woman's body becomes an environment that nurtures the development of another human being. This is life's cycle—from infancy to childhood to adolescence to pregnancy, returning to infancy. Each of these phases of life is vital to the cycle's continuation. Men and women leave the cycle only after the reproductive years are over, when they begin their journeys into later life. Until then, people who nourish and protect their bodies do so not only for their own sakes, but also for future generations.

Nutrition during Pregnancy

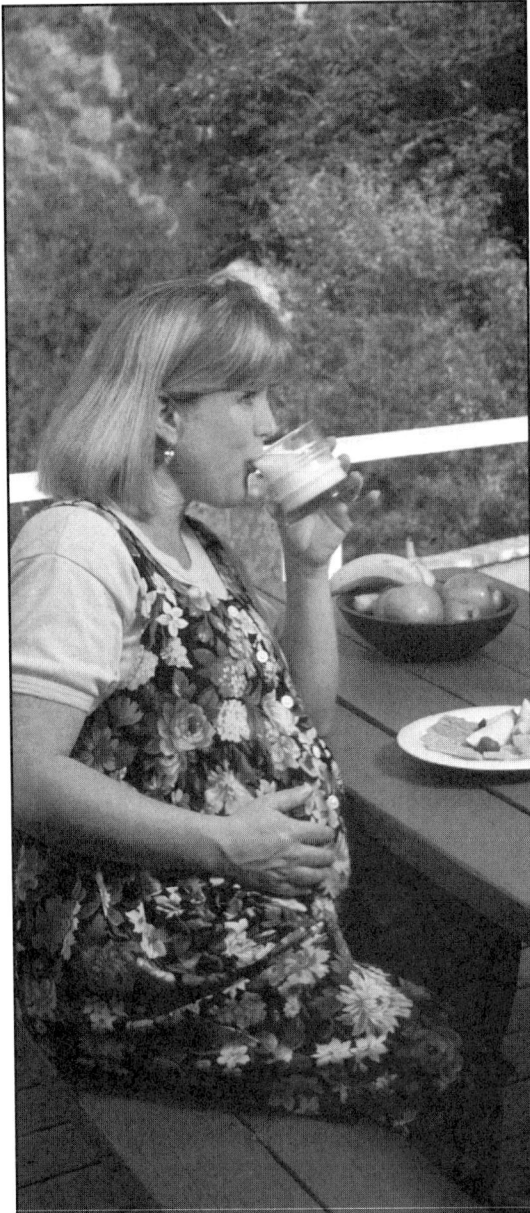

A whole new life begins at conception. Events follow quickly upon one another, and nutrition plays a supportive role in all of them. Nothing could be more extraordinary than the course of a pregnancy, yet this remarkable time is characterized as "normal." This chapter begins by describing the normal course of pregnancy and then proceeds to describe how nutrition supports that progress. The accompanying glossary defines terms used to describe periods in the lives of pregnant women and their newborn infants; Figure 2-1 places some of these terms on a time line.

conception: the union of the male spermatozoon and the female ovum; fertilization.

GLOSSARY *Pregnancy and Birth Terms*

These terms describe a woman's pregnancy status before, during, and after the birth of her child:

pregravid (pre-GRAV-id): before pregnancy (*pre* = before; *gravid* = pregnant).

gravid (GRAV-id): pregnant. A **gravida** (GRAV-ih-da) is a pregnant woman; **gravidity** is pregnancy. A woman during her first pregnancy is a **primigravida** (PRY-mee-gravida). A woman who has been pregnant two or more times is a **multigravida** (*primus* = first; *multi* = many).

postpartum: after childbirth (*post* = after; *partus* = birth).

These terms describe the number of a woman's pregnancies from the smallest to the largest number:

nullipara (nul-LIP-ah-ra): a woman who has borne no children; the adjective is **nulliparous** (nul-LIP-ah-rus) (*null* = none; *parere* = to bear).

primipara (pry-MIP-ah-ra): a woman who has borne, or is giving birth to, her first child; the adjective is **primiparous** (pry-MIP-ah-rus).

multipara (mul-TIP-ah-ra): a woman who has borne more than one infant, regardless of infant survival; the adjective is **multiparous** (mul-TIP-ah-rus).

Figure 2-1 Terms for Stages Surrounding Pregnancy and Birth

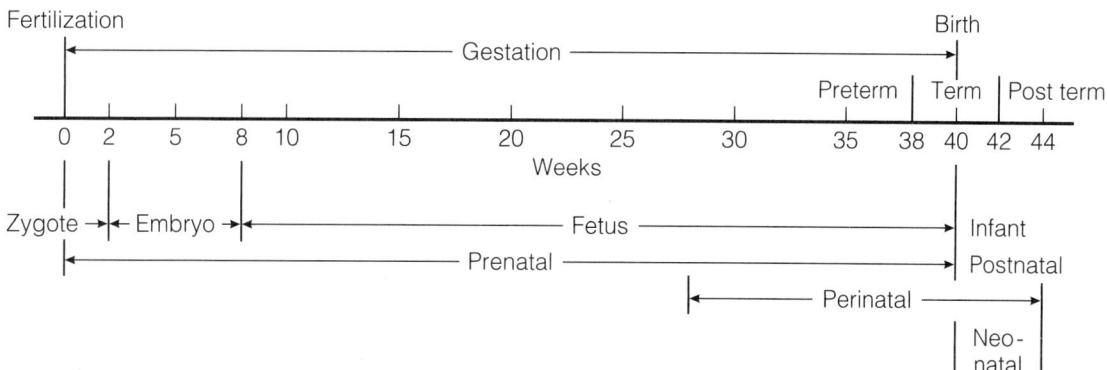

These terms describe the time surrounding birth:

prenatal: concerning the time before birth (*natal* = birth).

perinatal: concerning the time shortly before, during, or shortly after birth (*peri* = around).

neonatal: concerning the first four weeks after birth; a newborn infant is a **neonate** (*neos* = new).

postnatal: occurring after birth (*post* = after).

These terms describe the stages of intrauterine development:

zygote: the product of conception from zero to two weeks.

embryo: the developing organism from two weeks to eight weeks.

fetus: the developing organism from eight weeks to term (nine months).

These terms describe an infant's gestational age at birth:

gestation (jes-TAY-shun): the period from conception to birth; for human beings, gestation lasts from 38 to 42 weeks. Pregnancy is often divided into thirds, called **trimesters**.

preterm: an infant born prior to the 38th week of gestation; also referred to as a **premature** infant.

term: an infant born between the 38th and 42nd weeks of gestation.

post term: an infant born after the 42nd week of gestation.

Growth and Development during Pregnancy

Physiological events during pregnancy bring major changes to the mother's body, creating an entire new organ within it—the placenta—and producing a new human being. This section describes maternal physiology, placental development, and fetal growth, paying close attention to times of intense developmental activity.

Changes in Maternal Physiology

A woman's body changes dramatically during pregnancy. Her uterus and its supporting muscles increase in size and strength, her breasts grow and prepare to produce milk, and her blood volume increases by half to carry the additional nutrients and other materials. The normal weight gain during pregnancy is about 25 to 35 pounds. Increases in maternal tissues alone account for about two-thirds of this weight.

A woman's basal metabolism rises substantially during pregnancy as tissues are actively synthesized for the fetus and associated supportive structures. The maternal cardiovascular system, kidneys, and lungs work hard to deliver needed substances and to clear wastes. Over the nine months of gestation, all this activity is estimated to expend about 55,000 kcalories.[1]

Placental Development and Function

In the early days of pregnancy, the newly formed zygote is small and the nutrient-rich cells of the uterine lining provide adequate fuel and raw materials. Soon, though, a specialized system is required.

To continue nourishing the developing embryo and fetus, a spongy structure known as the placenta develops in the uterus. The placenta develops as an interweaving of fetal and maternal blood vessels embedded in the uterine wall (see Figure 2-2). Two associated structures also form. One is the amniotic sac, a fluid-filled balloonlike structure that houses the developing fetus. The other is the umbilical cord, a ropelike structure containing fetal blood vessels that extends from the fetus's "belly button" (the umbilicus) to the placenta. These three structures serve crucial roles during pregnancy and then are expelled from the uterus during childbirth.

The Blood Vessels within the Placenta. The fetal portion of the placenta develops when the membranes that surround the embryo invade the uterine wall and form villi. These villi contain embryonic blood vessels. The uterus responds by allowing pools of maternal blood to form around these villi so that maternal and fetal blood will be in close proximity.

zygote (ZY-goat): the product of the union of ovum and sperm; so called for the first two weeks after fertilization.

embryo (EM-bree-oh): the developing infant from two to eight weeks after conception.

fetus: the developing infant from eight weeks until term.

placenta (pla-SEN-tah): the organ that develops inside the uterus early in pregnancy, in which maternal and fetal blood circulate in close proximity so that materials can be exchanged between them.

Figure 2-2   The Placenta and Associated Structures

To understand how placental villi absorb nutrients without maternal and fetal blood interacting directly, think of how the intestinal villi work. The GI side of the intestinal villi is bathed in a nutrient-rich fluid (chyme). The intestinal villi absorb the nutrient molecules and release them into the body via capillaries. Similarly, the maternal side of the placental villi is bathed in nutrient-rich maternal blood. The placental villi absorb the nutrient molecules and release them to the fetus via fetal capillaries.

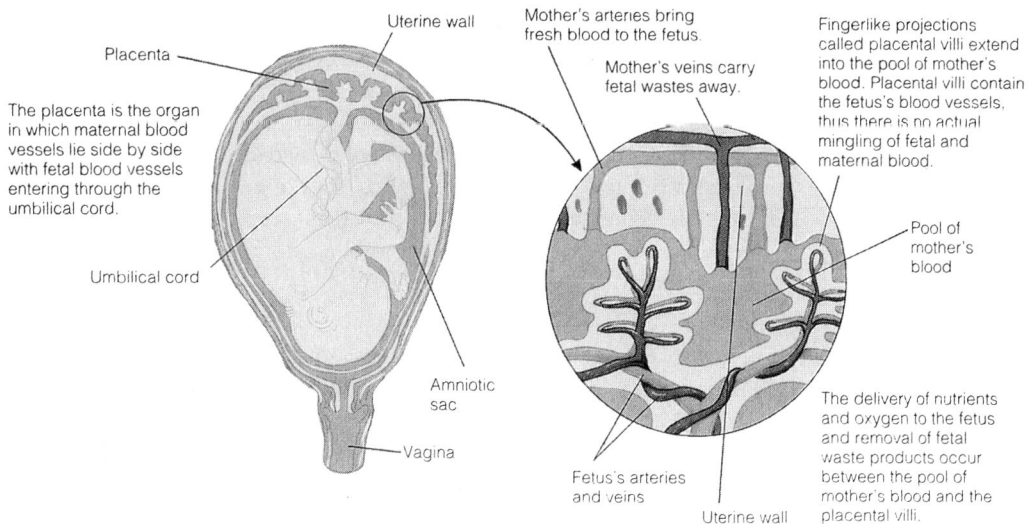

Uterine wall

Placenta

The placenta is the organ in which maternal blood vessels lie side by side with fetal blood vessels entering through the umbilical cord.

Umbilical cord

Amniotic sac

Vagina

Mother's arteries bring fresh blood to the fetus.

Mother's veins carry fetal wastes away.

Fetus's arteries and veins

Uterine wall

Fingerlike projections called placental villi extend into the pool of mother's blood. Placental villi contain the fetus's blood vessels, thus there is no actual mingling of fetal and maternal blood.

Pool of mother's blood

The delivery of nutrients and oxygen to the fetus and removal of fetal waste products occur between the pool of mother's blood and the placental villi.

amniotic (am-nee-OTT-ic) **sac:** the "bag of waters" in the uterus in which the fetus floats.

umbilical (um-BILL-ih-cul) **cord:** the ropelike structure through which the fetus's veins and arteries reach the placenta; the route of nourishment and oxygen into the fetus and the route of waste disposal from the fetus.

cardiac output: the amount of blood discharged per minute from the left or right ventricle of the heart.

Maternal blood enters the placenta through small artery branches of the uterus, bathes the villi, and leaves via the uterine veins. Fetal blood circulates from the umbilical artery, into the villi, and back through the umbilical vein. An exchange of materials takes place within the sponge-like endometrium, but no actual mingling of fetal and maternal blood occurs.

Blood flow to the placenta increases dramatically during pregnancy, accounting for as much as 25 percent of the maternal cardiac output by term. The rates of maternal and fetal blood flow influence the exchange of substances across the placenta. For example, as the flow of maternal blood increases, the delivery of oxygen to the fetus improves.

The Exchange of Nutrients and Wastes. The placenta transfers oxygen and nutrients to the fetus and returns carbon dioxide and other fetal waste products to the mother. Nutrients pass from maternal blood through the placenta into fetal blood; waste products move from fetal blood through the placenta into maternal blood. By exchanging oxygen, nutrients, and waste products, the placenta performs the respiratory, absorptive, and excretory functions that the fetus's lungs, gastrointestinal tract, and kidneys will provide after birth.

Water, oxygen, carbon dioxide, and electrolytes cross the placenta by passive diffusion. Glucose, calcium, phosphorus, magnesium, and other nutrients cross the placenta via active transport. Fetal accumulation of these nutrients increases throughout pregnancy, becoming greatest during the third trimester.

The Metabolism of the Placenta. The placenta is a versatile, metabolically active organ, producing some 60 enzymes. Much like muscles and other body tissues, the placenta uses energy fuels to support its work. It metabolizes glucose for its own energy needs, as well as actively pumping glucose into the fetal bloodstream. The placenta also synthesizes proteins and fatty acids from small precursors.

Glucose is the main energy source for fetal growth, and as mentioned, the placenta actively transfers it to the fetus. The placenta also stores glycogen, reflecting the vital importance of maintaining the glucose supply. Placental glycogen stores increase early in pregnancy, but decline during later pregnancy as they give up their glucose to the rapidly growing fetus. By the third trimester, the fetus is producing its own insulin. This insulin promotes fetal storage of glycogen as well as of protein and fat for use after delivery.

human chorionic gonadotropin (HCG): a hormone produced by the chorionic villi of the placenta that stimulates the secretion of the hormones progesterone and estrogen.

The Hormones of the Placenta. The placenta produces an array of hormones that maintain pregnancy and prepare the mother's breasts for lactation, the making of milk. Of the placental hormones, most notable are human chorionic gonadotropin (HCG) and placental lactogen. Early placental secretion of these hormones promotes placental production of the hormone progesterone from maternal cholesterol. Placental synthesis of the major estrogen of pregnancy, estriol, requires a precursor from the fetal adrenal glands. Together, the placenta and fetal adrenals produce the estriol required to maintain pregnancy.

HCG increases rapidly during early pregnancy, reaching its maximum concentration in the first eight to ten weeks, and then diminishing to a low but constant concentration at the end of the third month. Most pregnancy tests are based on the detection of this hormone in urine or plasma.

The secretion of placental lactogen starts at low levels and increases progressively. Late in pregnancy, high placental lactogen concentrations stimulate growth of the mammary glands. Placental lactogen also inhibits insulin's action on adipose and muscle tissue. Normally, insulin induces these tissues to take up amino acids, glucose, and fat, but lactogen keeps these fuels in circulation to support both the fetus and mammary gland activity.

A healthy placenta is essential for the developing fetus to attain its full genetic potential.[2] Maternal nutrition, in turn, is crucial to placental development. Maternal malnutrition limits placental development, which reduces placental blood flow. A diminished blood flow curtails the transfer of energy and essential nutrients to the fetus. Thus maternal malnutrition negatively affects the fetus first by impairing placental development and then by reducing the transfer of nutrients across the placenta.

Fetal Growth and Development

Fetal development begins with the fertilization of an ovum by a sperm. Three stages follow: the zygote, the embryo, and the fetus

The Zygote. The newly fertilized ovum, or zygote, begins to divide within a day after fertilization. It floats freely within the uterus for several days, receiving nourishment from the intrauterine fluid. As cell division continues, the uterine wall, or endometrium, prepares to receive the zygote.

Within two weeks, the zygote embeds itself in the endometrium—a process known as implantation. Cell division continues—each set of cells divides into many cells, and these cells sort themselves into layers. The outermost layer will form the placenta and amniotic sac; the innermost cells become the embryo.

The Embryo. The embryo accomplishes amazing developmental feats. The number of cells in the embryo doubles approximately every 24 hours at first; later the rate slows, and only one doubling occurs during the final ten weeks of pregnancy. The embryo's size changes very little up to the eighth week of pregnancy, but the events taking place during that time are momentous.

As the cells divide, they sort themselves into three layers—the outermost layer, or ectoderm; the middle layer, or mesoderm; and the innermost layer, the endoderm. From the ectoderm, the nervous system

placental lactogen: a hormone produced by the placenta during the second half of pregnancy that stimulates the metabolism of glucose to fat.

estriol (ESS-tree-ol): the major estrogen of pregnancy, produced by the placenta.

ovum: the female reproductive cell, capable of developing into a new organism upon fertilization; commonly called an egg; *ova* is the plural (*ovum* = egg).

sperm: the male reproductive cell, capable of fertilizing an ovum; formally called **spermatozoon** (SPER-mat-oh-ZOH-on), *spermatozoa* is the plural (*sperma* = seed; *zoon* = life).

endometrium (en-doe-MEE-tree-um): the membrane lining the inner surface of the uterus.

implantation: the stage of development in which the zygote embeds itself in the wall of the uterus and begins to develop; occurs during the first two weeks after conception.

ectoderm: the outermost layer of a developing embryo, which evolves into the nervous system and skin (*ecto* = outside; *derm* = skin).

mesoderm: the middle layer of a developing embryo, between the ectoderm and endoderm, which evolves into the muscular, skeletal, circulatory, and internal organ systems (*meso* = middle).

endoderm: the innermost layer of a developing embryo, which evolves into the glands and linings of the digestive, respiratory, and excretory systems (*endo* = inside).

and skin begin to develop; from the mesoderm, the muscles and internal organ systems; and from the endoderm, the glands and linings of the digestive, respiratory, and excretory systems.

At eight weeks, the 1 1/4-inch embryo has a complete central nervous system, a beating heart, a digestive system, well-defined fingers and toes, and the beginnings of facial features. Already an embryonic tail has formed and almost completely disappeared again.

The Fetus. During the next seven months, each organ grows to maturity according to its own schedule, with greater intensity at some times than at others. As Figure 2-3 shows, fetal growth is phenomenal: Weight increases from less than an ounce to about 7 1/2 pounds (3500 grams).

Figure 2-3 Stages of Embryonic and Fetal Development

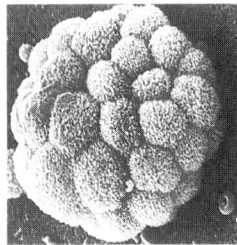

1. A newly fertilized ovum is about the size of the period at the end of this sentence. This zygote at less than one week after fertilization is not much bigger and is ready for implantation.

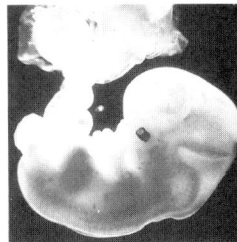

2. After implantation, the placenta develops and begins to provide nourishment to the developing embryo. An embryo five weeks after fertilization is about 1/2 inch long.

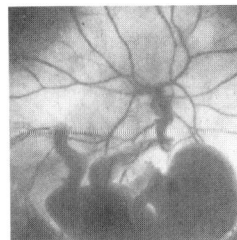

3. A fetus after 11 weeks of development is just over an inch long. Notice the umbilical cord and blood vessels connecting the fetus with the placenta.

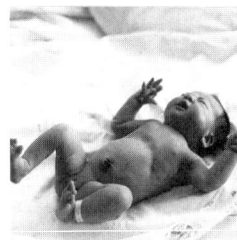

4. A newborn infant after nine months of development measures close to 20 inches in length. From eight weeks to term, this infant grew 20 times longer and 50 times heavier.

The gestation period, which normally lasts from 38 to 42 weeks, ends with the birth of the infant.

As mentioned, maternal malnutrition can impair fetal development by impairing placental development and by reducing the nutrients available for transfer across the placenta. In either case, malnutrition exerts its effects during critical periods of development. A word of introduction on the concept of critical periods is therefore in order.

Critical Periods

As development proceeds, intracellular activity becomes intense. Cell division requires that every organelle of the cell be duplicated—the mitochondria, the ribosomes, the vast and intricate network of intracellular membranes, the cell membrane, the nuclear membrane, the chromosomes with their extensive DNA blueprints for proteins, and numerous other intracellular bodies. To make all these new organelles requires synthesis and assembly of a multitude of different materials—DNA, RNA, proteins, lipids, and more. These syntheses are carried out by enzymes that themselves have been synthesized according to instructions from the genes—a process that requires extensive activity and coordination. Enzymes transcribe the relevant portions of DNA to make messenger RNA for delivery to the protein-making machinery of the cell, the ribosomes; transfer RNA carry amino acids to the messenger RNA, which dictates the sequence of attachment; enzymes link those amino acids together into new proteins; and enzymes and membrane proteins make or bring in new amino acids to make more proteins.

The number of molecular events taking place in a cell about to divide is astronomical, and the orderly sequence and completion of these events are crucial to normal development. Furthermore, any impairment of early cell multiplication may ultimately have greatly amplified effects, because each cell in an early stage of development is destined to become many cells later.

These times of intense development and rapid cell division are called critical periods—critical in the sense that the events scheduled for those times can occur only at those times. If cell division and the final cell number achieved in an organ are limited during a critical period, full recovery will not occur (see Figure 2-4).

critical periods: finite periods during development in which certain events may occur that will have irreversible effects on later developmental stages.

Increases in Cell Number and Size. The development of all organs occurs by the processes of cell division and cell growth. During cell division, cells increase dramatically in *number* (hyperplasia), and during cell growth, cells increase in *size* (hypertrophy). These two processes may occur simultaneously, may overlap, or may occur in tandem. Each developing organ follows its own unique schedule. For example, the heart is already well developed at 16 weeks, but the lungs remain immature until after 26 weeks.

hyperplasia: an increase in cell number.

hypertrophy: an increase in cell size.

The times of most obvious growth are when cells increase in size, but critical periods of cell division and differentiation may precede

Figure 2-4   The Concept of Critical Periods

Critical periods occur early in development. An adverse influence felt early can have a much more severe and prolonged impact than one felt later on.

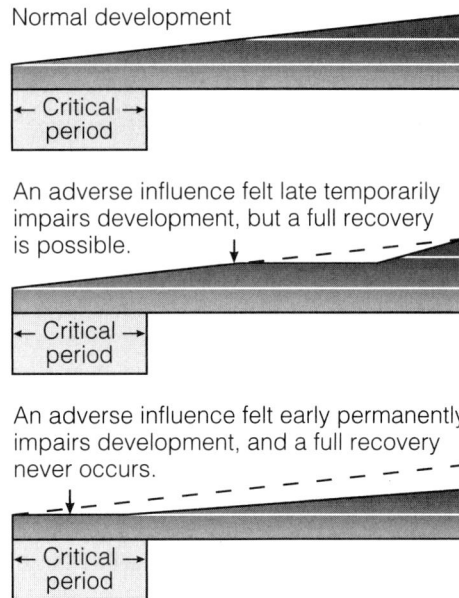

Normal development

← Critical → period

An adverse influence felt late temporarily impairs development, but a full recovery is possible.

← Critical → period

An adverse influence felt early permanently impairs development, and a full recovery never occurs.

← Critical → period

them. If the organ is to reach its full potential, nutrient availability and other environmental conditions must be optimal from the beginning. If cell division and the final cell number achieved in an organ are limited during a critical period, organ development cannot proceed normally and recovery is impossible. Early malnutrition can have irreversible effects, although they may not become fully apparent until the person reaches maturity.

Adverse Influences. Some research suggests that adverse influences at critical times set the stage for chronic diseases in adult life.[3] Undernutrition during "critical periods" may permanently alter body functions such as blood pressure, cholesterol metabolism, and immune functions that influence disease development.[4] Malnutrition during the critical period of pancreatic cell growth, for example, may be responsible for the development of diabetes mellitus in adulthood.[5] The pancreatic cells responsible for producing insulin (the beta cells) normally increase more than 130-fold between 12 weeks gestation and five months postnatal. Nutrition is a primary determinant of beta cell growth, and infants who have suffered prenatal malnutrition have significantly fewer beta cells than well-nourished infants. One hypothesis suggests that diabetes may develop from the interaction of inadequate nutrition early in life with abundant nutrition later in life: The mass of beta cells developed in lean times may be insufficient during times of overnutrition when the body needs more insulin.[6] Studies have found an association between low birthweight (an indicator of prenatal malnutrition) and glucose intolerance in adulthood.[7] Furthermore, the degree of intolerance appears to be influenced by the adults' body mass index values. Another study confirms the association between low birthweight and later glucose intolerance, but cautions that additional genetic and environmental factors are probably also involved.[8]

Each organ and tissue is most vulnerable to nutrient deficiencies or toxins during its own critical period. The critical period for neural tube development, for example, is from 17 to 30 days gestation.[9] The neural tube is the beginnings of the brain and spinal cord, key structures in the central nervous system. Recall that in the early weeks of embryo development, the nervous system and skin both derive from the ectoderm. Normally, by the fourth week of gestation, the ectoderm that forms the neural tube of the nervous system separates from the ectoderm that forms the skin. Neural tube development is most vulnerable to nutrient deficiencies or toxins at this critical time of differentiation. Any abnormal development of the neural tube or its failure to close completely can cause a major defect in the central nervous system. Chapter 3 presents research findings on the role of the vitamin folate in preventing neural tube defects.

Similarly, the phase of brain development most susceptible to malnutrition seems to be the growth spurt, when brain weight increases most rapidly. In human beings, the total number of brain cells rises rapidly during prenatal life and continues to increase after birth into the second year.[10] Undernutrition during this critical period may cause irreversible damage to the brain and adversely affect intelligence and behavior in later years.

neural tube defects: any of a number of defects in the orderly formation of the neural tube during early gestation. The brain and spinal cord develop from the neural tube, and the defects result in various central nervous system disorders.

The concept of critical periods is valid, but its simplicity may be somewhat misleading. Discrete developmental effects of single-nutrient deficiencies or toxicities are seldom seen in the real world, especially in human beings. Malnutrition does not occur in discrete episodes, uncomplicated by other factors. Because critical periods occur throughout pregnancy, a woman should continuously take good care of her health. That care should include achieving and maintaining a healthy body weight prior to conception and gaining sufficient weight during pregnancy to support a healthy infant.

Maternal Weight

Birthweight is the most reliable indicator of an infant's health. In general, higher birthweights present lower risks for infants. Two characteristics of the mother's weight influence an infant's birthweight: her weight for height prior to conception and her weight gain during pregnancy.

Weight before Conception

A woman's weight prior to conception influences fetal growth. Even with the same weight gain during pregnancy, underweight women tend to have smaller babies than heavier women.

Underweight is defined as BMI < 19.8.

The extensive problems associated with low-birthweight infants (5 1/2 pounds or less) and preterm births (prior to 38 weeks gestation) are discussed in Chapter 5 and Focal Point 5A.

Weight gain strategies:
- *Select energy-dense foods regularly.*
- *Eat at least three hearty meals a day.*
- *Eat large portions of foods and expect to feel full.*
- *Eat snacks between meals.*
- *Drink plenty of juice and milk.*
- *Exercise to build muscle.*

Overweight is defined as BMI > 26.0 to 29.0. Obese is defined as BMI > 29.0.

Weight loss strategies:
- *Select nutrient-dense foods regularly.*
- *Eat slowly.*
- *Select low-fat foods regularly.*
- *Limit concentrated sweets.*
- *Exercise to lose body fat.*

Focal Point 2B discusses medical complications during pregnancy that have nutrition implications.

gastric bypass surgery: a surgical procedure intended to limit food intake; a small segment of the upper stomach is closed off from the lower portion and attached to the small intestine, thus bypassing a major portion of the stomach.

Underweight. An underweight woman has a high risk of having a low-birthweight infant, especially if she is unable to gain sufficient weight during pregnancy. In addition, rates for preterm births and perinatal mortality are high for underweight women.[11] An underweight woman maximizes her chances of having a healthy infant by gaining sufficient weight prior to conception or by gaining extra pounds during pregnancy.

Weight gain prior to conception is best achieved by exercising to build muscle and eating energy-dense foods (foods that provide many kcalories in a small volume). Energy intake must be high enough to support both exercise and weight gain—about 700 to 1000 kcalories per day above normal energy intake. To increase food energy intake, an underweight woman can follow the dietary recommendations for pregnant women (described later in this chapter in Table 2-4). Additional tips for healthful weight gain are listed in the margin.

Overweight. Overweight also creates problems related to pregnancy and childbirth. Overweight women have an especially high risk of medical complications such as hypertension, gestational diabetes, and postpartum infections.[12] Compared with other women, overweight women are also more likely to have complications of labor and delivery.[13]

Infants of overweight women are likely to be born post term and to weigh more than 4000 grams (8.9 pounds). These women are less likely to have premature infants, but if they do, the infants may be large for their gestational age. Health care professionals need to be aware of the gestational age and not depend on size alone when caring for these infants.

Of greater concern than infant birthweight is the poor development of infants born to obese mothers. Recent research suggests that obesity may double the risk for neural tube defects in the infant.[14]

Weight loss dieting during pregnancy is never advisable. Overweight women should try to achieve a healthy body weight before becoming pregnant, avoid excessive weight gain during pregnancy, and postpone weight loss until after childbirth. Weight loss is best achieved by eating moderate amounts of nutrient-dense foods and exercising to lose body fat. Additional tips for healthful weight loss are listed in the margin.

For a few clinically obese people, gastric bypass surgery is a treatment option, but pregnancies following such surgery carry risks of their own. Nutrient deficiencies are common following gastric bypass surgery, and these deficiencies have been linked with neural tube defects. Women who have had gastric bypass surgery with adequate aftercare, including vitamin supplementation prior to and throughout pregnancy, have a lower incidence of neural tube defects, hypertension, and large newborns than women who have remained obese.

Weight Gain during Pregnancy

Women who plan to become pregnant, or who are already pregnant, often express concern about the weight gain that accompanies a healthy pregnancy. In a society such as ours, where body slimness is frequently equated with youth, vigor, and good looks, many prospective mothers may view the weight gain of pregnancy in a negative light. This is unfortunate. Maternal weight gain during pregnancy supports fetal growth and correlates closely with infant birthweight, which is a strong predictor of the health and subsequent development of the infant.

Recommended Weight Gains. Recommendations for weight gain during pregnancy have changed dramatically over the years, as shown in the margin. In the past, clinicians believed that restricting weight gain would reduce the risks of birth complications. Now they encourage weight gain, knowing that if the mother does not gain enough weight, she may give birth to an underweight baby with poor chances for survival.

Current recommendations for weight gains during pregnancy are based on a woman's prepregnant weight and offer a range that will support optimal fetal growth. The recommended gain for a woman who begins pregnancy at a healthy weight and is carrying a single fetus is 11.5 to 16 kilograms (25 to 35 pounds).[15] An underweight woman needs to gain between 12.5 and 18 kilograms (28 and 40 pounds), and an overweight woman, between 7 and 11.5 kilograms (15 and 25 pounds). Recommended weight gains for various prepregnant weights are shown in Table 2-1.

Historical look at weight gain recommendations:
- *1930s: <6.8 kg (15 lb).*
- *1940s: <9.1 kg (20 lb).*
- *1950s: <11.0 kg (24 lb).*
- *1960s: <12.5 kg (27.5 lb).*
- *1970s: 9.1 to 11.3 kg (20–25 lb).*
- *1980s: 11.5 to 16 kg (25–35 lb).*

Table 2-1 Recommended Weight Gains Based on Prepregnant Weight Status

Prepregnant Weight Status	Recommended Weight Gain
Underweight[a]	12.5 to 18.0 kg (28 to 40 lb)
Normal weight[b]	11.5 to 16.0 kg (25 to 35 lb)
Overweight[c]	7.0 to 11.5 kg (15 to 25 lb)
Obese[d]	6.8 kg minimum (15 lb minimum)

[a]Underweight defined as BMI <19.8.

[b]Normal weight defined as BMI 19.8 to 26.0.

[c]Overweight defined as BMI 26.0 to 29.0.

[d]Obese defined as BMI >29.0.

Source: Committee on Nutritional Status during Pregnancy and Lactation, Food and Nutrition Board, *Nutrition during Pregnancy* (Washington, D.C.: National Academy Press, 1990), pp. 10, 12.

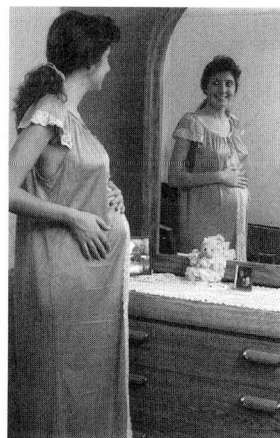

All pregnant women must gain weight—fetal growth and maternal health depend on it.

Some women should strive for gains at the upper end of the target range, notably adolescents who are still growing themselves. Short women (5 feet 2 inches and under) should strive for gains at the lower end of the target range. Women who are carrying twins should aim for a weight gain of 16 to 20.5 kilograms (35 to 45 pounds).

Weight Gain Patterns. For the normal-weight woman, weight gain ideally follows a pattern of 1.6 kilograms (3 1/2 pounds) during the first trimester and 0.44 kilograms (about 1 pound) per week thereafter. Underweight women should try to gain weight at a slightly faster rate, and overweight women, at a slightly slower rate. Figure 2-5 illustrates the rate of weight gain over the course of a pregnancy.

Components of Weight Gain. Weight gain during pregnancy supports both the growth of the fetus and associated tissues (placenta and amniotic fluid) and the expansion of maternal tissues. Table 2-2 shows the components of a typical 30-pound weight gain.

Risk of Inadequate Weight Gain. Normal weight gains during pregnancy vary dramatically from woman to woman, but certain maternal factors can predict a risk of an inadequate weight gain. Unmarried women, black and Hispanic women, low-income women, cigarette smokers, and women with little education may need additional nutrition counseling to ensure an adequate weight gain.[16]

Risk of Excessive Weight Gain. If a woman gains more than is recommended early in pregnancy, she should not restrict her energy intake later in order to lose weight. Women have been known to gain up to 60 pounds in pregnancy without ill effects. A large weight gain over a short period of time, however, indicates excessive fluid retention and

Table 2-2   Components of Weight Gain during Pregnancy

Development	Weight Gain (lb)
Infant at birth	7 1/2
Placenta	1 1/2
Increase in mother's blood volume to supply placenta	4
Increase in mother's fluid volume	4
Increase in size of uterus and supporting muscles	2
Increase in size of mother's breasts	2
Fluid to surround infant in amniotic sac	2
Mother's fat stores	7
Total	30

Source: ACOG Guide to Planning for Pregnancy, Birth, and Beyond (Washington, D.C.: American College of Obstetricians and Gynecologists, 1990), p. 109.

Figure 2-5 Recommended Prenatal Weight Gain Based on Prepregnancy Weight

Normal-weight women should gain about 3 1/2 lb in the 1st trimester and just under 1 lb/week thereafter, achieving a total gain of 25 to 35 lb by term; underweight women should gain about 5 lb in the 1st trimester and just over 1 lb/week thereafter, achieving a total gain of 28 to 40 lb by term; and overweight women should gain about 2 lb in the 1st trimester and 2/3 lb/week thereafter, achieving a total gain of 15 to 25 lb.

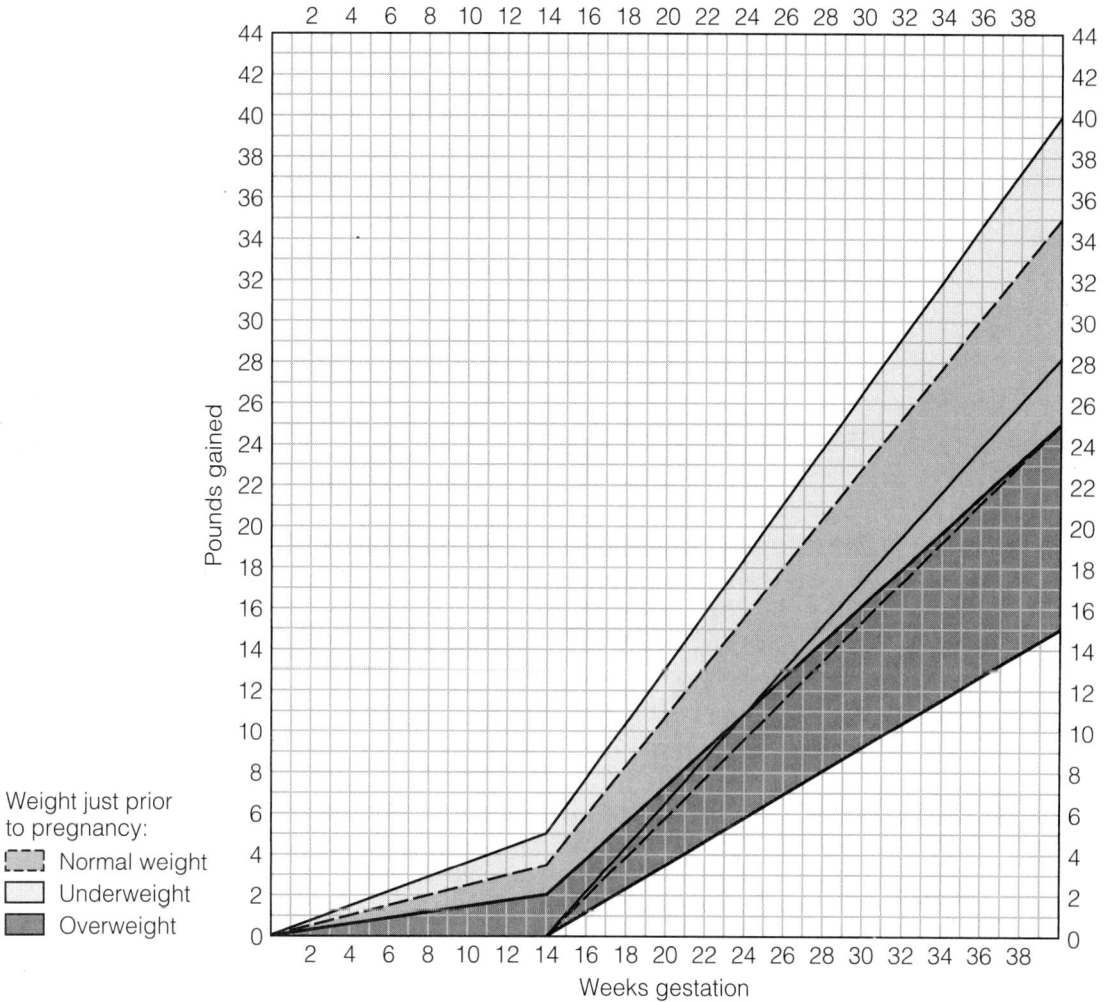

Weight just prior to pregnancy:
- ▪▪▪ Normal weight
- ☐ Underweight
- ▨ Overweight

may be the first sign of the serious medical complication preeclampsia (discussed in Focal Point 2B).

Weight Loss after Pregnancy

The pregnant woman loses some weight at delivery. In the following weeks, she loses more as her blood volume returns to normal and she sheds accumulated fluids. The typical woman does not, however, return to her prepregnancy weight. In general, the more weight a woman gains beyond what she needs for pregnancy, the more she will retain. Even with an average weight gain, though, most women tend to retain a couple of pounds with each pregnancy.[17]

For most women, pregnancy is a time of adjustment. A pregnant woman is undergoing a major transition and growing in many ways. Physically and emotionally, her needs are changing. If this is her first pregnancy, she senses that her lifestyle will have to change as she takes on the new responsibility of caring for a child. Ideally, she will be encouraged to develop this sense of responsibility by caring for herself during pregnancy. A large part of this care includes eating nutritious foods, staying physically active, and getting plenty of rest. The remainder of this chapter focuses on these topics and offers strategies to support a healthy pregnancy.

Energy and Nutrient Needs during Pregnancy

From conception to birth, all parts of the infant—bones, muscles, organs, blood cells, skin, and other tissues—are synthesized from nutrients in the foods the mother eats. For most women, nutrient needs during pregnancy and lactation are higher than at any other time. The better a woman takes care of herself nutritionally, the more successful her pregnancies are likely to be.

The need for some nutrients is greater than for others, as Figure 2-6 shows. A study of the figure reveals some of the key needs, and the following sections discuss them.

Energy and Energy Nutrients

Energy RDA during pregnancy (2nd and 3rd trimesters): + 300 kcal/day. Canadian RNI during pregnancy: + 100 to 300 kcal/day.*

A pregnant woman needs extra food energy, but only a little extra—300 kcalories above the allowance for nonpregnant women—and only during the second and third trimesters. A woman can easily obtain 300 kcalories from just one extra serving from each of the five food groups—a slice of bread, a serving of vegetables, an ounce of lean meat, a piece of fruit, and a cup of nonfat milk. Pregnant teenagers, underweight women, or physically active women may require more.

For a 2000-kcalorie daily intake, 300 kcalories represents about 15 percent more food energy than before pregnancy. The increase in nutrient needs is greater than this, so nutrient-dense foods should supply the 300 kcalories: foods such as whole-grain breads and cereals; legumes; dark green vegetables; citrus fruits; nonfat milk and milk products; and lean meats, fish, poultry, and eggs.

*For all Canadian RNI values during pregnancy, the lower value indicates recommendations for the first trimester, and the higher value indicates those for the second and third trimesters.

Figure 2-6 Comparison of Energy and Nutrient Recommendations for Nonpregnant and Pregnant Women

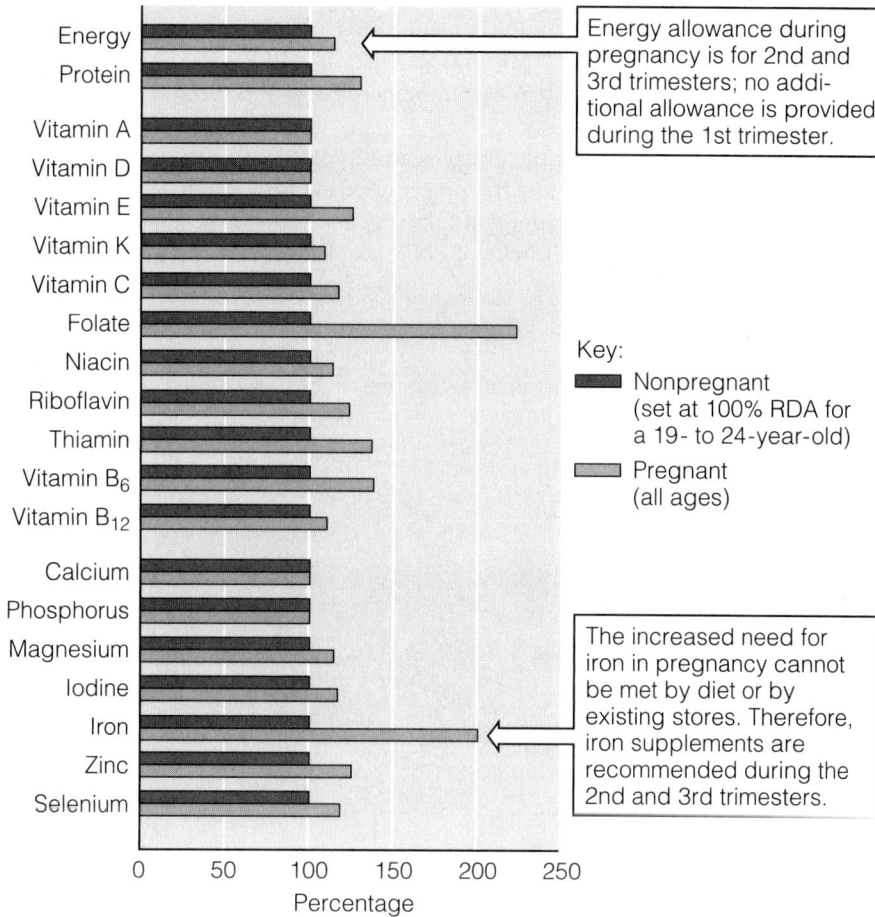

Energy allowance during pregnancy is for 2nd and 3rd trimesters; no additional allowance is provided during the 1st trimester.

Key:
■ Nonpregnant (set at 100% RDA for a 19- to 24-year-old)
▨ Pregnant (all ages)

The increased need for iron in pregnancy cannot be met by diet or by existing stores. Therefore, iron supplements are recommended during the 2nd and 3rd trimesters.

Note: For vitamin D, calcium, phosphorus, and magnesium, the 1997 recommendations differ from the 1989 RDA, but the *differences* between the values for nonpregnant and pregnant women are the same as those shown in this figure. For actual values, turn to the table on the inside front cover.

Protein. The expansion of maternal blood volume and the growth of the fetus, placenta, and maternal tissues require additional protein. The RDA for pregnancy is 10 grams per day higher than for nonpregnant women. Because people in the United States typically exceed the RDA, most women need not add 10 grams of protein to their diets. In fact, pregnant women in the United States—even those with low incomes who are not participating in food assistance programs—generally receive between 75 and 110 grams of protein a day.[18]

Protein needs during pregnancy can easily be met from dietary sources such as whole grains, meats, milk, and legumes (see the accom-

Protein RDA during pregnancy: +10 g/day.
Canadian RNI during pregnancy: +5 to 24 g/day.*

---

*For the first trimester, the RNI is an additional 5 grams/day; for the second trimester, it is an additional 20 grams/day; and for the third trimester, it is 24 grams/day.

panying photo). Obtaining enough protein need not pose a problem, even if the diet excludes all foods of animal origin. Pregnant vegetarian women who meet their energy needs by eating ample servings of protein-containing plant foods such as legumes, whole grains, nuts, and seeds meet their protein needs as well. Use of high-protein supplements during pregnancy can be harmful and is discouraged.

Carbohydrate. Ample carbohydrate (at least 100 grams per day) is necessary to spare the protein needed for growth. At least 50 percent of the total daily energy intake should derive from carbohydrate. For example, for a daily intake of 2000 kcalories, this represents at least 1000 kcalo-

Protein is most abundant in meats, meat alternates (legumes and tofu), and milk products. Grains and vegetables contain a little.

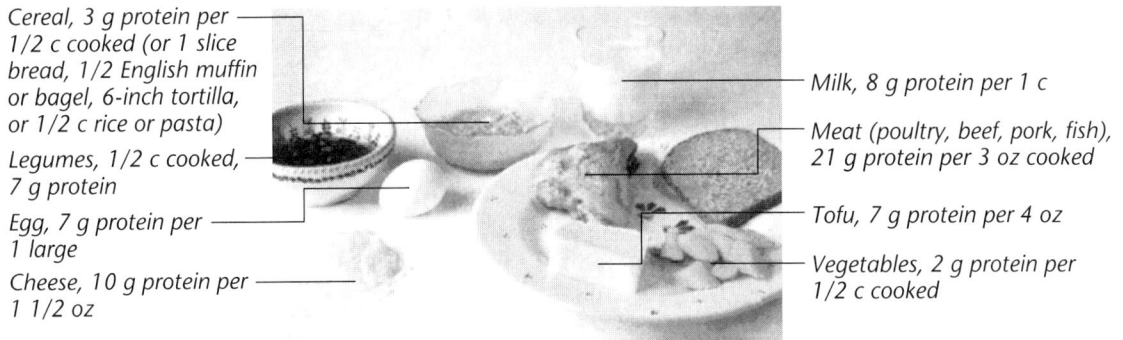

Cereal, 3 g protein per 1/2 c cooked (or 1 slice bread, 1/2 English muffin or bagel, 6-inch tortilla, or 1/2 c rice or pasta)

Legumes, 1/2 c cooked, 7 g protein

Egg, 7 g protein per 1 large

Cheese, 10 g protein per 1 1/2 oz

Milk, 8 g protein per 1 c

Meat (poultry, beef, pork, fish), 21 g protein per 3 oz cooked

Tofu, 7 g protein per 4 oz

Vegetables, 2 g protein per 1/2 c cooked

Note: Average gram values are adapted from the American Diabetes Association and the American Dietetic Association, *1995 Exchange Lists for Meal Planning.*

Carbohydrate-containing foods appear in several food groups: grain products, vegetables, fruits, and milk products.

Milk, 12 g carbohydrate per 1 c

Nonstarchy cooked vegetables (such as carrots, green beans, and broccoli), 5 g carbohydrate per 1/2 c
Starchy cooked vegetables (such as potatoes, legumes, and corn), 15 g carbohydrate per 1/2 c

Sugar (such as candies, syrups, and jellies), 5 g carbohydrate per 1 tsp

Fruit (such as small banana, apple, or orange); 15 g carbohydrate per fruit juices, 1/2 c; canned or fresh fruit, 1/2 c; or dried fruit, 1/4 c

Bread, 15 g carbohydrate per 1 slice (or 1/2 English muffin or bagel, 6-inch tortilla, 1/2 c rice, pasta or cooked cereal)

Small dessert, 15 g carbohydrate

Note: Average gram values are adapted from the American Diabetes Association and the American Dietetic Association, *1995 Exchange Lists for Meal Planning.*

ries, or 250 grams of carbohydrate. Foods that provide carbohydrate include milk, legumes, fruits, grains (breads, cereals, rice, and pasta), and vegetables, as the accompanying photo shows.

B Vitamins Associated with Energy and Protein Intake. The need for B vitamins increases in proportion to the increase in energy requirements. The RDA during pregnancy is set slightly above the nonpregnant woman's RDA for thiamin, riboflavin, and niacin. The usual intake of these nutrients is adequate for most pregnant women in the United States. Common food sources of these vitamins are shown in the accompanying photos.

Thiamin RDA during pregnancy: 1.5 mg/day.
Canadian RNI during pregnancy: +0.1 mg/day.

Riboflavin RDA during pregnancy: 1.6 mg/day.
Canadian RNI during pregnancy: +0.1 to 0.3 mg/day.

Niacin RDA during pregnancy: 17 mg NE/day.
Canadian RNI during pregnancy: +1 to 2 NE/day.

Thiamin-rich foods include meats (especially pork and ham), legumes such as black beans and green peas, sunflower seeds, and whole-wheat bread.

Green peas 0.23 mg per 1/2 c cooked

Watermelon 0.39 mg per melon wedge

Pork chop 0.75 mg per 3 oz broiled chop

Whole-wheat bread 0.12 mg per slice

Black beans 0.21 mg per 1/2 c cooked

Sunflower seeds (shelled) 0.1 mg per 2 tbs

Riboflavin is found in a variety of foods, but milk and milk products make the greatest contribution to most people's diets.

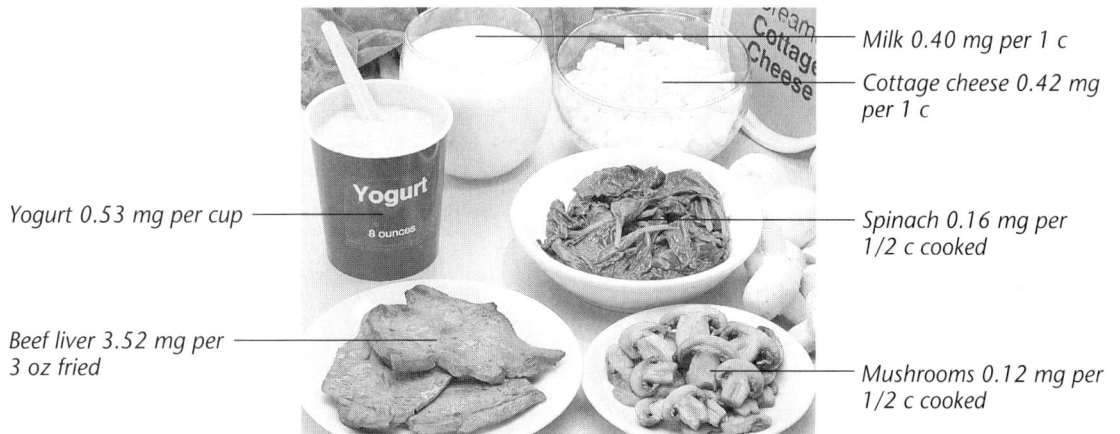

Milk 0.40 mg per 1 c

Cottage cheese 0.42 mg per 1 c

Yogurt 0.53 mg per cup

Spinach 0.16 mg per 1/2 c cooked

Beef liver 3.52 mg per 3 oz fried

Mushrooms 0.12 mg per 1/2 c cooked

Niacin contributions derive primarily from protein-rich foods such as tuna fish and chicken.

Mushrooms 1.1 mg per
1/2 c cooked

Pork chop 4.3 mg per
3 oz broiled chop

Baked potato 1.52 mg[a] per
whole small potato

Tuna (in water) 11.3 mg
per 3 oz

Chicken breast 10.8 mg
per 3 oz cooked

[a]Values are for preformed niacin, not niacin equivalents.

Vitamin B₆ appears in most protein-rich foods; some vegetables and fruits are good sources, too.

Chicken breast 0.51 mg
per 3 oz cooked

Spinach 0.22 mg
per 1/2 c cooked

Baked potato 0.32 mg per
whole small potato

Beef liver 2.2 mg
per 3 oz fried

Navy beans 0.15 mg
per 1/2 c cooked

Banana 0.66 mg
per whole banana

Vitamin B₆ RDA during pregnancy: 2.2 mg/day (or 0.016 mg/g dietary protein). Canadian RNI during pregnancy: 0.015 mg/g dietary protein.

Vitamin B_6 recommendations increase in parallel with protein recommendations. Fortunately, most protein-rich foods provide ample vitamin B_6, as the accompanying photo shows.

Substance abusers, pregnant adolescents, and women carrying more than one fetus need to pay attention to their vitamin B_6 intakes. For these women, a daily multivitamin supplement containing 2 milligrams vitamin B_6 is recommended.

Nutrients for Blood Production and Cell Growth

New cells are laid down at a tremendous pace as the fetus grows and develops. At the same time, the mother's red blood cell mass expands. All nutrients are important in these processes, but the needs for folate, vitamin B_{12}, iron, and zinc are especially great, due to their key roles in the synthesis of DNA and the synthesis of cells, including red blood cells.

Folate. Folate during pregnancy is particularly noteworthy because of its fundamental role in DNA synthesis and cell replication. A deficiency early in pregnancy could impair cell growth and replication, resulting in placental and fetal abnormalities.

The requirement for folate more than doubles during pregnancy, increasing from 180 micrograms to 400 micrograms per day. Pregnant women in the United States tend to consume less than the recommended amount, but deficiency is uncommon. It is fairly easy to obtain sufficient folate, without supplements, from a diet that includes fruits, juices, green vegetables, and whole-grain or fortified cereals (see the accompanying photo). In addition to being good sources of folate, these foods are low in kcalories, rich in other vitamins and minerals, and high in fiber.

When dietary intake is inadequate, daily supplementation with 300 micrograms folate is recommended. Supplementation may also be appropriate for adolescents, women who are carrying multiple fetuses, and women who use alcohol, drugs, or cigarettes.

Vitamin B_{12}. Normal cell division and protein synthesis during pregnancy also depend on the B vitamin that activates the folate enzyme—vitamin B_{12}. Consequently, the pregnant woman has a slightly greater need for vitamin B_{12}.

Generally, even modest amounts of meat, fish, eggs, or milk products together with body stores easily meet the increased need for vitamin B_{12} (see photo of food sources). Most pregnant women in the

Chapter 3 describes the relationships between folate and neural tube defects.

*Folate RDA during pregnancy: 400 μg/day.
Canadian RNI during pregnancy: +200 μg/day.*

*Vitamin B_{12} RDA during pregnancy: 2.2 μg/day.
Canadian RNI during pregnancy: +0.2 μg/day.*

Folate[a] suggests the word *foliage,* and indeed, leafy green vegetables are outstanding sources, as are other vegetables, legumes, and liver.

Cantaloupe 14 μg per small wedge (1/6 melon)

Liver 187 μg per 3 oz fried

Asparagus 131 μg per 1/2 c cooked

Spinach 109 μg per 1 c raw

Pinto beans 147 μg per 1/2 c cooked

Beets 68 μg per 1/2 c cooked

[a]As of 1998, most enriched breads, cereals, cornmeal, flour, pasta, rice, and other grain products will be fortified with 140 μg folate per 100 grams of food (about 1/2 cup cooked food or 1 slice bread).

Vitamin B₁₂ is found almost exclusively in foods derived from animals.

Sirloin steak 2.4 μg per 3 oz steak cooked

Chicken liver 16.5 μg per 3 oz cooked

Tuna (in water) 2.5 μg per 3 oz

Cottage cheese 1.6 μg per 1 c

Sardines 7.6 μg per 3 oz

United States have adequate intakes. Strict vegetarians, however, who exclude all foods of animal origin risk vitamin B₁₂ deficiency.[19] They may need to use soy milk that is fortified with vitamin B₁₂ (as well as calcium) or breakfast cereals fortified with vitamin B₁₂ (read the label). Alternatively, they may need daily supplements of 2 micrograms vitamin B₁₂ to prevent deficiency.

Iron RDA during pregnancy: 30 mg/day.
Canadian RNI during pregnancy: +0 to 10 mg/day.

Iron. Pregnant women need iron to support their enlarged blood volume and to provide for placental and fetal needs. The developing fetus draws on maternal iron stores to create stores of its own to last through the first four to six months after birth when milk, which is poor in iron, will be its sole food. Also, the blood losses inevitable at delivery, especially at a cesarean delivery, can further drain the mother's supply.* When these special needs are added to the body's usual requirements, the iron cost of pregnancy totals about 1000 milligrams.[20]

During pregnancy, the body makes several adaptations that help meet the exceptionally high need for iron. Menstruation, the major route of iron excretion in women, ceases, and iron absorption nearly triples due to an increase in the synthesis of transferrin, the body's iron-absorbing and iron-carrying protein. The hormones of pregnancy mobilize iron from storage sites, making it available in the blood. Without sufficient replacement, though, iron stores would quickly dwindle.

Iron deficiency anemia raises the risk of low birthweight, preterm birth, and perinatal mortality. The World Health Organization (WHO) estimates that half of the pregnant women worldwide have iron deficiency anemia. Some women, notably those who are poor, undereducated, and black or Hispanic, are more likely to develop iron deficiency than others.

*The average blood loss during a cesarean delivery is almost twice that occurring with the average vaginal delivery of a single fetus. Food and Nutrition Board, *Nutrition during Pregnancy* (Washington, D.C.: National Academy Press, 1990), p. 283.

Iron is most abundant in meats and legumes, with some fruits and vegetables contributing a little.

Swiss chard 2.0 mg per 1/2 c cooked

Clams 23.8 mg per 3 oz steamed

Navy beans 2.3 mg per 1/2 c cooked

Beef steak 2.9 mg per 3 oz cooked

Tofu 6.7 mg per 1/2 c

Dried figs 0.6 mg per 1/4 c

The best dietary sources of iron are meat, fish, and poultry, and thanks to the MFP factor, the absorption of iron from nonmeat foods is enhanced. Vegetarians can obtain iron from plant foods such as legumes, dark green, leafy vegetables, iron-fortified cereals, and whole-grain breads and cereals, but the iron is not as well absorbed as iron from meats. Fortunately, vitamin C in fruits and vegetables can enhance iron absorption from foods eaten at the same meal. Eggs are also a valuable source for those who include them. Few women enter pregnancy with adequate iron stores, so a daily iron supplement of 30 milligrams is recommended during the second and third trimesters for all pregnant women.[21]

Zinc. Zinc is required for DNA and RNA synthesis and thus for protein synthesis and cell development. Low blood zinc is a significant predictor of low birthweight.[22] The recommendation for zinc for pregnant women is slightly higher than for nonpregnant women.

Zinc RDA during pregnancy: 15 mg/day.
Canadian RNI during pregnancy: +6 mg/day.

Zinc is similar to iron in that meat is its richest food source. Half the zinc in the average diet comes from meat, fish, and poultry. Zinc from plant sources such as cereals and legumes is not well absorbed. In vegetarian diets, the abundant fibers and binders in foods of plant origin limit its absorption. Soy products, which are commonly used as meat alternates, also interfere with zinc absorption. Consequently, vegetarian diets are often limited in zinc quantity, bioavailability, or both. Perhaps the best advice to vegetarians regarding zinc is to eat meat alternates such as black-eyed peas, pinto beans, and kidney beans instead of soy products; to eat a variety of nutrient-dense foods; and to maintain an adequate energy intake.[23] Oysters, crabmeat, and shrimp are rich sources of zinc for those vegetarians who include seafood.

Typical zinc intakes for pregnant women are lower than recommendations, but routine supplementation is not advised.[24] Women taking iron supplements (more than 30 milligrams per day), however, may need zinc supplementation because large doses of iron can interfere with the body's absorption and use of zinc.

Zinc-containing foods are mostly found in the meat group.

Yogurt 2.2 mg per cup

Black beans 1.0 mg per 1/2 c cooked

Crabmeat 6.5 mg per 3 oz steamed

Green peas 1.0 mg per 1/2 c

Beef steak 5.6 mg per 3 oz cooked

Oysters 155 mg per 3 oz steamed

Nutrients for Bone Development

The nutrients involved in building the skeleton are in great demand during pregnancy. To build a bone or a tooth, cells first lay down a matrix of the protein collagen and then fill it with crystals of minerals. Insufficient intakes of calcium, phosphorus, vitamin D, magnesium, or fluoride during pregnancy may result in abnormal fetal bone development. Recommendations for these five nutrients reflect the 1997 Dietary Reference Intakes (DRI).*

Calcium and Phosphorus. Adaptive physiological changes during pregnancy favor calcium retention. Intestinal calcium absorption more than doubles early in pregnancy, and the mother's bones store the mineral. During the last trimester, as the fetal bones begin to calcify, a dramatic shift of calcium across the placenta occurs. Whether calcium added to the mother's bones early in pregnancy is withdrawn to provide sufficient calcium to the fetus later in gestation is unclear. In the final weeks of pregnancy, over 300 milligrams a day are transferred to the fetus. Recommendations to ensure an adequate calcium intake during pregnancy are aimed at conserving maternal bone mass while supplying fetal needs.

Because bones are still actively depositing minerals until about age 25, adequate calcium intake is especially important for young women.[25]

Calcium DRI during pregnancy:
1300 mg/day (14 to 18 yr).
1000 mg/day (19 to 50 yr).

✶ **Healthy People 2000**

Increase calcium intake so at least 50% of pregnant women consume three or more servings daily of foods rich in calcium.

---

*The DRI encompass several sets of standards, including an RDA that serves as a goal for nutrient intake for individuals. To distinguish this value from the previous RDA, margin notes refer to the 1997 values as DRI.

Pregnant women under age 25 who receive less than 600 milligrams of dietary calcium daily need to increase their intake of milk, cheese, yogurt, and other calcium-rich foods. Alternatively, and less preferably, they may need a daily supplement of 600 milligrams calcium.

Most pregnant women tend to drink more milk than other women, but still their intakes typically fall below recommendations. This is particularly true for black, Hispanic, and Native American women. These ethnic groups have a high prevalence of lactose intolerance and customarily limit their milk consumption. They should be encouraged to try calcium-rich foods that contain relatively little lactose such as yogurt, fermented cheese, or lactose-free milk.

Vegetarians who exclude milk products need calcium-fortified foods, such as soy milk. It is worth noting that not all soy milk is fortified with calcium; the calcium- and vitamin D-fortified type is recommended.

Diets rich in cereals and vegetables (typical of vegetarian diets) contain abundant fibers, phytates, and oxalates—compounds that impair calcium absorption even when calcium intakes meet recommendations. Together, the low calcium intakes of strict vegetarians, the abundance of calcium binders in vegetarian diets, and the high calcium needs of pregnancy jeopardize the calcium status of these women. They need a reliable calcium source in their diets—either a calcium-fortified food or a calcium supplement. Within physiological limits, the body partially compensates for low calcium intakes by increasing absorption.

Phosphorus is closely linked with calcium metabolism. Because phosphorus intakes usually exceed recommendations, getting enough does not present a problem for pregnant women. In fact, too much phosphorus may be more likely and more of a problem in that high phosphorus intakes interfere with calcium absorption. Additives in processed foods, including soft drinks, can add significant amounts of phosphorus to the diet.

Vitamin D. Vitamin D plays a vital role in calcium absorption and utilization. It stimulates calcium absorption from the GI tract, it stimulates calcium retention by the kidneys, and it helps withdraw calcium from bones into the blood. In these three ways, vitamin D raises blood calcium. Consequently, severe maternal vitamin D deficiency interferes with normal calcium metabolism, resulting in rickets in the fetus and osteomalacia in the mother.

The body can synthesize vitamin D, with the help of sunlight, from a precursor that the body makes from cholesterol. Given enough time in the sun, people need no vitamin D from foods. This is fortunate, because only a few foods supply significant amounts of vitamin D, notably those derived from animals: egg yolks, liver, fatty fish, butter, and fortified milk. The fortification of milk with vitamin D is the best guarantee that people will meet their needs and underscores the importance of milk in a well-balanced diet.

Vitamin D deficiency is common in other parts of the world, but in the United States and Canada, exposure to sunlight and vitamin D-fortified milk is usually sufficient to provide the recommended amount of vitamin D during pregnancy. Pregnant women who do not drink

lactose intolerance: a condition that results from inability to digest the milk sugar lactose; characterized by bloating, gas, abdominal discomfort, and diarrhea. Lactose intolerance differs from milk allergy, which is caused by an immune reaction to the protein in milk.

Phosphorus DRI during pregnancy:
1250 mg/day (14 to 18 yr).
700 mg/day (19 to 50 yr).

rickets: the vitamin D deficiency disease in children characterized by inadequate mineralization of bone (manifested in bowed legs or knock-knees, outward-bowed chest, and knobs on ribs).

osteomalacia (os-tee-o-mal-AY-shuh): a vitamin D deficiency disease characterized by softening of the bones; symptoms include bending of the spine and bowing of the legs. The disease occurs most often in adult women who have low calcium intakes and little exposure to sun, and who go through repeated pregnancies and periods of lactation (*osteo* = bone; *mal* = bad).

enough fortified milk or exposure to sunlight may need supplements of 5 micrograms vitamin D daily. Routine supplementation, however, is not recommended because of the toxicity risk. Toxicity symptoms reflect an excess of blood calcium, which tends to precipitate in soft tissues, forming stones, especially in the kidneys where calcium is concentrated in the effort to excrete it. Calcification may also harden the blood vessels and is especially dangerous in the major arteries of the heart and lungs, where it can cause death.

Because practically all milk in the United States and Canada is fortified, vegetarians who include milk in their diets are protected from deficiency. Vegetarians who avoid milk, eggs, and fish may receive enough vitamin D from daily exposure to sunlight or from fortified soy milk. If their vitamin D status is inadequate, though, they may need supplements of 10 micrograms daily. Without adequate sunshine, fortification, or supplementation, a diet of plant foods alone cannot meet vitamin D needs.[26]

Magnesium. The recommendation for magnesium during pregnancy is slightly higher than for nonpregnant women because of its integral role in bone formation. Magnesium participates in the activation of vitamin D and in the release of the parathyroid hormone that acts on the kidneys and bones to raise blood calcium.

Magnesium is commonly found in such foods as grains, seafoods, and green vegetables. Pregnant women typically consume less than the amount recommended for magnesium, but the consequences of inadequate intakes are unclear. Some research suggests that women who take magnesium supplements during pregnancy have fewer complications.[27] Routine supplementation, however, is not recommended.

Fluoride. Mineralization of the fetal teeth begins in the fifth month. Some studies report reduced tooth decay in children born to women taking fluoride supplements, but supplementation is not routinely recommended for pregnant women.[28] Fluoride crosses the placenta, but the placenta may not defend well against excess fluoride. Instead, women are encouraged to drink fluoridated water, or physicians may prescribe supplements, not to exceed 1 milligram fluoride per day.

Other Nutrients

The nutrients mentioned so far are those most intensely involved in cell production and bone mineralization. Of course, other nutrients are also needed during pregnancy to support the growth and health of both fetus and mother.

Vitamin A. Vitamin A deficiency during pregnancy has been associated with fetal growth retardation, preterm birth, and low birthweight, but such deficiencies are uncommon in the United States. Women in the United States typically have enough vitamin A in their diets and liver stores to support normal fetal growth. Therefore, dietary recommendations for vitamin A do not increase for women during pregnancy. Supplementation is both unnecessary and potentially harmful.

Vitamin E. Pregnant women in the United States typically have vitamin E intakes below the RDA, which is slightly higher than for nonpregnant women. Vitamin E-rich foods include vegetable oils, wheat germ, whole grains, and nuts.

Vitamin E RDA during pregnancy: 10 mg α-TE/day. Canadian RNI during pregnancy: +2 mg α-TE/day.

Vitamin E crosses the placenta in the last 8 to 10 weeks of pregnancy, when the fetus is accumulating fat. In the event of a premature birth, the infant may need vitamin E supplementation (Chapter 5 provides more details). There is no evidence, however, that maternal supplementation would improve the health risks in premature infants.

Vitamin C. One of vitamin C's many roles in the body is to help form the fibrous protein collagen. Collagen gives structure to bones, cartilage, muscle, and blood vessels. Plasma vitamin C normally falls during pregnancy, largely because of hormonal changes, blood volume expansion, and increased needs. The placenta transfers vitamin C into fetal blood against a concentration gradient. At term, fetal vitamin C plasma concentration is 50 percent greater than maternal concentration. The RDA for nonpregnant women increases by 10 milligrams during pregnancy, and most pregnant women in the United States meet this recommendation.

Vitamin C RDA during pregnancy: 70 mg/day. Canadian RNI during pregnancy: +10 mg/day.

Some women have a greater need for vitamin C: women who are carrying more than one fetus and women who use illicit drugs, smoke cigarettes, or drink alcohol. These women need to include vitamin C-rich fruits and vegetables in their diets. Otherwise they may need a daily supplement of 50 milligrams vitamin C.

Iodine. Iodine is an integral part of two hormones released by the thyroid gland that regulate body temperature, metabolic rate, reproduction, growth, the making of blood cells, nerve and muscle function, and more. Maternal iodine deficiency impairs fetal development, causing the extreme and irreversible mental and physical retardation known as cretinism. Cretinism affects approximately 6 million people worldwide.[29] A person with cretinism may have an IQ as low as 20 and many physical abnormalities. To prevent fetal damage, iodine deficiencies need to be corrected before conception.

cretinism (CREE-tin-ism): an iodine deficiency disease characterized by mental and physical retardation.

The recommended intake of iodine during pregnancy is slightly higher than for nonpregnant women. Iodine needs are easily met by consuming seafood, vegetables grown in iodine-rich soil, and (in iodine-poor areas) iodized salt. In the United States, labels state whether salt is iodized; in Canada, all table salt is iodized.

Iodine RDA during pregnancy: 175 μg/day. Canadian RNI during pregnancy: +25 μg/day.

During pregnancy, no special recommendations are made for intakes of the additional nutrients not listed in the main RDA table (shown on the inside front cover). It is assumed that the ranges recommended for adults will cover pregnant women as well.

Nutrient Supplements for Pregnant Women

Women who make wise food choices during their pregnancies can meet most of their nutrient needs. For women who do not eat adequately, daily *multi*vitamin-mineral supplements are recommended. Multivitamin-mineral supplements are also recommended for women in high-risk groups: adolescents, women who are carrying multiple fetuses, cigarette smokers, and alcohol and drug abusers. Table 2-3 lists nutrient recommendations for multivitamin-mineral supplements.

In general, supplements should be taken between meals or at bedtime to enhance absorption. Calcium supplements are an exception; they should be taken with meals to enhance absorption and limit interactions with iron and zinc supplements.

Specific Nutrient Supplements. Specific nutrient supplements may be appropriate for some women under special circumstances. Table 2-3 includes those recommendations as well.

Iron Supplements. Iron is the only nutrient for which supplements are routinely recommended during pregnancy. A balanced diet alone cannot supply the iron a pregnant woman needs. Iron supplementation is needed to accommodate the expansion in red blood cell mass that begins at the end of the first trimester.[30] During the second and third trimesters, all pregnant women are advised to take 30 milligrams of ferrous iron daily.* They are also advised to eat an iron-rich diet that contains meat, fish, poultry, and vitamin C-rich fruits and vegetables to enhance iron absorption. Women who have iron deficiency anemia may need to take iron supplements that provide 60 to 120 milligrams a day until iron status returns to normal.†

Iron absorption is enhanced by taking supplements between meals or at bedtime on an empty stomach. Supplements should be taken with liquids other than milk, tea, or coffee, because these beverages inhibit iron absorption; water will do fine. There is no benefit to taking iron supplements with orange juice. Vitamin C does not enhance iron absorption from supplements as it does from foods. (Vitamin C helps iron absorption by converting the insoluble ferric iron in foods to the more soluble ferrous iron; supplemental iron is already in the ferrous form.)

Iron supplements can cause such side effects as heartburn, nausea, upper abdominal discomfort, constipation, and diarrhea. These problems may be alleviated by taking the iron supplements at bedtime.

The combination of a well-balanced diet and judicious supplementation of key nutrients can support the nutrient demands of pregnancy. Wise food choices are essential to prenatal care.

---

*Doses are given in terms of elemental iron; 30 milligrams of ferrous iron is provided by 150 milligrams ferrous sulfate, 300 milligrams ferrous gluconate, or 100 milligrams ferrous fumarate.
†In pregnancy, iron deficiency anemia is defined as a hemoglobin value below 11 g/dL during the first and third trimesters, and below 10.5 g/dL during the second trimester.

Table 2-3   Nutrient Recommendations for Supplements

Multivitamin-Mineral Supplements

Nutrient	Amount	Nutrient	Amount
Vitamin B_6	2 mg	Iron	30 mg
Folate	300 μg	Zinc	15 mg
Vitamin C	50 mg	Copper	2 mg
Vitamin D	5 μg	Calcium	250 mg

Supplements for Special Circumstances

Nutrient	Amount	Circumstance
Vitamin B_{12}	2 μg	Complete vegetarians
Vitamin D	10 μg	Complete vegetarians, others with low intakes of vitamin D-fortified milk
Calcium	600 mg	Women under age 25 whose calcium intake is less than 600 mg/day
Zinc	15 mg	Women taking >30 mg iron/day to treat anemia
Copper	2 mg	Women taking >30 mg iron/day to treat anemia

Source: Committee on Nutritional Status during Pregnancy and Lactation, Food and Nutrition Board, *Nutrition during Pregnancy* (Washington, D.C.: National Academy Press, 1990), pp. 20–21.

Food Choices and Fitness during Pregnancy

Meeting the heightened energy and nutrient needs of pregnancy without overconsuming food energy is a challenge: The pregnant woman must seek out foods of high nutrient density. A balance similar to that of the Food Guide Pyramid (presented in Chapter 1) is recommended, but with at least one extra serving from each of the five food groups (see Table 2-4).

For most women, appropriate food choices include nonfat milk and milk products; lean meats, fish, and poultry; eggs; legumes; dark green and other vegetables; citrus and other fruits; and whole-grain breads and cereals. If a woman's diet is already meeting nutrient needs at the start of pregnancy, then she can easily adjust it to meet the increased demands; but if a woman has not been eating well, she may need help. Some women need only a little counseling and motivation to get the pregnancy on the right nutrition track, as the next section describes. Others may need the support of a maternal and infant assistance program, some of which are described in Chapter 3.

A pregnant woman's nutrition choices support both her health and her developing infant's growth.

Table 2-4 Daily Food Guide for Women during Pregnancy

Food Group	Number of Servings
Breads, cereals, and other grain products	7 to 11
Vegetables	4 to 5
Fruits	3 to 4
Meat and meat alternatives	3
Milk and milk products	3 to 4

This sample meal plan follows the Daily Food Guide for pregnant women and provides about 2500 kcalories (50% from carbohydrate, 20% from protein, and 30% from fat).

Breakfast

2 medium bran muffins

2 tsp butter or margarine

1 c vanilla yogurt

1/2 c fresh strawberries

1 c orange juice

Midmorning snack

1 medium apple

Lunch

Sandwich (2 oz ham, 1 oz swiss cheese, 2 slices rye bread, 2 tsp mayonnaise, lettuce)

1 1/4 c salad (lettuce, tomatoes, carrots)

1 tbs salad dressing

1 c low-fat milk

Dinner

Chicken cacciatore

   4 oz chicken

   3/4 c stewed tomatoes

1 c rice

3/4 c summer squash

1 1/2 c salad (spinach, mushrooms, onions)

1 tbs salad dressing

2 slices Italian bread

2 tsp butter or margarine

1 c low-fat milk

Evening snack

1 c low-fat milk

3 oatmeal cookies

Diets Tailored to Individual Preferences

Perhaps at no other time in life is the motivation to take care of oneself so great as it is during the unique time of pregnancy. The health and well-being of two individuals are inseparable at this time. Many women willingly abstain from smoking and drinking alcohol, and pay more attention to their eating habits than ever before. The wise health professional will offer guidance to support this motivation and enhance the chances of an optimal pregnancy and birth.

No two women are alike, and few carefully follow the diet pattern presented in Table 2-4. An unknowing teenager driven by peer pressure might eat hamburgers, fries, and colas. A busy mother with three children and too many other demands might skip dinner and drink three

glasses of wine in the evening. The newly converted member of a religious cult might be persuaded to give up all but a very few foods. Less
extreme examples are more common: One person eats too much meat
at the expense of vegetables, whereas another drinks too much milk at
the expense of iron-rich foods.

Beliefs and Superstitions. Individual food practices may be not only
haphazard, but irrational and rigid. Unusual beliefs, superstitions, and
dietary practices have surrounded childbirth since the beginning of
time. Some women, even today, in various parts of the world, believe
that overconsumption or underconsumption of a craved food imparts
physical or behavioral peculiarities to the infant. For example, they may
believe that a strawberry-shaped birthmark will appear if a woman
stares at strawberries, eats too many strawberries, or does not satisfy a
craving for strawberries.[31]

Cravings and Aversions. Some women develop cravings for, or aversions to, particular foods and beverages during pregnancy. These likes
and dislikes are fairly common, but they do not seem to reflect real
physiological needs. In other words, a woman who craves pickles does
not necessarily need salt, nor does a woman who craves chocolate need
caffeine or fat. Similarly, cravings for ice cream are common in pregnancy, but do not signify a calcium deficiency. Food cravings and aversions that arise during pregnancy are most likely due to psychological
factors or hormone-induced changes in sensitivity to taste and smell.

Vegetarian Diets during Pregnancy

Like other pregnant women, pregnant vegetarian women are usually
motivated to do their best to ensure successful pregnancies. As long as
this motivation leads to practices consistent with health, both mother
and fetus benefit. In general, well-planned vegetarian diets that are adequate in food energy and include milk products and eggs can support a
healthy pregnancy.[32] In contrast, severely restricted vegetarian diets can
be inadequate in energy, protein, vitamins, and minerals and thereby
pose significant risks for both mother and fetus. Table 2-5 provides a
vegetarian eating pattern for pregnant women.

Many vegetarian women willingly eat more liberal diets than usual
throughout their pregnancies. For example, they might include additional protein-rich foods and milk products in their diets. Many vegetarian women are well nourished, with nutrient intakes from diet alone
exceeding the RDA for all vitamins and minerals except iron, which is
low for most women.

Vegetarian women who restrict themselves to an exclusively plant-
based diet generally have low food energy intakes and are thin; for pregnant women, this can be a problem. As mentioned earlier, women with
low prepregnancy weights and low weight gains during pregnancy jeopardize a healthy pregnancy.

Table 2-5 Daily Food Guide for Vegetarians during Pregnancy

Food Group	Number of Servings	Serving Sizes
Breads, cereals, and other grain products	7 or more	1 slice bread
		1/2 bun, bagel, or English muffin
		1/2 c cooked cereal, rice, or pasta
		1 oz dry cereal
Vegetables	4 or more[a]	1/2 c cooked or 1 c raw
Fruits	3 or more	1 piece fresh fruit
		3/4 c fruit juice
		1/2 c canned or cooked fruit
Legumes and other meat alternates	3 to 4	1/2 c cooked beans
		4 oz tofu or tempeh
		8 oz soy milk
		2 tbs nuts or seeds (these tend to be high in fat, so use sparingly)
		1 egg or 2 egg whites
Milk and milk products	3 to 4[b]	1 c low-fat or nonfat milk
		1 c low-fat or nonfat yogurt
		1 1/2 oz low-fat cheese

Note: Vegetarians need to eat sufficient quantities of a variety of foods to help ensure an adequate intake of energy and nutrients. See Table 2-3 on p. 63 for supplement recommendations.

[a]Include 1 cup of dark green vegetables daily to help meet iron requirements.

[b]People who do not use milk or milk products: use soy milk fortified with calcium, vitamin D, and vitamin B_{12}.

In addition to being low in energy, poorly planned vegetarian diets can be low in protein, vitamins, and minerals. Earlier discussions mentioned nutrients of particular concern for vegetarians.

For the most part, it appears that a vegetarian diet favors a healthy pregnancy if it provides adequate energy; includes milk and milk products; and contains a wide variety of legumes, cereals, fruits, and vegetables. Vegetarian diets that exclude all foods of animal origin may require supplementation with vitamin B_{12}, vitamin D, calcium, and iron, or the addition of foods fortified with these nutrients. For the strict vegetarian, appropriate food choices include calcium- and vitamin D-fortified soy milk and tofu, legumes, nuts, seeds, whole-grain breads and cereals, vegetables (especially dark green, leafy vegetables), and fruits.

Alleviating Maternal Discomfort

Nausea, constipation, and heartburn are common during pregnancy. A few simple strategies can help avert them (see Table 2-6).

Nausea. Not all women have uneasy stomachs in the early months of pregnancy, but many do. The nausea of "morning" (actually, any time)

Table 2-6 Strategies to Alleviate Maternal Discomforts

To alleviate the nausea of pregnancy:

- On waking, arise slowly.
- Eat dry toast or crackers.
- Chew gum or suck hard candies.
- Eat small, frequent meals.
- Avoid foods with offensive odors.
- When nauseated, drink no citrus juice, water, milk, coffee, or tea.

To prevent or alleviate constipation:

- Eat foods high in fiber (fruits, vegetables, and whole-grain cereals).
- Exercise regularly.
- Drink at least eight glasses of liquids a day.
- Respond promptly to the urge to defecate.
- Use laxatives only as prescribed by a physician; do not use mineral oil, because it interferes with absorption of fat-soluble vitamins.

To prevent or relieve heartburn:

- Relax and eat slowly.
- Chew food thoroughly.
- Eat small, frequent meals.
- Drink liquids between meals.
- Avoid spicy or greasy foods.
- Sit up while eating; elevate the head while sleeping.
- Wait an hour after eating before lying down.
- Wait two hours after eating before exercising.

sickness ranges from mild queasiness to debilitating nausea and vomiting. The hormonal changes early in pregnancy seem to be responsible for the sensitivities to the appearance, texture, or smell of foods. Traditional strategies for alleviating nausea are listed in Table 2-6, but some women benefit most from simply eating the foods they want when they feel like eating.[33] They may also find comfort in a cleaner, quieter, and more temperate environment.[34]

Constipation and Hemorrhoids. As the hormones of pregnancy alter muscle tone and the growing fetus crowds intestinal organs, an expectant mother may experience constipation. She may also develop hemorrhoids (swollen veins of the rectum). These can be painful, and straining during bowel movements may cause bleeding. She can gain relief by following the strategies listed in Table 2-6.

Heartburn. Heartburn is another common complaint during pregnancy. As the growing fetus puts increasing pressure on the mother's stomach, acid may back up into the lower esophagus and create a burning sensation near the heart. Heartburn may be especially troublesome at night, when the woman is lying down; she may find relief by sleeping with her torso slightly raised. Table 2-6 lists additional tips to help relieve heartburn.

Nutrition during Labor

When a woman goes into labor, she does not cease to have dietary needs. If she gives birth within two hours or so, she can simply wait until after the event to eat again. She may, however, give birth after 20, or even more than 40, hours. Giving birth is an athletic event, and nutritional preparation for it is not unlike the preparation for competition. She will find labor easier if she is well nourished and well hydrated at the start. This requires eating enough food to provide energy during labor without eating more than the digestive system can handle. As in other strenuous events, blood supply to the digestive tract receives low priority, resulting in indigestion and, possibly, vomiting.

During a long labor, it is wise to eat periodically, and many health care professionals encourage women to take snacks with them when they go to the birthing center or hospital. During the heavier stages of labor, few women care about food, but they may find ice chips refreshing.

The combination of wise food choices and regular physical activity promotes good health. With this in mind, a few words about fitness during pregnancy are in order.

Fitness during Pregnancy

It is not unusual today to see pregnant women jogging, swimming, participating in aerobic dance classes, walking energetically, or doing other exercises with the same vigor as their nonpregnant peers. An active, physically fit woman experiencing a normal pregnancy can continue to exercise throughout pregnancy, adjusting the intensity and duration as the pregnancy progresses.

A woman who has not been physically active before pregnancy should consult with her health care provider before beginning an exercise program. Exercise programs that gradually progress from mild to moderate intensity are likely to benefit her cardiorespiratory and muscular fitness.

Staying active throughout pregnancy can improve fitness, prevent gestational diabetes, facilitate labor, and reduce stress.[35] Women who exercise during pregnancy report fewer discomforts throughout their pregnancies.[36] Regular exercise develops the strength and endurance a woman needs to carry the extra weight gain through pregnancy and to labor through an intense delivery. It also maintains the habits that help a woman lose excess weight and get back into shape after the birth.

A pregnant woman should avoid sports in which she might fall or be hit by other people or objects. For example, playing tennis with one person on each side of the net is safer than a fast-moving game of racquetball in which the two competitors can collide. Swimming is ideal because it allows the body to remain cool and move freely with the water's support. Table 2-7 provides guidelines for exercise during pregnancy developed by the American College of Obstetricians and Gynecologists

Pregnant women can enjoy the benefits of physical activity.

Table 2-7   Exercise Guidelines for Pregnancy

- Get regular exercise (at least three times a week). Avoid spurts of heavy exercise followed by long periods of no activity.

- Avoid brisk exercise in hot, humid weather or when you are sick with a fever (such as the flu).

- Avoid jerky, bouncy, or high-impact motions. Activities that call for jumping, jarring motions or quick changes in direction may strain your joints and cause pain. Low-impact exercise is best. A wooden floor or a tightly carpeted surface reduces impact and gives you sure footing.

- Wear a bra that fits well and gives lots of support to help protect your breasts.

- Wear the proper shoes for the activity to be sure your feet are well cushioned and to give your body good support. There are shoes designed just for walking, running, aerobics, or tennis.

- Avoid deep knee bends, full sit-ups, double leg raises (in which you raise and lower both legs together), and straight-leg toe touches.

- After 20 weeks of pregnancy, avoid exercises that require lying with the back on the floor for more than a few minutes.

- Always begin with 5 minutes of slow walking or stationary cycling with low resistance to warm up your muscles. Intense exercise should not last longer than 15 minutes.

- Follow intense exercise with 5 to 10 minutes of gradually slower activity that ends with gentle stretching in place. To reduce the risk of injuring the tissue connecting your joints, do not stretch as far as you possibly can.

- The extra weight you are carrying will make you work harder as you exercise at a slower pace. Measure your heart rate at peak times of activity.

- Get up from the floor slowly and gradually to avoid feeling dizzy or fainting. Once standing, walk in place briefly.

- Drink water often before, during, and after exercise to be sure your body gets enough fluids. Take a break in your workout to drink more water if necessary.

- If you did not exercise two to three times a week before getting pregnant, you should start with physical activity of very low intensity and, bit by bit, move to higher levels.

- Stop exercising and consult your doctor or nurse if you get any of these symptoms, and they are unusually severe: pain, vaginal bleeding, dizziness or feeling faint, shortness of breath, irregular or rapid heart beat, difficulty walking, or pain in your back or pubic area.

Source: Reprinted with permission. American College of Obstetricians and Gynecologists, *Planning for Pregnancy, Birth, and Beyond* (Washington, D.C.: ©ACOG, 1995), pp. 88–90.

(ACOG).[37] Several of the guidelines listed prevent excessively high internal body temperature and dehydration, both of which can harm fetal development. To this end, pregnant women should also stay out of saunas, steam rooms, and hot whirlpools.

The expectant mother needs support in thinking of herself as a worthwhile and important person with a new and challenging task that she can and will perform well. She may still be working out her relationship with her mate, and he and she both know that an infant will affect that relationship profoundly. There is a need for sensitive communication and understanding on both parts in this time of transition.

With all her planning and adjusting, a woman can still relax and enjoy her pregnancy. With the help and guidance of health professionals, and the support of family and friends, she can feel healthy and confident as she awaits the birth of her infant. She has been given the chance to nurture a new life. It is a serious and challenging task, but it can be a rewarding one.

All women should be advised of the importance of eating a healthy diet during pregnancy. Some women may need intensive instruction and perhaps even supplements to correct for any inadequacies revealed through nutrition assessment. The next chapter examines high-risk pregnancies, including nutrition assessments and interventions.

CHAPTER TWO NOTES

1. Committee on Nutritional Status during Pregnancy and Lactation, Food and Nutrition Board, *Nutrition during Pregnancy* (Washington, D.C.: National Academy Press, 1990), p. 98.
2. W. W. Hay and coauthors, Workshop summary: Fetal growth: Its regulation and disorders, *Pediatrics* 99 (1997): 585–591.
3. G. R. Goldberg and A. M. Prentice, Maternal and fetal determinants of adult diseases, *Nutrition Reviews* 52 (1994): 191–200.
4. D. J. P. Barker, Growth in utero and coronary heart disease, *Nutrition Reviews* 54 (1996): S1–S7.
5. T. J. Wilkin, Early nutrition and diabetes mellitus, *British Medical Journal* 306 (1993): 283–284.
6. C. N. Hales and D. J. P. Barker, Type 2 (non-insulin-dependent) diabetes mellitus: The thrifty phenotype hypothesis, *Diabetologia* 35 (1992): 595–601.
7. C. N. Hales and coauthors, Fetal and infant growth and impaired glucose tolerance at age 64, *British Medical Journal* 303 (1991): 1019–1022; S. Robinson and coauthors, The relation of fetal growth to plasma glucose in young men, *Diabetologia* 35 (1992): 444–446.
8. J. T. E. Cook and coauthors, Association of low birth weight with β cell function in the adult first-degree relatives of non-insulin-dependent diabetic subjects, *British Medical Journal* 306 (1993): 302–305.
9. Committee on Nutritional Status during Pregnancy and Lactation, 1990, pp. 412–419.
10. E. M. Widdowson, Nutrition and cell and organ growth, in *Modern Nutrition in Health and Disease*, 8th ed., eds. M. E. Shils, J. A. Olson, and M. Shike (Philadelphia: Lea & Febiger, 1994), pp. 728–739.
11. R. L. Goldenberg and T. Tamura, Prepregnancy weight and pregnancy outcome, *Journal of the American Medical Association* 275 (1996): 1127–1128.
12. Goldenberg and Tamura, 1996.
13. M. M. Werler and coauthors, Prepregnant weight in relation to risk of neural tube defects, *Journal of the American Medical Association* 275 (1996): 1089–1092.
14. Werler and coauthors, 1996; G. M. Shaw, E. M. Velie, and D. Schaffer, Risk of neural

tube defect—Affected pregnancies among obese women, *Journal of the American Medical Association* 275 (1996): 1093–1096.

15. Committee on Nutritional Status during Pregnancy and Lactation, 1990, pp. 10–11.

16. Committee on Nutritional Status during Pregnancy and Lactation, 1990, p. 118.

17. Committee on Nutritional Status during Pregnancy and Lactation, 1990, p. 229.

18. Committee on Nutritional Status during Pregnancy and Lactation, 1990, p. 384.

19. D. R. Miller and coauthors, Vitamin B_{12} status in a macrobiotic community, *American Journal of Clinical Nutrition* 51 (1991): 524–529.

20. Committee on Nutritional Status during Pregnancy and Lactation, 1990, p. 282.

21. Committee on Nutritional Status during Pregnancy and Lactation, 1990, pp. 272–298.

22. Y. H. Neggers and coauthors, A positive association between maternal serum zinc concentration and birth weight, *American Journal of Clinical Nutrition* 51 (1990): 678–684.

23. J. Freeland-Graves, Mineral adequacy of vegetarian diets, *American Journal of Clinical Nutrition* 48 (1988): 859–862.

24. Life Sciences Research Office, Federation of American Societies for Experimental Biology, Executive summary from the third report on nutrition monitoring in the United States, *Journal of Nutrition* 126 (1996): 1907S–1936S; Committee on Nutritional Status during Pregnancy and Lactation, 1990, pp. 299–317.

25. R. R. Recker and coauthors, Bone gain in young adult women, *Journal of the American Medical Association* 268 (1992): 2403–2406.

26. C. Lamberg-Allardt and coauthors, Low serum 25-hydroxyvitamin D concentrations and secondary hyperparathyroidism in middle-aged white strict vegetarians, *American Journal of Clinical Nutrition* 58 (1993): 684–689.

27. D. Schendel and coauthors, Prenatal magnesium sulfate exposure and the risk for cerebral palsy and mental retardation among very low-birth-weight children aged 3 to 5 years, *Journal of the American Medical Association* 276 (1996): 1805–1810; L. Spatling and G. Spatling, Magnesium supplementation in pregnancy: A double-blind study, *British Journal of Obstetrics and Gynaecology* 95 (1988): 120–125.

28. Committee on Nutritional Status during Pregnancy and Lactation, 1990, pp. 299–317.

29. N. S. Scrimshaw, Nutrition and health from womb to tomb, *Nutrition Today*, March–April 1996, pp. 55–67.

30. R. A. Chez, Nutritional factors in pregnancy affecting fetal growth and subsequent infant development, in *Textbook of Pediatric Nutrition*, 2nd ed., eds. R. M. Suskind and L. Lewinter-Suskind (New York: Raven Press, 1993), pp. 1–7.

31. C. A. Bryant and coauthors, *The Cultural Feast: An Introduction to Food and Society* (St. Paul: West, 1985), pp. 149–186; C. A. Rinzler, *Feed a Cold, Starve a Fever: A Dictionary of Medical Folklore* (New York: Facts on File, 1991), p. 150; Food and Nutrition Board, *Alternative Dietary Practices and Nutritional Abuse in Pregnancy: Summary Report* (Washington, D.C.: National Technical Information Service, 1982), p. 3.

32. Position of The American Dietetic Association: Vegetarian diets, *Journal of the American Dietetic Association* 93 (1993): 1317–1319.

33. M. Erick, Battling morning (noon and night) sickness: New approaches for treating an age-old problem, *Journal of the American Dietetic Association* 94 (1994): 147–148.

34. M. Erick, Hyperolfaction and hyperemesis gravidarum: What is the relationship? *Nutrition Reviews* 53 (1995): 289–295.

35. K. G. Dewey and M. A. McCrory, Effects of dieting and physical activity on pregnancy and lactation, *American Journal of Clinical Nutrition* (supplement) 59 (1994): 446S–453S.

36. B. Sternfeld and coauthors, Exercise during pregnancy and pregnancy outcome, *Medicine and Science in Sports and Exercise* 27 (1995): 634–640.

37. American College of Obstetricians and Gynecologists, *ACOG Technical Bulletin 189: Exercise During Pregnancy and the Postpartum Period* (Washington, D.C.: American College of Obstetricians and Gynecologists, 1994).

An Overview of Genetics and Conception

Chapter 2 described how nutrition supports conception and fetal development. To fully appreciate the importance of this relationship, it is necessary to understand the basics of inheritance and conception. The accompanying glossary defines related terms.

Inheritance

An ovum—the gamete produced by a female—carries within it a set of 23 chromosomes containing all the genetic information necessary to make a human being. Each chromosome bears along its length thousands of genes, and each gene consists of coded instructions for making a single working protein—an enzyme, a structural protein, a muscle protein, or some other body protein. Each protein determines, or helps determine, traits for the new person, such as eye color, hair color and texture, maximum height, susceptibility to various diseases, potential for muscular development, and some aspects of temperament. The exact combination of these and many other characteristics makes each individual unique. These inherited characteristics define the range within which nutrition and other environmental influences can affect a person's development.

A spermatozoon—the gamete produced by a male—also contains 23 chromosomes bearing sets of instructions for the same characteristics. When the ovum and spermatozoon merge at fertilization to form a single cell, the chromosomes from each parent line up side by side within that cell. Now there are not 23 chromosomes, but 23 *pairs*—46 chromosomes in all (see Figure FP2A-1).

After this union, the zygote splits into two new cells, and those cells split again. All 46 chro-

GLOSSARY

birth defects: congenital abnormalities (present in an individual from birth), often caused by somatic mutations.

chromosomes: the bodies within each cell that contain the genetic material (deoxyribonucleic acid, or DNA).

dominant gene: a gene that produces a product whose effect is expressed in the appearance or functioning of the organism.

gamete: a mature male or female reproductive cell; the spermatozoon or ovum (*gamein* = to marry).

genes: the basic units of hereditary information, made of DNA, that are passed from parent to offspring in the chromosomes of the gametes. Each gene codes for a protein.

germ cells: an informal term for the cells in the ovary and testis that produce gametes (*germinate* = to begin life).

gonads: the primary sex organs containing the germ cells; testes in the male and ovaries in the female (*gone* = seed).

mutagens: agents or events that cause genetic mutations (*mutare* = to change; *genesis* = to produce).

mutation: a change in a cell's genetic material (DNA).

recessive gene: a gene that produces no product, or a product whose effect is not expressed when the product from a dominant gene is present.

somatic cells: the nonreproductive cells or tissues of the body (*soma* = body).

teratogens (teh-RAT-oh-gens): agents that cause somatic mutations that lead to abnormal fetal development and birth defects (*terat* = monster).

Figure FP2A-1 Gametes and Chromosomes

Human body cells contain 23 pairs of chromosomes, and gametes contain one member of each pair, or 23 single chromosomes. In these cells, only two pairs are shown. These pairs separate when the gametes are formed, and each gamete receives only one member of each pair (2 chromosomes, in this diagram).

When the gametes join in fertilization to form a zygote, new pairs of chromosomes are formed. Each contains one member from the male and one from the female. Thus the zygote has a full set of 46 chromosomes, like the parent cells (4 are shown in this diagram).

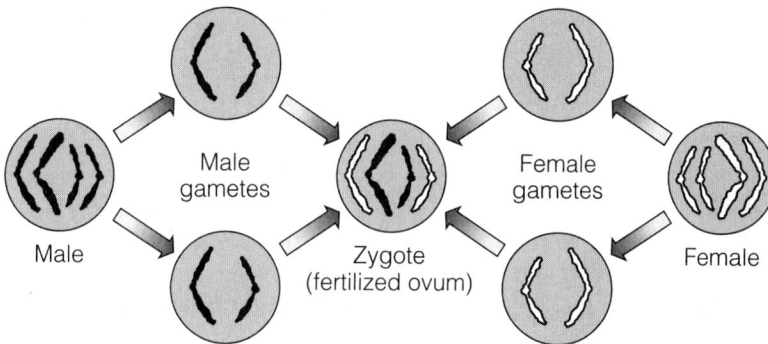

mosomes are faithfully copied at each division so that every new cell inherits the entire set of 46 chromosomes. (To be strictly accurate, each somatic cell—that is, every cell in the body except the germ cells—contains the full complement of 46 chromosomes.) When the individual has become a sexually mature adult, divisions of germ cells in the gonads produce cells with just 23 chromosomes once again. These are, of course, the gametes, the special cells through which the individual contributes one member of each pair of chromosomes to the *next* generation. Thus a child receives some chromosomes from its father, and some from its mother.

An offspring always, therefore, inherits two sets of genes, or instructions for making each piece of molecular machinery—one from each parent. In many cases, the two genes code for identical products. In some cases, the two genes code for two products, both are produced, and the resulting trait is a blend. In a few cases, one of the two genes may produce no product or a defective product that does not function—but the other produces a product that functions. (The inheritance of two sets of instructions gives the new individual two chances to "get it right," in a sense.)

Occasionally two defective genes come together in an individual, and the result is the malfunction or total absence of some gene product. One such case may present no problem if the characteristic governed by that gene product is not essential to life or health. Another such case may cause a major biochemical defect in an organism, known as an inborn error of metabolism (featured in Focal Point 5C). Many such cases are lethal, leading to spontaneous abortions and stillbirths.

The foregoing discussion makes clear why some of the new individual's traits may be identical to the father's and some to the mother's and why some are intermediate between the two. Some brand-new traits also emerge; for example, traits determined by genes on chromosomes donated by the grandparents—genes that were hidden but not expressed in one or the other parent.

New traits may also arise from unique combinations of gene products occurring for the first time in this new individual.

The simplest genetic arrangement to understand is the type governed by a single gene pair—the arrangement for eye color, for example. A gene on one chromosome from the mother and a gene on the same location on the corresponding chromosome from the father govern this characteristic. The two genes may or may not be identical; the combination determines the actual eye color. If the father's gene specifies blue-eye pigment, but the mother's codes for brown pigment, then the eyes will be brown, whether blue-eye pigment is present or not; this is because the dominant gene's product, brown pigment, literally outshadows the recessive gene's product. Thus the person who inherits one copy of each gene will have brown eyes.

Another ovum may not carry the message for brown eyes but may contain the recessive message for blue eyes, just as the spermatozoon does. (Remember, the mother also inherited two genes for eye color; she can pass along either one.) Now the products of the normally recessive genes are visible, and the person is blue-eyed. Two brown-eyed people can thus have a blue-eyed child. When this happens, although both parents possessed a dominant, brown-eye gene, both also carried a blue-eye gene and passed that one along to their offspring. This is one way in which "new" traits emerge in new generations.

An infant's gender is determined by a variation on this theme. The father's gametes determine the sex of the infant. One of the 23 pairs of chromosomes, known as the X-Y, or sex, chromosomes, carries the sex-determining genes. People can inherit either two Xs, in which case they are female, or an X and a Y, in which case they are male. No one ever gets two Y chromosomes, because mothers, being female, have two Xs and can only donate Xs to their offspring. Males donate X chromosomes to half of their offspring, on the average, and Y chromosomes to half—thus girls and boys are born in equal numbers. In the race to fertilize the ovum, the spermatozoon that swims best or that gets through the ovum's outer covering first is the winner— and is the one that determines the infant's gender (see Figure FP2A-2).

Nutrition and Inheritance

With this basic understanding, you can see how nutrition relates to conception and heredity. Genes are normally copied with amazing accuracy. This is important, for they contain instructions that are crucial to the correct making of an organism. A set of parental genes is copied each time a gamete is made. The genes are copied from the parent's germ cell into the set of 23 chromosomes that are packaged in the gamete. Then the male and female gametes unite to form a zygote with 46 chromosomes. After that, all 46 chromosomes are copied over and over again as the zygote divides and redivides to form the billions of cells of the adult organism.

As a cell duplicates its genetic material in preparation for dividing to form two cells, every now and then it makes a mistake—a mutation. Such mistakes arise in the average gene about once in every 100 million copies. Most errors are immediately corrected by an intracellular system of comparing copies with the original, destroying defective copies, and recopying. A few mistakes slip by, however, and mutations may be transmitted to new cells. Once transmitted (inherited), a mutation will be as faithfully copied as the original was, and will be transmitted to all future offspring of that cell. The only ways mutations are ever lost from a cell line is by changing back (an event even rarer than the mutations' genesis) or by the cells' dying out without reproducing.

Mutagens, including both chemical and physical agents, cause mutations to arise in the copying of the genetic material. Chemical mutagens include tars in tobacco smoke, toxins produced by bacteria, many pesticides, heavy metals, and many drugs, both illegal and legal (prescription and over-the-counter). Among physical mutagens are several forms of radiation, including the sun's ultraviolet rays and radioactivity. (Irradiated food, however, does not contain mutagens; it has been exposed to radiation, but does not contain any substances that give

Figure FP2A-2 Sex Inheritance

One of the 23 pairs of chromosomes, known as the X-Y, or sex, chromosomes, carries the sex-determining genes. Males contain an X and a Y; females contain two Xs. Males donate X chromosomes to half of their offspring, on the average, and Y chromosomes to half—thus boys and girls are born in equal numbers.

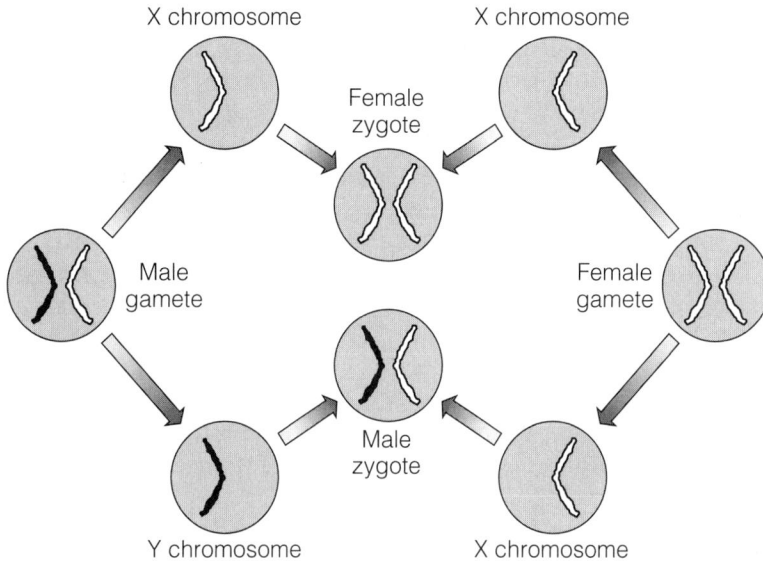

off radiation.) Mutagens can cause not only mutations leading to inborn errors, but also others leading to birth defects, cancer, possibly atherosclerosis and diabetes, and probably other diseases as well. The special class of mutagens that cause fetal malformations are known as *teratogens.*

The earlier in development a harmful influence exerts itself, the more devastating the consequences are likely to be. Mutations in germ cells are especially harmful, for gametes are copied from these cells. Such mutations are passed on to every cell of the offspring's body *and from generation to generation* through the germ cell line. They are usually recessive; typically, the gene altered by mutation produces an inactive product or no product. An individual who inherits such a defective gene exhibits a normal trait only if the defective gene is paired with a normal gene contributed by the other parent. (Such an individual is said to be a carrier of

the mutation.) When two carriers conceive a child, and each contributes the defective member of its pair of genes for that trait to the child, the child cannot produce a normal product (such as an enzyme), and so has an inborn error of metabolism (see Focal Point 5C). Mutations in the germ cell line may persist invisibly for many generations before their effects become apparent. Should enough mutations accumulate within the human genetic material, they could condemn the human race to ever-increasing disabilities. For this reason, more than any other, both women and men should, throughout their reproductive lives, avoid contact with any chemical or physical agent that causes mutations. (It is for this reason, too, that accidents involving mutagenic radiation, such as the one that took place at Chernobyl in April 1986, are so frightening.)

Teratogens cause mutations, not in the germ cell line, but in somatic cells during an individual's development. These are also harmful, but *they affect only that individual* and are not passed on to future generations. They can be devastating, however. If mutations occur during the first two weeks of a zygote's development, they are likely to be lethal, because each cell in a developing zygote is destined to become a large part of the body of an adult. Major abnormalities induced in whole organ systems preclude survival. Lesser abnormalities lead to birth defects. Chapter 3 and its Focal Point describe how teratogenic effects of alcohol and drugs arise.

Nearly all mutations are harmful. Exceptions are known, but 999 out of 1000 mutations are disadvantageous. Therefore all *mutagens* can be classed as harmful.

Although inborn errors and many birth defects are known to be caused by mutations, it is never possible to look back in time and say when those mutations arose. They are, however, most likely to arise during times of cell division, which means that nearly all periods of early life are especially vulnerable—for nearly all involve cell division in the germ cell line. Boys and men make new spermatozoa continuously from puberty on. In young women, as soon as they are pregnant, zygotic and embryonic cell division proceeds rapidly. Female fetuses develop all the ova they will release for a lifetime. Young men and women should therefore at all times avoid exposure to mutagenic radiation and contamination of food and water that might threaten the integrity of their genetic material and young offspring. In addition to the recommendations pre-

sented in Chapter 2 to eat a diet that supports a healthy pregnancy—one that is adequate, balanced, and varied—all the foods and beverages selected should be free of contamination.

Nutrition and Conception

Once the gametes are produced, they may then meet, join, and begin the production of a new individual. The meeting and joining are, of course, fertilization, or conception. Sexual intercourse from 48 hours before to 15 hours after ovulation is most likely to result in conception. The average life spans of the spermatozoon (48 hours) and the unfertilized ovum (10 to 15 hours) determine this time frame. During ovulation, the ovary discharges an ovum into the oviduct. Over the next several days, the ovum works its way down the oviduct into the uterus. The short life span of the ovum requires fertilization to occur in the oviduct (one of the two tubes connecting the ovaries to the uterus). As Chapter 3 explains, poor nutrition can impair fertility and even prevent conception.

Reproduction is an astonishing process. That human bodies package, into cells too small to see, entire libraries of all the information necessary to make human beings inspires wonder. To produce gametes as nearly perfect as possible, and to permit their union in the safest conditions, men and women need to treat their own bodies with respect.

Medical Disorders during Pregnancy

Medical disorders during pregnancy can threaten the lives of both mother and fetus. If diagnosed and treated early, though, most conditions can be managed to ensure a healthy outcome—a strong argument for early prenatal care. The health problems featured in this discussion have special relationships with nutrition. In two cases—diabetes and hypertension—the condition may have developed before pregnancy, or it may have first appeared during pregnancy. In the third case—phenylketonuria, an inborn error of metabolism—the condition has been present since the mother's birth but has critical ramifications during her pregnancy. Preexisting medical problems are more easily managed when they are under control before pregnancy—a good reason for women to discuss pregnancy plans with their physicians. Careful management of the disorder during pregnancy helps to minimize potential problems.

Diabetes in Pregnancy

Blood insulin begins to rise soon after conception, and the cells respond by storing energy nutrients to provide for the developing fetus. Later in pregnancy, insulin remains high, but the cells become insulin resistant. Hormones that antagonize the action of insulin rise.* This hormonal shift signals the body to stop storing energy fuels and allows the fetus to rapidly take up energy nutrients. Because pregnancy stresses the glucose regulatory system in these ways, women with diabetes should expect control to become more difficult during pregnancy.

Preexisting Diabetes. Women with diabetes who are contemplating pregnancy should know that poorly controlled diabetes presents risks for both mother and infant. Without proper management of maternal diabetes, women face a high infertility rate, and those who do conceive may experience episodes of severe hypoglycemia or hyperglycemia, spontaneous abortion, and pregnancy-related hypertension. The amniotic sac may retain excess fluid, which stimulates premature labor and delivery. Infants may be larger than normal and suffer congenital abnormalities and other complications such as severe hypoglycemia or respiratory distress, both of which can be fatal. Women with diabetes should receive preconception care that aims to achieve blood glucose control before conception, and continued prenatal care to maintain blood glucose control throughout pregnancy. They must be careful, however; aggressive efforts to control blood glucose raise the likelihood of hypoglycemic episodes.[1]

For the pregnant woman with diabetes, a diet tailored to meet the increased demands of pregnancy *and* carefully coordinated with insulin therapy is central to her care. Diet plans provide adequate energy to support weight gain and frequent small meals to prevent hypoglycemia. A bedtime snack is recommended, for two reasons. For one, a bedtime snack provides glucose during the night, preventing nocturnal hypoglycemia. For another, it provides glucose for the developing fetus, preventing the buildup of ketone bodies. Ketone body accumulation is a serious complication of uncontrolled diabetes that can cause stillbirth.

Gestational Diabetes. Women who never had diabetes before may develop it during pregnancy (gestational diabetes). Gestational diabetes usually appears later in pregnancy when the tissues become insulin resistant, with subsequent return to normal glucose tolerance after childbirth. Almost one-third of all women with gestational diabetes, however, develop non-insulin-dependent diabetes later in life, especially if they are overweight. For this reason, health care professionals advise these women to avoid excessive

*The hormones that antagonize the action of insulin during late pregnancy are placental lactogen, cortisol, prolactin, and progesterone.

weight gain during pregnancy and to achieve a healthy weight thereafter.

Without proper management, gestational diabetes can lead to complications such as a high-birthweight infant and difficult delivery; rates of cesarean delivery are also exceptionally high (almost 30 percent).[2] To ensure that the problems of gestational diabetes are dealt with promptly, during the first prenatal examination, health care professionals look for the risk factors listed in Table FP2B-1. Women with one or more of the factors listed receive glucose tolerance tests to evaluate their risk.[3]

All pregnant women should have urine tests for glucose and ketone bodies at each checkup. Some physicians give a glucose tolerance test routinely between 24 and 28 weeks of gestation.

An important aspect of nutrition management in gestational diabetes is prevention of excessive weight gain, but weight reduction diets are not normally recommended during pregnancy. Women with gestational diabetes should not reduce their total carbohydrate intakes but should select foods rich in complex carbohydrates, such as legumes, vegetables, and grain products, and limit their intakes of concentrated sweets. Dietary recommendations encourage three meals a day plus two snacks, each containing protein, carbohydrate, and moderate fat. Diet alone may control gestational diabetes, but insulin therapy may be required if blood glucose fails to normalize. Preliminary research indicates that coordinating insulin doses with postprandial, rather than preprandial, glucose concentrations improves glycemic control.[4]

Hypertension in Pregnancy

Hypertension complicates pregnancy and affects its outcome in different ways, depending on when it first develops and on how severe it becomes.[5] Hypertension can be a preexisting chronic condition that develops before a woman becomes pregnant or a transient condition that

Table FP2B-1   Risk Factors for Gestational Diabetes

- Previous gestational diabetes.
- History of large infants (9 lb or more).
- Family history of diabetes.
- Symptoms of diabetes (glycosuria, recurrent urinary tract infections).
- Obesity or excessive weight gain.
- Age 30 or older.
- Complications in previous pregnancies (spontaneous abortions, unexplained stillbirths).

develops during the pregnancy and subsides after childbirth. In some cases, hypertension warns of the ominous disorder preeclampsia.

Preexisting Chronic Hypertension. More than 85 percent of the women who have hypertension before they become pregnant have uncomplicated pregnancies. Still, their risks for problem pregnancies are high compared with women who do not have high blood pressure.[6] In addition to the complications associated with hypertension (heart attack and stroke), high blood pressure increases the risks of a low-birthweight infant or the separation of the placenta from the wall of the uterus before the birth, resulting in stillbirth. Ideally, before a woman with hypertension becomes pregnant, her blood pressure will be under control.

Transient Hypertension of Pregnancy. Some women first develop hypertension during the second half of pregnancy. Most often, the rise in blood pressure is mild and does not affect the pregnancy adversely.[7] Blood pressure usually returns to normal during the first few weeks after childbirth. This transient hypertension of pregnancy differs from the pregnancy-induced hypertension that accompanies preeclampsia.*

---

*The Working Group on High Blood Pressure in Pregnancy, convened by the National High Blood Pressure Education Program of the National Heart, Lung, and Blood Institute has suggested abandoning the term "pregnancy-induced hypertension" because it fails to differentiate between the mild, transient hypertension of pregnancy and the life-threatening hypertension of preeclampsia.

Preeclampsia. Hypertension may signal the onset of preeclampsia, a condition characterized not only by high blood pressure but by protein in the urine and fluid retention (edema). Preeclampsia usually occurs with first pregnancies and almost always after 20 weeks gestation, most often near term.[8] Symptoms typically regress within 48 hours of delivery.

Preeclampsia affects almost all the mother's organs—the circulatory system, liver, kidneys, and brain. Blood flow through the vessels that supply oxygen and nutrients to the placenta diminishes. For this reason, preeclampsia often retards fetal growth. In some cases, the placenta separates from the uterus, resulting in stillbirth.

Preeclampsia can progress rapidly to eclampsia—a condition characterized by convulsions. Maternal mortality during pregnancy and childbirth is extremely rare in most developed countries, but eclampsia is the most common cause.

Preeclampsia demands prompt medical attention. Treatment focuses on regulating blood pressure and preventing convulsions. If preeclampsia is mild, bed rest and medication may be sufficient to control the symptoms until fetal growth is complete enough to sustain life. If preeclampsia develops early and is severe, induced labor or cesarean delivery may be necessary, regardless of gestational age. The infant will be preterm, with all the associated problems, including poor lung development and special care needs. Table FP2B-2 lists the warning signs of preeclampsia.

Prenatal care includes monitoring weight gains and maternal blood pressure throughout pregnancy. A sudden, large weight gain in the second or third trimester alerts health care professionals to excessive fluid retention (edema). Blood pressure of 140/90 millimeters mercury during the second half of pregnancy in a woman who has not previously exhibited hypertension indicates high blood pressure. So does a rise in systolic blood pressure of 30 millimeters or in diastolic blood pressure of 15 millimeters on at least two occasions more than six hours apart. By this rule, an apparently "normal" blood pressure of 120/85 would be high for a woman whose normal value was 90/70.

Several approaches have been proposed to

Table FP2B-2   Symptoms of Preeclampsia

- Hypertension.
- Protein in the urine (proteinuria).
- Upper abdominal pain.
- Severe and constant headaches.
- Dizziness.
- Blurred vision.
- Swelling, especially of the face.
- Sudden weight gain (1 lb/day).
- Fetal growth retardation.

prevent preeclampsia, including sodium restriction, calcium supplementation, and magnesium sulfate injections. Sodium restriction does not reduce the incidence or severity of preeclampsia. Furthermore, a sodium-restricted diet tends to lower the intakes of fat, protein, calcium, and energy; limit maternal weight gain; and reduce maternal fat stores—all unwanted side effects in women whose nutrition status may already be compromised.[9] Until and unless the kidneys prove unable to handle sodium and edema is evident, sodium restriction is not recommended.

Several studies have reported an inverse relationship between calcium intake and preeclampsia.[10] Furthermore, research has determined that calcium supplementation (1500 to 2000 milligrams per day) during pregnancy can lower blood pressure.[11] Such findings are promising, but additional research is needed to determine the exact relationship between calcium, preeclampsia, and other forms of gestational hypertension. At this time, evidence is insufficient to recommend calcium supplements routinely to prevent pregnancy-related hypertension. Furthermore, clinicians caution that calcium supplementation may have risks of its own, including the development of kidney stones.[12]

To prevent seizures during labor, physicians may give magnesium sulfate injections to women with preeclampsia.[13] Such treatment is both effective and superior to other anticonvulsive drugs. Exposure to magnesium sulfate reduces the risk for cerebral palsy and possibly mental retardation.[14]

Phenylketonuria (PKU)

Phenylketonuria (PKU) is an inherited disorder that can damage the developing nervous system if left untreated. Like all inborn errors, PKU is a recessive disorder; that is, it appears only when a person inherits two genes that code for PKU—one from each parent. This can occur even if nei-ther parent has PKU, because both may be carriers (see Figure FP2B-1). A carrier is a person who inherits one PKU gene and one normal gene. The carrier may be unaware of having the PKU gene, for the symptoms are usually mild or absent.

Preexisting PKU. All newborns in the United States are routinely screened for PKU so that a di-

Figure FP2B-1 The Inheritance of PKU

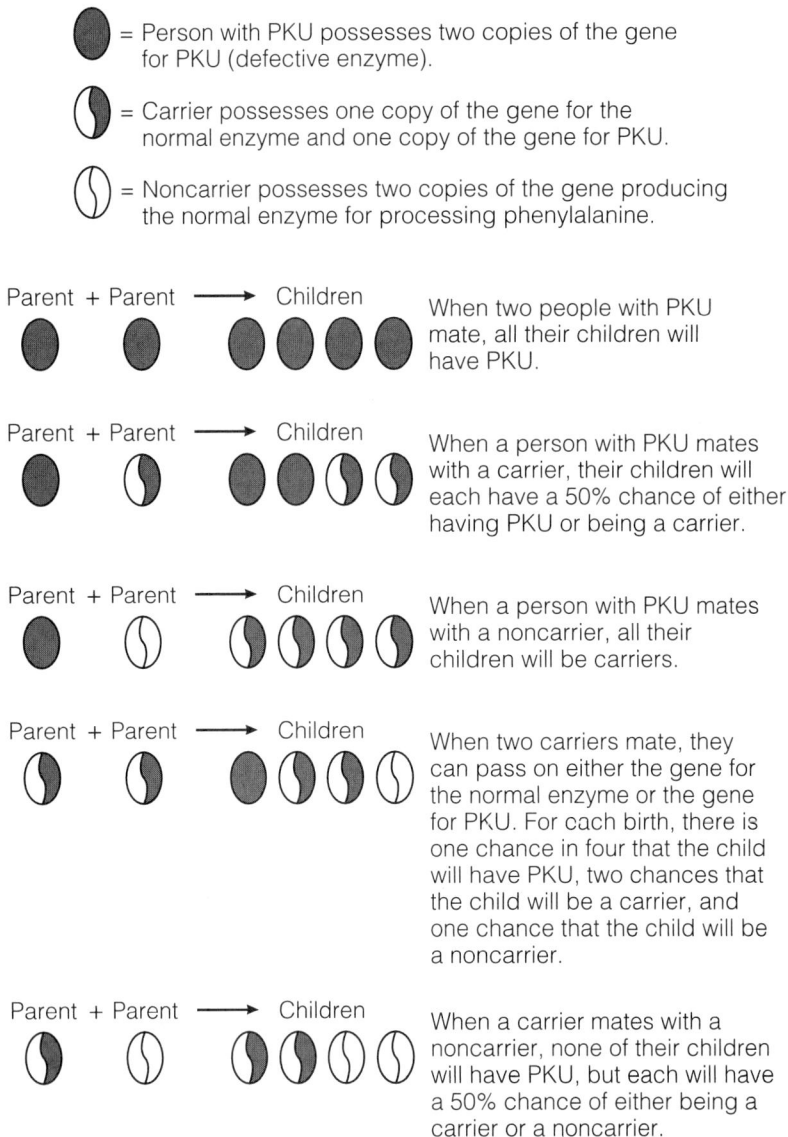

= Person with PKU possesses two copies of the gene for PKU (defective enzyme).

= Carrier possesses one copy of the gene for the normal enzyme and one copy of the gene for PKU.

= Noncarrier possesses two copies of the gene producing the normal enzyme for processing phenylalanine.

Parent + Parent → Children

When two people with PKU mate, all their children will have PKU.

Parent + Parent → Children

When a person with PKU mates with a carrier, their children will each have a 50% chance of either having PKU or being a carrier.

Parent + Parent → Children

When a person with PKU mates with a noncarrier, all their children will be carriers.

Parent + Parent → Children

When two carriers mate, they can pass on either the gene for the normal enzyme or the gene for PKU. For each birth, there is one chance in four that the child will have PKU, two chances that the child will be a carrier, and one chance that the child will be a noncarrier.

Parent + Parent → Children

When a carrier mates with a noncarrier, none of their children will have PKU, but each will have a 50% chance of either being a carrier or a noncarrier.

Note: Focal Point 2A discusses inheritance in more detail.

agnosis can be made early. PKU infants are given special formulas and must follow a diet that restricts the amino acid phenylalanine. Focal Point 5C describes the diet and how, without strict adherence to the diet, a healthy baby develops mental retardation.

Maternal PKU. The risks for a PKU mother primarily affect her baby. If she has not carefully followed the prescribed diet, her blood concentrations of phenylalanine and its metabolic by-products rise. In adults, these toxic levels do not cause irreversible brain damage, but in a fetus, blood concentrations rise higher than the mother's, and fetal development is impaired. She may experience spontaneous abortion; or her infant is likely to suffer mental retardation, microcephaly, congenital heart disease, and low birthweight.[15]

If implemented early enough, dietary control of maternal PKU may protect the fetus.[16] Women who become pregnant without planning for

---

GLOSSARY

carrier: an individual who possesses one dominant and one recessive gene for a recessive trait, such as an inborn error of metabolism. Such a person may show no signs of the trait but can pass it on.

diabetes (DEY-uh-BEET-eez) **mellitus** (MELL-ih-tus or mell-EYE-tus): a metabolic disorder characterized by altered glucose regulation and utlization, usually caused by insufficient or relatively ineffective insulin.

eclampsia: a severe stage of preeclampsia characterized by convulsions.

edema (eh-DEEM-uh): the swelling of body tissue caused by excessive amounts of fluid in the interstitial spaces; seen in preeclampsia (among other conditions).

gestational diabetes: the appearance of abnormal glucose tolerance during pregnancy, with subsequent return to normal after birth.

glycosuria: glucose in the urine.

hyperglycemia: an abnormally high blood glucose concentration.

hypertension: higher-than-normal blood pressure.

hypoglycemia: an abnormally low blood glucose concentration.

inborn error of metabolism: an inherited flaw evident as a metabolic disorder or disease present from birth.

insulin (IN-suh-lin): a hormone secreted by special cells in the pancreas in response to (among other things) increased blood glucose concentration. The primary role of insulin is to control the transport of glucose from the bloodstream into the cells.

insulin resistance: the condition of having a normal amount of insulin producing a subnormal effect.

ketone (KEE-tone) **bodies:** the product of the incomplete breakdown of fat when glucose is not available in the cells. The combination of elevated ketone bodies in the blood (ketonemia) and in the urine (ketonuria) is **ketosis**.

PKU, phenylketonuria (FEN-el-KEY-toe-NEW-ree-ah): an inborn error of metabolism in which phenylalanine, an essential amino acid, cannot be converted to tyrosine. Alternative metabolites of phenylalanine (phenylketones) accumulate in the tissue, causing damage, and overflow into the urine.

postprandial: following a meal.

preeclampsia: a condition characterized by hypertension, fluid retention, and protein in the urine.

preexisting chronic hypertension: an ongoing condition of high blood pressure that developed before a specified time (in this case, pregnancy).

preprandial: prior to a meal.

transient hypertension of pregnancy: high blood pressure that develops in the second half of pregnancy and resolves after childbirth, usually without affecting the outcome of pregnancy.

their pregnancies expose their fetuses to high phenylalanine concentrations during a critical time of early development. A woman with PKU who departed from her restricted diet as a child must resume the low-phenylalanine diet at least one to two months prior to conception and maintain it throughout her pregnancy. As mentioned in Focal Point 5C, many physicians recommend adherence to a restricted diet throughout life. Resuming a low-phenylalanine diet is not easy. The special formulas are costly, inconvenient, and unpalatable to an adult who has been eating foods freely. Many women have forgotten that they were ever on a special diet as a child or may never have understood why.

The low-phenylalanine diet for maternal PKU must meet the energy, protein, vitamin, and mineral needs of pregnancy. Special PKU formulas for the pregnant PKU woman are now available. Although dietary control does not ensure a successful outcome of pregnancy, a low-phenylalanine diet used from before conception throughout pregnancy seems promising for women with PKU. Their children are more likely to have higher birthweights, larger head circumferences, fewer malformations, and higher IQs than children of women who begin diet therapy during their pregnancy or not at all.[17]

Given the genetic risks and fetal abnormalities associated with PKU, a woman with PKU needs genetic and medical counseling. She must consider the possible consequences of pregnancy, her ability to follow the special diet, and the options of contraception to prevent pregnancy or adoption if she wants children.

In recent years, new techniques and methods have greatly enhanced early detection and treatment of diseases. Women with chronic conditions can deliver healthy infants today, whereas only a few years ago pregnancy was not even an option for many of them.

FOCAL POINT 2B NOTES

1. A. M. Ferris and E. A. Reece, Nutritional consequences of chronic maternal conditions during pregnancy and lactation: Lupus and diabetes, *American Journal of Clinical Nutrition* 59 (1994): 465S–473S.

2. C. D. Naylor and coauthors, Cesarean delivery in relation to birth weight and gestational glucose tolerance: Pathophysiology or practice style? *Journal of the American Medical Association* 275 (1996): 1165–1170.

3. *ACOG Guide to Planning for Pregnancy, Birth, and Beyond* (Washington, D.C.: American College of Obstetricians and Gynecologists, 1990), pp. 128–140.

4. M. deVeciana and coauthors, Postprandial versus preprandial blood glucose monitoring in women with gestational diabetes mellitus requiring insulin therapy, *New England Journal of Medicine* 333 (1995): 1237–1241.

5. B. M. Sibai, Treatment of hypertension in pregnant women, *New England Journal of Medicine* 335 (1996): 257–265.

6. F. G. Cunningham and M. D. Lindheimer, Hypertension in pregnancy, *New England Journal of Medicine* 326 (1992): 927–932.

7. Cunningham and Lindheimer, 1992.

8. *ACOG Guide to Planning for Pregnancy, Birth, and Beyond*, pp. 130–131.

9. B. J. A. vanBuul and coauthors, Dietary sodium restriction in the prophylaxis of hypertensive disorders of pregnancy: Effects on the intake of other nutrients, *American Journal of Clinical Nutrition* 62 (1995): 49–57.

10. Calcium supplementation prevents hypertensive disorders of pregnancy, *Nutrition Reviews* 50 (1992): 233–236.

11. H. C. Bucher and coauthors, Effect of calcium supplementation on pregnancy-induced hypertension and preeclampsia: A meta-analysis of randomized controlled trials, *Journal of the American Medical Association* 275 (1996): 1113–1117.

12. T. F. Ferris, Pregnancy, preeclampsia, and the endothelial cell, *New England Journal of Medicine* 325 (1991): 1439–1440.

13. M. J. Lucas, K. J. Leveno, and G. Cunningham, A comparison of magnesium sulfate with phenytoin for the prevention of eclampsia, *New England Journal of Medicine* 333 (1995): 201–205.

14. D. E. Schendel and coauthors, Prenatal magnesium sulfate exposure and the risk for cere-

bral palsy or mental retardation among very low-birth-weight children aged 3 to 5 years, *Journal of the American Medical Association* 276 (1996): 1805–1810.

15. The Maternal Phenylketonuria Collaborative Study: A status report, *Nutrition Reviews* 52 (1994): 390–393; Committee on Genetics, American Academy of Pediatrics, Maternal phenylketonuria, *Pediatrics* 88 (1991): 1284–1285.

16. P. B. Acosta, Phenylketonuria—Impact of nutrition support on reproductive outcomes, *Nutrition Today*, January–February 1991, pp. 43–47.

17. Committee on Genetics, 1991; Acosta, 1991.

High-Risk Pregnancies

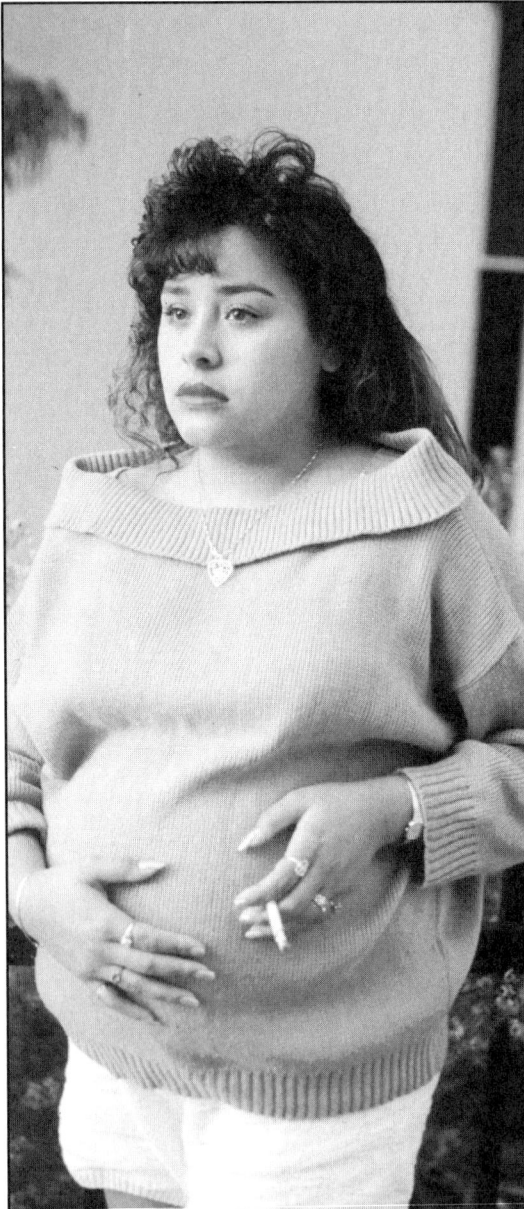

As Chapter 2 explained, between the moment of conception and the moment of birth, innumerable events determine the course and outcome of fetal development and, ultimately, the health of the newborn infant. Maternal diet plays a vital role, but it is only one of several influencing factors; paternal diet also plays a role in that it may influence fertility. Some influencing factors, such as genetics, also play important roles in fetal development, but they are beyond a couple's control and cannot be changed. Other factors, such as maternal health and age, may not be changeable, but their associated problems may be managed in ways that support a healthy pregnancy. Still other factors, such as maternal *and* paternal smoking and alcohol use, are personal behaviors that can be changed to improve the likelihood of a successful pregnancy.

This chapter begins with a description of how maternal health and age influence a pregnancy and how malnutrition and lifestyle habits such as smoking and drug use impair fetal growth and development. Such pregnancies are risky to the life and health of both the mother and infant. Table 3-1 identifies several characteristics of a high-risk pregnancy. A woman with none of these risk factors is said to have a low-risk pregnancy. The more factors that apply, the higher the risk.

All pregnancies, especially high-risk pregnancies, need careful assessment and special management, including dietary intervention. The chapter includes a discussion of nutrition assessment during pregnancy and closes with a look at government efforts to provide assistance to pregnant women in the United States.

As you will see, the nutrition habits and lifestyle choices people make can influence the course of a pregnancy they are not even thinking of at the time. Many couples are willing to change lifestyle behaviors to promote a healthy pregnancy provided they have the knowledge and the means to do so. They may be motivated to make these changes if they understand how their behaviors influence prenatal development and the health of the child long after birth.

Young adults can prepare themselves for a healthy pregnancy by taking care of themselves today.

high-risk pregnancy: a pregnancy characterized by indicators that make it likely the birth will be surrounded by problems such as premature delivery, difficult birth, retarded growth, birth defects, and early infant death.

low-risk pregnancy: a pregnancy characterized by indicators that make a normal outcome likely.

The Mother's Health and Age

Normal weight gain and adequate nutrition support the health of the mother and growth of the infant. Conversely, maternal diseases detract from growth and health. If discovered early, many diseases can be controlled—a prime reason early prenatal care is recommended. Focal Point 2B described the development and treatment of certain diseases during pregnancy.

Maternal age also influences the course of a pregnancy. Compared with women of the physically ideal childbearing age of 20 to 25, both younger and older women face more complications of pregnancy, as the next paragraphs describe.

Table 3-1 High-Risk Pregnancy Factors

Factor	Condition That Raises Risk
Maternal weight	
Prior to pregnancy	Prepregnancy BMI either <19.8 or >26.0
During pregnancy	Insufficient or excessive pregnancy weight gain
Maternal nutrition	Nutrient deficiencies or toxicities; eating disorders
Socioeconomic status	Poverty, lack of family support, low level of education, limited food available
Lifestyle habits	Smoking, alcohol or other drug use
Age	Teens 15 years or younger; women 35 years or older
Previous pregnancies	
Number	Many previous pregnancies (3 or more to mothers under age 20 and 4 or more to mothers age 20 and over)
Interval	Short intervals between pregnancies (<1 year)
Outcomes	Previous history of problems
Multiple births	Twins or triplets
Birthweight	Low- or high-birthweight infants
Maternal health	
High blood pressure	Development of pregnancy-related hypertension
Diabetes	Development of gestational diabetes
Chronic diseases	Diabetes; heart, respiratory, and kidney disease; certain genetic disorders; special diets and drugs

Pregnancy in Adolescents

Many adolescents become sexually active before age 19. With this sexual activity comes the responsibility of contraception. Nine out of ten sexually active couples who do not practice contraception become pregnant within a year.[1]

Approximately one million adolescent girls face pregnancies each year in the United States. About half of them give birth. Put another way, about one out of every four babies is born to a teenager, and more than a tenth of these mothers are younger than 15.[2]

Nourishing a growing fetus adds to a teenager's nutrition burden, especially if her growth is still incomplete.[3] The timing of adolescent growth and sexual maturation varies greatly among teenagers, but in general, females attain physiological maturity about three years after menarche (their first menstrual period). Pregnancy before the end of these three years, while a teenage girl is still growing, jeopardizes the health of both mother and infant. Simply being young increases the risks of pregnancy complications independently of important socioeconomic factors.[4]

Maternal Health. Common complications among adolescent mothers include iron deficiency anemia (which may reflect poor diet and inadequate prenatal care) and prolonged labor (which reflects physical immaturity of the mother). On a positive note, maternal mortality is lowest for mothers under age 20.

Infant Health. Pregnant teenagers have higher rates of stillbirths, preterm births, and low-birthweight infants than do adult women.[5] Many of these infants suffer physical problems, require intensive care, and die within the first year. The care of infants born to teenagers costs our society an estimated $1 billion annually. Because teenagers have few financial resources, they cannot pay these costs. Furthermore, their low economic status contributes significantly to the complications surrounding their pregnancies. Clearly, teenage pregnancy is a major public health problem.

Nutrition Recommendations. Young mothers (13 to 16 years of age) tend to have smaller newborns than older mothers (17 to 25 years of age) even when prepregnancy weights and pregnancy weight gains are similar. Weight gains need to cover the expected growth of both the adolescent and her fetus if she is to give birth to an infant of optimal birthweight. To support the needs of both mother and fetus, physically immature teenagers are encouraged to strive for the highest weight gains recommended for pregnancy.

For a teen who enters pregnancy at a healthy body weight, a weight gain of approximately 35 pounds is recommended; this minimizes the risk of delivering a low-birthweight infant.[6] Gaining less may limit fetal growth.[7] Pregnant teenagers can use the food guide presented in Table 2-4 (on p. 64), making sure to select at least four servings of milk or milk products daily.

Little information is available on the specific energy and nutrient needs of the pregnant adolescent. The RDA for pregnancy do not differ for women of various ages. Chapter 2 presented the nutrient needs of pregnancy and recommended food intakes. Figure 3-1 compares the RDA for adolescent girls with those of pregnancy.

Many pregnant teens eat the same amounts and kinds of foods as pregnant adults. Some, however, continue to follow typical teenage patterns of depending on fast-food meals, skipping meals, and paying little attention to food selections. Those who do will not have the nutrient stores they need for their own growth plus pregnancy. Adolescents who understand the nutrient demands of pregnancy are more likely to have sufficient energy intakes and weight gains than those with little nutrition education.

Adolescent Needs and Prenatal Care. The plight of a pregnant teenager is often compounded by a multitude of social and economic problems. Most are from low-income families. Often they have several children during their teen years, and this further increases their pregnancy risks. Compared with the first pregnancies of adolescents, subse-

Figure 3-1 Comparison of Nutrient RDA of Nonpregnant Adolescents and Pregnant Women

This figure uses adolescents 15 to 18 as the nonpregnant standard. For many of the nutrients, the RDA do not differ for adolescents 15 to 18 and those 11 to 14. For several, however, they do, and it is important to note that:

■ For vitamin K, vitamin C, vitamin B₆, folate, magnesium, and selenium, the RDA increase slightly for older teens, so increased needs during pregnancy are more dramatic for younger teens.

■ For protein, the RDA decreases slightly for older teens, so increased needs during pregnancy are more dramatic for older teens.

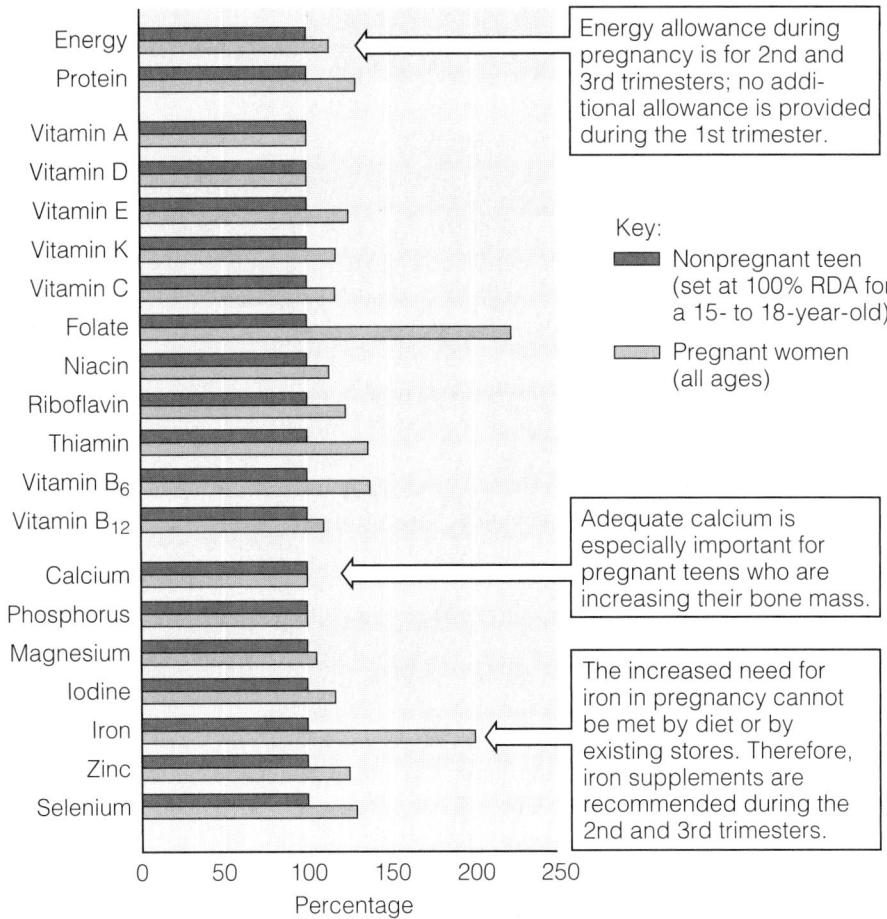

Note: For vitamin D, calcium, phosphorus, and magnesium, the 1997 recommendations differ from the 1989 RDA, but the *differences* between the values for nonpregnant and pregnant adolescents are similar to those shown in this figure. For actual values, turn to the table on the inside front cover.

quent pregnancies have higher rates of low birthweight, mortality, and morbidity.

Pregnant teenagers often find themselves without resources or access to prenatal care. They urgently need programs addressed to all their problems, including medical attention, nutrition guidance, and emotional support. Continued schooling is most important to their future.

Many communities offer programs to improve the health of adolescent mothers and their infants. The goals of such programs are to keep pregnant teenagers enrolled in school, reduce the rate of subsequent unwanted pregnancies, and encourage father and family involvement.

Without the appropriate economic, psychosocial, and physical support, a young mother will not be able to care for herself during her pregnancy and for her child after the birth.[8] To improve her chances for a successful pregnancy and healthy infant, she must seek prenatal care. Unfortunately, only about half of all teenage mothers receive prenatal care in the first trimester of pregnancy. Among the reasons teenagers do not seek prenatal care immediately are their fears of losing confidentiality, concerns about costs, and failure to recognize the early signs of pregnancy.

Counseling the pregnant teenager is a challenge. Teenagers typically turn a deaf ear to lectures, so it is extremely important to establish rapport. Open-ended questions will encourage the young woman to talk about herself. Once she has revealed something of herself, the counselor can use this information to employ nutrition counseling. For example, if she says she eats meals away from home, the counselor can provide her with tips on how to get the most nutrients for the money she spends. If she enjoys snacking, she will welcome ideas for nutritious snacks. She needs respect and support for her good judgment whenever possible. With emotional support and adequate prenatal care, she will look and feel better throughout the pregnancy, and her infant will have a better chance of being born healthy and staying that way.

Pregnancy in Older Women

In the last three decades, as many women have pursued their education and careers, they have delayed childbearing. As a result, the number of first births to women 35 and older has increased dramatically.

Maternal Health. Each year, out of every 1000 pregnant women over the age of 35, 994 have healthy pregnancies.[9] Most of the complications associated with later childbearing reflect chronic conditions that accompany aging, such as hypertension and diabetes, which can complicate an otherwise healthy pregnancy. These complications often result in the need for cesarean delivery, which is twice as common in pregnant women over the age of 35 as among younger women. For all these reasons, maternal mortality rates are higher in women over 35 than in younger women.

Infant Health. Infants of older mothers face problems of their own. Birth defects, preterm births, growth retardation, and death are common among infants born to women over 35. Because 1 out of 50 pregnancies in older women produces an infant with genetic abnormalities, obstetricians routinely screen women older than 35. For a 40-year-old mother, the risk of having a child with Down syndrome, for example, is about 1 in 100 compared with 1 in 300 for a 35-year-old and 1 in 10,000

Down syndrome: a genetic abnormality that causes mental retardation, short stature, and flattened facial features.

for a 20-year-old. Fetal mortality is twice as high for women 35 years and older than for younger women.[10] Simply being older seems to carry a risk of its own. Why this is so remains a bit of a mystery. One possibility is that the uterine blood vessels of older women cannot fully adapt to the increased demands of pregnancy.

Malnutrition and Pregnancy

fertility: the capacity of a woman to produce a normal ovum periodically and of a man to produce normal spermatozoa; the ability to reproduce.

Good nutrition clearly plays a supportive role in the course of a pregnancy. In contrast, malnutrition interferes with the ability to conceive, the likelihood of implantation, and the subsequent development of a fetus should conception and implantation occur.

Malnutrition and Fertility

Reports from years past or from distant countries tell us of the starvation and malnutrition that accompany political struggles such as wars or natural events such as droughts. Closer to home, we know of homeless people who are hungry or young women who are starving themselves for the sake of fashion. As you read this section, keep in mind that the reasons for starvation may differ, but the consequences are the same: weight loss, malnutrition, and poor health.

Fertility depends both on the health of the ovum and sperm and on their environments. Severe nutritional deprivation can reduce fertility. Furthermore, both men and women appear to lose sexual interest during times of starvation. Starvation arises predictably during famines, wars, and drought, but can also occur amidst peace and plenty. Many women who diet excessively and exercise intensely are starving and suffering from malnutrition.

The influence of severe malnutrition on fertility was first reported during World War II when food supplies were severely restricted: the number of births declined dramatically below the expected rate. When food became available again, birthrates climbed, demonstrating that starvation-caused infertility is reversible. In areas that had adequate food supplies, fertility and birthrates remained normal, even under similar conditions of war and weather.

amenorrhea (a-MEN-oh-REE-ah): the absence, temporary or permanent, of menstrual periods; normal before puberty, after the menopause, during pregnancy, and during lactation; otherwise abnormal; the adjective is *amenorrheic* (*a* = not; *men* = month; *rhea* = flow).

Infertility in Women. One precondition of fertility is a woman's ability to ovulate. Most women will either develop amenorrhea (stop menstruating altogether) or have irregular menstrual cycles whenever food intake is severely limited.

Exactly how amenorrhea develops is unclear, but contributing factors seem to include restricted energy intake, low body weight, depleted fat stores, altered metabolism, and psychological stress. Whatever the underlying mechanism, the amenorrhea of an underweight woman may be viewed as an adaptive response to curtail reproduction. Many animals respond to food shortages and stress with hormonal changes that suppress ovulation and thereby fertility. Thus animals do not use reproductive energies to create offspring when the environment is not conducive to their survival. Sufficient weight gain usually initiates or restores regular menstrual cycles.

Many girls and women diet so excessively or exercise so intensely that they are underweight and amenorrheic (Focal Point 7A).

Infertility in Men. Another precondition of fertility is a man's ability to produce enough viable sperm. Food deprivation in men has several effects on sperm: It reduces their numbers, alters their motility, and shortens their life spans.

Subclinical vitamin C deficiency may be responsible for nonspecific sperm agglutination—a clumping of the sperm that causes infertility in men. Men who are infertile because of sperm agglutination have low serum vitamin C concentrations. With supplementation, their serum and semen vitamin C concentrations rise, and the percentage of agglutinated sperm falls to below the level that distinguishes fertility from infertility.[11]

In summary, there appears to be a nutritional threshold below which hormonal secretions diminish, causing both women and men to become infertile. Between the optimal nourishment that produces a healthy infant and the starvation that causes infertility, an intermediate zone exists where nourishment is adequate enough to support fertility, but not enough to support implantation, placental development, and fetal development.

Malnutrition and Implantation

Conditions within the uterus at the time of conception can be either hospitable or hostile. With an appropriate balance of hormones and nutrients, a zygote can implant successfully and begin normal development. When nutrients are absent or when drugs or other toxins are present, the endometrium may reject the zygote, or the zygote may be unable to attach to the lining. Without implantation, the zygote will be lost, possibly even before the woman knows she is pregnant. A woman cannot control all factors influencing her body's chemistry during the time surrounding implantation, but ideally, her nutrition will have been optimal and her exposure to any drugs or environmental contaminants minimal at this time.

Malnutrition and Placental Development

Malnutrition prior to and around conception prevents the placenta from developing fully.[12] Most notably, a low protein intake suppresses placental growth.[13] A poorly developed placenta cannot deliver optimum nourishment to the fetus, and the infant will be born small and possibly with physical and cognitive abnormalities. If this small infant is a female, she may in turn develop poorly and will have an elevated risk of having a miscarriage or a stillbirth, or giving birth to an infant in need of neonatal intensive care.

Malnutrition and Fetal Development

Malnutrition includes both undernutrition and overnutrition, and both nutrient deficiencies and excesses during pregnancy can have adverse effects. The accompanying glossary defines some adverse effects of preg-

Table 3-2   Effects of Malnutrition on Pregnancy and Fetal Development

Deficiencies	Possible Effect
Energy	Low infant birthweight
PEM	Fetal growth retardation
Folate	Miscarriage and neural tube defect
Thiamin	Infantile beriberi (congestive heart failure), brain lesions
Vitamin A	Congenital malformations
Vitamin D	Low infant birthweight, altered composition of dental enamel, neonatal hypocalcemia, congenital rickets
Iron	Stillbirth, premature birth, and low infant birthweight
Iodine	Cretinism (varying degrees of mental and physical retardation in the infant)
Zinc	Low infant birthweight and congenital malformations, fetal growth retardation

Excesses	Possible Effect
Vitamin A	Microcephaly (small head); hydrocephalus (water on the brain); spontaneous abortion
Vitamin D	Hypercalcemia
Iodine	Congenital goiter

nancy, and Table 3-2 summarizes some of the effects of specific nutrient deficiencies and excesses on pregnancy and fetal development. In general, consequences of malnutrition during pregnancy include:

■ Fetal growth retardation.

■ Congenital malformations, birth defects.

■ Spontaneous abortion and stillbirth.

■ Premature birth.

■ Low infant birthweight.

Birthweight is most frequently used as a predictor of an infant's survival and health. Malnutrition, coupled with low birthweight, is the underlying or associated cause of more than half of all the deaths of children under four years of age.[14]

Nutrient Deficiencies. Nutrient deficiencies during pregnancy can often be seen in the immediate outcome, but they also can cause defects that become apparent years later. Furthermore, the damage might be such that a developing infant will never fully recover. For example, mothers who are severely iodine deficient during pregnancy give birth to infants with mental retardation. The irreversibility of such a condition is obvious when abundant, nourishing food fed after the critical time fails to remedy the deficit. Clearly, optimal nutrition during pregnancy enhances the likelihood of a healthy outcome.

GLOSSARY *ADVERSE OUTCOMES OF PREGNANCY*

congenital malformations: abnormalities present at birth.

miscarriage: known medically as a spontaneous abortion, the separation of the developing fetus and the placenta from the inner wall of the uterus, resulting in the unintentional termination of the pregnancy, occurring before the beginning of the 20th week of gestation.

neonatal death: death of a newborn within the first month after birth.

respiratory distress: severe impairment of pulmonary function due to unexpanded lungs or poor blood flow to the lungs; symptoms include abnormally rapid breathing, abnormally rapid heartbeat, blue skin tone, and grunting during exhalation.

stillbirth: birth of a dead fetus.

Nutrient Toxicities. The pregnant woman who is constantly being told to eat well and take care of herself may mistakenly assume that more is better regarding vitamin and mineral supplements. This is simply not true. All the nutrients can be harmful when taken in excess.[15] A pregnant woman can obtain most of the vitamins and minerals she needs by eating nutritious foods and should take supplements only on the advice of a registered dietitian or a physician. To prevent toxicities, health care professionals need to provide counseling on the proper use of supplements.

Problem Nutrients

Much of our understanding about malnutrition's influence on pregnancy comes from studies that have focused on diet in general. When one nutrient is lacking in the diet, chances are good that other nutrients are lacking as well. Single-nutrient deficiencies rarely occur in human beings, but each nutrient deficiency can impair a pregnancy's outcome in a specific way.

Folate Inadequacies. Several research studies have focused on the relationship between folate and neural tube defects.[16] As Chapter 2 mentioned, the neural tube is the beginning of the central nervous system. Any abnormal development of the neural tube or its failure to close completely constitutes a defect. Neural tube defects commonly cause serious disabilities and infant mortality.

One of the most common types of neural tube defects is spina bifida, a disorder characterized by incomplete closure of the spinal cord and its bony encasement. The membranes covering the spinal cord often protrude as a sac, which may rupture and lead to meningitis, a life-threatening inflammation of the membranes. Spina bifida is accompanied by varying degrees of paralysis, depending on the extent of spinal cord damage. Mild cases may not even be noticed, but severe cases lead to death. Common problems associated with spina bifida include club-foot, dislocated hip, kidney disorders, curvature of the spine, muscle weakness, mental handicaps, and motor and sensory losses.

In the United States, neural tube defects occur in approximately 1 of every 1000 newborns, affecting some 2500 to 3000 infants each year.* Many other pregnancies with neural tube defects end in abortion or stillbirth.

Note: 0.4 milligrams = 400 micrograms

Recent evidence suggests that folate supplements (0.4 milligrams per day) taken one month before conception and continued throughout the first trimester can prevent neural tube defects.[17] For this reason, the Public Health Service has recommended that all women of child-bearing age who are capable of becoming pregnant should take 0.4 milligrams of folate daily. This amount of folate is easy to obtain from a diet that includes plenty of fruits and vegetables, but supplements offer women a convenient way to ingest enough folate regularly and continuously enough to benefit pregnancies. Most over-the-counter multivitamin supplements contain 0.4 milligrams of folate; prenatal supplements usually contain at least 0.8 milligrams. A woman who has previously had an infant with a neural tube defect may be advised by her physician to take folate supplements in doses 10 times larger—4 milligrams daily. The risks associated with high doses of folate are not all known, but folate supplements can mask the pernicious anemia of a vitamin B_{12} deficiency. For this reason, quantities of 1 milligram or more require a prescription.

To deliver folate to the U.S. population, the Food and Drug Administration (FDA) has mandated fortification of grain products. This decision carefully weighed the benefits of fortification against the risks of overconsumption. On the one hand, an adequate folate intake is expected to reduce the incidence of neural tube defects by 50 percent. On the other hand, if vitamin B_{12} deficiency is masked by folate and left untreated, irreversible nerve damage may occur. The FDA regulation requires manufacturers to fortify foods to provide 140 micrograms of folate per 100 grams of foods, which should increase average daily intakes by 100 micrograms. Fortified foods include pasta, flour, rolls, buns, farina, grits, cornmeal, and rice.

When selections are made wisely, even the **minimum** *recommended number of servings of two fruits and three vegetables can provide enough folate: orange juice at breakfast, a fresh spinach salad with garbanzo beans for lunch, cantaloupe for a midafternoon snack, and broccoli with dinner, for example. The photo on p. 55 features the folate contents of selected foods.*

The exact role of folate in neural tube defect prevention is unclear. Folate absorption is similar in women with a history of neural tube defect births and women with a normal pregnancy history.[18] A difference is apparent, however, in that those with a history of neural tube defects have higher concentrations of homocysteine, a metabolic intermediate that requires folate for its conversion to the amino acid methionine.[19] The resulting methionine shortage during the critical period of neural tube development may be responsible for the defect. Furthermore, because most women with a severe folate deficiency do not give birth to

*Worldwide, some 300,000 to 400,000 infants are born with neural tube defects.

infants with a neural tube defect, another factor must be involved. Researchers speculate that "folate deficiency must act on an underlying nutrient-sensitive genetic defect to yield a defective newborn."[20]

Insufficient folate has been associated with other poor pregnancy outcomes as well. Women have twice the risk of preterm delivery and low infant birthweight when their folate intakes and serum concentrations are low.[21]

Vitamin A Toxicity. Excessive vitamin A poses a teratogenic risk. One study reported that among women who took more than 10,000 IU of supplemental vitamin A daily, approximately 1 out of every 57 infants was born with a malformation of the cranial nervous system that was attributable to high vitamin A intake. The time of intake that appeared most damaging was before the seventh week.[22] For this reason, vitamin A is not given as a supplement in the first trimester of pregnancy unless there is specific evidence of deficiency, which is rare.

teratogenic (ter-AT-oh-jen-ik): causing abnormal fetal development and birth defects (*terato* = monster; *genic* = to produce).

Practices Incompatible with Pregnancy

Besides malnutrition, which presents many hazards to pregnancy, a variety of lifestyle factors can have adverse impacts, and some may be teratogenic. People who are planning to have children can make the choice to practice healthy behaviors. Table 3-3 summarizes the effects of harmful substances on fetal development. The following sections describe the major ones.

Smoking and Smokeless Tobacco Use

Smoking cigarettes or chewing tobacco at any time exerts harmful effects, and pregnancy dramatically magnifies the hazards of these practices. Unfortunately, an estimated 20 percent of pregnant women smoke, with higher rates for unmarried women, teenagers, and those with a limited education. Smokers tend to eat less nutritious foods during their pregnancies than do nonsmokers, which in turn impairs fetal nutrition.[23] Smoking during pregnancy also directly harms the mother and the development of the placenta, the embryo, the fetus, and the infant and child later in life. Smoking increases the risk of retarded development and complications at birth. Mislocation of the placenta, premature separation of the placenta, and vaginal bleeding are almost twice as frequent among women who smoke more than one pack of cigarettes per day than among nonsmokers.

Smoking restricts the blood supply to the growing fetus and so limits oxygen and nutrient delivery and waste removal. Tobacco smoke contains hundreds of compounds that are harmful, including nicotine and carbon monoxide. Carbon monoxide in the blood of a smoking mother may deprive the developing fetus of the oxygen necessary for optimal growth. The primary cause of most smoking-related fetal deaths is oxygen deprivation.

Table 3-3 Effects of Potentially Harmful Substances on Pregnancy and the Fetus

Substance	Effect on Fetus
Cigarette smoke	Low birthweight, increased incidence of spontaneous abortion and preterm birth; nervous system disturbances, increased incidence of sudden infant death syndrome (SIDS), fetal death
Caffeine	Central nervous system stimulant, increased incidence of spontaneous abortion
Medications	
Salicylates (large doses)	Pulmonary hypertension and neonatal bleeding
Acetaminophen	Renal failure
Anticonvulsants	Growth retardation and mental retardation
Oral progestogens, androgens, and estrogens	Masculinization and advanced bone age
Tetracyclines	Inhibition of bone growth, discoloration of teeth
Illicit drugs	
Marijuana	Short-term irritability at birth
Heroin and methadone	Drug addiction and acute narcotic withdrawal symptoms (tremors, excessive, high-pitched crying, and disturbed sleep), low birthweight
Cocaine	Greater incidence of placental abruption and spontaneous abortion, uncontrolled jerking motion, paralysis, depressed interactive behavior, poor organizational response to environmental stimuli
Phencyclidine (PCP)	Facial malformations, tremors, low birthweight
Alcohol	Fetal alcohol syndrome: distorted facial features, low birthweight, nervous system disturbances, cleft palate
Heavy metals	
Lead	Spontaneous abortion, stillbirth, low birthweight, neurobehavioral deficits
Mercury	Central nervous system damage

The risk of spontaneous abortion and neonatal death increases directly with increasing levels of maternal smoking. A positive association also exists between maternal smoking and sudden infant death syndrome (SIDS).[24] This relationship is found for frequency, quantity, and even postnatal exposure to cigarette smoke (passive smoking). The U.S. surgeon general's report concludes that "maternal smoking can be a direct cause of fetal or neonatal death in an otherwise normal infant."[25]

The U.S. surgeon general also states that of all *preventable* causes of low birthweight in the United States, smoking has the greatest impact.[26] The more a mother smokes, the smaller her infant will be. On the average, infants of mothers who smoke weigh 200 grams (7.1 ounces, or almost 1/2 pound) less than those born to nonsmoking mothers. The deficit reflects a disproportionate lack of lean tissue—that is, body composition as well as size is abnormal. This low birthweight reflects a small-for-gestational-age profile, not preterm birth, indicating that maternal smoking directly slows fetal growth rate. Furthermore, smoking during pregnancy may even harm the intellectual and behavioral development of the child later in life.[27]

Research shows that tobacco chewing by pregnant women also affects the fetus adversely. Even in the absence of carbon monoxide inhalation, the constituents of tobacco absorbed into the bloodstream, which include lead, are damaging. (Focal Point 6A examines the toxic effects of lead poisoning.) Infants of mothers who chew tobacco, like those whose mothers smoke tobacco, have lower birthweights and a higher risk of fetal death than infants born to women who do not use tobacco.

The risks of smoking cigarettes and chewing tobacco during pregnancy should be taught from junior high school through college. Any woman who smokes cigarettes or chews tobacco and is considering pregnancy or who is already pregnant should be urged to quit.

Alcohol and Drug Use

Alcohol consumption during pregnancy can cause irreversible mental and physical retardation of the fetus—fetal alcohol syndrome (FAS). Of the leading causes of mental retardation, FAS is the only one that is totally *preventable*. As a consequence, the U.S. surgeon general has issued a statement that pregnant women should drink absolutely no alcohol. Fetal alcohol syndrome is the subject of Focal Point 3.

In addition to the effects seen in the FAS child, alcohol prevents the birth of some children. Spontaneous abortions occur more frequently in women who drink more than two alcoholic beverages daily than in women who do not drink.

Medicinal Drugs. Drugs other than alcohol can also cause complications during pregnancy, problems in labor, and serious birth defects. For these reasons, pregnant women should not take any medicines without consulting their physicians. Drug labels warn, "As with any drug, if you are pregnant or nursing a baby, seek the advice of a health professional before using this product." For aspirin and ibuprofen, an additional warning immediately follows: "It is especially important not to use aspirin (or ibuprofen) during the last three months of pregnancy unless specifically directed to do so by a doctor, because it may cause problems in the unborn child or excessive bleeding during delivery."

An example of a drug whose effects are particularly well known is Accutane, a synthetic relative of vitamin A. Generically known as isotretinoin, Accutane is effective in treating severe cystic acne that has been unresponsive to other treatments. Accutane bears a warning label that advises women to use an effective form of contraception beginning at least one month before the inception and continuing until one month

Labels warn women of the drug's potential for harm.

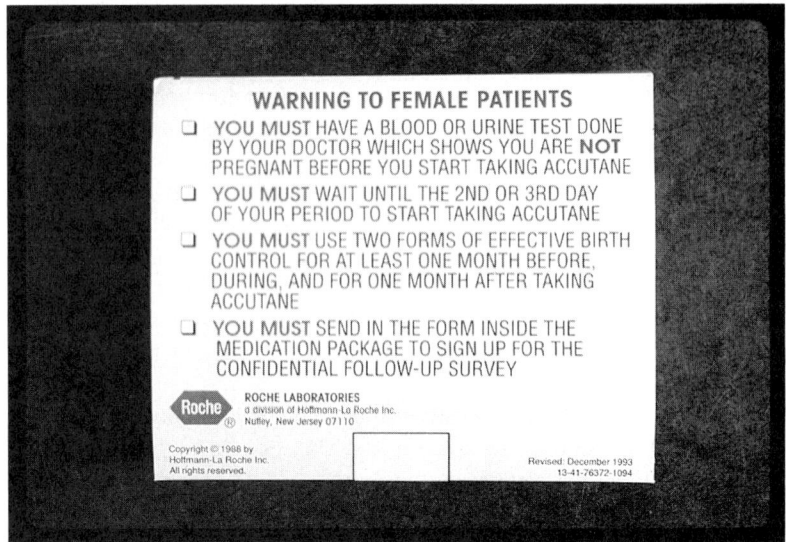

WARNING TO FEMALE PATIENTS

❑ YOU MUST HAVE A BLOOD OR URINE TEST DONE BY YOUR DOCTOR WHICH SHOWS YOU ARE **NOT** PREGNANT BEFORE YOU START TAKING ACCUTANE

❑ YOU MUST WAIT UNTIL THE 2ND OR 3RD DAY OF YOUR PERIOD TO START TAKING ACCUTANE

❑ YOU MUST USE TWO FORMS OF EFFECTIVE BIRTH CONTROL FOR AT LEAST ONE MONTH BEFORE, DURING, AND FOR ONE MONTH AFTER TAKING ACCUTANE

❑ YOU MUST SEND IN THE FORM INSIDE THE MEDICATION PACKAGE TO SIGN UP FOR THE CONFIDENTIAL FOLLOW-UP SURVEY

Roche ROCHE LABORATORIES
a division of Hoffmann-La Roche Inc.
Nutley, New Jersey 07110

Copyright © 1988 by
Hoffmann-La Roche Inc.
All rights reserved.

Revised: December 1993
13-41-76372-1094

after the termination of Accutane's use. Red stickers warn that it can cause spontaneous abortions and birth defects. Pharmacists add labels that caution, "Do not become pregnant while using Accutane."

Because infants born to women taking Accutane have exhibited major birth defects, dermatologists must ascertain whether women consulting them for acne are sexually active, and if so, whether they are using effective contraception methods. A pregnancy test two weeks prior to the onset of treatment is imperative. Some dermatologists require monthly pregnancy tests while Accutane is taken.

A list of specific drugs contraindicated prior to pregnancy is not available. However, this one example of Accutane illustrates the potent effect that a drug taken near the time of conception, even one approved by the Food and Drug Administration and prescribed by a physician, can have on the outcome of a pregnancy. The effectiveness and blood clearance rates of different drugs vary. Even a short-term, low dose of a potent drug can expose a fetus to irreparable harm. Other factors influencing the nature and severity of the harm to the fetus include the time and duration of administration. Obviously, more caution is required when a woman is sexually active and conception is a possibility than when she is sexually abstinent or practicing contraception.

Among the drugs to discontinue if a woman wants to become pregnant are, of course, the oral contraceptives. Once a woman stops taking oral contraceptives, she can expect a delay of one to three months before regular ovulation and menstruation resume. If she waits for her menstrual cycle to resume before becoming pregnant, she can calculate her due date. After the initial delay, conception rates become the same for these women as for others not using contraceptives. Oral contraceptives cause no fetal damage, however, even if taken after conception.

Illicit Drugs. The recommendation to avoid drugs during pregnancy includes illicit drugs, of course. Unfortunately, use of illicit drugs, such as cocaine and marijuana, is common among some pregnant women.*

Drugs of abuse, such as cocaine, easily cross the placenta and impair fetal development. Furthermore, they are responsible for preterm births, low-birthweight infants, and sudden infant deaths. If these newborns survive, their cries and behaviors at birth are abnormal, and their cognitive and motor development later in life is often impaired.[28] They may be hypersensitive or underaroused; those who test positive for drugs suffer the greatest effects of toxicity and withdrawal.[29]

Much research focuses on the effects of maternal drug ingestion on conception and fetal development, but few studies have examined the fetal effects of paternal exposure to drugs. Pregnancy begins with the union of an ovum and a sperm, and it stands to reason that adverse influences on the male might affect conception. Male rats given methadone prior to mating with untreated females sire pups with low birthweights and altered behavior patterns. Neonatal mortality is also high. In human males, lead, anesthetic gases, cigarettes, and caffeine are associated with such adverse effects as spontaneous abortions, neonatal mortality, and low birthweight.[30] Drugs may alter sperm formation, maturation, or motility, and this can affect fertility. In addition, paternal drug ingestion may alter male hormone activity, damage the sperm, or, if the drug or its metabolites pass in the ejaculate, affect the intrauterine environment or the newly fertilized ovum.

methadone: a synthetic analgesic drug widely used in the treatment of heroin abuse.

Other Potentially Harmful Practices

Some factors, such as environmental contaminants, can be seriously damaging to the fetus, but are largely out of a person's control. Other factors, such as caffeine and sugar substitutes, are fully within a person's control, but not particularly damaging. In all cases, expectant mothers need to understand the risks.

Environmental Contaminants. Evidence of exposure to environmental contaminants such as lead and mercury has been detected in the amniotic fluid of pregnant women.[31] Infants and young children of these mothers show signs of impaired cognitive development.[32] For this reason, it is particularly important that pregnant women receive foods and beverages grown and prepared in environments free of contamination.

Focal Point 6A describes how lead toxicity impairs a child's development.

Caffeine. Caffeine crosses the placenta, and the developing fetus has a limited ability to metabolize it. For this reason, pregnant women may wonder whether they should give up coffee, tea, and colas because of their caffeine contents. Research studies have not proven that caffeine (even in high doses) causes birth defects in human infants (as it does in

*Estimated prevalence of marijuana use among pregnant women is 17 percent, and for cocaine use, >6 percent. Committee on Nutritional Status during Pregnancy and Lactation, Food and Nutrition Board, *Nutrition during Pregnancy* (Washington, D.C.: National Academy Press, 1990), p. 48.

animals), but limited evidence suggests that moderate-to-heavy use may lower infant birthweight.[33] (Heavy caffeine use was defined as more than 300 milligrams per day—the equivalent of two to three cups of coffee.) All things considered, it might be most sensible to limit caffeine consumption to the equivalent of a cup of coffee or two 12-ounce cola beverages a day.

Sugar Substitutes. Artificial sweeteners have been extensively investigated and found to be safe for use during pregnancy.[34] (Women with phenylketonuria should not use aspartame, as Focal Point 5C explains.) It would be prudent for pregnant women to use sweeteners in moderation and within an otherwise nutritious and well-balanced diet.

Nutrition Assessment of Pregnant Women

A pregnant woman's physiology is different from that of a nonpregnant woman, and it changes throughout the pregnancy. Health care professionals use special sets of standards to assess nutrition status and fetal development during pregnancy.

Historical Information

Health, diet, social, and drug histories provide valuable clues about a pregnant woman's health and eating habits. Table 3-1 (on p. 86) lists several factors that alert health care professionals to the possibility of a high-risk pregnancy.

Prenatal care should include a dietary assessment to determine the adequacy of the woman's usual dietary intake. When taking a diet history, the assessor needs to be alert to any dietary practices and food patterns that may create nutrient deficiencies or toxicities. Specifically, does the diet history reveal:

■ A diet that severely restricts energy intake?

■ A diet that excludes one or more food groups?

■ A diet that includes nonfood items?

■ A diet high in nutrient-poor, high-kcalorie foods?

■ The use of self-prescribed vitamin and mineral supplements in amounts above the RDA?

If the answer to any of these questions is yes, then nutrition counseling is in order.

Anthropometric Data

The assessor can learn much about fetal development from a simple anthropometric measurement: weight. The mother's prepregnancy weight and the amount and pattern of her weight gain during pregnancy are central to assessing the progress of a pregnancy.

Weight for Height before Conception. Ideally, the mother's prepregnancy height and weight measurements were recorded in her medical chart during a recent preconceptional visit. Alternatively, her weight at the first prenatal visit or self-reported weight can be used with discretion. (Assessors need to keep in mind that many women underestimate their weights by about 3 pounds and overestimate their heights by about 1 inch.) Prepregnancy weight is then assessed by calculating the BMI (see p. 27), which is used to determine the recommended weight gain.

Weight Gain during Pregnancy. At the beginning of each prenatal visit, the assessor weighs the pregnant woman and records the measurement in both the medical record and on a prenatal weight gain chart (see Figure 3-2). By plotting measurements on a chart, the assessor can compare the woman's weight against target weights and rates of gains.

In general, steady weight gains represent lean and fat tissue, whereas abrupt gains usually represent excessive fluid retention. A sudden or large weight gain, especially in the second and third trimesters, may indicate the onset of preeclampsia, a complication that demands immediate medical attention.

Weight Gain and the Duration of Pregnancy. In addition to maternal weight for height prior to conception and maternal weight gain during pregnancy, duration of pregnancy is also predictive of infant birthweight: The longer the gestation, the more the infant will weigh at birth. Consequently, weight gain during pregnancy is most meaningful when compared to the length of gestation. For example, a 10-pound gain is appropriate at 20 weeks gestation, but not at 40. Similarly, a 3 1/2-pound gain is appropriate for the entire first trimester, whereas a 3 1/2-pound gain *per month* is appropriate during the second trimester, as the weight gain grid in Figure 2-5 on p. 49 shows.

Other Measures. The assessor can also gather evidence of fetal and placental growth from the medical record, which may include several indirect measures such as the symphysis–fundus measure, human placental lactogen concentration, and ultrasound measurement. The symphysis–fundus measure indicates uterine (and fetal) growth; the hormone concentration correlates with fetal weight; and the ultrasound measurement accurately determines fetal size, which is often used to estimate age.

Physical Examinations

Physical examination of a pregnant woman may find common signs of malnutrition, similar to those seen in other malnourished adults (see Chapter 1). It may also uncover problems that develop because of the pregnancy, such as edema.

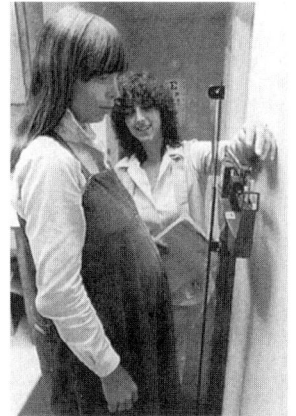

Every prenatal exam includes charting the weight gain.

Focal Point 2B describes preeclampsia and its relationships with nutrition.

symphysis–fundus measure: the distance from the junction of the pubic bones on midline in front to the uppermost part of the uterus.

Figure 3-2 Nutrition Assessment: Prenatal Weight Gain Grid

A prenatal weight gain grid plots the rate of weight gain during pregnancy.

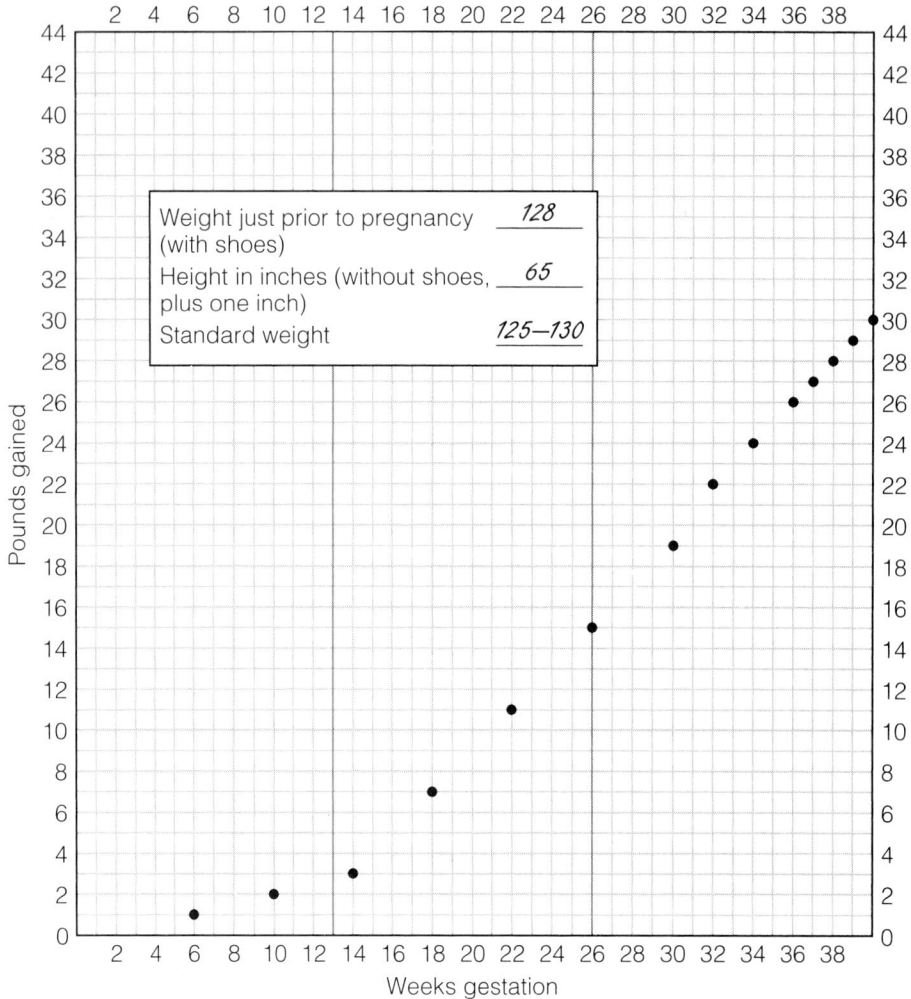

> Weight just prior to pregnancy   *128*
> (with shoes)
> Height in inches (without shoes,   *65*
> plus one inch)
> Standard weight   *125–130*

edema (eh-DEE-ma): local or generalized swelling caused by an excessive accumulation of tissue fluids (*edema* = swelling).

The combination of edema, high blood pressure, and protein in the urine signals the onset of an abnormal condition of pregnancy known as preeclampsia (discussed in Focal Point 2B).

Edema. Physical examination may reveal edema of the ankles or legs, which worsens on standing, or a generalized edema, which does not change with body position. In both cases, edema indicates excessive fluid retention.

Some edema during pregnancy is expected and normal—provided that it is not accompanied by high blood pressure or protein in the urine. Edema develops in response to the normal physiological changes that accompany pregnancy: elevated estrogen, which promotes water retention, and low serum albumin, which lowers osmotic pressure.

Biochemical Analyses

As Chapter 2 described, plasma volume increases, hormones alter metabolism, and kidney function changes during pregnancy. These changes in maternal physiology complicate the analyses of laboratory tests that measure nutrients or their metabolites in plasma or urine. The increase in plasma volume, for example, dilutes the concentrations of nutrients, making them appear low when, in fact, they may have actually increased, decreased, or remained the same. Hemodilution is illustrated in Figure 3-3.

hemodilution (HE-mow-dye-LOO-shun): a relatively reduced concentration of red blood cells caused by an increase in the volume of blood plasma.

Other factors also influence nutrition status during pregnancy. The rate of transfer from the mother to the fetus varies from nutrient to nutrient. It also varies over the course of pregnancy. For these reasons, health care professionals must use a special set of norms to evaluate the results of laboratory tests performed on pregnant women. Otherwise, laboratory values might appear to indicate a deficiency when, in fact, they reflect a normal state of pregnancy. Unfortunately, though, standards for pregnancy are not available for many nutrients.

Furthermore, laboratory tests to assess most vitamins and minerals are expensive to conduct routinely. Hemoglobin (or hematocrit) and possibly serum ferritin are the only laboratory tests routinely used during prenatal care. These tests help in assessing iron status.

Iron. As Chapter 1 explained, the first stage of iron deficiency is characterized by depletion of iron stores (serum ferritin). In most pregnant women, though, iron stores are low or depleted, regardless of supplementation. Hemoglobin levels normally decline because of the expansion of plasma volume (hemodilution). A further decline signals the onset of an iron deficiency. At this stage, hemoglobin is low, but not so

Figure 3-3 Hemodilution

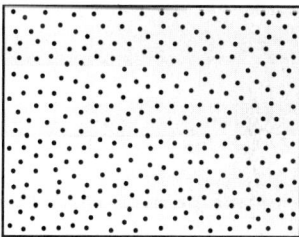

At the start of pregnancy, a woman may have a hemoglobin concentration of 14 g/100 ml and a plasma volume of 2 L, so the total hemoglobin in the blood amounts to 280 g.

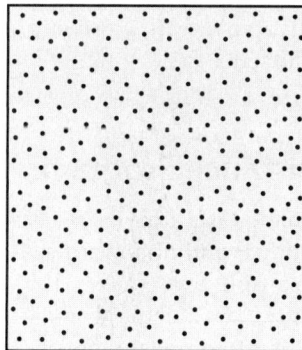

At the end of pregnancy, hemoglobin concentration may have fallen to 10 g/100 ml, but plasma volume has expanded to 3 L, so total hemoglobin in the blood is now 300 g, more than before.

Table 3-4   Nutrition Assessment: Anemia Criteria for Nonpregnant and Pregnant Women

| | Nonpregnant Women | | Pregnant Women |
	Normal	Anemia	Anemia
Hemoglobin	≥13.5 g/dL	≤12.0 g/dL	≤11.0 g/dL (1st trimester) ≤10.5 g/dL (2nd trimester) ≤11.0 g/dL (3rd trimester)
Serum ferritin	>18 µg/L		<12 µg/L

low as to indicate anemia. Table 3-4 provides the criteria for defining anemia during pregnancy.

Protein. Protein deficiency is difficult to assess during pregnancy. Some blood proteins, including albumin, decrease with the increase in plasma volume (hemodilution). Other blood proteins increase during pregnancy in response to elevated estrogen levels. Assessors need to be aware of these normal physiological changes when assessing protein status.

Laboratory values alone do not provide enough information to reach conclusions about a pregnant woman's nutrition status. Health care professionals must give sufficient emphasis to other assessment measures before making recommendations. For example, low serum zinc is typical for pregnant women (due to hemodilution) and usually needs no intervention. But zinc supplementation may be warranted for a pregnant woman with low serum zinc, an inadequate dietary zinc intake, and a history of poor pregnancy outcomes.

How extensively should maternal assessment be pursued? The answer varies, depending on factors such as those listed in Table 3-1, but all pregnancies require a minimal nutrition assessment. If such an evaluation reveals possible problems, then additional tests should be conducted.

Assistance Programs

Pregnancy is a time of increased need. Nutrient needs increase; so do emotional and financial needs. If any of these needs are not met, the course and outcome of the pregnancy may be compromised. Frequently, low-income women receive little, if any, prenatal care. They cannot afford private care or may be unaware of the free health services they may be eligible for in their communities.

Teens, who may at first be unaware that they are pregnant, or who are trying to hide the fact that they are, may fail to seek help until late in pregnancy. Many women do not eat properly while they are pregnant because they are not aware of the importance of nutrition, they do not know how to eat well, or they have too little money to purchase the food they need. Women can deal with all these situations provided they can learn of the options available to them.

Several resources are available for anyone who may be in need of information or assistance:

- The Agriculture Extension Service provides many educational services and materials, including nutrition, food budgeting, and shopping information.
- Community hospitals and health clinics employ registered dietitians who can assist with meal planning and health care during pregnancy.
- The Special Supplemental Food Program for Women, Infants, and Children (WIC) provides nutrition information and low-cost nutritious foods to low-income pregnant women and their children.
- The Food Stamp Program or other federal assistance programs may be available to some.

Health care professionals can be invaluable in guiding clients to the resources available.

Food Assistance Programs

WIC (the Special Supplemental Food Program for Women, Infants, and Children) provides nutrition education and nutritious foods to low-income pregnant women and their young children. WIC provides eggs, milk, cereal, juice, cheese, legumes, peanut butter, and infant formula to infants, children up to age five, and pregnant and breastfeeding women who qualify financially and are at medical or nutritional risk. The program is both remedial and preventive: Services include health care referrals, nutrition education, and food packages or vouchers for specific foods to supply nutrients known to be lacking in the diets of the target population.

Prenatal WIC participation can effectively reduce infant mortality, low birthweight, and newborn medical costs.[35] These benefits are even greater for teenagers than for others in the WIC program.

For every $1 spent on WIC, an estimated $3 in medical costs are saved. In 1992, participation in WIC reduced first-year medical expenses for infants by $1.19 billion.[36] Chapter 5 presents additional information on the benefits of WIC for infants.

Nutrition Counseling

Nutrition counseling is an essential component of prenatal care.[37] Anyone counseling a pregnant woman on nutrition should listen for clues about dietary beliefs or superstitions and cravings or aversions. If any of these dietary practices interfere with an adequate intake, then she needs a word of warning. It is extremely important not to alienate her, so that she may readily accept such advice. Suggestions to the diet adviser follow.

See p. 57 for the iron contents of selected foods.

When counseling a pregnant woman, support her in all she is doing well, and work within the framework of her beliefs and prejudices. If she is severely restricting kcalories, express approval of her desire to do what she perceives as beneficial, and try to help her realize that adequate food energy is indispensable to her health and the growth of her fetus. Help her select nutrient-dense foods that will provide enough energy to support normal growth without adding unnecessarily to her body fat. If she is omitting food groups, accept this choice, and help her find foods that will supply the nutrients normally provided by those groups. For example, a person who eats no meat can select iron-rich foods such as legumes and combine them with vitamin C-rich fruits and vegetables to enhance iron absorption. Encourage the woman whose diet is kcalorie rich and nutrient poor to continue enjoying food and to maintain an adequate energy intake—but to do so using foods that will provide more vitamins, minerals, and fiber. The person who replaces two soft drinks each day with two glasses of low-fat milk significantly improves the nutrient density of an otherwise high-kcalorie, low-nutrient diet. Encourage such a person to select nutrient-dense snacks such as hard-boiled eggs, yogurt, and fresh fruits and vegetables every day, and provide her with a list of healthy snack foods (see Table 6-6 in Chapter 6).

If a woman is taking self-prescribed supplements, for whatever reason, congratulate her for making the effort to do the best she can to obtain all the nutrients she needs. Indicate that she is correct in realizing that she has added nutrient needs, but that she will do better to obtain her nutrients from foods. Advise her to take supplements only as prescribed by her health care provider.

Support the woman in all that she is doing well during her pregnancy, and without being pushy, try to make her realize why a healthy body and mind are important to her own well-being and to that of her developing fetus. Skillful evaluation of a pregnant woman's diet is essential in determining possible risks due to alternative dietary practices.

As changes that improve her nutrition status gradually become part of her daily routine, the pregnant woman will feel and look better. Her motivation will begin to come from within, and the job of the health counselor will become easier.

Whether a pregnancy is planned or unplanned, the potential for many factors to harm or benefit the pregnancy demonstrates what a pivotal time this is. If all goes well, and it usually does, a pregnancy's full potential can be realized. Fortunately, control of many factors influencing pregnancy is within reach for most women.

CHAPTER THREE NOTES

1. *Contraception: Choosing a Method* (Rockville, Md.: American College Health Association, 1989).

2. U.S. Department of Health and Human Services, *Health United States 1992 and Healthy People 2000 Review*, August 1993, p. 17.

3. M. Story and I. Alton, Nutrition issues and adolescent pregnancy, *Nutrition Today* 30 (1995): 142–151.

4. A. M. Fraser, J. E. Brockert, and R. H. Ward, Association of young maternal age with adverse reproductive outcomes, *New England Journal of Medicine* 332 (1995): 1113–1117.

5. Fraser, Brockert, and Ward, 1995; J. M. Rees, S. A. Lederman, and J. L. Kiely, Birthweight associated with lowest neonatal mortality: Infants of adolescent and adult mothers, *Pediatrics* 98 (1996): 1161–1166.

6. M. L. Hediger and coauthors, Rate and amount of weight gain during adolescent pregnancy: Associations with maternal weight-for-height and birth weight, *American Journal of Clinical Nutrition* 52 (1990): 793–799.

7. Committee on Nutritional Status during Pregnancy and Lactation, Food and Nutrition Board, *Nutrition during Pregnancy* (Washington, D.C.: National Academy Press, 1990), pp. 1–23; J. M. Rees and coauthors, Weight gain in adolescents during pregnancy: Rate related to birth-weight outcome, *American Journal of Clinical Nutrition* 56 (1992): 868–873.

8. Position of The American Dietetic Association: Nutrition care for pregnant adolescents, *Journal of the American Dietetic Association* 94 (1994): 449–450.

9. F. G. Cunningham and K. J. Leveno, Childbearing among older women—The message is cautiously optimistic, *New England Journal of Medicine* 333 (1995): 1002–1004.

10. R. C. Fretts and coauthors, Increased maternal age and the risk of fetal death, *New England Journal of Medicine* 333 (1995): 953–957.

11. E. B. Dawson, W. A. Harris, and W. J. McGanity, Effect of ascorbic acid on sperm fertility, *Federation Proceedings* 42 (1983): 531; E. R. Gonzalez, Sperm swim singly after vitamin C therapy, *Journal of the American Medical Association* 249 (1983): 2747, 2751; W. A. Harris, T. E. Harden, and E. B. Dawson, Apparent effect of ascorbic acid medication on semen metal levels, *Fertility and Sterility* 32 (1979): 455–459.

12. W. W. Hay and coauthors, Workshop summary: Fetal growth: Its regulation and disorders, *Pediatrics* 99 (1997): 585–591.

13. M. E. Symonds and L. Clarke, Nutrition–environment interactions in pregnancy, *Nutrition Research Reviews* 9 (1996): 135–148.

14. D. L. Pelletier, The potentiating effects of malnutrition on child mortality: Epidemiologic evidence and policy implications, *Nutrition Reviews* 52 (1994): 409–415.

15. Committee on Dietary Allowances, Food and Nutrition Board, *Recommended Dietary Allowances* (Washington, D.C.: National Academy Press, 1989), p. 14.

16. C. Bower, Folate and neural tube defects, *Nutrition Reviews* 53 (1995): S33–S38.

17. Committee on Genetics, American Academy of Pediatrics, Folic acid for the prevention of neural tube defects, *Pediatrics* 92 (1993): 493–494.

18. B. A. Davis and coauthors, Folic acid absorption in women with a history of pregnancy with neural tube defect, *American Journal of Clinical Nutrition* 62 (1995): 782–784.

19. J. B. Ubbink, Is an elevated circulating maternal homocysteine concentration a risk factor for neural tube defect? *Nutrition Reviews* 53 (1995): 173–175.

20. V. Herbert, Folate and neural tube defects, *Nutrition Today*, November–December 1992, pp. 30–33.

21. T. O. Scholl and coauthors, Dietary and serum folate: Their influence on the outcome of pregnancy, *American Journal of Clinical Nutrition* 63 (1996): 520–525.

22. K. J. Rothman and coauthors, Teratogenicity of high vitamin A intake, *New England Journal of Medicine* 333 (1995): 1369–1373.

23. F. M. Haste and coauthors, Nutrient intakes during pregnancy: Observations on the influence of smoking and social class, *American Journal of Clinical Nutrition* 51 (1990): 29–36.

24. H. S. Klonoff-Cohen and coauthors, The effect of passive smoking and tobacco exposure through breast milk on sudden infant death syndrome, *Journal of the American Medical Association* 237 (1995): 795–798; E. A. Mitchell and coauthors, Smoking and the sudden infant death syndrome, *Pediatrics* 91 (1993): 893–896; K. C. Schoendorf and J. L. Kiely, Relationship of sudden infant death syndrome to maternal smoking during and after pregnancy, *Pediatrics* 90 (1992): 905–908; M. G. Bulterys, S. Greenland, and J. F. Kraus, Chronic fetal hypoxia and sudden infant death syndrome: Interaction between maternal smoking and low hematocrit during

pregnancy, *Pediatrics* 86 (1990): 535–540; B. Haglund and S. Cnattingius, Cigarette smoking as a risk factor for sudden infant death syndrome: A population-based study, *American Journal of Public Health* 80 (1990): 29–32.

25. U.S. Department of Health and Human Services, Public Health Service, *The Health Benefits of Smoking Cessation: A Report of the Surgeon General* (Washington, D.C.: Government Printing Office, 1990): 279–366.

26. U.S. Department of Health and Human Services, 1990.

27. C. D. Drews and coauthors, The relationship between idiopathic mental retardation and maternal smoking during pregnancy, *Pediatrics* 97 (1996): 547–553; D. L. Olds, C. R. Henderson, Jr., and R. Tatelbaum, Intellectual impairment in children of women who smoke cigarettes during pregnancy, *Pediatrics* 93 (1994): 221–227; D. M. Fergusson, L. J. Horwood, and M. T. Lynskey, Maternal smoking before and after pregnancy: Effects on behavioral outcomes in middle childhood, *Pediatrics* 92 (1993): 815–822.

28. L. Fetters and E. Z. Tronick, Neuromotor development of cocaine-exposed and control infants from birth through 15 months: Poor and poorer outcomes, *Pediatrics* 98 (1996): 938–943; C. A. Chiriboga and coauthors, Neurological correlates of fetal cocaine exposure: Transient hypertonia of infancy and early childhood, *Pediatrics* 96 (1995): 1070–1077; S. D. Azuma and I. J. Chasnoff, Outcome of children prenatally exposed to cocaine and other drugs: A path analysis of three-year data, *Pediatrics* 92 (1993): 396–402; M. J. Corwin and coauthors, Effects of in utero cocaine exposure on newborn acoustical cry characteristics, *Pediatrics* 89 (1992): 1199–1203; L. N. Eisen and coauthors, Perinatal cocaine effects on neonatal stress behavior and performance on the Brazel-ton Scale, *Pediatrics* 88 (1991): 477–480; M. Mirochnick and coauthors, Circulating catecholamine concentrations in cocaine-exposed neonates: A pilot study, *Pediatrics* 88 (1991): 481–485.

29. Corwin and coauthors, 1992; L. C. Mayes and coauthors, Neurobehavioral profiles of neonates exposed to cocaine prenatally, *Pediatrics* 91 (1993): 778–783.

30. L. F. Soyka and J. M. Joffe, Male-mediated drug effects on offspring, *Progress in Clinical and Biological Research* 36 (1980): 49–66.

31. M. Lewis and coauthors, Prenatal exposure to heavy metals: Effect on childhood cognitive skills and health status, *Pediatrics* 89 (1992): 1010–1015.

32. Lewis and coauthors, 1992; M. W. Shannon and J. W. Graef, Lead intoxication in infancy, *Pediatrics* 89 (1992): 87–90.

33. Committee on Nutritional Status during Pregnancy and Lactation, 1990, pp. 397–399; T. S. Hinds and coauthors, The effect of caffeine on pregnancy outcome variables, *Nutrition Reviews* 54 (1996): 203–207.

34. Position of The American Dietetic Association: Use of nutritive and nonnutritive sweeteners, *Journal of the American Dietetic Association* 93 (1993): 816–821.

35. P. A. Buescher and coauthors, Prenatal WIC participation can reduce low birth weight and newborn medical costs: A cost–benefit analysis of WIC participation in North Carolina, *Journal of the American Dietetic Association* 93 (1993): 163–166.

36. S. Avruch and A. P. Cackley, Savings achieved by giving WIC benefits to women prenatally, *Public Health Reports* 110 (1995): 27–34.

37. S. M. Yu and R. T. Jackson, Need for nutrition advice in prenatal care, *Journal of the American Dietetic Association* 95 (1995): 1027–1029.

Fetal Alcohol Syndrome

Some women give birth to infants who are not perfectly healthy, and often this is beyond the woman's control. Much of an infant's development is genetically determined, and so some errors are inherited. One such disorder, PKU, is highlighted in Focal Point 5C. Prevention is impossible for such a defect, but it can be controlled with dietary treatment. The subject of this discussion is fetal alcohol syndrome (FAS), a defect for which the reverse is true. FAS can only be prevented; there is no treatment. To prevent FAS, an expectant mother need only refuse to drink alcoholic beverages. The one characteristic that all mothers of FAS infants have in common is alcohol consumption during pregnancy.

As Chapter 3 mentioned, drinking alcohol during pregnancy may result in fetal alcohol syndrome, a cluster of symptoms that includes:[1]

- *Prenatal and postnatal growth retardation.* Infants with FAS are smaller than healthy infants and experience little postnatal catch-up growth, even with adequate nutrition and care.
- *Impairment of the brain and nerves, with consequent mental retardation, poor coordination, and hyperactivity.* FAS is the third most common cause of mental retardation in the United States. In addition to their mental disabilities, FAS children often have behavioral problems. They are irritable and fussy as infants and difficult to control as children.
- *Abnormalities of the face and skull (see Figure FP3-1).* The most obvious symptoms of FAS are the abnormal facial features, as illustrated in the photograph. These facial abnormalities reflect inadequate development of the brain, as well as parts of the face.

    This point is important. The FAS child's physical features are different from those of normal children, but these features alone do not pose a handicap. The extent of brain damage is mirrored in the facial structure; the more severe the facial deformities, the more severe the mental function impairment. Alcohol is responsible for both.

- *Increased frequency of major birth defects.* These include cleft palate, heart defects, and defects in ears, genitals, and urinary system. Many FAS infants have physical malformations that impair their ability to suck effectively. Sucking is an infant's natural response to hunger. It develops facial muscles and enables the infant to obtain nutrients from either a bottle of formula or a mother's breast. This serious feeding problem can easily lead to malnutrition, which further limits the growth of the infant, who is destined never to reach full potential.

Tragically, the damage evident at birth persists—children with FAS never fully recover.[2]

Of every 10,000 children born in the United States, some 6 or 7 suffer health problems because their mothers drank alcohol during preg-

The most obvious symptoms of FAS are the abnormal facial features, but the most tragic ones are the mental disabilities.

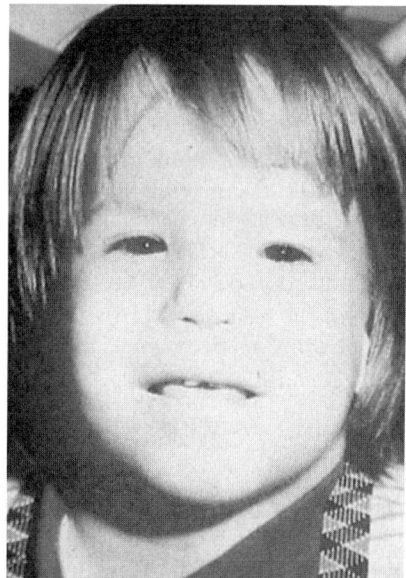

Figure FP3-1 Typical Facial Characteristics of FAS

The severe facial abnormalities shown here are outward signs of the severe mental impairments within. The internal organs also suffer irreversible damage that, although hidden, may create major problems for a child's health.

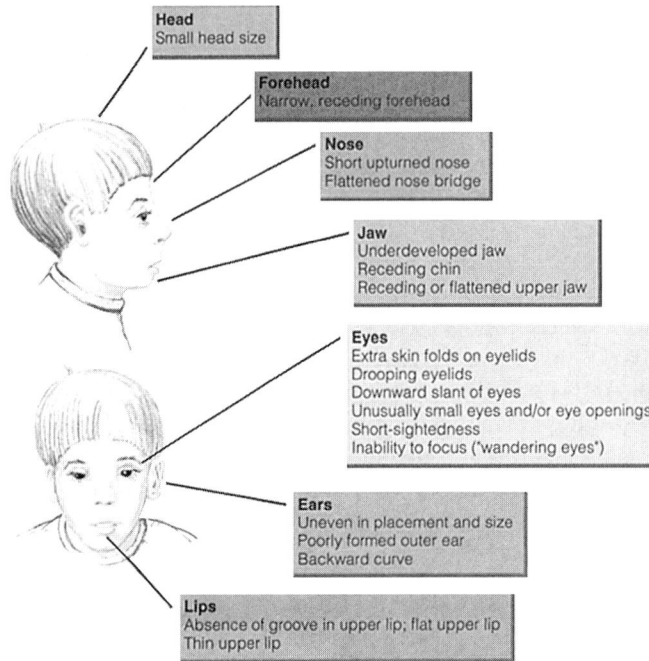

Head
Small head size

Forehead
Narrow, receding forehead

Nose
Short upturned nose
Flattened nose bridge

Jaw
Underdeveloped jaw
Receding chin
Receding or flattened upper jaw

Eyes
Extra skin folds on eyelids
Drooping eyelids
Downward slant of eyes
Unusually small eyes and/or eye openings
Short-sightedness
Inability to focus ("wandering eyes")

Ears
Uneven in placement and size
Poorly formed outer ear
Backward curve

Lips
Absence of groove in upper lip; flat upper lip
Thin upper lip

Source: Adapted from J. O. Beattie, Alcohol exposure and the fetus, *European Journal of Clinical Nutrition* 46 (1992): S7 – S17.

nancy—a sixfold increase over the past 15 years.[3] In addition, many infants are born with the less serious, yet still significant, damage some clinicians describe as *fetal alcohol effects*, or FAE (see the glossary).[4] Some children with FAE have no outward signs; others may be abnormally short or have minor facial abnormalities. Often FAE goes undiagnosed even when these children develop problems in the early school years. Learning disabilities, behavioral abnormalities, and motor impairments are common symptoms of FAE.

This Focal Point asks two questions: How much alcohol causes FAS? When during pregnancy is the damage done? For now, abstinence from alcohol is the best policy for pregnant women both because alcohol consumption during pregnancy has such severe consequences, and because FAS can only be prevented—it cannot be treated.[5] Further, because the most severe damage occurs around the time of conception— *before a woman may even realize that she is pregnant*—a woman planning to conceive should also abstain from alcohol.

Drinking during Pregnancy

When a person drinks an alcoholic beverage, the alcohol moves rapidly from the digestive system into the blood. The blood travels to the liver, where the alcohol can be metabolized. There is a limit, however, to the amount of alcohol the liver can process in a given time. The potential

GLOSSARY

fetal alcohol syndrome (FAS): the cluster of symptoms seen in an infant or child whose mother consumed excess alcohol during pregnancy, including retarded growth, impaired development of the central nervous system, and facial malformations.

fetal alcohol effects (FAE): a subclinical version of FAS, with hidden defects including learning disabilities, behavioral abnormalities, and motor impairments.

for fetal damage arises when the liver receives more alcohol than it can handle.

When a woman drinks during pregnancy, she causes damage in two ways: directly, by intoxication, and indirectly, by malnutrition. When alcohol crosses the placenta, fetal blood alcohol rises until it reaches an equilibrium with maternal blood alcohol. The mother may not even appear drunk, but the fetus may be poisoned. The fetus's body is small, its detoxification system is immature, and alcohol remains in fetal blood long after it has disappeared from maternal blood. Alcohol interferes with many developmental events, reducing the number of cells produced and damaging those that are produced.[6]

How Much Alcohol Is Too Much?

A pregnant woman need not have an alcohol abuse problem to give birth to a baby with FAS. She need only drink in excess of the liver's capacity to detoxify alcohol. About four drinks a day dramatically worsens the risk of having physical malformations. Even one to two drinks a day threatens to retard growth.

In addition to total alcohol intake, drinking patterns play an important role. Most FAS studies report their findings in terms of average intake per day, but people usually drink more heavily on some days than on others. For example, a woman who drinks an *average* of 1 ounce of alcohol (2 drinks) a day may not drink at all during the week but then might have 10 drinks on a Saturday night, exposing the fetus to toxic quantities of alcohol. Whether drinking a certain number of drinks during binges or

spreading them out over several days causes more damage depends on the frequency of the binges, the quantity consumed, and the stage of fetal development at the time of each drinking episode.[7]

An occasional drink may be innocuous, but researchers have been unable to say how much alcohol is safe to consume during pregnancy.[8] For this reason, health care professionals urge women to stop drinking alcohol as soon as they realize they are pregnant, or better, as soon as they *plan* to become pregnant.[9] Why take any risk? Only the woman who abstains completely is sure of protecting her infant from FAS.

No health care professional condones alcohol use during pregnancy, but some prefer not to issue a broad edict against alcohol consumption. They believe that an absolute statement to all women is less effective than one that targets those at highest risk: heavy drinkers. Preventive efforts need to focus on urging heavy drinkers to quit drinking during pregnancy. For most other women, the advice might be to reduce drinking to a minimum.

Other health risks also require attention. The mother's socioeconomic status, age, emotional stability, diet, drug use, smoking habits, genetic makeup, and prenatal care all play at least as great a role as an alcoholic beverage or two.

The difference between a "no drinks" position and an "occasional drinks" position is one of philosophy. All agree, however, that heavy drinking is hazardous to the fetus. Undeniably, a risk is associated with *any* drinking during pregnancy; the minimal alcohol intake position simply states that at some level of consumption, that risk is small enough to be negligible.

Characteristic facial features may diminish with time, but children with FAS typically continue to be short and underweight for their age.

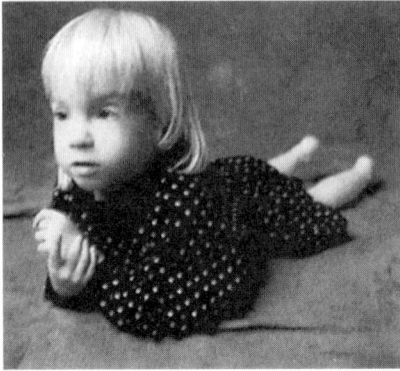

When Is the Damage Done?

At no time is excessive drinking safe. Even if a woman stops drinking at conception, her prior alcohol abuse will affect the development of her fetus. One study showed this clearly—infants born to women who had a history of alcohol abuse *but abstained during pregnancy* weighed less than infants born to women who did not consume alcohol.[10] Perhaps the earlier alcohol abuse led to maternal liver dysfunction and nutrient deficits that impaired fetal growth and development. This is not to say there is no point in abstaining during pregnancy. The other half of the story is that infants born to those alcohol abusers who abstained from drinking during pregnancy weighed more than infants born to women who had a similar history of alcohol abuse and continued drinking during pregnancy.

The first month of pregnancy is a critical period of fetal development. Because pregnancy confirmation usually requires five to six weeks, a woman may not even realize she is pregnant during that critical first month. Therefore, it is advisable for women who are trying to conceive, or who suspect they might be pregnant, to curtail their alcohol intakes to ensure a healthy start.

The type and extent of abnormality observed in an FAS infant depend on the developmental events occurring at the times of alcohol exposure. During the first trimester, developing organs such as the brain, heart, and kidneys may be malformed. During the second trimester, the risk of spontaneous abortion increases. During the third trimester, body and brain growth may be retarded.

Male alcohol ingestion may also affect fertility and fetal development. Animal studies have found smaller litter sizes, lower birthweights, reduced survival rates, and impaired learning ability in the offspring of males consuming alcohol prior to conception.[11] One human study found an association between paternal alcohol intake one month prior to conception and low infant birthweight.[12] (Paternal alcohol intake was defined as an average of two or more drinks daily or at least five drinks on one occasion.) This relationship was independent of either parent's smoking and of the mother's use of alcohol, caffeine, or other drugs.

Maternal Malnutrition

One of the health hazards associated with alcohol abuse is malnutrition. Alcohol depresses appetite because it produces euphoria, so that heavy drinkers usually eat poorly, if at all. Alcohol also attacks the digestive tract lining, causing pain and malabsorption of nutrients.

Many alcoholic women can easily drink 100 grams of ethanol a day. This amount is roughly equivalent to eight beers, a pint of whiskey, or a bottle of wine. Alcohol provides about 7 kcalories per gram, so the daily kcaloric intake from alcohol alone can amount to 700 kcalories. The woman who derives so many kcalories from a source that has no nutritive value finds it difficult to obtain the many essential nutrients she needs to maintain her health and support the fetus's growth.

Not only do alcohol abusers suffer malnutrition from lack of food, but even if they eat well, the direct effects of alcohol take their toll. Alcohol hinders the absorption, alters the metabolism, and increases the excretion of many nutrients, so that malnutrition can occur even in a well-nourished drinker.

Children born with FAS must live with the long-term consequences of prenatal brain damage.

Thus, an alcoholic pregnant woman harms her unborn child not only by consuming alcohol but also by not consuming food. This combination enhances the likelihood of malnutrition and a poorly developed infant. It is important to realize, however, that malnutrition is not the cause of FAS. It is true that mothers of FAS children often have unbalanced diets and nutrient deficiencies. It is also true that malnutrition may augment the clinical signs seen in these children, but it is the *alcohol* that is the determining factor. An adequate diet alone will not prevent FAS if alcohol is abused during the pregnancy.

Counseling and Prevention

In view of these findings, it is important to advise women not to drink during pregnancy. Everyone should know of the potential dangers. Heavy drinkers who are sexually active need effective contraception to prevent pregnancy.

The health care professional who realizes a pregnant woman drinks heavily and has no intention of quitting faces a difficult challenge. Drinking is more important to this woman than the possibility of harming an infant she has never touched and does not yet love. The task is to inform her that alcohol causes birth defects, and make it clear that the connection is well established. She should understand that the only way to be sure of preventing this kind of retardation is to abstain from drinking altogether throughout the pregnancy. The decision to abstain or not is hers, however. Pushing the woman to make the decision will alienate her. She will thereafter be less inclined to confide in those trying to help her. This will also limit opportunities to guide her in other nutrition choices.

If she does decide to quit drinking, at least during the pregnancy, she deserves a pat on the back. Let her know she has made a wise decision and that the sooner she abstains, the less the risk of FAS.

At the other extreme is the anxious woman who is usually quite careful of her health but who had a cocktail with friends during the first month of her pregnancy, before she even knew she was pregnant. Now she fears her infant will be born with mental and physical defects. In such a situation, the chances are very small that harm will have been done. The woman should be reassured that a small amount of alcohol probably will not harm the developing infant. Many drinkers, even heavy drinkers, deliver normal babies. Furthermore, and even more important, the episode is in the past, and nothing can be done to change it. A pregnancy can be a long, suspenseful experience, especially if parents fear an imperfect outcome. The health care professional should minimize the things that cannot be changed and emphasize what can be done to ensure the best possible outcome.

Of the leading causes of mental retardation, FAS is the only one that is totally preventable. Health care education on the potential dangers of alcohol use during pregnancy should begin *before* conception. FAS information should be included in all classes that discuss birth control, sexual intercourse, or pregnancy. All containers of beer, wine, and liquor carry the warning: Women should not drink alcoholic beverages during pregnancy because of the risk of birth defects. Everyone should hear the message loud and clear: Don't drink alcohol prior to conception or during pregnancy.

FOCAL POINT 3 NOTES

1. Committee on Substance Abuse and Committee on Children with Disabilities, American Academy of Pediatrics, Fetal alcohol syndrome and fetal alcohol effects, *Pediatrics* 91 (1993): 1004–1006; J. O. Beattie, Alcohol exposure and the fetus, *European Journal of Clinical Nutrition* 46 (1992): S7–S17.

2. H. L. Spohr, J. Willms, and H. C. Steinhausen, Prenatal alcohol exposure and long-term developmental consequences, *Lancet* 341 (1993): 907–910.

3. Update: Trends in fetal alcohol syndrome—United States, 1979–1993, *Morbidity and Mortality Weekly Report* 44 (1995): 249–251.

4. J. M. Aase, K. L. Jones, and S. K. Clarren, Do we need the term "FAE"? *Pediatrics* 95 (1995): 428–430.

5. Committee on Nutritional Status during Pregnancy and Lactation, Food and Nutrition Board, *Nutrition during Pregnancy* (Washington, D.C.: National Academy Press, 1990), pp. 390–411.

6. Beattie, 1992.

7. Effects of alcohol on fetal and postnatal development, in *Alcohol and Health* (Washington, D.C.: U.S. Department of Health and Human Services, 1993), pp. 203–232.

8. Beattie, 1992.

9. Committee on Substance Abuse and Committee on Children with Disabilities, 1993; *The Surgeon General's Report on Nutrition and Health* (Washington, D.C.: U.S. Government Printing Office, 1988), p. 72.

10. R. E. Little and coauthors, Decreased birth weight in infants of alcoholic women who abstained during pregnancy, *Journal of Pediatrics* 96 (1980): 974–977.

11. When Dad drinks: Can his liquor intake impair his future offspring? *Scientific American*, February 1990, p. 23; L. F. Soyka and J. M. Joffe, Male-mediated drug effects on offspring, *Progress in Clinical and Biological Research* 36 (1980): 49–66.

12. R. E. Little and C. F. Sing, Father's drinking and infant birth weight: Report of an association, *Teratology* 36 (1987): 59–65.

Breast Milk, Infant Formula, and Lactation

C hildbirth marks the end of pregnancy and the beginning of a new set of parental responsibilities, decisions, and behaviors. Many of these focus on care of the newborn. Newborns arrive without an instruction manual, and people become parents without any formal training. Parents must gather information from a variety of sources to determine what to do and how and when to do it. Translating this information into action requires making many adjustments and learning a number of new behaviors. This chapter and its accompanying Focal Point describe the responsibility of choosing what milk to feed an infant, the factors that influence the choice, and the behaviors parents must adopt to carry it out successfully.

The Choice: Breast Milk or Infant Formula?

Before the end of her pregnancy, a woman will need to decide whether to feed her infant breast milk, infant formula, or both. These options are the only recommended foods for an infant during the first four to six months of life.

In many countries around the world, a woman breastfeeds her newborn without considering the alternatives or consciously making a decision. In other parts of the world, a woman feeds her newborn formula simply because she knows so little about breastfeeding. She may have misconceptions or feel uncomfortable about a process she has never seen or experienced. In each of these settings, mothers can benefit from knowledge about both alternatives before deciding what best meets their own needs and the needs of their infants.

To learn about infant feeding practices, a pregnant woman can read any of the many publications available. Other good sources of information are certified lactation consultants, other health care professionals, and mothers who have successfully breastfed and formula-fed their infants. In addition to these resources, a woman examines her personal values and those of the society in which she lives. Each contributes to the decision-making process. The following discussion examines factors influencing the decision whether to feed an infant breast milk or formula, and the Focal Point that follows the chapter offers suggestions as to how to feed, in either case.

A Historical Overview

Prior to the eighteenth century, breast milk was the only source of nourishment for a newborn infant; satisfactory substitutes for breast milk were nonexistent. If a mother could not or did not want to breastfeed her infant, she hired a wet nurse. Deplorable sanitary conditions, as well as prevailing beliefs at the time, precluded even the thought of using cow's milk to feed an infant. (In those days, people believed that an infant drinking cow's milk would take on the characteristics of the animal. Similarly, they assumed that the characteristics of a wet nurse

certified lactation consultants: consultants, often registered nurses, who specialize in helping new mothers to establish a healthy breastfeeding relationship with their newborns.

Appendix F provides a list of nutrition resources.

wet nurse: a lactating woman who breastfeeds another woman's infant.

would pass through her breast milk to the infant. Consequently, much care was given to selecting a kind-hearted wet nurse.)

Around the mid-1700s, women began to adopt the practice of feeding infants a mixture of bread and flour soaked in water once the first tooth had erupted. By the late 1700s cow's milk was considered an option when breast milk was unavailable. A major problem at the time was the lack of an appropriate feeding vessel for the infant. (Cow's milk was first fed, logically, by way of a cow's horn.) By the end of the nineteenth century, the advantages of glass bottles and the need for cleanliness were recognized.

Formula feeding gained popularity as the "modern," efficient, and practical way to feed an infant at the turn of the twentieth century. For the infants of women who were unable to breastfeed or who had died in childbirth, infant formulas became a healthy alternative. Formula feeding coincided with the rise in hospital births and the associated practice of separating mothers from infants after delivery. Other twentieth-century developments also influenced the shift from breast to formula. An essential step was the development of the technology required to analyze breast milk and to create an infant formula that was similar. Systems to purify water supplies and to control pathogenic organisms allowed formula feeding to develop as a safe method of nourishing infants. Simultaneously, the advertising and fashion industries grew bolder in glamorizing women's bodies to a point of distorting the natural function of breasts. To view the breasts as sex objects is to forget that their primary function is to provide milk to a suckling infant.

Breastfeeding steadily declined in the United States until less than one in five mothers of newborns breastfed. During the 1970s, the trend reversed until, by the 1990s, just over half of all mothers were choosing to breastfeed.[1] About a third of these women continue to breastfeed for at least five to six months. As might be expected, the incidence of breastfeeding progressively declines throughout the first year of the infant's life. In developing countries, the initial incidence of breastfeeding is much higher than in developed countries, and the decline over the first year is much smaller.

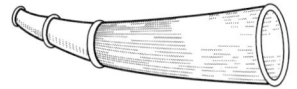

Cow's horn

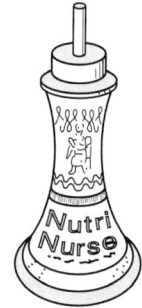

Eighteenth-century German pewter nursing bottle

Nineteenth-century American glass nursing bottle

✶ Healthy People 2000

Increase to at least 75% the proportion of mothers who breast-feed their babies in the early weeks and to at least 50% the proportion who continue breastfeeding until their babies are five to six months old.

Breastfeeding: A Learned Behavior

Although lactation is an automatic physiological process, breastfeeding is a learned behavior that is most successful in a supportive environment. Health care professionals play an important role in providing

lactation (lack-TAY-shun): maternal secretion of milk for a suckling offspring.

Table 4-1 Ten Steps to Successful Breastfeeding

To promote breastfeeding, every maternity facility should:

- Develop a written breastfeeding policy that is routinely communicated to all health care staff.
- Train all health care staff in the skills necessary to implement the breastfeeding policy.
- Inform all pregnant women about the benefits and management of breastfeeding.
- Help mothers initiate breastfeeding within 1/2 hour of birth.
- Show mothers how to breastfeed and how to maintain lactation, even if they need to be separated from their infants.
- Give newborn infants no food or drink other than breast milk, unless medically indicated.
- Practice rooming-in, allowing mothers and infants to remain together 24 hours a day.
- Encourage breastfeeding on demand.
- Give no artificial nipples or pacifiers to breastfeeding infants.[a]
- Foster the establishment of breastfeeding support groups and refer mothers to them at discharge from the facility.

[a]Compared with nonusers, infants who use pacifiers breastfeed less frequently and stop breastfeeding at a younger age. C. G. Victora and coauthors, Pacifier use and short breastfeeding duration: Cause, consequence, or coincidence? *Pediatrics* 99 (1997): 445–453.

Source: United Nations Children's Fund and World Health Organization, *Barriers and Solutions to the Global Ten Steps to Successful Breast-feeding*, 1994.

"There is a reason behind all these things in nature."
Aristotle

encouragement and accurate information on breastfeeding.[2] Of women who do breastfeed, those who receive early and repeated information and support breastfeed their infants longer than others. Table 4-1 lists 10 steps health care professionals can take to promote successful breastfeeding among new mothers.

Fathers also play an important role in encouraging breastfeeding.[3] One study reported that most of those fathers whose partners planned to breastfeed supported that decision and respected breastfeeding women. By comparison, fathers whose partners planned to bottle-feed believed that breastfeeding would make the breasts ugly and interfere with sexual relations. Clearly, educating fathers could change attitudes and promote breastfeeding.

In societies where few women breastfeed, appropriate breastfeeding etiquette remains undefined. A woman faces conflict, confusion, and frustration. Must she retreat to a private place to nurse? What if she cannot find such a place in a public setting? A hungry infant is impatient, and a mother must act quickly. With abundant role models, a consensus defines accepted behaviors, thus offering a nursing mother guidance and confidence. Many public buildings now provide "baby rooms" with tables for changing diapers and comfortable chairs for nursing. These rooms are open to both mothers and fathers in response to current parenting needs.

Parents in today's society also have to coordinate work and family. All mothers are working women—many of them with jobs outside the home. A social system that provides extended, paid maternity leaves, breaks during the workday to nurse infants or pump breasts, and workplace child care promotes breastfeeding as a feasible option.

Focal Point 4 describes how to express breast milk, using a variety of pumps.

Bonding

Some breastfeeding proponents argue that breastfeeding encourages bonding. The affection of a mother for her infant depends more on the time spent in close physical contact than on the method of feeding. There is evidence, however, to suggest that mothers allowed early extended contact with their newborns are more likely to breastfeed and to continue to do so for a longer duration than mothers denied contact. This is not to say that breastfeeding mothers bond better than mothers who feed their infants formulas, but that early, prolonged contact facilitates both bonding and breastfeeding.

Controversy surrounds the question whether early and extended mother–child contact is crucial to bonding. The notion that during a sensitive period just after birth a bond develops that has a profound effect on the parent–child relationship is based on observations of farm animals. In animals, the bond between a mother and her offspring forms within the first few minutes of life. If the two are separated during this crucial time, the mother later rejects the offspring. She refuses to nurse and exhibits physically hostile behaviors toward the newborn. A brief initial contact, even if followed by a separation, is sufficient to establish the relationship. A variety of visual, auditory, and olfactory cues establishes the maternal acceptance of her offspring. Most likely, postpartum hormonal conditions influence maternal acceptance.

In human beings, factors such as odor also play an important role in breastfeeding. Two out of three infants placed between a washed breast and an unwashed breast spontaneously selected their mothers' unwashed breasts.[4] Of course, in human beings, mothering behaviors are more complex than can be explained by such physiological cues and hormones alone. The actual birth in human beings is preceded by months of maternal thought and anticipation. Many women begin the bonding process during pregnancy. They may pat their bellies when they feel a kick, attend childbirth classes, read books, seek advice from friends, and decorate the baby's room. A woman enters motherhood with memories of her own relationship with her mother. She has learned her culture's patterns of mother–child interactions. Whether the infant was planned or wanted also affects the mother's maternal behaviors, as does her relationship with the father.

The notion that events at the time of birth can influence the bonding process has altered birth practices in this country. In the first half of the twentieth century, mother and newborn were separated at birth and

bonding: a process that occurs immediately after birth in which a parent and infant form an affectionate attachment.

cared for in the maternity and nursery wards, respectively. In more recent years, the trend has been to allow mother and newborn to lie together in a warm and quiet room, often joined by the father. Many health care professionals encourage breastfeeding mothers to nurse their infants immediately after delivery.

Such family-centered birth experiences may indeed promote loving relationships, but the theory is difficult to prove. As is often the case in research involving human beings, designing a methodologically sound and ethical study is close to impossible. In this instance, the results would require subjective interpretations of significance. If one mother picks up her infant more often than another, who is to determine the significance and effects of that behavior?

Convenience

In a society that embraces microwave ovens and frozen foods, convenience plays a large part in the many decisions parents make about feeding themselves and their children. The lifestyles, attitudes, and habits of each individual dictate what is defined as convenient.

Breastfeeding is a simple, yet elegant, procedure—the natural way to nourish infants since the beginning of time. Once a woman learns the correct technique, she becomes confident and usually finds the experience enjoyable. In a few short weeks, the mother and infant adjust to the feeding process, and the milk supply is established. Many women find that breastfeeding is as easy as breathing. It can be done anywhere, anytime, as long as mother and infant are together. Breast milk is sterile and always at the appropriate temperature. The breastfeeding mother is free of cleaning bottles and mixing formulas.

Breastfeeding may not be convenient, however, for a mother whose schedule does not easily permit her to be with her infant at feeding times. A woman who works outside the home may find it less troublesome to feed her infant formula. Hot cycles on a dishwasher make bottle cleaning easy; and measuring, mixing, and pouring infant formula is easier than preparing any other meal. Ready-to-pour formulas that do not need mixing, although more expensive, are available for busy mothers who can afford them. Formula feeding also allows the father and other family members an opportunity to feed the infant.

The working mother who is motivated to continue breastfeeding when she returns to work may be able to find ways to do so. She can breastfeed exclusively up to that time; then, depending on her schedule and location, she can continue to breastfeed some feedings and supplement with expressed breast milk or formula for others. Breastfeeding manuals provide a multitude of suggestions to help her work out the details. Breastfeeding mothers may find that working out such details may be well worth the effort. At least one study reports that breastfed infants suffer fewer and less severe illnesses, and their mothers miss work less often than those feeding formula.[5]

Before making the decision of whether to feed an infant breast milk or formula, parents may want to examine the characteristics of each. The following sections present this information, making comparisons where appropriate.

Characteristics of Breast Milk

From the nutrition standpoint, breast milk is the preferred choice for infants. The American Dietetic Association advocates breastfeeding "because of the nutritional and immunologic benefits of human milk for the infant; the physiological, social, and hygienic benefits of the breastfeeding process for the mother and infant; and the economic benefits to the family and health care system."[6] The American Academy of Pediatrics and the Canadian Pediatric Society have issued this joint statement: "Breastfeeding is strongly recommended for full-term infants, except in the few instances where specific contraindications exist." The recommendations in favor of breast milk arise from its unique nutrient composition and nonnutrient protective factors that promote optimal infant health and development. Alternatively, infant formula can be used to meet the nutrient needs of infancy because it imitates breast milk's nutrient composition as closely as possible. Table 5-4, in Chapter 5, compares the nutrient composition of the two.

Because breast milk best meets an infant's nutrient needs, it is used as a standard for estimating those needs. The infant's nutrition needs are examined in detail in Chapter 5; the following examination of breast milk composition lays the groundwork for that discussion and shows how breast milk best meets a newborn infant's needs.

Nutrient Composition

Breast milk excels as a source of nutrients for the young infant.[8] The following paragraphs describe its unique nutrient composition, comparing it to milk-based infant formula where appropriate.

Carbohydrate. The carbohydrate in breast milk and milk-based infant formula is the disaccharide lactose. Lactose is a major constituent of breast milk, second only to water. In addition to being easily digested, lactose facilitates calcium absorption.

Other carbohydrates in breast milk include oligosaccharides and bifidus factors. Each of these nonlactose carbohydrates is present only in small concentrations, but plays a critical role in defending the infant against harmful bacteria. Oligosaccharides (short chains of monosaccharides) interfere with the binding of pathogenic bacteria or their toxins to the epithelial cells of the intestine. Bifidus factors (nitrogen-containing sugars) promote the growth of special bacteria that secrete acidic compounds that inhibit the growth of pathogenic bacteria (described further on p. 125).

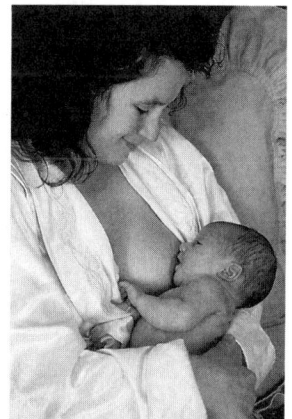

Breastfeeding is a natural extension of pregnancy—of the mother's body nourishing the infant.

renal solute load: a measure of the concentration of all dissolved substances in the urine that result from the feeding of a milk, formula, or diet.

alpha-lactalbumin (LACT-al-BYOO-min): the chief protein in human breast milk.

casein (CAY-seen): the chief protein in cow's milk.

Protein. The total protein concentration of breast milk is less than that of cow's milk, but this deficit is actually beneficial because a low protein concentration contributes to a low renal solute load. Because the solute-concentrating capacity of the kidneys is limited during infancy, a low renal solute load eases the work of the kidneys. Solutes in excess of the infant's need must be excreted via the kidneys, a process that requires water. If water is unavailable, then the kidneys must concentrate the urine. Water balance during early infancy is critical; infants can easily become dehydrated. A low renal solute load is therefore beneficial to the infant's immature kidneys and delicate water balance.

The unique protein quality of breast milk also offers benefits. Compared with cow's milk, breast milk contains a greater proportion of the protein alpha-lactalbumin; cow's milk contains a larger proportion of the protein casein. Alpha-lactalbumin is efficiently digested and absorbed, whereas casein produces tough, hard-to-digest curds in the infant's intestine. Alpha-lactalbumin is also richer in sulfur-containing amino acids than is casein. One of these sulfur-containing amino acids, methionine, is an essential amino acid, and another, cysteine, may be essential for premature infants. The availability of cysteine is especially important because the enzymes responsible for synthesizing cysteine from methionine are inactive in early infancy. Breast milk contains all the essential amino acids in appropriate amounts, plus nonprotein nitrogen compounds as well.

Lipids. The total lipid content of breast milk varies considerably, due partly to differences in sampling techniques, but also to individual differences between milks from different women or from time to time within the same woman.[9] The total fat present in breast milk is not a reflection of the total fat content of the maternal diet, except in severely malnourished women. Such is not the case with the fatty acid composition, which changes in response to maternal diet.

The lipids in breast milk, cow's milk, and infant formulas provide the main source of energy in the infant's diet. The unique lipid composition of breast milk offers other nutritional advantages as well. The infant digests the fat in breast milk more completely than the fat in cow's milk, thanks to the acid-resistant lipase enzymes present in breast milk. Another advantage is the unique configuration of the triglycerides on hydrolysis by the lipase enzymes. Figure 4-1 illustrates how enzymes dismantle triglycerides at the 1 and 3 positions of the glycerol molecule, leaving a monoglyceride with the fatty acid attached at the 2 position. In breast milk, palmitic acid usually occupies the 2 position, whereas in cow's milk, stearic acid is often in the 2 position. Infants absorb monoglycerides with palmitic acid in the 2 position more efficiently than either free fatty acids or monoglycerides with stearic acid in the 2 position.[10]

Triglycerides constitute 98 percent of the lipids present in breast milk; free fatty acids, cholesterol, and phospholipids make up the remainder. Breast milk contains more than adequate concentrations of the essential fatty acid linoleic acid. Breast milk contains about 10 times

Figure 4-1 Triglyceride Hydrolysis and Fatty Acid Configuration

Palmitic acid occupies the 2 position in the monoglycerides of breast milk, whereas stearic acid occupies that position in cow's milk. After hydrolysis, monoglycerides with palmitic acid (a 16-carbon saturated fatty acid) are more efficiently absorbed than free fatty acids or monoglycerides with stearic acid (an 18-carbon saturated fatty acid).

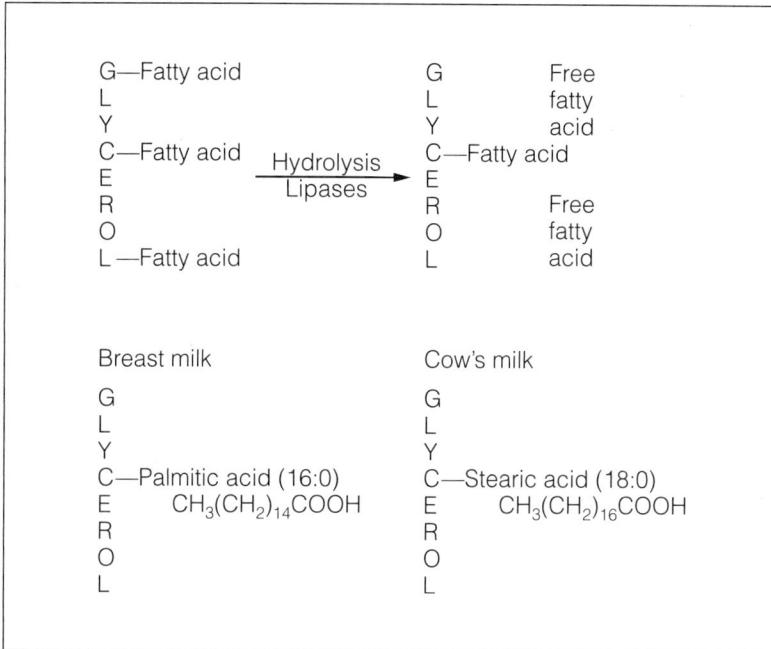

```
G—Fatty acid              G          Free
L                         L          fatty
Y                         Y          acid
C—Fatty acid  Hydrolysis  C—Fatty acid
E            ──────────►  E
R             Lipases     R          Free
O                         O          fatty
L —Fatty acid             L          acid


Breast milk               Cow's milk

G                         G
L                         L
Y                         Y
C—Palmitic acid (16:0)    C—Stearic acid (18:0)
E     CH₃(CH₂)₁₄COOH       E     CH₃(CH₂)₁₆COOH
R                         R
O                         O
L                         L
```

as much cholesterol as infant formula. Cholesterol is essential for normal myelination in the central nervous system. Many questions remain unanswered concerning exactly how much cholesterol and other lipids might be appropriate for an infant. At the present time, the American Academy of Pediatrics recommendations discourage fat restriction during infancy, even for those with a family history of atherosclerosis.

The energy-nutrient composition of breast milk differs dramatically from the energy contributions of a healthy diet for adults. Protein contributes only about 6 percent of the total kcalories in breast milk, whereas protein in the adult diet contributes about twice that percentage (see Figure 4-2). Carbohydrate in breast milk contributes only two-thirds of the proportion recommended for adults. Fat contributes over 40 percent of the kcalories in breast milk, whereas adults are advised to derive no more than 30 percent of their daily energy from fat. Most remarkably, breast milk's cholesterol content is 10 times that recommended for adults. Yet for infants, breast milk is nature's most nearly perfect food, providing a clear lesson in the necessity of appreciating that people at different stages of life have different nutrient needs.

Figure 4-2 Percentage of Energy-Yielding Nutrients in Human Milk and in Recommended Adult Diets

The proportions of energy-yielding nutrients in human breast milk differ from those recommended for adults.

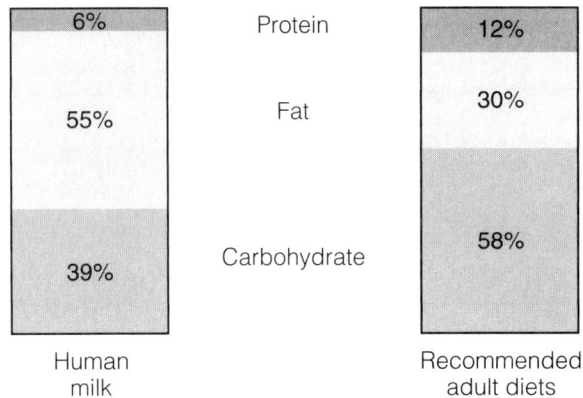

Human milk		Recommended adult diets
6%	Protein	12%
55%	Fat	30%
39%	Carbohydrate	58%

Vitamins and Minerals. In general, breast milk from healthy mothers contains all the vitamins and minerals needed for normal infant growth, with the possible exception of vitamin D, which can be obtained easily with daily exposures to sunlight. The infant's vitamin and mineral requirements and supplement recommendations are discussed, nutrient by nutrient, in Chapter 5.

A note on electrolytes is also in order. Breast milk has a low electrolyte content, which offers at least two nutritional advantages. First, the sodium concentration is low. Whether a high sodium intake during infancy predisposes infants to the later development of hypertension is unclear, but in some adults, an association between high sodium intake and high blood pressure does exist. A second advantage of the low electrolyte content of breast milk is the low renal solute load produced. The benefits of a low renal solute load were discussed earlier with respect to protein.

The nutritional advantages of breast milk go beyond the nutrient content of the milk. For example, zinc is present in low quantities in breast milk, but its absorption is exceptionally efficient because of zinc-binding proteins that enhance its bioavailability. Similarly, breast milk contains an amylase enzyme that facilitates starch digestion. The presence of this enzyme explains why the first solid food normally fed to infants is a grain food, usually cereal.

amylase (AM-ih-lace): an enzyme that hydrolyzes amylose (a form of starch).

Protective Factors

Parallel to the nutritional benefits of breast milk in promoting infant health are the nonnutritional benefits. Not only is breast milk sterile, but it actively fights disease in a number of ways. Such protection is most valuable during the first year, when the infant's immune system is not fully prepared to mount a response against infection.

Colostrum. During the first two to three days after delivery, before the onset of true lactation, the breasts produce colostrum. Colostrum is a premilk substance that contains mainly serum with antibodies and white blood cells.

Colostrum differs from mature breast milk in nutrient composition and immunological factors. Compared with mature breast milk, colostrum is higher in protein, fat-soluble vitamins, carotenoids, minerals, and immunological factors. During the first two weeks of lactation, the concentrations of proteins, fat-soluble vitamins, minerals, and immunoglobulins decrease, while energy, lactose, fat, and water-soluble vitamins increase. These changes reflect changing infant needs.

Colostrum (like breast milk) helps protect the newborn infant from infections against which the mother has developed immunity. Researchers have proposed that colostrum may also contain a growth-promoting substance.

Like colostrum, breast milk is also rich in substances that protect the infant from infection. Bifidus factors, immunoglobulins, lipases, lysozyme, and lactoferrin are among the protective factors in breast milk.

Bifidus Factors. Microbial growth factors, referred to as bifidus factors, are present in both colostrum and breast milk. These bifidus factors benefit the infant by favoring the growth of harmless bacteria in the digestive tract. The presence of "friendly" *Lactobacillus bifidus* in the digestive tract inhibits the growth of potentially harmful *Escherichia coli* and other bacteria, thereby protecting the infant from intestinal infection.

Immunoglobulins. Breast milk, and especially colostrum, is also rich in immunoglobulins that protect the infant's digestive tract against antigens. Most often, the mother and infant are exposed to the same antigens. When disease-causing antigens enter the mother's body (usually from the small intestine or respiratory tract), her immune system is called into action. Her immune cells travel to the mammary glands, where they produce antibodies to fight the antigen. In this way, the infant receives protection against the specific infectious agents that are in the surrounding environment.

Of all the antibodies found in colostrum and breast milk, secretory IgA is the most abundant. The infant receives secretory IgA and other antibodies with each feeding of breast milk. Secretory IgA is resistant to protein digestion, so it can function in the infant's digestive tract. There, secretory IgA binds with harmful bacteria, preventing their entry into the blood. Secretory IgA, in conjunction with lactoferrin and other factors, accounts for the lower incidence of intestinal infection among breastfed infants compared with formula-fed infants. Even when an infant becomes infected with a disease-causing microorganism, secretory IgA helps prevent the accompanying diarrhea.[11]

colostrum (co-LAHS-trum): the secretion from the breast before the onset of true lactation; also referred to as the "first milk."

antibodies: large proteins produced by the immune system in response to the invasion of the body by foreign molecules (usually antigens).

bifidus (BIFF-id-us or by-FEED-us) **factors:** factors in colostrum and breast milk that favor the growth of the "friendly" bacteria *Lactobacillus bifidus* in the infant's intestinal tract, so that other, less desirable intestinal inhabitants will not flourish.

immunoglobulins: proteins that are capable of acting as antibodies.

antigens: substances foreign to the body that elicit either the formation of antibodies, an inflammation reaction, or both, from immune system cells.

secretory IgA: one of several types of antibodies produced by the human immune system; the predominant antibody of breast milk.

lipase (LYE-pase): an enzyme that hydrolyzes lipids (fats).

Lipase Activity. The bifidus factors and immunoglobulins in colostrum and breast milk are not alone in offering the breastfed infant antimicrobial protection. The lipase activity of breast milk may also be responsible for some antimicrobial activity. There appears to be a relationship between the formation of fatty acids and monoglycerides in breast milk and antimicrobial activity. Milks rich in free fatty acids and monoglycerides exhibit stronger activity against certain viruses, compared with milks rich in triglycerides. (Cow's milk has a high triglyceride and low monoglyceride content.) This finding supports the use of monoglycerides in the manufacture of infant formulas, instead of triglyceride-rich vegetable oils, as is the current practice.

lysozyme (LYE-so-zime): an enzyme that breaks up cell wall structures, including those of bacteria.

Lysozyme. Breast milk is richer in the enzyme lysozyme than are other milks. Lysozyme breaks apart bacterial cell walls, thus inhibiting bacterial growth in the digestive tract of breastfed infants.

lactoferrin (lak-toh-FERR-in): a factor in breast milk that binds iron and keeps it from supporting the growth of the infant's intestinal bacteria.

Lactoferrin. Another factor that contributes to the antimicrobial activity of breast milk is the iron-binding protein lactoferrin. Lactoferrin captures iron, making it less available to bacteria in the intestinal tract, thus inhibiting their growth. Also, lactoferrin may be responsible for the high bioavailability of the little iron present in breast milk; by comparison, cow's milk contains little lactoferrin and iron bioavailability is low. The low bioavailability of iron from infant formulas is compensated for by the manufacture of iron-fortified formulas, which are effective in preventing iron deficiency. In addition to its iron-binding capacities, lactoferrin appears active in other, unrelated functions: It participates in the immune system and as a growth factor.[12]

Intestinal Growth Factor. An epidermal growth factor that stimulates growth of intestinal cells has also been discovered in breast milk. This growth factor allows damaged cells to be replaced more rapidly than normal. The efficient replacement of damaged cells helps keep the infant's digestive tract resistant to infection.

Health Advantages of Breast Milk

Consistent with the presence of many antibacterial factors in breast milk is convincing evidence that breast milk protects infants against gastrointestinal infections. Most notable is that this protection persists beyond the actual duration of breastfeeding.[13] In developing countries, where the incidence of infection is especially high, the infection rate among breastfed infants is low. In countries where poor sanitation is prevalent, failure to breastfeed an infant who lives in a house without piped water and a toilet incurs twice the risk of perinatal mortality as for a breastfed infant living in a house with good sanitation.

In addition to protection against infection, breast milk may offer protection against the development of allergies as well. Compared with formula-fed infants, breastfed infants show a lower incidence of allergic reactions, such as asthma, recurrent wheezing, and skin rash.[14] This protection is especially noticeable among infants with a family history of allergy.

One other potential advantage of breast milk is the wide spectrum of flavors it introduces to an infant; this early introduction of flavors may make the acceptance of a variety of foods later in life more likely.[15] The flavor compounds of foods eaten by lactating mothers pass into their breast milk. When infants drink the breast milk, they experience the full array of flavors from their mother's diet.[16] Research has found that young animals prefer the foods, and readily accept unfamiliar foods, their mothers have eaten during lactation. Such findings have led researchers to question whether formula-fed infants may be missing early flavor experiences that could influence future food choices.

Characteristics of Infant Formulas

A woman who breastfeeds for one year can wean her infant to cow's milk, bypassing the need for infant formula. However, a woman who decides to feed her infant formula from birth, to wean to formula after a short time, or to substitute formula for breastfeeding on occasion must select an appropriate infant formula and learn to prepare it. A variety of infant formulas are available, and the selection must be made carefully. The use of iron-fortified formula throughout the first year of life helps ensure an adequate iron intake. Chapter 5 offers help with the question of what type of formula is appropriate, and the Focal Point that follows this chapter describes how to feed formula.

Appendix D provides a table comparing the composition of infant formulas available in the United States.

Formula manufacturers attempt to copy the nutrient composition of breast milk as closely as possible. The mother who feeds her infant formula can feel confident that the quantity she gives her infant is adequate. She can see the bottle being emptied and can replenish it as needed. The supply is limited only by finances.

Formulas prepared under sanitary conditions minimize exposure to disease-causing organisms. In developed countries, preventive medical care (such as vaccinations) and public health measures (such as purified water) help protect infants from infections. Safety and sanitation can be achieved with either mode of feeding by the informed mother whose water supply is reliable.

Making the Decision

Some women believe that the breast is best and that infant formulas are unnatural. They are convinced that breastfeeding offers the most benefits and best nourishment to their newborns. They find suckling an infant at their breast to be a most rewarding experience. Breastfeeding is compatible with their views on mothering and fits into their work and home arrangements.

Some women consider breastfeeding old-fashioned and formula feeding more convenient. They are uncomfortable with breastfeeding;

they may attempt to breastfeed, but find they are too nervous or too distracted to successfully nurse an infant. Other women find breastfeeding just does not work with their lifestyles and schedules. These mothers have valid reasons for making their choices, and their feelings need to be honored. Bearing and nurturing an infant involves much more than merely pouring in nutrients.

The decision whether to breastfeed is best made by the parents with the advice of their health care providers. Many friends, relatives, and strangers will probably voice their opinions as well. A mother will ultimately need to rely on well-informed sources for advice and information. She can obtain help from registered dietitians, lactation consultants, instructors at natural childbirth classes, and maternity nurses.

Quite likely they will encourage her to breastfeed. Many health care professionals and dietitians believe breastfeeding is sufficiently important to warrant that every effort be made to do so, even if only for a short time. Breastfeeding during the most critical first few weeks or months provides the infant with valuable immunological protection and other special advantages. The mother can then shift to formula, knowing she has given her infant those benefits.

If this discussion appears biased in favor of breastfeeding, it is because breastfeeding offers many benefits to both mother and infant (see Table 4-2), and every pregnant woman should seriously consider it. Still, there are valid reasons for not breastfeeding, and formula-fed infants grow and develop into healthy children. After all, the primary goal is to provide the infant optimal nourishment in a relaxed and loving environment.

The responsibilities and decisions continue once a mother decides how she will feed her infant. If she decides to feed infant formula, she will need to learn how to select the appropriate formula and prepare it. Focal Point 4 provides the needed information. If she decides to breastfeed, she will need to choose foods wisely to promote optimal milk production.

Lactation

Much attention is focused on nutrition for pregnant women, whereas nutrition for lactating women is often neglected. But the lactating mother has unique nutrient needs and concerns, and her care and feeding deserve special attention. First, a description of the physiology of lactation sets the foundation for understanding the special nutrient needs of a breastfeeding mother.

Physiology of Lactation

mammary glands: glands of the female breast that secrete milk.

Lactation is the natural extension of pregnancy—of the mother's body nourishing the infant. The mammary glands secrete milk for this purpose.

Table 4-2 Benefits of Breastfeeding

For infants:

- Provides a favorable balance of nutrients with high bioavailability.
- Provides hormones that promote physiological development.
- Improves cognitive development.
- Protects against a variety of infections.
- Reduces the risk of sudden infant death syndrome (SIDS).
- Protects against some chronic diseases, such as diabetes.[7]
- Makes food allergies less likely.

For mothers:

- Contracts the uterus.
- Lengthens birth intervals.
- Conserves iron stores (by prolonging amenorrhea).
- Reduces risk of breast cancer.
- Protects bone density.
- Saves money and offers convenience.

The mammary glands, stimulated by estrogen during puberty, develop a system of ducts, lobes, and alveoli. This system remains fairly inactive until pregnancy. During pregnancy, hormones stimulate a proliferative stage of mammary gland activity. Estrogen promotes growth and branching of the duct system, while progesterone stimulates development of the alveoli (see Figure 4-3). During pregnancy, the elevated concentrations of these hormones support mammary gland development, while inhibiting the actual secretion of milk. These two hormones re-

ducts: narrow tubular vessels that drain the lobes of the mammary gland into the tip of the nipple.

lobes: segments of the mammary gland.

alveoli (al-VEE-oh-lie): the milk-producing cells of the mammary gland; the singular is **alveolus.**

Figure 4-3 Breast Development from Puberty to Lactation

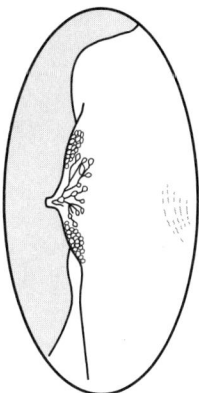

During puberty, a system of ducts, lobes, and alveoli develops.

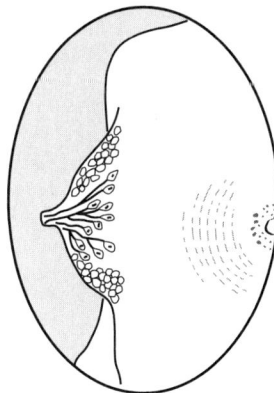

This system remains inactive until pregnancy.

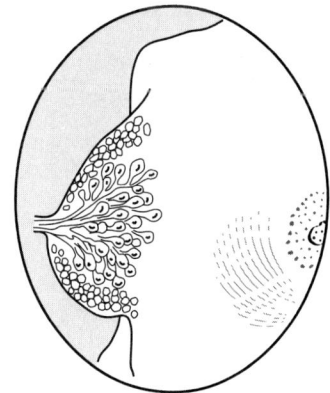

During pregnancy, growth proliferates, with ductal branching and lobular-alveolar development proceeding at a spectacular rate, yet in an orderly fashion.

prolactin (pro-LAK-tin): a hormone secreted from the anterior pituitary gland that acts on the mammary glands to initiate and sustain milk production (*pro* = promote; *lacto* = milk).

oxytocin (OK-si-TOE-sin): a hormone secreted from the posterior pituitary gland that stimulates the uterus to contract and the mammary glands to eject milk (*oxy* = quick; *tocin* = childbirth).

prolactin-inhibiting hormone: a hormone secreted from the hypothalamus that acts on the anterior pituitary gland to regulate the release of prolactin.

turn to their basal concentrations within a few days postpartum, thus allowing milk production to begin.

The Hormones of Lactation. As illustrated in Figure 4-4, the hormones prolactin and oxytocin finely coordinate lactation. Prolactin is responsible for milk production. Throughout pregnancy, the blood concentration of prolactin increases. Upon delivery, its concentration rises substantially in preparation for infant feeding.

High prolactin concentrations signal the release of prolactin-inhibiting hormone, which ensures that prolactin concentrations and milk production do not exceed the need. This is an example of hormone regulation by negative feedback—prolactin secretion is turned off by its own high concentrations.

In contrast, infant suckling turns off the secretion of prolactin-inhibiting hormone and signals prolactin secretion and milk production. The consequence of these hormonal interactions is that prolactin concentrations remain high and milk production continues as long as the infant is nursing. The infant's demand for milk signals the mammary glands to supply milk.

Figure 4-4 Prolactin and Oxytocin Activity

An infant suckling at the breast stimulates the pituitary to release the hormones prolactin and oxytocin.

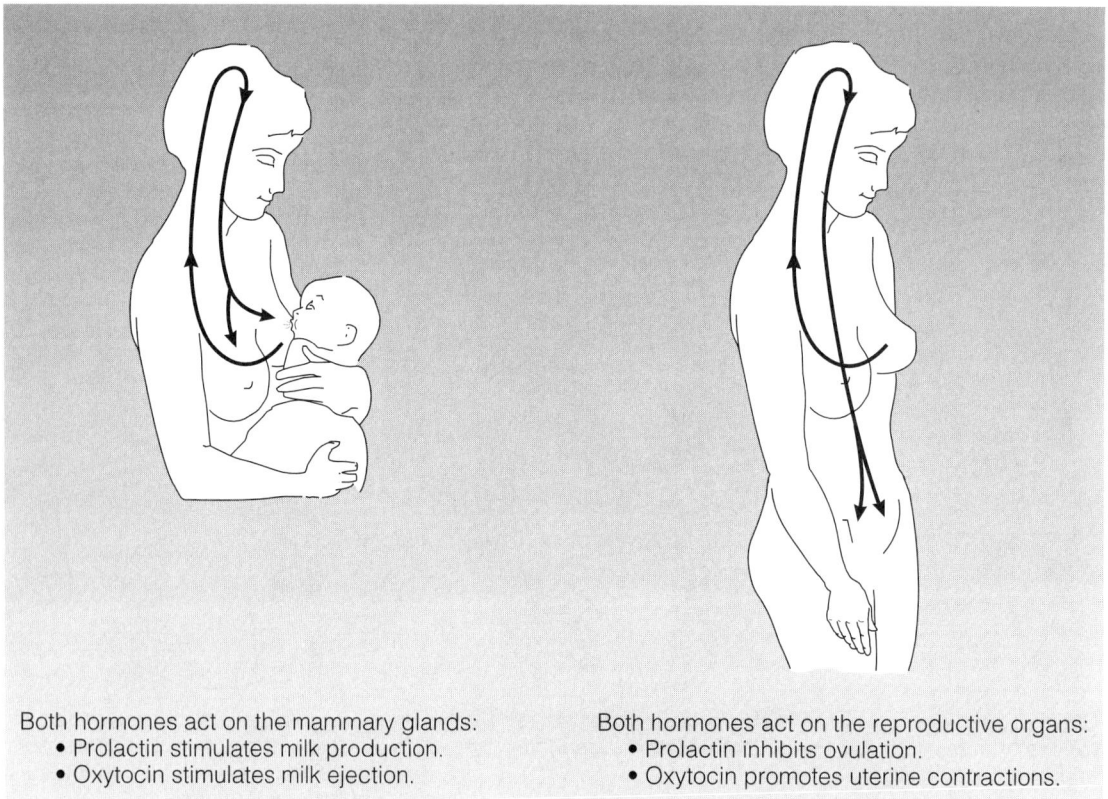

Both hormones act on the mammary glands:
- Prolactin stimulates milk production.
- Oxytocin stimulates milk ejection.

Both hormones act on the reproductive organs:
- Prolactin inhibits ovulation.
- Oxytocin promotes uterine contractions.

In breastfeeding mothers, concentrations of prolactin are extremely high for the first three months but then decline to near-normal levels, even with continued milk production. In nonbreastfeeding mothers, prolactin concentrations decline to prepregnant levels within the first two to three weeks postpartum.

The hormone oxytocin causes the myoepithelial cells surrounding the alveoli to contract, thus initiating the ejection of milk into the ducts, a response known as the let-down reflex. (The mother feels this as a contraction of the breast, followed by the flow of milk and relief of pressure.) After the birth, the elevated progesterone concentrations of pregnancy (which turn off oxytocin) decline. The stretching of the cervix during childbirth signals the release of oxytocin, which causes two organs to react: the uterus to contract and the mammary glands to eject milk.

An infant's suckling also elicits oxytocin to eject milk from the mammary glands. At first, the stimulus for let-down is the infant's suckling. Later, when the reflex is well established, the sound of the infant's crying may be enough to trigger it. An efficient let-down reflex is essential to successful lactation. Emotional upset, pain, and fatigue may inhibit the let-down reflex. By relaxing and eating well, the nursing mother promotes easy let-down of milk and greatly enhances her chances of successful lactation.

myoepithelial cells: contractile cells of the mammary glands.

let-down reflex: the reflex that forces milk to the front of the breast when the infant begins to nurse.

The Release of Milk. From early to late in a nursing session, the character of milk changes. The milk released first, known as the foremilk, provides most of the nutrients the infant receives. The draught reflex, which occurs later during a nursing session, draws milk from the hindmost milk-producing glands of the breast after the foremilk has been released. The mother feels this as a tingly sensation within the breast. Hindmilk has a higher fat content than foremilk, leading researchers to speculate that its appearance late in a breastfeeding session permits the infant to satisfy its sucking need and receive sufficient nutrients from the foremilk before achieving satiety.

foremilk: the milk at the front of the breast that is released early in a nursing session; it is low in fat and high in nutrients.

draught reflex (DRAFT reflex): the reflex that moves the hindmilk toward the nipple after the infant has drawn off the foremilk.

hindmilk: the milk released late in a nursing session; it is higher in fat than foremilk.

Postpartum Amenorrhea. Women who breastfeed their infants experience prolonged postpartum amenorrhea. An infant's suckling serves the dual purpose of not only promoting milk production, but also of causing lactational anovulation; both of these effects are mediated by the hormone prolactin. Prolactin inhibits the release of the hormones responsible for ovulation, which is beneficial because it postpones pregnancy and allows time for the replenishment of maternal nutrient reserves.

Physically, the new mother's body is still undergoing major changes. Hormones are shifting from a state of pregnancy to one of lactation or nonpregnancy. Some nutrient stores may be low or depleted. A new mother rarely sleeps more than four hours at a time. Even with optimal nourishment and adequate rest, she will not be back to "100 percent" for at least a year. Repeated pregnancies at intervals of less than one year deplete nutrient reserves. Whether contraception is passive or active,

postpartum amenorrhea: the normal temporary absence of menstrual periods immediately following childbirth.

lactational anovulation: the normal suppression of ovulation during lactation.

it seems best to avoid pregnancy until the mother has had time to re-adjust to nonpregnancy, restore nutrient banks, and regain her vigor. The reproductive system needs a rest before being called into active duty again.

The uterus returns to its normal size about six weeks after delivery. Menstruation usually occurs four to eight weeks postpartum in nonlac-tating women. Women need not be concerned if menstruation is delayed for three or four months, because variation is quite normal. The duration of postpartum amenorrhea is as individual as the women themselves. The delayed onset of ovulation, and the consequent delay of menstruation during lactation may be a pleasant bonus for some women, but the absence of menstrual periods does not always protect against pregnancy. An ovum may be released at any time, so to avoid pregnancy a couple must use some form of contraception. Breastfeeding mothers, however, should not use oral contraceptive agents. Most oral contraceptives contain estrogen, which reduces milk volume and the protein content of breast milk.[17] Mothers can continue breastfeeding without harmful effects after menstruation resumes.

The effect of nutrition on fertility is discussed in Chapter 3.

Maternal nutrition status plays a role in altering plasma prolactin concentration and, therefore, in postpartum amenorrhea and anovula-tion. When diets of undernourished lactating women are supplemented to provide the needed energy, protein, vitamins, and minerals to support lactation, plasma prolactin concentrations decline, and the duration of postpartum amenorrhea is shortened. Thus, the length of post-partum amenorrhea depends on both the infant's sucking behavior and the mother's nutrition status. Prolonged lactation and poor nutrition lengthen the period of postpartum amenorrhea and consequently reduce fertility. Whether it is wise to improve maternal nutrition and shorten lactational amenorrhea is a topic of much debate.[18]

The hormonal events preceding the first menstruation postpartum are rarely like those of an ordinary interpregnancy cycle. Fewer than 20 percent of postpartum women have a normal ovulatory cycle before their first menstruation. Furthermore, conditions within the uterus are unlikely to support a pregnancy even if an ovum is fertilized. Given a 25 percent probability of conception in a normal menstrual cycle, it is estimated that only 5 percent of lactating women who engage in unpro-tected intercourse before their first menstruation are likely to conceive. Indeed, the incidence of conception during lactational amenorrhea in developing countries is less than 10 percent. Such statistics seen in *popu-lations* bespeak the effectiveness of lactation as a contraceptive influ-ence, but do not make it acceptable as a contraceptive method for an *individual*.

Unfortunately, no simple or reliable procedure to detect the onset of fertility has been developed. Many biological factors influence the vari-ability in duration of postpartum infertility. Hormonal responses to the frequency and vigor of sucking are responsible for much of the varia-tion. Women who frequently nurse their infants have elevated prolactin concentrations. The number of feedings as well as the total time at the breast also influence prolactin concentrations. As mothers begin to sup-plement their infants' diets and nurse less often and for shorter periods, prolactin concentrations decline and ovarian activity resumes. Most lactating women will start to menstruate and ovulate prior to complete weaning.

The End of Lactation. The woman who wants to stop lactating may be given an injection of estrogen or a large dose of vitamin B_6 to inhibit milk production. Such measures hasten the end of lactation but are not necessary. Without the stimulation of suckling, milk production will eventually stop.

Lactation for Adoptive Mothers. Amazingly, women who have never been pregnant can breastfeed an infant.[19] Breastfeeding an adopted baby requires patience and perseverance, but most women who try are reasonably successful. One study reports that one-fourth of adoptive mothers produced enough milk to eliminate infant formulas altogether, while most of the others needed to supply about half of the infants' intakes with formula. Beyond the nourishment and protection of breast milk, these mothers most appreciated the bonding experience. Focal Point 4 provides instructions for how to breastfeed an adopted baby.

Some health conditions interfere with lactation or prohibit breastfeeding. The following paragraphs describe these situations.

Maternal Health

If a woman has an ordinary cold, she can go on nursing without worry. If susceptible, the infant will catch it from her anyway. (Thanks to immunological protection, a breastfed baby may be less susceptible than a formula-fed baby would be.) If a woman has a communicable disease such as tuberculosis or hepatitis that could threaten the infant's health, then mother and baby must be separated; mothers can pump their breasts several times a day and feed breast milk by bottle. If a woman has a chronic disease such as cancer that could drain her energy and strain her emotional health, then her physician may advise against breastfeeding.

HIV Infection and AIDS. For mothers with HIV infections, advice differs depending on context. The virus can be transmitted through breast milk, but the risk of infecting the infant by breastfeeding is low.[20] Where safe alternatives are available, the Centers for Disease Control and the American Academy of Pediatrics recommend that HIV-positive women not breastfeed their infants. In developing countries, however, the feeding of inappropriate or contaminated formulas is the cause of 1.5 million infant deaths each year, so WHO and UNICEF urge mothers to breastfeed irrespective of HIV infection. For these infants, the protection of being breastfed outweighs the risk of HIV transmission.[21]

Diabetes. Women with insulin-dependent diabetes mellitus (IDDM) may need careful monitoring and counseling to ensure successful lactation.[22] Women with IDDM need to adjust their energy intakes and insulin doses to meet the heightened needs of lactation. Maintaining good glucose control helps to initiate lactation and support milk production.[23]

Breast Health. Some women fear that breastfeeding will cause their breasts to sag. The breasts do swell and become heavy and large immediately after the birth, but even when they are producing enough milk to nourish a thriving infant, they eventually shrink back to their prepregnant size. Given proper physical support, diet, and exercise, breasts return to their former shape and size after weaning. Breasts change their shape as the body ages, but breastfeeding does not accelerate this process.

Whether lactation protects women from later breast cancer is an area of active research.[24] Some research suggests no association between breastfeeding and breast cancer, whereas other research suggests a protective effect.[25] Any relation between lactation and the risk of breast cancer is most apparent for premenopausal women who were young when they breastfed and who breastfed for a long time.[26] Whether the physical and hormonal events of lactation predispose a woman to breast cancer is unclear.

Breast Implants. During the 1960s, 1970s, and 1980s, some 1 million women received silicone breast implants for either cosmetic or reconstructive reasons. The incidence of breastfeeding among these women is unknown, but questions have been raised about the health of infants breastfed from silicone-implanted breasts. First, do implants leak chemicals into breast milk? Second, do implants cause an immunological disease in the mother that might release compounds into breast milk? Research studies have yet to be conducted to ascertain the answers to these questions. Some children breastfed by mothers with silicone implants suffer gastrointestinal problems;[27] others report "no problems." As of this time, there is no contraindication to women with silicone implants breastfeeding their infants.[28]

Nutritious foods support successful lactation.

Maternal Nutrition during Lactation

During lactation, as during pregnancy, the mother requires sufficient nutrient intakes and stores to support both the infant's growth and her own health. If she does not eat well throughout pregnancy and lactation, her health may be compromised—in some instances, to a greater extent than that of her child. In addition, lactation is likely to be unsuccessful.

Water

Water is the major nutrient in breast milk. Total milk volume varies with infant age, but not with maternal fluid intake, as might be expected. Despite previous misconceptions, a mother who drinks more fluid does not produce more breast milk.[29] To protect herself from dehydration, however, a lactating woman needs to drink plenty of fluids (at least two quarts of fluid a day). A sensible rule of thumb is to drink a glass of milk, water, or juice at each meal and each time the baby nurses.

Energy and Energy Nutrients

Energy from maternal diet and tissue reserves supports both lactation and maternal health and activities. A nursing mother produces about 25 ounces of milk per day, with considerable variation from woman to woman and in the same woman from time to time, depending primarily on the infant's demand for milk.[30] To produce milk, a woman needs extra energy—almost 650 kcalories a day above her regular need during the first six months of lactation. To meet this energy need, the woman is advised to eat an extra 500 kcalories of foods each day and let the fat reserves she accumulated during pregnancy provide the rest. Some research suggests that many women need less energy for milk production than is currently recommended; other research findings confirm current recommendations.[31] Severe energy restriction, however, hinders milk production. Most women need at least 1800 kcalories a day to receive all the nutrients required for successful lactation.[32]

Energy RDA during lactation: +500 kcal/day.

Carbohydrate. Maternal diet has no effect on the carbohydrate content of breast milk. The carbohydrate is always lactose, and the concentration is always about 70 grams per liter.

Protein. Maternal nutrition may have no effect on breast milk's protein, either. In general, the protein concentration of breast milk in malnourished women is similar to that of well-nourished women. The protein RDA during lactation reflects both the nonlactating RDA and estimates from data on milk composition and milk volume. It adds 15 grams per day during the first six months of lactation and 12 grams thereafter, well within most women's intakes.

Protein RDA during lactation:
*   *65 g/day (1st 6 months).*
*   *62 g/day (2nd 6 months).*
Canadian RNI during lactation:
*   *+20 g/day.*

Lipids. Maternal dietary intake alters the fatty acid composition of breast milk, but not the total fat concentration or milk volume. Consequently, the omega-3 fatty acid docosahexaenoic acid (DHA) concentrations of breast milk increase with maternal consumption of oils rich in DHA. This, in turn, raises the infant's consumption of DHA, one of the most abundant structural lipids in the brain. Breastfed infants score slightly better than formula-fed infants on visual and developmental tests, and researchers speculate that higher brain concentrations of DHA may be responsible.[33] Cholesterol concentrations in breast milk are unaffected by maternal fat and cholesterol consumption.

Vitamins and Minerals

A question often raised is whether a mother's milk may lack a nutrient if she fails to get enough in her diet. The answer differs from one nutrient to the next, but in general, nutritional inadequacies reduce the *quantity*, not the *quality*, of breast milk. Women can produce milk with adequate carbohydrate, protein, fat, and most minerals, even when their own supplies are limited.[34] For these nutrients and for the vitamin folate as well, milk quality is maintained at the expense of maternal stores. Nutrients in breast milk are most likely to decline in response to prolonged inadequate intakes of the vitamins—especially vitamins B_6, B_{12}, A, and D.[35] Figure 4-5 repeats Figure 2-6 from Chapter 2, adding the nutrient needs of lactating women.

Figure 4-5 Comparison of Energy and Nutrient Recommendations for Nonpregnant, Pregnant, and Lactating Women

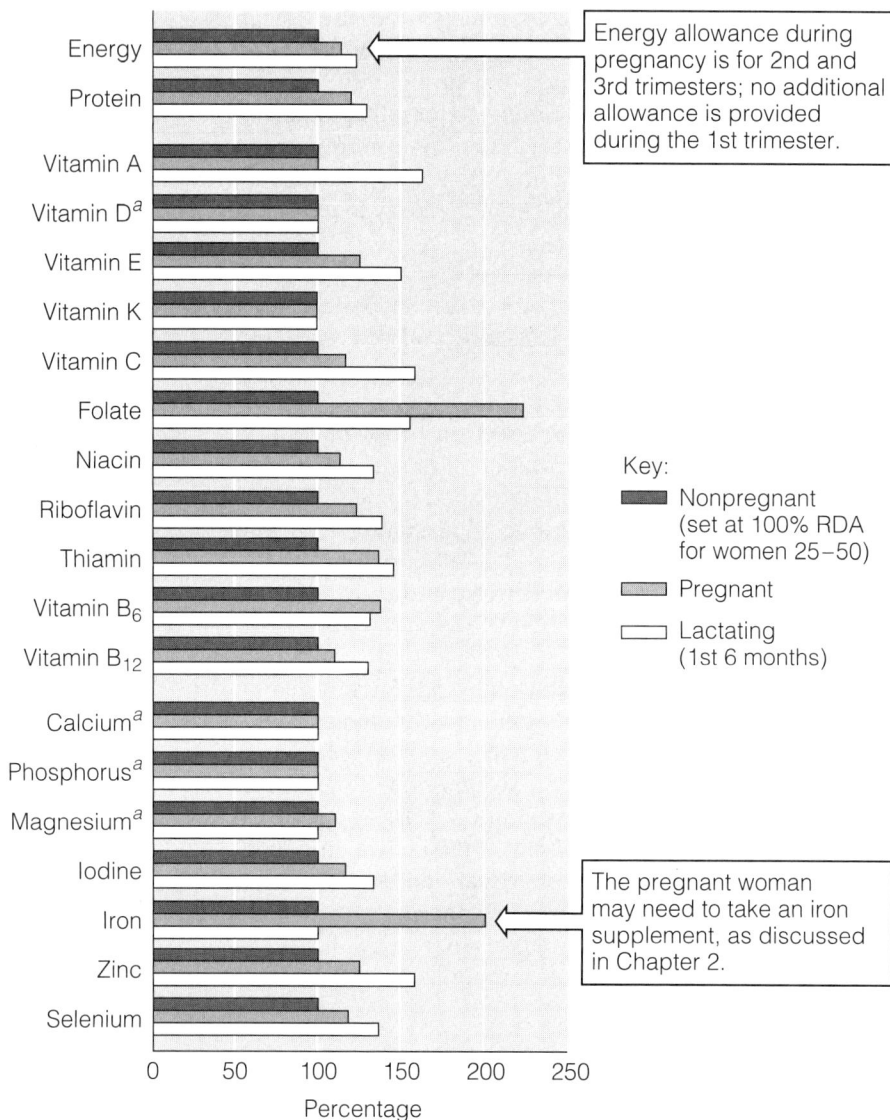

Energy allowance during pregnancy is for 2nd and 3rd trimesters; no additional allowance is provided during the 1st trimester.

Key:
- Nonpregnant (set at 100% RDA for women 25–50)
- Pregnant
- Lactating (1st 6 months)

The pregnant woman may need to take an iron supplement, as discussed in Chapter 2.

[a]Values reflect 1997 DRI.

Fat-Soluble Vitamins. Breast milk composition may change with maternal dietary excesses and deficiencies of the fat-soluble vitamins, depending on the vitamin. The vitamin A concentration in breast milk reflects the mother's intakes and stores. Therefore, vitamin A deficiency is rare in infants breastfed by well-nourished mothers. To maintain maternal vitamin A stores and sufficient concentrations in breast milk, women need an extra 400 to 500 micrograms RE daily.

Vitamin D in breast milk also correlates directly with maternal vitamin D status. Still, its concentration is minimal even with adequate maternal dietary intake. For this reason, recommendations do not change during lactation.

Vitamin E in breast milk also varies directly with maternal intake and stores. The RDA increases slightly during lactation to account for the vitamin E in breast milk.

Breast milk vitamin K concentrations are insufficient to meet the infant's needs, even when the maternal diet is adequate.[36] Consequently the RDA does not change during lactation.

Water-Soluble Vitamins. Like the fat-soluble vitamins, water-soluble vitamins in breast milk reflect maternal intake to varying extents. Marginal deficiencies and daily fluctuations have little, if any, influence on breast milk composition, but severely vitamin-deficient mothers produce vitamin-deficient breast milk. For example, lactating women who consume a strict vegetarian diet produce vitamin B_{12}-deficient milk. The RDA for most of the vitamins are greater during lactation than during pregnancy; the RDA for vitamin D and vitamin B_{12} remain at their pregnancy level during lactation; the RDA for vitamin B_6 and folate are lower during lactation than during pregnancy.

Minerals. In general, maternal dietary intake of minerals does not influence their concentrations in milk. This is most apparent in the calcium, phosphorus, and magnesium contents of breast milk, which remain fairly constant, regardless of maternal dietary intake. Diet is important, however, because a low calcium intake promotes mobilization of calcium from maternal bone stores. A lactating woman whose daily diet lacks calcium-rich foods may find herself later in life with weakened bones.

Breast milk iron concentration also remains fairly constant whether the mother takes an iron supplement or suffers iron deficiency anemia. The mother's body is designed to deliver iron to the infant, no matter what the cost to her health.

Vitamin A RDA during lactation:
  1300 μg RE /day (1st 6 months).
  1200 μg RE /day (2nd 6 months).
Canadian RNI during lactation:
  +400 μg RE /day.

Vitamin D DRI during lactation:
  5 μg /day.

Vitamin E RDA during lactation:
  12 mg α-TE /day (1st 6 months).
  11 mg α-TE /day (2nd 6 months).
Canadian RNI during lactation:
  +3 mg α-TE /day.

✦ Healthy People 2000

Increase calcium intake so at least 50% of lactating women consume three or more servings daily of foods rich in calcium.

Nutrient Supplements for Lactating Women

Most lactating women can obtain all the nutrients they need from a well-balanced diet without taking vitamin and mineral supplements; some, however, may need iron supplements. Maternal iron stores dwindle when the fetus takes iron to meet its own needs for the four to six months after birth. In addition, childbirth may have incurred blood losses. A woman may therefore need iron supplements during lactation, not to augment the iron in her breast milk, but to replenish her depleted iron stores. When iron stores become depleted, the symptoms of iron deficiency anemia (such as weakness, fatigue, and headaches) become evident. A new mother trying to care for her infant cannot do so optimally if she is tired and weak. Her compromised emotional and physical health will eventually interfere with successful lactation. Women who restrict their dietary intakes of certain food groups, such as vegetarians, may also need supplements, as the section below explains.

Food Choices and Fitness during Lactation

Ideally, the mother who chooses to breastfeed her infant will continue to eat nutrient-dense foods throughout lactation. An adequate diet is needed to support the stamina, patience, and self-confidence that nursing an infant demands. Table 2-4 on p. 64 offers a daily food guide to meeting nutrient needs during pregnancy that can also be used during lactation.

Diets Tailored to Individual Preferences

To receive all the nutrients necessary to support lactation, a woman needs to select several servings from each of the five food groups. Women who restrict their intakes may fall short in meeting their nutrient needs.

Vegetarian Diets during Lactation. In general, well-planned vegetarian diets that are adequate in food energy and include milk products and eggs can support successful lactation.[37] Vegetarian diets that exclude all foods of animal origin may require supplementation with vitamin B_{12}, calcium, and vitamin D or the addition of foods fortified with these nutrients.[38]

A strict vegetarian needs a regular source of vitamin B_{12}-fortified foods or a supplement that provides 2.6 micrograms daily. Appropriate food choices for those excluding milk products include calcium- and vitamin D-fortified soy milk and tofu, legumes, nuts, seeds, and vegetables (especially dark green, leafy vegetables). A woman who cannot meet her calcium needs through diet alone may need 600 milligrams of supplemental calcium daily, taken with meals. Women who do not receive sufficient dietary vitamin D or enough exposure to sunlight may need a supplement that provides 10 micrograms daily.

Weight Loss during Lactation. After the birth of the infant, many women are in a hurry to lose the extra body fat they accumulated during pregnancy. Opinions differ about whether breastfeeding helps with this postpartum weight loss. One study reports that the amount of weight lost does not depend on whether a woman breastfeeds her infant.[39] Another study suggests that breastfeeding enhances weight loss initially, especially fat loss from the lower body, but not thereafter.[40] Still another study indicates that weight loss is significant only if breastfeeding continues for at least six months.[41] In general, most women lose 1 to 2 pounds a month during the first four to six months of lactation; some may lose more, and others may maintain or even gain weight.[42] Regardless of prepregnancy weight, the more weight a woman gains during pregnancy, the more weight she loses following delivery (when measured at six weeks and one year).[43] Neither the quality nor the quantity of breast milk is adversely affected by moderate weight loss.[44]

Breastfeeding can actually aid in restoring a woman's figure to its nonpregnant state. As mentioned earlier, the infant's suckling stimulates the nerves controlling the muscles of the uterus, causing them to contract. The contractions expel any tissue remaining after the birth and return the uterus to its normal size. Breastfeeding also facilitates the mobilization of fat stored during pregnancy; this breakdown of fat is especially apparent from the woman's thighs. The energy costs of milk production can also help a lactating mother lose weight. A woman who chooses nutrient-dense foods during lactation will experience a gradual weight loss, even though her energy intake is greater than normal. Breastfeeding does not, however, promote rapid, easy, weight loss and should not be expected to do so. Disappointment arising from such false expectations may lead some women to stop nursing their infants earlier than they had originally intended.

Food Assistance Programs. Low-income women may be eligible for the Special Supplemental Nutrition Program for Women, Infants, and Children (WIC). Ideally, these women receive the educational information and nutritious foods to support breastfeeding, but this may not always be the case. Although WIC promotes breastfeeding, it also provides infant formula. Some research shows a negative correlation between WIC participation and breastfeeding.[45] To encourage more women to breastfeed, and to breastfeed longer, WIC counselors emphasize prenatal counseling and breastfeeding advice.

Fitness during Lactation

Women often exercise to lose weight and improve fitness, and this is compatible with breastfeeding.[46] Studies have found that physically active lactating women compensate for their high energy expenditures by increasing their energy intakes.[47] Intense physical activity can raise the lactic acid concentration of breast milk, which influences the milk's taste. Infants appear to prefer milk produced prior to exercise (which

A brisk walk through the neighborhood provides an opportunity for physical activity and fresh air.

has a lower lactic acid content).[48] For this reason, mothers may want to breastfeed their infants before exercise or express their milk before exercise for use afterward.

A healthy diet and regular physical activity go a long way to support successful lactation. In contrast, some substances impair milk production or enter breast milk and interfere with infant development. The next section describes the practices that are incompatible with breastfeeding, and then the following section closes the chapter with a few words on the proper care of a lactating mother.

Practices Incompatible with Breastfeeding

A variety of lifestyle factors can have adverse effects on breastfeeding. Women who are planning to breastfeed can make the choice to practice healthy behaviors.

Smoking. Cigarette smoking reduces milk volume, so smokers may produce too little milk to meet their infants' energy needs. One study of lactating women found that infants of smoking mothers gained less weight than infants of nonsmoking mothers.[49] Furthermore, infant exposure to passive smoke negates the protective effect breastfeeding offers against sudden infant death syndrome (SIDS) and increases the risks dramatically.[50]

Alcohol. Alcohol easily enters breast milk. One study showed that the alcohol concentration of breast milk peaks within one hour after ingestion.[51] In this study, even small amounts of alcohol (equivalent to a can of beer) consumed by lactating women significantly reduced their infants' intakes of breast milk. The researchers suggest three possible reasons, acting separately or together. For one, the alcohol may have altered the flavor of the breast milk and thereby the infants' acceptance of it. For another, because infants metabolize alcohol inefficiently, even low doses may be potent enough to suppress infant feeding behavior. Third, the alcohol may have reduced milk production; alcohol is known to inhibit the release of the hormone oxytocin.

In the past, alcohol has been recommended to mothers to facilitate lactation despite a lack of scientific evidence that it does so. The research summarized here suggests that alcohol actually hinders breastfeeding. An occasional glass of wine or beer is considered within safe limits, but in general, lactating women should consume little or no alcohol.

Medicinal Drugs. Many drugs are compatible with breastfeeding, but some medicines are contraindicated, either because they suppress lactation or because they are secreted into breast milk and can harm the infant.[52] As a precaution, a nursing mother should consult with her physician prior to taking any drug. The physician may be able to prescribe a low dose or a short-acting form of the drug.

Table 4-3 presents a list of medical drugs that are contraindicated during breastfeeding and shows their effects on infants and lactation. In addition, drugs containing radioactivity (such as iodine-131) require cessation of breastfeeding for a period of time that depends on the specific drug. (To maintain lactation during times of cessation, the mother should express her milk and discard it.) If a woman must take a medication that is regarded as generally safe, she can minimize any effects by taking the drug while or immediately after the infant nurses. This will produce the lowest concentration of the drug in the breast milk at the next feeding.

Illicit Drugs. Illicit drugs such as marijuana and cocaine are harmful to the physical and emotional health of both the mother and the nursing infant. Breast milk can deliver such high doses of illicit drugs as to cause irritability, tremors, and hallucinations in infants. Table 4-4 lists the effects of drugs of abuse on infants and lactation.

Environmental Contaminants. Environmental contaminants, such as DDT, PCBs, and methylmercury can find their way into breast milk.[53] Inuit mothers living in Arctic Québec who eat seal and beluga whale blubber have concentrations of DDT and PCBs in their breast milk two to ten times greater than those found in breast milk from women in southern Québec.[54] The impact of contaminated breast milk on infant development is unclear, however. Preliminary studies indicate that the children of these Inuit mothers are developing normally. Researchers speculate that the abundant omega-3 fatty acids of the Inuit diet may protect against any damage to the central nervous system.

Table 4-3 Medical Drugs That Are Contraindicated during Breastfeeding

Drug	Registered Name	Indications	Reported Sign or Symptom in Infant or Effect on Lactation
Bromocriptine	Parlodel	Amenorrhea, female infertility, suppression of lactation, Parkinson's disease	Suppression of lactation
Cyclophosphamide[a]	Cytoxan	Cancer	Possible immune suppression; abnormally small number of white blood cells; unknown effect on growth or association with carcinogenesis
Cyclosporine	Sandimmune	Organ transplants	Possible immune suppression; unknown effect on growth or association with carcinogenesis
Doxorubicin[b]	Adriamycin	Cancer	Possible immune suppression; unknown effect on growth or association with carcinogenesis
Ergotamine	Cafergot	Migraine headaches	Vomiting, diarrhea, convulsions (doses used in migraine medications)
Lithium	Eskalith	Manic episodes	One-third to one-half therapeutic blood concentration in infants
Methotrexate	Methotrexate	Cancer, psoriasis, rheumatoid arthritis	Possible immune suppression; abnormally small number of white blood cells; unknown effect on growth or association with carcinogenesis

[a]Data not available for other cytotoxic agents.

[b]Drug is concentrated in breast milk.

Source: Committee on Drugs, American Academy of Pediatrics, The transfer of drugs and other chemicals into human milk, *Pediatrics* 93 (1994): 137–150.

Caffeine. Caffeine use during lactation may make a breastfed infant irritable and wakeful. As during pregnancy, caffeine consumption should be moderate—say, the equivalent of one to two cups of coffee a day.

Care of the Lactating Mother

Life is hectic, to say the least, for any new mother, whether she is breastfeeding or formula feeding. The breastfed infant demands frequent feedings in the early weeks, stimulating milk production. The nursing mother may at first feel as if she is doing little else. Finding time to prepare meals for herself and other family members is often difficult in the beginning.

Table 4-4 Drugs of Abuse That Are Contraindicated during Breastfeeding

Drug	*Reported Sign or Symptom in Infant or Effect on Lactation*
Amphetamine[a]	Irritability, poor sleeping pattern
Cocaine	Cocaine intoxication
Heroin	Tremors, restlessness, vomiting, poor feeding
Nicotine (smoking)	Shock, vomiting, diarrhea, rapid heart rate, restlessness; decreased milk production
Phencyclidine (PCP)	Potent hallucinogen

[a]Drug is concentrated in breast milk.

Source: Committee on Drugs, American Academy of Pediatrics, The transfer of drugs and other chemicals into human milk, *Pediatrics* 93 (1994): 137–150.

Nutrition plays a significant role in successful lactation, affecting both the physical and mental well-being of the mother. Thus, it is important that she eat well. To do so, she needs help and support. She cannot expect, nor should she be expected, to prepare family meals, breast-feed her infant, attend to other responsibilities, and get the rest she so urgently needs, without help from family members and friends. Her main priorities at first are to feed and care for herself and her newborn infant.

Family members can help by shopping and preparing food. Meals may be simple for a while, but they can still be nutritious. Casseroles, salads, and soups are easy to prepare. They also provide many nutrients, especially when foods from most of the food groups are used in the recipe.

Rest is vital to successful lactation and an important feature of the road to recovery. Giving birth to an infant is an exhausting and stressful experience. Friends and family members can help by caring for the infant (and other children) for a few hours while mother rests or spends some time away from home. Mothers can benefit by napping when the infant naps.

Once postpartum bleeding has stopped and the mother is feeling stronger, physical activity and fresh air will improve her strength, well-being, and self-esteem. Exercise classes that provide free babysitting are offered throughout the country today. If the mother can afford this luxury, these classes provide an excellent opportunity to do something just for herself, with the added bonus of a break from the infant and the social stimulation of adult company.

Nursing women are frequently the recipients of unsolicited, "friendly" advice on how to care for themselves and their newborns, including what they should or should not eat during lactation. Aside from creating unnecessary confusion for the mothers, this advice is often incorrect. Each mother and infant couple is unique, and what is true for one nursing mother may not be for another. For example, one mother may find that when she eats garlic, her infant becomes irritable, whereas another mother has no such experience. Infant reactions to substances in mother's milk are matters that require individual detective work.

In conclusion, successful lactation requires adequate nutrition and rest. This, plus the support of all those who care, will help to enhance the well-being of mother and infant.

The initial care of a healthy newborn involves such simple tasks as providing clean diapers, warm clothing, a place to sleep, nourishment, and love. With each interaction, the parent and infant relationship develops. During each of several feedings every day, the parent has the opportunity to nurture the infant physically, emotionally, and mentally. With each day, their relationship continues to grow and change, just as the infant does. Chapter 5 presents the nutrition needs of infancy.

CHAPTER FOUR NOTES

1. Life Sciences Research Office, Federation of American Societies for Experimental Biology, Executive summary from the third report on nutrition monitoring in the United States, *Journal of Nutrition* 126 (1996): 1907S–1936S; Committee on Nutritional Status during Pregnancy and Lactation, Food and Nutrition Board, *Nutrition during Lactation* (Washington, D.C.: National Academy Press, 1991) pp. 28–49; National Center for Health Statistics, *Health: United States, 1992* (Hyattsville, Md.: U.S. Public Health Service, 1993).

2. A. Wright, S. Rice, and S. Wells, Changing hospital practices to increase the duration of breastfeeding, *Pediatrics* 97 (1996): 669–675.

3. J. P. Sciacca and coauthors, Influences on breast-feeding by lower-income women: An incentive-based, partner-supported educational program, *Journal of the American Dietetic Association* 95 (1995): 323–328; G. L. Freed, J. K. Fraley, and R. J. Schanler, Attitudes of expectant fathers regarding breast-feeding, *Pediatrics* 90 (1992): 224–227.

4. H. Varendi, R. H. Porter, and J. Winberg, Does the newborn baby find the nipple by smell? *Lancet* 344 (1994): 989–990.

5. R. Cohen, M. B. Mrtek, and R. G. Mrtek, Comparison of maternal absenteeism and infant illness rates among breast-feeding and formula-feeding women in two corporations, *American Journal of Health Promotion* 10 (1995): 148–153.

6. Position of The American Dietetic Association: Promotion of breast-feeding, *Journal of the American Dietetic Association* 97 (1997): 662–666.

7. M. J. Heinig and K. G. Dewey, Health advantages of breast feeding for infants: A critical review, *Nutrition Research Reviews* 9 (1996): 89–110.

8. A. C. Goedhart and J. G. Bindels, The composition of human milk as a model for the design of infant formulas: Recent findings and possible applications, *Nutrition Research Reviews* 7 (1994): 1–23.

9. M. T. Ruel and coauthors, Validation of single daytime samples of human milk to estimate the 24-h concentration of lipids in urban Guatemalan mothers, *American Journal of Clinical Nutrition* 65 (1997): 439–444.

10. V. P. Carnielli and coauthors, Effect of dietary triacylglycerol fatty acid positional distribution on plasma lipid classes and their fatty acid composition in preterm infants, *American Journal of Clinical Nutrition* 62 (1995): 776–781.

11. J. N. Walterspiel and coauthors, Secretory anti-*Giardia lamblia* antibodies in human milk: Protective effect against diarrhea, *Pediatrics* 93 (1994): 28–31.

12. B. Lönnerdal and S. Iyer, Lactoferrin: Molecular structure and biological function, *Annual Review of Nutrition* 15 (1995): 93–110; J. C. Fleet, A new role for lactoferrin: DNA binding and transcription activation, *Nutrition Reviews* 53 (1995): 226–231.

13. J. S. Forsyth, Is it worthwhile breastfeeding? *European Journal of Clinical Nutrition* 46 (1992): 19S–25S.

14. U. M. Saarinen and M. Kajosaari, Breastfeeding as prophylaxis against atopic disease: Prospective follow-up study until 17 years old, *Lancet* 346 (1995): 1065–1069; A. L. Wright and coauthors, Relationship of infant feeding to recurrent wheezing at age 6 years, *Archives of Pediatrics & Adolescent Medicine* 149 (1995): 758–763.

15. J. A. Mennella and G. K. Beauchamp, Early flavor experiences: When do they start? *Nutrition Today*, September–October 1994, pp. 25–31.

16. J. A. Mennella and G. K. Beauchamp, Maternal diet alters the sensory qualities of human milk and the nursling's behavior, *Pediatrics* 88 (1991): 737–744.

17. Committee on Drugs, American Academy of Pediatrics, The transfer of drugs and other chemicals into human milk, *Pediatrics* 93 (1994): 137–150.

18. K. G. Dewey, Does maternal supplementation shorten the duration of lactational amenorrhea? *American Journal of Clinical Nutrition* 64 (1996): 377–378.

19. T. Waterston, Any questions, *British Medical Journal* 310 (1995): 780.

20. L. A. Guay and coauthors, Detection of human immunodeficiency virus type 1 (HIV-1) DNA and p24 antigen in breast milk of HIV-1-infected Ugandan women and vertical transmission, *Pediatrics* 98 (1996): 438–444.

21. R. F. Black, Transmission of HIV-1 in the breast-feeding process, *Journal of the American Dietetic Association* 96 (1996): 267–274.

22. A. M. Ferris and E. A. Reece, Nutritional consequences of chronic maternal conditions during pregnancy and lactation: Lupus and diabetes, *American Journal of Clinical Nutrition* 59 (1994): 465S–473S; A. M. Ferris and coauthors, Perinatal lactation protocol and outcome in mothers with and without insulin-dependent diabetes mellitus, *American Journal of Clinical Nutrition* 58 (1993): 43–48.

23. C. M. van Beusekom and coauthors, Milk of patients with tightly controlled insulin-dependent diabetes mellitus has normal macronutrient and fatty acid composition, *American Journal of Clinical Nutrition* 57 (1993): 938–943.

24. United Kingdom National Case-Control Study Group, Breast feeding and risk of breast cancer in young women, *British Medical Journal* 307 (1993): 17–20.

25. K. B. Michels and coauthors, Prospective assessment of breastfeeding and breast cancer incidence among 89,887 women, *Lancet* 347 (1996): 431–436; United Kingdom National Case-Control Study Group, 1993.

26. P. A. Newcomb and coauthors, Lactation and a reduced risk of premenopausal breast cancer, *New England Journal of Medicine* 330 (1994): 81–87.

27. J. L. Levine and N. T. Ilowite, Sclerodermalike esophageal disease in children breast-fed by mothers with silicone breast implants, *Journal of the American Medical Association* 271 (1994): 213–216.

28. C. M. Berlin, Silicone breast implants and breast-feeding, *Pediatrics* 94 (1994): 547–549; J. A. Flick, Silicone implants and esophageal dysmotility—Are breast-fed infants at risk? *Journal of the American Medical Association* 271 (1994): 240–241.

29. L. B. Dusdieker and coauthors, Prolonged maternal fluid supplementation in breast-feeding, *Pediatrics* 86 (1990): 737–740.

30. K. G. Dewey and coauthors, Maternal versus infant factors related to breast milk intake and residual milk volume: The DARLING Study, *Pediatrics* 87 (1991): 829–837; Committee on Nutritional Status during Pregnancy and Lactation, 1991, pp. 1–19.

31. M. A. Guillermo-Tuazon and coauthors, Energy intake, energy expenditure, and body composition of poor rural Philippine women throughout the first 6 mo of lactation, *American Journal of Clinical Nutrition* 56 (1992): 874–880; C. Frigerio and coauthors, A new procedure to assess the energy requirements of lactation in Gambian women, *American Journal of Clinical Nutrition* 54 (1991): 526–533; J. M. A. van Raaij and coauthors, Energy cost of lactation, and energy balances of well-nourished Dutch lactating women: Reappraisal of the extra energy requirements of lactation, *American Journal of Clinical Nutrition* 53 (1991): 612–619.

32. Committee on Nutritional Status during Pregnancy and Lactation, 1991, p. 232.

33. Heinig and Dewey, 1996; M. Makrides and coauthors, Fatty acid composition of brain, retina, and erythrocytes in breast- and

formula-fed infants, *American Journal of Clinical Nutrition* 60 (1994): 189–194.

34. Committee on Nutritional Status during Pregnancy and Lactation, 1991, p. 140.

35. Committee on Nutritional Status during Pregnancy and Lactation, 1991, p. 140.

36. Committee on Nutritional Status during Pregnancy and Lactation, 1991, p. 126.

37. Position of the American Dietetic Association: Vegetarian diets, *Journal of the American Dietetic Association* 93 (1993): 1317–1319.

38. B. L. Specker, Nutritional concerns of lactating women consuming vegetarian diets, *American Journal of Clinical Nutrition* 59 (1994): 1182S–1186S.

39. S. Potter and coauthors, Does infant feeding method influence maternal postpartum weight loss? *Journal of the American Dietetic Association* 91 (1991): 441–446.

40. F. M. Kramer and coauthors, Breast-feeding reduces maternal lower-body fat, *Journal of the American Dietetic Association* 93 (1993): 429–433.

41. K. G. Dewey, M. J. Heinig, and L. A. Nommsen, Maternal weight-loss patterns during prolonged lactation, *American Journal of Clinical Nutrition* 58 (1993): 162–166.

42. Committee on Nutritional Status during Pregnancy and Lactation, 1991, pp. 1–19.

43. Potter and coauthors, 1991.

44. L. B. Dusdieker, D. L. Hemingway, and P. J. Stumbo, Is milk production impaired by dieting during lactation? *American Journal of Clinical Nutrition* 59 (1994): 833–840.

45. J. B. Schwartz and coauthors, Does WIC participation improve breast-feeding practices? *American Journal of Public Health* 85 (1995): 729–731.

46. K. G. Dewey and M. A. McCrory, Effects of dieting and physical activity on pregnancy and lactation, *American Journal of Clinical Nutrition* 59 (1994): 446S–453S; K. G. Dewey and coauthors, A randomized study of the effects of aerobic exercise by lactating women on breast-milk volume and composition, *New England Journal of Medicine* 330 (1994): 449–453.

47. Dewey and coauthors, 1994; C. A. Lovelady, B. Lönnerdal, and K. G. Dewey, Lactation performance of exercising women, *American Journal of Clinical Nutrition* 52 (1990): 103–109.

48. J. P. Wallace, G. Inbar, and K. Ernsthausen, Infant acceptance of postexercise breast milk, *Pediatrics* 89 (1992): 1245–1247.

49. F. Vio, G. Salazar, and C. Infante, Smoking during pregnancy and lactation and its effects on breast-milk volume, *American Journal of Clinical Nutrition* 54 (1991): 1011–1016.

50. H. S. Klonoff-Cohen, The effect of passive smoking and tobacco exposure through breast milk on sudden infant death syndrome, *Journal of the American Medical Association* 273 (1995): 795–798.

51. J. A. Mennella and G. K. Beauchamp, The transfer of alcohol to human milk: Effects on flavor and the infant's behavior, *New England Journal of Medicine* 325 (1991): 981–985.

52. Committee on Drugs, American Academy of Pediatrics, 1994.

53. Committee on Environmental Health, American Academy of Pediatrics, PCBs in breast milk, *Pediatrics* 94 (1994): 122–123.

54. E. Dewailly and coauthors, Inuit exposure to organochlorines through the aquatic food chain in Arctic Québec, *Environmental Health Perspectives* 101 (1993): 618–620.

How to Feed Infants

As Chapter 4 explained, the feeding options for infants during their first four to six months are limited to breast milk, infant formula, or both. This Focal Point provides a "how to" lesson on each of these methods of infant feeding.

How to Breastfeed

Most healthy women who want to breastfeed can do so with little preparation. The mother-to-be may find it reassuring to learn that 95 percent of all women who try are successful; physical obstacles are rare. The size and shape of a woman's breasts do not affect her ability to breastfeed an infant; neither does it matter what her age is or whether she is thin, normal weight, or obese.

Newborn infants readily adapt to breastfeeding. In fact, fetuses evince sucking behaviors before birth. Newborns are prepared to suckle immediately after birth and demonstrate a rooting reflex that orients them toward a nipple. Nursing the infant immediately after birth facilitates successful lactation.

If the mother is unable to nurse immediately (for example, if she is sedated), the newborn should not be fed. Nutrients are less critical at this time than the infant's first impression. If given formula first, the newborn may develop an attachment to the bottle and then resist the breast later.

Positioning the Infant. Beginning at the first feeding, the mother needs to learn how to relax and position herself so that she and the infant will be comfortable. The position must also allow the infant to nurse without obstructing breathing. The mother squeezes the areola, the colored halo around the nipple, between two fingers, slipping enough of it into the infant's mouth to promote good pumping action (see Figure FP4-1). If the infant is to successfully milk the mammary glands, the nipple must rest well back on the infant's tongue. The infant's lips and gums pump the areola, thus releasing milk from the mammary glands into the ducts that lie beneath the areola. The sucking and swallowing reflexes work together. The infant's tongue and jaw suck milk from the breast, and the swallow

Figure FP4-1 Infant's Grasp on Mother's Breast

The mother squeezes the areola, slipping enough of it into the infant's mouth to promote good pumping action. The infant's lips and gums pump the areola, releasing milk from the mammary glands into the milk ducts that lie beneath the areola.

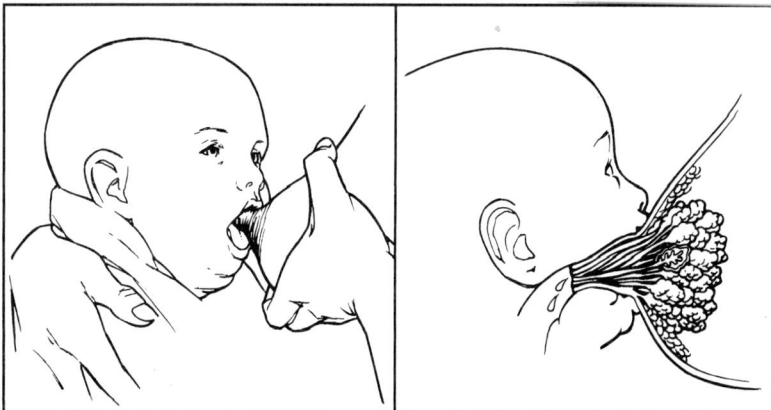

follows. To break the suction, the mother can slip a finger between the infant's mouth and her breast.

The Let-Down Reflex. The let-down reflex forces milk to the front of the breast when the infant begins to nurse, allowing the milk to flow. Let-down has to occur for the infant to obtain milk easily, and the mother needs to relax for let-down to occur. This means that at a time when the stress response might be more natural, she must will the relaxation response. Willed relaxation first requires that a person assume a comfortable position in a calm environment and then maintain a passive attitude toward interruptions. It may take several feeding sessions for the mother to learn how to respond to cries of hunger before she can relax enough to achieve let-down promptly and fully.

Duration of a Feeding Session. The infant sucks half the milk from the breast within the first two minutes, and 80 to 90 percent of it within four minutes. Sucking on one breast, however, is encouraged for 8 to 12 minutes. The sucking itself,

as well as the complete removal of milk from the breast, stimulates lactation. After ten minutes or so, the mother offers the other breast to finish satisfying the infant's hunger. Nursing sessions start on alternate breasts to ensure that each breast is emptied regularly. This pattern maintains the same supply and demand for each breast and thus prevents either breast from overfilling. When the infant finishes nursing on the first breast, the mother allows the infant to expel any swallowed air and then offers another chance to nurse.

Frequency of Feeding Sessions. Approximately six to eight feedings a day, when the infant cries with hunger, promote optimal milk production and infant growth. The mother encourages the infant to nurse the full 15 to 20 minutes per feeding. Some infants fall asleep in less time and may need to be aroused to continue feeding. Feeding intervals vary with each infant, but by

Infants need a few moments during their feedings to relax and release swallowed air.

A mother breastfeeds her infant while sitting in a relaxed, comfortable position.

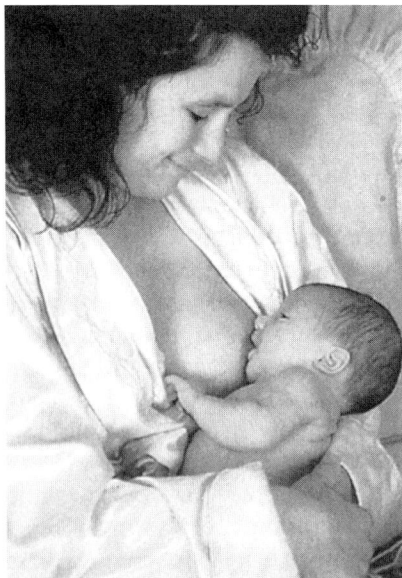

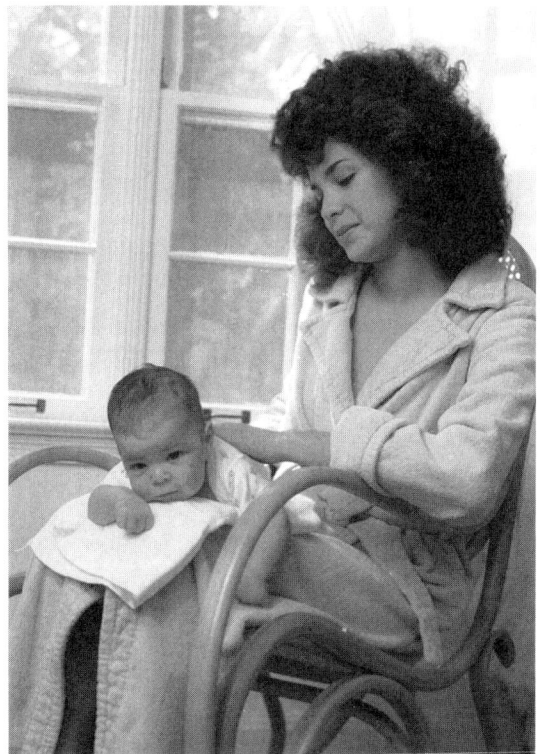

six weeks should be at least two hours apart. If they are less than two hours apart, the mother may begin to feel like a human pacifier, and her milk supply, at first unable to meet the demand, may come to exceed it. If a feeding interval exceeds four hours, the mother may need to express some milk to relieve pressure and maintain the demand. Figure FP4-2 illustrates methods of expressing milk. Until lactation is well established, the infant should be encouraged to feed

Figure FP4-2 Methods of Expressing Milk

Before expressing breast milk, gently massage the breasts to encourage the let-down reflex.

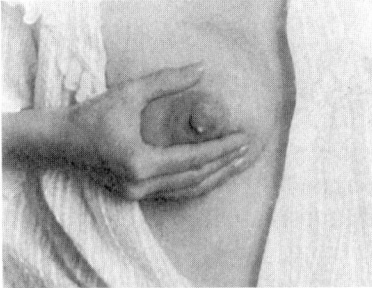

Hand Expression

Expressing breast milk by hand requires no equipment, but it does take practice to perfect the technique. With the thumb above the nipple and the fingers below, gently push inward toward the chest wall and then squeeze the thumb and fingers together toward the nipple. Continue this procedure while rotating the position of the hand around the breast.

Manual Pumps

Expressing breast milk by manual pump is simple and economical. The pump consists of two plastic cylinders (one fits inside the other) and uses a piston-type suction action. A funnel that fits over the nipple catches the milk into a bottle.

Electric Pumps

Expressing breast milk by electric pump is easy and effective, but expensive. Hospitals and medical supply companies often rent electric pumps. Many electric pumps express milk from both breasts simultaneously.

Small electric or battery-powered pumps are also available and cost less than the larger models.

regularly and not be allowed to sleep through a feeding.

Breastfeeding Twins What if a mother is breast-feeding twins? As far as milk supply is concerned, a woman need have no problem breast-feeding twins. The more milk the infants drink, the more milk the breasts produce. Most mothers of twins agree that the problem lies in finding time to feed two—whether breast milk or formula. Each infant has a unique "hunger clock," and every mother has to work out her own system. Breastfeeding twins is no more difficult than any other task involving twins.

Breastfeeding Adopted Infants. What if a mother wants to breastfeed an adopted infant? With plenty of support and determination, she has a reasonable chance of success.[1] In preparation, she will need to stimulate the nipples with a breast pump and put the infant to the breast frequently (every two hours initially) to stimulate lactation. Until lactation is fully established, she can attach a tube to her chest that delivers

Mothers of twins who want to breastfeed can usually find a way to do so.

formula when the infant suckles at the breast. This procedure helps the infant grow accustomed to the breast and signals the breasts to produce milk. As the mother's supply increases, the supplemental formula is reduced.

Substituting Infant Formula on Occasion. A mother might want to know if she can skip an occasional feeding, substituting a bottle of formula. To avoid suppressing lactation, the mother will need to express her milk; that way, the breast receives the message to continue producing milk. This may not be necessary once lactation is well established.

A mother who wants to skip one or two feedings daily—for example, if she works outside the home—can substitute formula for those feedings and continue to breastfeed at other feedings. Or the infant can be fed, in a bottle, the breast milk she has expressed and frozen on previous occasions. Breast milk can be kept refrigerated for up to 48 hours or frozen for up to six weeks. Safe storage of breast milk maintains the protein structure of the milk and slows bacterial growth.[2]

If the mother feels that she is spending immense amounts of time breastfeeding, she should remind herself that the mother who is feeding formula is also spending time washing bottles, preparing formula, and feeding her infant. Breastfeeding is less time consuming after the first few weeks. The mother is encouraged to remember the advantages and to enjoy this time with her infant while it lasts.

Troubleshooting. Most problems associated with breastfeeding can be resolved. Many new mothers experience sore nipples during the initial days of breastfeeding. Sore nipples need to be treated kindly, but nursing can continue. Air and sunlight between feedings help to heal them. Before let-down, the infant must suck hard on the nipple to receive milk. For this reason, a mother will want to nurse on the less-sore breast first. Then, when the milk lets down and is freely flowing, she can switch to the sore nipple. When the fast-flowing, early milk from that sore breast is gone, she can switch back and satisfy the infant's hunger and sucking need.

Engorgement. Engorgement is common before lactation is established, when the schedule is changed as in weaning, or when a feeding is missed. The breasts become so full and hard that the infant cannot grasp the nipple and the mother is very uncomfortable. A gentle massage or warming the breasts (with a heating pad or in a shower) helps initiate let-down and release some of the accumulated milk. The best solutions are to pump out some milk, to use a nipple shield that will help the infant grasp the nipple, and to allow the infant to nurse. The breasts will, in time, get smaller and softer, while still producing ample milk.

Inverted Nipples. A mother can also manage to nurse with an inverted nipple. An inverted nipple folds inward toward the breast when the areola is pressed between two fingers. Pushing the areola toward the chest wall—manually or with a shield—everts the nipple, making it available for infant sucking.

Clogged Ducts. An undrained duct can make a hard, uncomfortable lump in the breast. By massaging the lump while the infant is nursing, the mother can move the milk toward the nipple, where it will join the main supply.

Mastitis. Infection of a breast, known as mastitis, is best managed by *continuing to breastfeed*. By drawing off the milk, the infant helps to relieve pressure in the infected area. The infant is safe because the infection is *between* the milk-producing glands, not inside them. Neither milk quality nor quantity is affected by such an infection.[3]

Gaining Confidence. Most importantly, if the infant is irritable and wakeful, the mother may fear that her milk supply is inadequate. The infant's normal small bowel movements may suggest to her that the infant is underfed, but a health care professional can reassure her that breast milk contains little indigestible material and therefore little waste. The stress of worrying, itself, can inhibit lactation. All infants cry. The mother's ability to relax and set her fears aside will better support lactation than will anxiety about inadequate milk production.

Weaning the Infant. To wean an infant, the mother gradually introduces small amounts of formula or milk and solid foods to the infant while she continues to breastfeed. (The type of food or milk and the age of weaning are discussed fully in Chapter 5.) As the infant consumes more solid foods and formula or milk, breast milk consumption will decrease. Because the breasts need time to adjust their supply to the diminishing demand for milk, the key to a comfortable weaning is to do it *gradually*. The less breast milk an infant drinks, the less the mother produces until finally weaning is complete—that is, the infant is eating solid foods, drinking milk or formula, and not receiving any breast milk.

The mother should allow several weeks for complete weaning. The first step is to replace any one breastfeeding with formula or milk and solid foods. After a few days, a second breastfeeding is replaced in the same way. This process continues until all breastfeedings have been replaced. Such a gradual schedule will allow the breasts to adjust with little discomfort until they are no longer producing milk.

How to Feed Infant Formula

Formulas in the United States and Canada are available in a variety of physical forms. Liquid concentrate formulas are relatively inexpensive and easy to prepare by mixing with equal parts water. Powdered formulas are the least expensive and lightest for travel. Labels provide preparation instructions. Ready-to-feed formulas are the easiest and most expensive; premixed and sterile formula is poured directly into clean bottles or disposable bottle liners.

Preparing Infant Formula. To avoid bacterial contamination, caregivers must apply the rules of safe formula preparation. They must clean all

Parents may want to consult with their health care providers before selecting from the variety of infant formulas available.

To check the temperature of infant formula, sprinkle a few drops on the wrist or back of the hand.

bottles, caps, nipples, and utensils used in preparing formula.* Liquid concentrate and powdered formulas are prepared using cooled, sterile water. The ratio of formula to water must be carefully measured to ensure the correct nutrient density. Opened cans of liquid concentrate and ready-to-feed formulas must be covered, refrigerated, and consumed within 48 hours or thrown away. Powdered formula, once prepared, should also be used within 48 hours.

Warming Infant Formula. Infants will drink cold formula, but most prefer it warm. To warm formula, the caregiver places the bottle in a larger container of hot water for a few minutes. Before offering the bottle of formula to the infant, the caregiver shakes the bottle and sprinkles a few drops of formula on the wrist or back of the hand to check that the temperature is not too hot. Microwave ovens are not recommended for heating formulas; they tend to heat unevenly. The drops may feel warm, but the sip an infant takes may cause a burn. Anyone who does use a microwave oven to heat formula should shake the bottle vigorously to equalize the temperature throughout the formula before testing it.

Positioning the Infant. Close contact during feeding is important. Infant and caregiver should both be comfortable and relaxed, with the infant cradled on an incline so that its head is higher than its body. The nipple hole should be large enough to allow one swallow of milk to flow each time the infant sucks; if the hole is too small, enlarge it with a sterile needle; if too large, replace the nipple. The caregiver tilts the bottle so that the nipple is full of formula, not air, while the infant is sucking. Now and then the infant should be held upright and given a gentle pat on the back to help eliminate any bubbles of air.

Duration of a Feeding Session. Infants generally feed for about 15 minutes and should not be forced to empty the bottle. Like making children clean their plates, this practice promotes obesity. Formula left in a bottle after feeding should be discarded.

Frequency of Feeding Sessions. The feeding schedule can vary, but is best if adjusted to the infant's expressed hunger needs at first, within reason. Some infants need to feed more fre-

*To protect infants from bacterial infections, it is important to keep all bottles, bottle caps, and nipples clean and free of contamination. Some pediatricians recommend sterilizing feeding equipment in boiling water for at least 5 minutes and using only boiled water for mixing formulas; others recommend using hot cycles on a dishwasher or scrubbing equipment with a bottle brush and hot, soapy water.

The infant thrives on infant formula offered with affection.

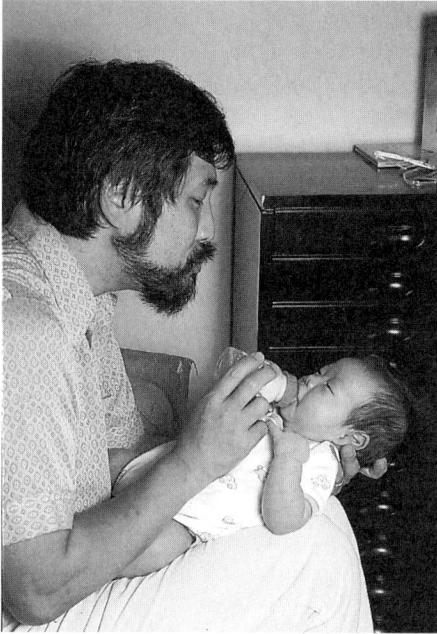

An extreme example of nursing bottle tooth decay. This child was frequently put to bed sucking on a bottle filled with apple juice, so the teeth were bathed in carbohydrate for long periods of time—a perfect medium for bacterial growth. The upper teeth have decayed all the way to the gum line.

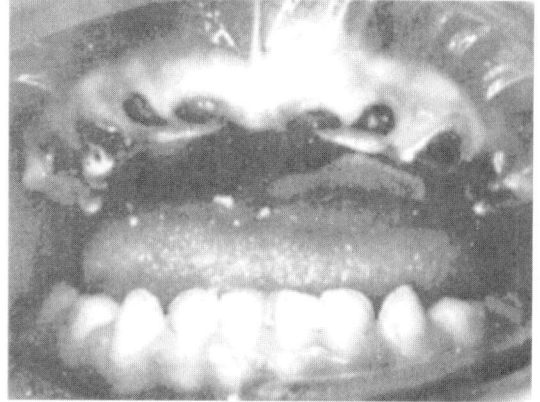

quently than others. Most infants enjoy bottles; the sucking provides stimulation and satisfaction, as well as nutrients.

Nursing Bottle Tooth Decay

Infants cannot be allowed to sleep with bottles, because of the potential damage to developing teeth. Salivary flow, which normally cleanses the mouth, diminishes as the infant falls asleep. Prolonged sucking on a bottle of formula, milk, or juice bathes the upper teeth in a carbohydrate-rich fluid that nourishes decay-producing bacteria. (The tongue covers and protects most of the lower teeth, but they, too, may be affected.) The result is extensive and rapid tooth decay. To prevent tooth decay, no infant should be put to bed with a bottle of nourishing fluid.

✷ Healthy People 2000

Increase to at least 75% the proportion of parents and caregivers who use feeding practices that prevent nursing bottle tooth decay.

GLOSSARY

areola (ah-REE-oh-la): the colored portion of the mammary gland that surrounds the nipple.

engorgement: overfilling of the breasts with milk so that they become swollen and hard.

mastitis: inflammation of the breast, most common in women during lactation; caused by infectious bacteria entering through a crack or abrasion of the nipple.

nursing bottle tooth decay: extensive tooth decay due to prolonged tooth contact with formula, milk, fruit juice, or other carbohydrate-rich liquid offered to an infant in a bottle.

relaxation response: the opposite of the stress response; the normal state of the body.

rooting reflex: a reflex that causes an infant to turn toward whichever cheek is touched, in search of a nipple.

stress response: the body's response to a physical or psychological threat, mediated by nerves and hormones.

FOCAL POINT 4 NOTES

1. T. Waterston, Any questions, *British Medical Journal* 310 (1995): 780.
2. M. Hamosh and coauthors, Breastfeeding and the working mother: Effect of time and temperature of short-term storage on proteolysis, lipolysis, and bacterial growth in milk, *Pediatrics* 97 (1996): 492–498.
3. N. Zavaleta and coauthors, Effect of acute maternal infection on quantity and composition of breast milk, *American Journal of Clinical Nutrition* 62 (1995): 559–563.

Nutrition during Infancy

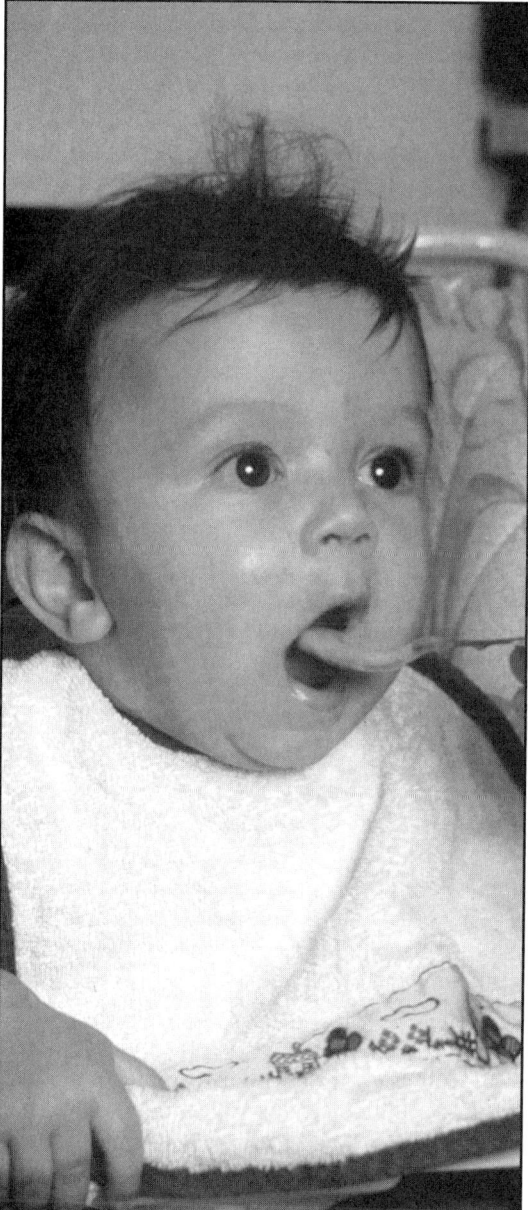

T he first year of life is a time of phenomenal growth and development. To attain full potential, the infant needs an abundant supply
of nutrients. The infant's high nutrient needs and developing maturity
define the foods most appropriate for each stage of the first year. Nutrition during that critical first year is the focus of this chapter.

Growth and Development during Infancy

Physical growth and development involve not only the progressive increase in size of a living being but also the changes that accompany this
increase. A variety of interrelated factors—genetics, nutrition, and environment—influence how growth and development progress.

In many ways, growth and development go hand in hand: As something grows, it also develops. Development involves psychological and
physiological changes, including the attainment of sensory and motor
skills.

Growth and development are not uniform. Each body system has its
own unique schedule of growth and development, varying in rate, pattern, and duration.

Growth

An infant grows more rapidly during the first year of life than ever
again, as Figure 5-1 shows. In the United States, the average, healthy infant weighs 3.3 kilograms (7 1/4 pounds) at birth.* The infant's birthweight doubles by about five months of age and triples by one year. (If
an adult, starting at 150 pounds, were to do this, the person's weight
would increase to 450 pounds in a single year.) The infant's length
changes more slowly than weight, increasing about 25 centimeters
(10 inches) from birth to one year—or about 50 percent. This tremendous growth reflects the differing growth patterns of all the internal
organs.

Development

The course of development in the first year is remarkable. Externally, an
observer can note that at birth infants can hardly see and cannot roll
over; at a year, infants can crawl and are beginning to walk and talk. At
birth, infants can only suck; at a year, infants can hold a spoon and feed
themselves. Internal physiological changes parallel the external ones.
The internal changes, especially the development of the gastrointestinal
tract and kidneys, are of particular relevance to nutrition. These internal
changes enable the infant to ingest, digest, absorb, and process more
and more complex foods, from breast milk or infant formula alone at
the start of the year to foods from all food groups at the end of one year.

*Fiftieth percentile, National Center for Health Statistics growth chart.

Figure 5-1   Weight Gain of Human Infants in Their First Five Years of Life

In the first year, an infant's birthweight may triple, but over the following several years, the rate of weight gain gradually diminishes.

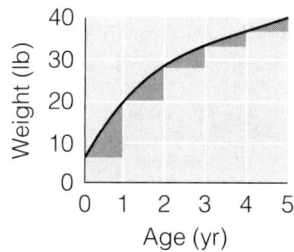

Readiness of the Gastrointestinal Tract. The full-term, newborn infant is capable of sucking and swallowing at birth, although somewhat inefficiently. Within a few days, however, the infant is better able to coordinate sucking, swallowing, and breathing, so this process becomes more efficient. Until about four months of age, all solid objects entering the infant's mouth are pushed out by the action of the tongue; this is how the infant manages to obtain milk from the breast or bottle: sucking the nipple in, drawing the milk out of it and swallowing the milk, and then pushing the nipple out. This tongue movement is called the extrusion reflex; it also prevents foods from being swallowed for the first four to six months of age. Only after the extrusion reflex disappears is the infant able to push food to the back of the mouth for swallowing. Food in the back of the mouth touches the soft palate of the roof of the mouth, which triggers the swallowing reflex. This is why the nipple must be well inside the infant's mouth when breastfeeding or feeding infant formula from a bottle.

extrusion reflex: the reflex that causes an infant to push food out, rather than swallowing it.

   Spoon feeding of solid foods promotes proper swallowing movements. In contrast, the use of an infant feeder, which allows an infant to suck pureed food from a bottlelike container, discourages the maturation of swallowing movements. Sometime after about four months of age, the infant is able to sit alone and has greater control of head movement than previously. Thus the infant can sit in a high chair and is capable of giving signs of satiety, such as turning the head away. Table 5-6 later in this chapter lists feeding skills and appropriate foods during infancy.

   The main role of the stomach is to physically and chemically prepare foods for digestion. As Figure 5-2 illustrates, the infant's stomach is much smaller, shaped differently, and functionally immature compared with the older child's stomach. During the first few months of life, foods move more slowly and are mixed less efficiently in the stomach than they are in later infancy. The high energy density of breast milk or infant formula helps compensate for the stomach's small capacity and prolonged emptying time and meets the high energy requirements of the infant.

Figure 5-2 Infant and Child Stomachs Compared

The stomach of the normal newborn is more horizontal than that of the older child. The stomach becomes curved and upright as the child assumes a more erect posture. The subdivisions of fundus, body, and antrum are not well defined for the first year of life. The vigorous mixing action of the antrum, which is most involved with emptying solids, is apparent in older children but is minimal in the first months of infancy. These differences in stomach anatomy partly explain why solid foods are digested less readily at this time.

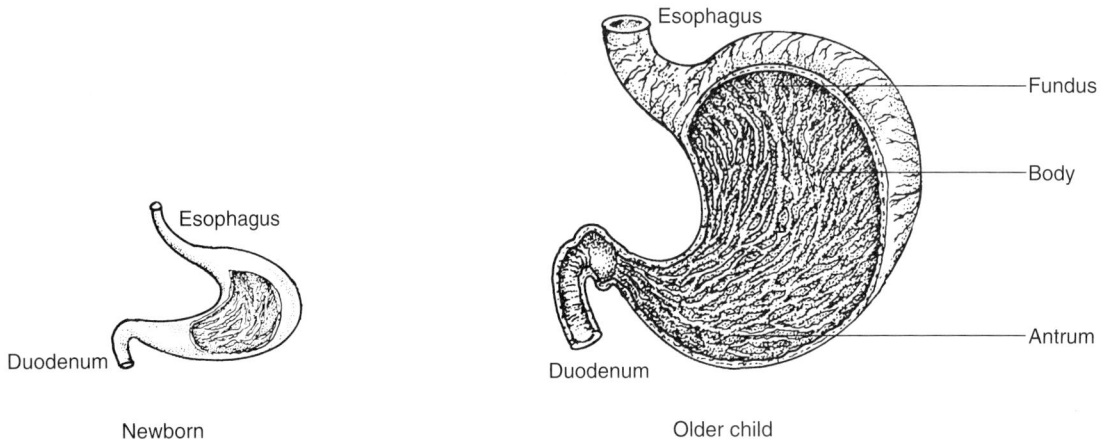

Newborn

Older child

The concentrations of gastric secretions, such as hydrochloric acid and the enzyme pepsin, are low initially, which prevents some types, as well as large amounts, of food from being digested.[1] This explains why the large, tough curds of the protein casein in cow's milk do not readily break down. Similarly, enzyme secretions of the pancreas and small intestine are low in early infancy, so the digestion of solid foods is inefficient.

Readiness of the Renal System. The limited capacity of the infant's immature renal system during the first few months of life further justifies delaying the introduction of solid food. The immature renal system has only a limited ability to filter out waste products or to excrete sodium. As the kidneys mature, they can handle higher solute loads. Until then, foods with a high renal solute load incur too great an obligatory water loss and cause dehydration, a life-threatening condition. To protect against dehydration and to compensate for the higher renal solute load, water should be offered to the infant regularly once solid food is added to the diet.[2]

Adequate nutrition during infancy plays a crucial role in normal growth and development. The rapid growth and metabolism of the infant demand an ample supply of *all* the nutrients, but the next section emphasizes the energy-yielding nutrients and those vitamins and minerals most critical to the growth process.

Energy and Nutrient Needs during Infancy

Before any discussion of nutrient needs for the infant can begin, it must be acknowledged that the information on which nutrient recommendations for infants are based is incomplete at this point. Much remains to be learned, even about protein and energy. Estimates of vitamin and mineral needs for the infant are based on even more uncertainty. Despite this, health care providers must have standards for nutrient adequacy on which timing of introduction of supplemental foods for the infant can be based. Because of large variations in growth rate, activity level, size, metabolic rate, environment, and other factors, it is impossible to establish a single standard applicable to all infants. For this reason, recommendations are often expressed as ranges. The standard used in establishing nutrient allowances for infants is the average amount of nutrients consumed by thriving infants breastfed by healthy, well-nourished mothers. Therefore, much of the discussion of infant nutrient needs that follows is a discussion of the nutrient contents of breast milk.[3]

In the past, the RDA used 750 mL/day as the reference value for mean breast milk volume for the first six months of life. The 1997 DRI are based on a mean breast milk volume of 780 mL/day.

During the first year of life, the infant's need for most nutrients, in proportion to body weight, is more than double that of the adult. Figure 5-3 shows this to be true by comparing a five-month-old infant's needs with those of an adult male. The sections that follow describe the healthy infant's needs, nutrient by nutrient, and the bases on which requirements for individual nutrients are established. Focal Point 5A discusses the special nutrient needs of the premature infant.

Energy and Energy Nutrients

Energy needs per unit of body weight peak during the first year of life because of the infant's rapid basal metabolic rate and growth. This may not be obvious, for a newborn requires only about 650 kcalories per day, whereas most adults require about 2000 kcalories per day, but to examine the energy needs per unit of body weight is to see the great differences. Infants require about 100 kcalories per kilogram body weight per day; most adults require fewer than 40. If the energy needs of the infant were superimposed on the adult, a 170-pound adult would require over 7000 kcalories a day.

The infant uses energy for three major purposes—to support basal metabolism, growth, and activity. The proportion between these components of energy expenditure changes throughout the first year. Figure 5-4 compares the infant's uses of energy at three different periods within the first year. Note how the infant's daily energy needs for growth are greatest during the early months of life, then decline by the infant's first birthday, as the energy needs for activity rise. Note also that basal metabolic needs do not change; the energy requirement for basal metabolism for the healthy infant or child is about double that of the adult on a per-pound basis.[4]

The Committee on Dietary Allowances recommends an average of 108 kcalories per kilogram body weight per day for infants up to six

At six months, the energy saved by slower growth is spent on increased activity.

Figure 5-3 Nutrient RDA of a Five-Month-Old Infant and an Adult Male Compared on the Basis of Body Weight

Because infants are small, they need smaller total amounts of the nutrients than adults do, but when comparisons are based on body weight, infants need over twice as much of many nutrients. Infants use large amounts of energy and nutrients, in proportion to their body size, to keep all their metabolic processes going.

Infant's metabolism:

- Heart rate: 120 to 140 beats per minute.
- Respiration rate: 20 per minute.
- Energy needs: 45 kcalories per pound (100 kcalories per kilogram) body weight.

Adult's metabolism:

- Heart rate: 70 to 80 beats per minute.
- Respiration rate: 12 to 14 per minute.
- Energy needs: <18 kcalories per pound (<40 kcalories per kilogram) body weight.

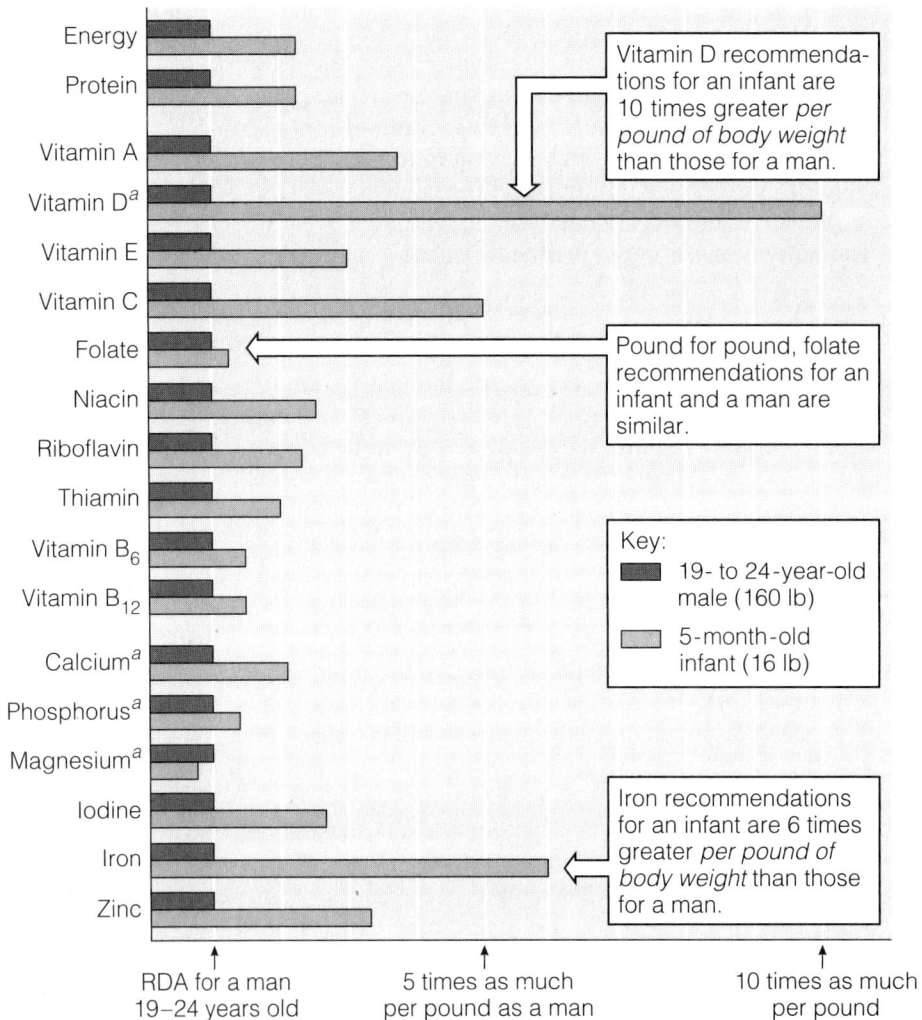

Vitamin D recommendations for an infant are 10 times greater *per pound of body weight* than those for a man.

Pound for pound, folate recommendations for an infant and a man are similar.

Key:
- 19- to 24-year-old male (160 lb)
- 5-month-old infant (16 lb)

Iron recommendations for an infant are 6 times greater *per pound of body weight* than those for a man.

RDA for a man 19–24 years old | 5 times as much per pound as a man | 10 times as much per pound

Energy, Protein, Vitamin A, Vitamin D*a*, Vitamin E, Vitamin C, Folate, Niacin, Riboflavin, Thiamin, Vitamin B₆, Vitamin B₁₂, Calcium*a*, Phosphorus*a*, Magnesium*a*, Iodine, Iron, Zinc

*a*Values reflect DRI.

Figure 5-4   Estimated Energy Requirements of Infants during the First Year

Infant energy needs vary according to age and weight. As the infant grows older and larger, the growth rate diminishes, and the activity level increases.

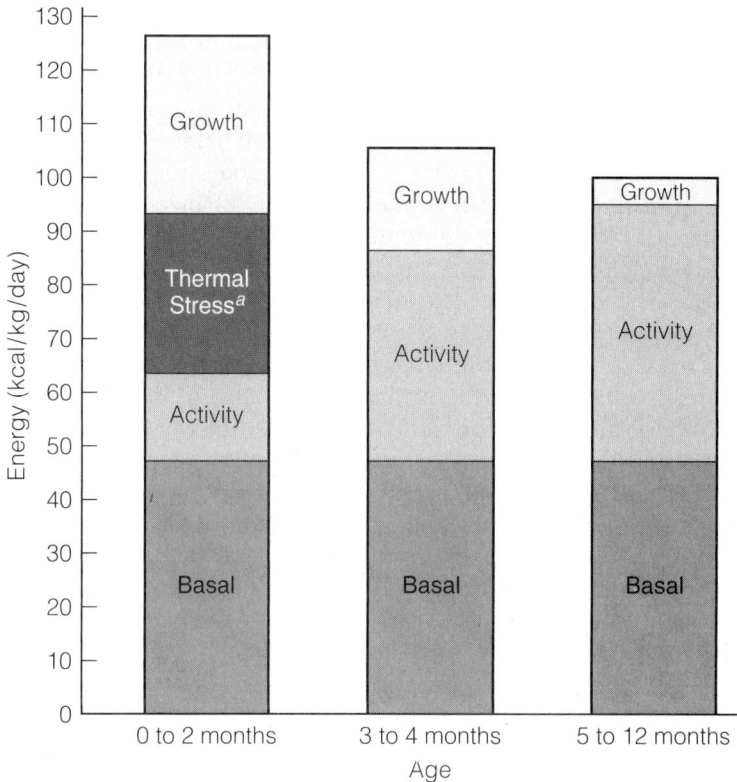

[a]The newborn infant's thermoregulatory system requires extra energy at first to adapt from womb temperature to room temperature.

Source: Adapted from data in E. M. Widdowson, Nutrition, in *Scientific Foundation of Pediatrics*, 2nd ed. (Baltimore: University Park Press, 1981), pp. 41–53, as cited by M. Gracey and F. Falkner, *Nutritional Needs and Assessment of Normal Growth* (New York: Raven Press, 1985), pp. 23–40.

months of age, and 98 kcalories per kilogram per day for infants 6 to 12 months old.[5] Actual food energy intakes, however, vary widely among both breastfed and formula-fed infants. Food energy intakes of breastfed infants in North America decrease at a faster rate and at an earlier age than current recommendations.[6] One study reported that formula-fed infants meet recommendations, consuming about 100 to 120 kcalories per kilogram body weight per day during the first six months of life. In contrast, breastfed infants' energy intakes fell below recommendations, ranging from about 100 to 110 kcalories per kilogram per day early on, and only 70 to 80 kcalories per kilogram by six months of age.[7] Research shows that growth patterns of breastfed infants differ from those of formula-fed infants; after the first two to three

Energy RDA during infancy:
650 kcal/day (0 to 6 mo).
850 kcal/day (6 to 12 mo).
Canadian RNI during infancy:
600 kcal/day (0 to 4 mo).
900 kcal/day (5 to 12 mo).

months, breastfed infants tend to gain less weight than formula-fed infants. Whether the more rapid weight gain of formula-fed infants is advantageous or disadvantageous is unclear.[8]

Protein. No single nutrient is more essential to growth than protein. All of the body's cells and most of its fluids contain protein; it is the basic building material of the body's tissues. Table 5-1 lists the nine essential amino acids; the absence of any one can stunt growth.[9] Other amino acids are conditionally essential (also listed in Table 5-1). That is, an amino acid may be essential for an infant under certain conditions, such as illness, prematurity, or inborn errors of metabolism. For example, premature infants and some term infants require cysteine, tyrosine, and taurine.[10]

Dietary protein is needed to replace daily losses of amino acids. An adequate protein intake supplies all the essential amino acids in amounts sufficient for maintenance and growth; this definition considers protein quality as well as quantity.[11] Older infants in North America typically obtain protein primarily from breast milk or infant formula; meat, fish, and eggs; and some foods of vegetable origin, such as infant cereal. These foods supply all the essential amino acids in reasonably high concentrations relative to their food energy.

The protein RDA for infants are the same as the amounts of protein provided by the quantities of breast milk that ensure adequate rates of growth.[12] Infants consuming 150 to 200 milliliters of breast milk per kilogram body weight per day receive adequate protein.

Protein recommendations are designed for healthy, full-term infants. Infection, premature birth, illness, and genetic disorders all increase protein requirements. Protein needs are also greater when energy intake is inadequate, as dietary protein is used for energy.

Protein contributes a small percentage of the energy in breast milk— about 6 to 7 percent. Breast milk in the quantity normally consumed provides approximately 2 grams protein per kilogram body weight per day for the average 7 1/2-pound infant. Therefore, the American Academy of Pediatrics (AAP) recommends that infant formulas provide a comparable amount, with a protein efficiency ratio equal to or greater than that of casein.

Excess dietary protein can cause problems, especially in the small infant, when amino acids build up in the blood. The ingestion of excess protein stresses the kidneys and liver, which have to metabolize and excrete the excess nitrogen. Signs of protein overload include acidosis, dehydration, diarrhea, elevated blood ammonia, elevated blood urea, and fever. Such problems are not common but have been observed in infants fed inappropriate foods, such as nonfat milk or concentrated formula.

Carbohydrate. Carbohydrate provides a readily available source of energy for all human beings and, in doing so, spares protein from being used for energy. The brain relies almost exclusively on carbohydrate for its energy. Breast milk provides about 39 percent of its energy from

---

Protein malnutrition during infancy and early childhood can profoundly impair growth and development, as Focal Point 5A explains.

Focal Point 5C discusses the unique nutrient needs of infants with inborn errors of metabolism.

Protein RDA during infancy:
13 g/day (0 to 6 mo).
14 g/day (6 to 12 mo).
Canadian RNI during infancy:
12 g/day (0 to 12 mo).

150 to 200 mL breast milk contains between 1.7 and 2.2 g protein.

Assuming infants consume approximately 750 mL/day (approximately equal to 25 oz or about 3 c), protein intake = 8.25 g protein/day.

protein efficiency ratio (PER): a measure of protein quality assessed by determining how well a given protein supports weight gain in laboratory animals.

Table 5-1 Amino Acids

Essential amino acids:	*Conditionally essential amino acids:*
■ Histidine.	■ Cysteine.
■ Isoleucine.	■ Tyrosine.
■ Leucine.	■ Taurine.
■ Lysine.	
■ Methionine.	
■ Phenylalanine.	
■ Threonine.	
■ Tryptophan.	
■ Valine.	

carbohydrate; infant formula, about 42 percent. Figure 5-5 illustrates the energy–nutrient balance of both breast milk and standard infant formula.

In most instances, lactose is by far the major carbohydrate consumed by infants. Lactose makes up about 90 percent of the total carbohydrate in breast milk and contributes most of the carbohydrate in standard infant formulas.[13]

Fat. Fat supports the normal development of the central nervous system, helps to maintain body temperature, and cushions the vital organs. Dietary fat is a concentrated source of energy, imparts flavor and satiety, provides essential fatty acids, and serves as a vehicle for the absorption of the fat-soluble vitamins. Fat contributes about 55 percent of the energy in breast milk, and this percentage is considered ideal for

Figure 5-5 Percentages of Energy-Yielding Nutrients in Human Milk and in Infant Formula

The proportions of energy-yielding nutrients in human breast milk and formula differ slightly.

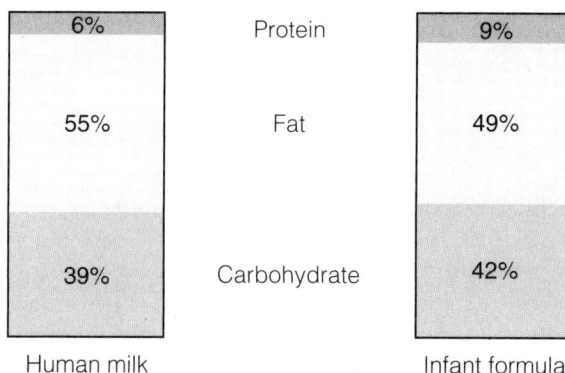

Human milk: Protein 6%, Fat 55%, Carbohydrate 39%. Infant formula: Protein 9%, Fat 49%, Carbohydrate 42%.

*Cholesterol in breast milk:
20 to 30 mg/100 mL.
Breastfed infants consume
approximately 200 mg
cholesterol/day.*

linoleic acid (lin-oh-LAY-ick): an essential fatty acid with 18 carbons and two double bonds (18:2).

linolenic acid (lin-oh-LEN-ick): an essential fatty acid with 18 carbons and three double bonds (18:3)

omega-3 fatty acid: a polyunsaturated fatty acid in which the first double bond is three carbons away from the methyl (CH_3) end of the carbon chain.

eicosapentaenoic (EYE-cossa-PENTA-ee-NO-ic) **acid** (EPA): an omega-3 polyunsaturated fatty acid with 20 carbons and five double bonds (20:5); synthesized from linolenic acid.

docosahexaenoic (DOE-cossa-HEXA-ee-NO-ic) **acid** (DHA): an omega-3 polyunsaturated fatty acid with 22 carbons and six double bonds (22:6); synthesized from linolenic acid.

infants. As Chapter 4 mentioned, approximately 98 percent of the fat is in the form of triglycerides; the remainder is phospholipids, cholesterol, and free fatty acids.

Breast milk is a relatively rich source of cholesterol.[14] Formula manufacturers replace cow's milk fat with vegetable oils, which lowers the cholesterol content of infant formulas. Clinical studies show no adverse effects of low-cholesterol formulas, but further research is needed regarding the long-term effects.

Linoleic acid, an omega-6 fatty acid, cannot be made by the body but is indispensable to body function. It is therefore an essential fatty acid. Linoleic acid contributes about 7 percent of the total energy in breast milk; in infant formula, it contributes about 10 percent. Infants develop deficiency symptoms, such as dermatitis and failure to thrive, when linoleic acid provides less than 0.1 percent of the total daily energy intake, but such deficiencies are rare.

Another essential fatty acid, linolenic acid, is a member of the omega-3 fatty acid family. Linolenic acid is the best known of the omega-3 fatty acids, but two others, eicosapentaenoic acid (EPA) and docosahexaenoic acid (DHA), have gained worldwide attention in the last decade or so. Research on the omega-3 fatty acids has focused on the prevention of heart disease, but interest is now turning to other roles, including those in growth and development. Most of this research has focused on DHA, an omega-3 fatty acid found abundantly in both the retina of the eye and the brain. About half of the DHA accumulates in the brain before birth and half, after birth, emphasizing the importance of lipids during pregnancy and infancy.[15]

Breast milk contains significant quantities of linoleic acid, linolenic acid, EPA, and DHA; infant formula contains only linoleic acid and linolenic acid.[16] Because breastfed infants receive more DHA than formula-fed infants, research is under way to determine the physiological significance of this difference. Researchers have analyzed the fatty acid composition of brains taken from previously healthy infants who died of sudden infant death syndrome (SIDS).[17] They found that the brains of the breastfed infants contained significantly more DHA than the brains of the formula-fed infants. The longer the infants were breastfed prior to death, the more DHA their brains contained. Results of another study showed that infants' visual acuity improved when DHA was added to formula.[18] These findings suggest that including all the omega-3 fatty acids in infant formula may be desirable for optimum development.[19]

Water

One of the most essential nutrients for the infant, as for anyone, is water. The younger the infant, the greater the percentage of body weight is water. The water in an infant's body is easily lost, because compared with the water in an adult's body, a larger percentage of an infant's body water is located in the interstitial (extracellular) and vascular spaces. During early infancy, breast milk or infant formula normally provides enough water for a healthy infant to replace water losses from the skin, lungs, feces, and urine. Under normal conditions, the AAP advises little or no supplemental water for infants before the introduction of solid foods.[20] However, conditions that cause fluid loss, such as sweating, diarrhea, or vomiting, may justify offering additional water. Adults must remember that infants may cry from thirst as well as hunger. When water is needed, allow infants to drink it until their thirst is quenched.

All infants need supplemental water once they are eating solid foods. Foods with a high protein content, such as meats and eggs, impose a high renal solute load on the kidneys. Additional water eases the burden on the kidneys. Without supplemental water, the kidneys are stressed, and dehydration becomes a threat. Foods such as fruits and vegetables present the kidneys with a low renal solute load.

Vitamins

The extraordinary growth of an infant during the first year of life may not be obvious, because each day brings only gradual changes. Yet internally, the metabolic machinery is fast at work creating new body parts—with the assistance of the vitamins.

Vitamin A. The average retinol content of breast milk is about 50 micrograms per 100 milliliters. Assuming a milk intake of 750 milliliters per day, the average intake of vitamin A for the infant is 375 micrograms per day.[21]

In the United States and other developed countries, few vitamin A deficiencies develop during infancy. They occur only in infants with impaired fat absorption. The earliest clinical sign of vitamin A deficiency is impaired dark adaptation in the retina, which is difficult to detect in infants. As the deficiency progresses, failure to thrive, apathy, anemia, dry skin, and corneal changes appear. In developing countries, vitamin A deficiencies occur far more commonly in infancy, though they take an especially terrible toll in early childhood.

retinol: the active form of vitamin A found in milk.

Vitamin A RDA during infancy:
375 μg RE/day (0 to 12 mo).
Canadian RNI during infancy:
400 μg RE/day (0 to 12 mo).

Vitamin D. Vitamin D is similar in structure, metabolism, and mechanism of action to steroid hormones, and the active form to which it is converted in the body is actually a hormone.[22] Nevertheless, for convenience and historical reasons, it is still called vitamin D. A major function of vitamin D is to promote calcium absorption and transport.

Research suggests that breast milk is unsuitable as a standard from which to estimate the vitamin D needs of infants; it provides less than they need, and the difference is made up by sunlight. Breastfed infants who are deprived of sufficient sunlight require supplemental vitamin D (see the section on supplements for infants, later in this chapter).[23]

Vitamin D DRI during infancy:
5 μg/day.

Vitamin E. Vitamin E is a fat-soluble antioxidant that protects the lipids and other vulnerable components of the cells and their membranes from oxidative destruction. Breast milk contains 1.3 to 3.3 milli-

Vitamin E RDA during infancy:
  3 mg α-TE/day (0 to 6 mo).
  4 mg α-TE/day (6 to 12 mo).
Canadian RNI during infancy:
  3 mg α-TE/day (0 to 12 mo).

Vitamin K RDA during infancy:
  5 μg/day (0 to 6 mo).
  10 to 20 μg/day (6 to 12 mo).

grams vitamin E per liter and is assumed to provide an adequate intake for nursing infants. By comparison, infant formula contains about 10 milligrams vitamin E per liter.

Plasma vitamin E in newborn infants is less than that of older children and adults.[24] The smaller the infant, the lower the vitamin E concentration at birth. Full-term, breastfed infants attain adult values of the vitamin soon after birth, as do infants who consume adequate quantities of commercial infant formula.

Vitamin K. The newborn infant presents a unique case when it comes to vitamin K nutrition.[25] At birth, the intestinal tract of the newborn is sterile and thus lacks vitamin K-producing bacteria. At the same time, plasma prothrombin concentrations fall, to reduce the likelihood of fatal blood clotting during the stress of birth. Prothrombin synthesis depends on vitamin K, and prothrombin concentrations climb back to adult levels as milk is consumed and the intestinal tract gradually develops a population of vitamin K-producing bacteria. If prothrombin concentrations do not rise, hemorrhagic disease of the newborn may appear. The AAP therefore recommends that a single dose of vitamin K be given to infants at birth; see the section "Nutrient Supplements for Infants," later in this chapter.

The B Vitamins. If infants consume adequate amounts of infant formula or breast milk from healthy women, then their daily requirements for the B vitamins will be met. Deficiencies may occur if requirements increase—for example, due to severe injury or malabsorption.

Two of the B vitamins, vitamin B_{12} and folate, deserve special mention for their crucial roles in growth and development. Both vitamin B_{12} and folate perform essential roles in DNA synthesis and thus support growth. The decreased DNA synthesis that results from either a vitamin B_{12} or folate deficiency causes all replicating cells in the body to have a megaloblastic (giant cell) appearance. This is usually observed in blood cells and is called megaloblastic anemia.

megaloblastic anemia: anemia in which the red blood cells are immature (and therefore large).

Vitamin B_{12} deficiency is rare in infants, because they are usually born with stores sufficient for the first year of life and then receive additional amounts from breast milk or infant formula. Thus vitamin B_{12} deficiency does not occur in infants who are breastfed by women with adequate serum vitamin B_{12}. Daily outputs of vitamin B_{12} in breast milk range from 0.2 to 0.8 micrograms. Vitamin B_{12} deficiency has been observed, however, in infants of women who consume strict vegetarian diets, as described in a later section.

Vitamin B_{12} RDA during infancy:
  0.3 μg/day (0 to 6 mo).
  0.5 μg/day (6 to 12 mo).
Canadian RNI during infancy:
  0.3 μg/day (0 to 4 mo).
  0.4 μg/day (5 to 12 mo).

Folate RDA during infancy:
  25 μg/day (0 to 6 mo).
  35 μg/day (6 to 12 mo).
Canadian RNI during infancy:
  25 μg/day (0 to 4 mo).
  40 μg/day (5 to 12 mo).

Folate depends on vitamin B_{12} to properly fulfill its role in DNA synthesis. Folate deficiency is the most common cause of megaloblastic anemia in infants and children. At birth, an infant's serum folate is three times the maternal concentration, but infant stores are limited, and the rapid growth of the infant depletes these stores. By two weeks of age, serum folate falls below adult values and stays there for several months.[26]

The folate needs of infants are adequately met by breast milk or infant formula, but not by goat's milk, which has a notoriously low folate content. Breast milk and cow's milk contain about 50 micrograms of folate per liter.

Vitamin C. The vitamin C in breast milk ranges between 30 and 80 milligrams per liter, depending on the dietary intake of the mother. Infant formulas in the United States contain between 55 and 61 milligrams of vitamin C per liter.[27]

Vitamin C deficiency is uncommon in the United States, but scurvy does occur in infants fed exclusively cow's milk for the first 6 to 12 months of life. Symptoms of scurvy in infants include anorexia, diarrhea, failure to gain weight, irritability, and increased susceptibility to infection. As the disease worsens, hemorrhages under the skin and failure of spontaneous leg movements occur. Dramatic improvement of symptoms is seen with the administration of 25 milligrams of vitamin C four times a day.[28]

Minerals

The minerals are as actively involved in the growth and development of infants as are the vitamins. All are essential, and each serves a unique function. Those discussed here are particularly important to infant nutrition.

Calcium and Phosphorus. Not surprisingly, calcium needs are great during infancy, a period of rapid skeletal growth and mineralization. During infancy, the calcium content of the body increases faster relative to body size than at any other time in the life span. By nature's design, milk is both the best source of calcium and the main component of the infant's diet. Breast milk contains about 300 milligrams of easily absorbed calcium per liter (two-thirds of which, or 200 milligrams, is absorbed efficiently).[29] Standard infant formula contains more calcium (400 to 500 milligrams per liter), but it is less readily absorbed (less than one-half retention).[30] The full-term infant's calcium needs are fully met by either breastfeeding or formula feeding.[31]

Excessive dietary phosphorus relative to calcium may contribute to calcium loss from bone.[32] The calcium-to-phosphorus ratio of breast milk is 1.5:1; the ratio of calcium to phosphorus in commercial infant formulas ranges between 1.3:1 and 1.5:1, amounts compatible with the AAP recommendations.

Sodium. Sodium, the principal electrolyte of the extracellular fluid, is critical to the maintenance of fluid balance. Sodium also helps maintain acid–base balance and is essential to nerve transmission and muscle contraction. Breast milk provides 7 milliequivalents of sodium per liter, whereas infant formulas contain between 7 and 11 milliequivalents per liter.[33] Infants' average sodium intake is 120 milligrams per day. Sodium

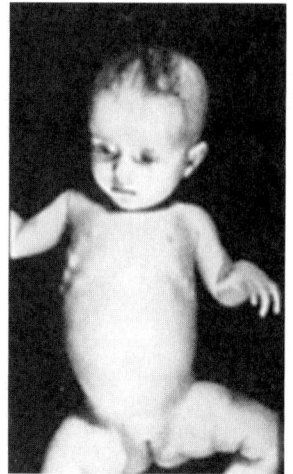

Infant scurvy. This is the characteristic "scorbutic pose," with legs bent and thighs rotated open. The infant's joints are painful, and she will cry if made to move.

Vitamin C RDA during infancy:
  30 mg/day (0 to 6 mo).
  35 mg/day (6 to 12 mo).
Canadian RNI during infancy:
  20 mg/day.

Calcium DRI during infancy:
  210 mg/day (0 to 6 mo).
  270 mg/day (6 to 12 mo).

Phosphorus DRI during infancy:
  100 mg/day (0 to 6 mo).
  275 mg/day (6 to 12 mo).

milliequivalent: the concentration of electrolytes in a volume of solution.

Sodium (estimated safe in-takes) during infancy:
  120 mg/day (0 to 6 mo).
  200 mg/day (6 mo to 1 yr).

intakes of infants who are breastfed or fed formula are within the rec-ommended range.

Potassium. Potassium is the principal electrolyte *inside* the body cells. The infant's potassium requirement for growth exceeds that of sodium, as reflected in the sodium-to-potassium ratio of 0.6:1 in breast milk.[34] Infant formulas contain slightly more potassium than breast milk. Con-sistent with the infant's limited renal concentrating capacity, upper lim-its for the electrolyte content of infant formula have been set by the AAP. If blood potassium rises three to four times above normal, the heart stops beating. Years ago, breastfed infants died when given potas-sium supplements for colic.

Iron RDA during infancy:
  6 mg/day (0 to 6 mo).
  10 mg/day (6 mo to 1 yr).
Canadian RNI during infancy:
  0.3 mg/day (0 to 4 mo).
  7 mg/day (5 mo to 1 yr).

Iron. During gestation, the fetus receives high priority when it comes to available iron, often at the expense of the mother. In fact, iron stores of newborn infants show negligible differences between infants of iron-deficient, anemic mothers and infants of iron-sufficient mothers.

Infant iron status may depend on the way the attending health care professional clamps the umbilical cord at birth. When clamping of the cord is delayed slightly—about one minute after birth—while the in-fant is level with, or below, the vaginal opening, greater blood flow from the placenta to the infant permits greater iron transfer.[35] The timing of cord clamping and the positioning of the infant at birth are controver-sial and require careful consideration. The transfer of too much blood can be as critical as too little.

The infant continues to receive high priority for iron during lacta-tion. Poor iron status of the mother does not reduce the iron concentra-tion of her breast milk. Breast milk contains relatively small amounts of iron (0.5 milligrams per liter), but this iron has a high bioavailability (50 percent). In contrast, iron-fortified infant formula contains about 12 milligrams of iron per liter, but this iron has low bioavailability (about 4 percent). The presence of lactoferrin in breast milk and a lower phosphate and protein content may explain the enhanced absorption of iron from breast milk.[36] The full-term, newborn infant arrives en-dowed with enough iron to last for at least the first four to six months, assuming that breast milk or iron-fortified formula also is provided.

Chapter 4 offers a discussion of lactoferrin.

Iron Deficiency. Rapid growth demands iron. At about four to six months, the infant begins to need more iron than stores plus breast milk or iron-fortified formula can provide. Iron deficiency becomes pos-sible, and in fact likely, unless iron-fortified cereal or iron supplements are introduced.

The depletion of iron reserves during later infancy is due to the large amount of iron needed to maintain an almost constant hemoglobin concentration against a rapidly expanding blood volume.[37] The use of cow's milk during infancy also contributes to iron deficiency anemia among infants. Infants who consume cow's milk have increased gas-trointestinal blood loss, which, in turn, impairs iron status.[38]

Research suggests that socioeconomic status is a factor in the development of iron deficiency. In 1980, researchers found that 30 percent of infants seen in a U.S. Public Health Service clinic were anemic, and 25 percent were iron deficient, compared with 8 percent who were anemic and 11 percent who were iron deficient in a private practice.[39] Recent research shows an improvement in the iron status of low-income infants and suggests that participation in the Special Supplemental Food Program for Women, Infants, and Children (WIC) is a primary contributing factor.[40] Formula-fed infants enrolled in this program receive iron-fortified formula.

Within the past 15 years or so, the prevalence of iron deficiency anemia among all infants in the United States has declined impressively.[41] The changes that have led to this reduced prevalence include the following:

■ More infants are initially breastfed.

■ More infants receive iron-fortified formula.

■ Fewer infants receive whole cow's milk during the first year of life than in the past.

Iron deficiency is still the leading cause of anemia, however, despite a better understanding of how to prevent it and improved methods to detect it.

The question of whether iron deficiency influences infant behaviors has been extensively explored in recent years. Iron-deficient infants display symptoms of irritability, anorexia, and poor weight gain. Many studies also reveal that behavioral disturbances tend to persist. Focal Point 5B explores this topic.

The length of this discussion on iron accentuates the importance of this nutrient to the infant, especially to the older infant. Because iron deficiency is the most common nutrient deficiency of infants in the United States, Canada, and developing countries as well, iron nutrition should be a high priority for all health professionals involved in the care of young children.

Zinc. Zinc needs of infants are not well defined. A deficiency of this nutrient impairs growth, and growth velocity is the main determinant of infant zinc requirements.[42] Zinc needs therefore are highest during early infancy and decline with advancing age.

The zinc concentration of breast milk declines as lactation progresses, ranging from as much as 4 milligrams per liter during the first month; falling to about 1.5 milligrams per liter by month 3; and then falling further still to about 0.5 milligrams per liter by month 12.[43] Despite this marked decline in breast milk zinc, breastfed infants' intakes remain adequate, as long as maternal zinc intakes are adequate.[44]

The secretion of zinc into breast milk appears to be tightly controlled, but with time may be modified by maternal zinc status. One group of researchers hypothesized that declines in growth rate (com-

Zinc RDA during infancy:
 5 mg/day.
Canadian RNI during infancy:
 2 mg/day (0 to 4 mo).
 3 mg/day (5 mo to 1 yr).

pared with rates shown on reference growth curves) among infants who are breastfed for longer than 4 months may be partly attributed to inadequate zinc intakes.[45] After three months, weight and length gains of zinc-supplemented infants were significantly greater than for controls. The researchers suggest that the improvements in growth rates observed in this study might have resulted from correction of a marginal zinc deficiency.

Less zinc is available from formula than from breast milk because much of the zinc in the formula is bound to casein, the cow's-milk protein that is believed to inhibit zinc absorption.[46] Zinc availability from soy-based formulas is possibly lower still because of the presence of phytates.[47] To compensate for lower zinc bioavailability, infant formulas contain more zinc than breast milk.

Nutrient Supplements for Infants

Pediatricians may routinely prescribe supplements containing vitamin D, iron, and fluoride. Table 5-2 offers a schedule of supplements during infancy.

In addition, the AAP recommends that a single dose of vitamin K be given to infants at birth.[48] In many states, this preventive dose of vitamin K is required by law. Reports of an association between childhood cancer and intramuscular vitamin K administration to infants in Britain are, according to the AAP, unproven and unlikely.[49]

Vitamin K:
Intramuscular dose:
0.5 to 1.0 mg.
Oral dose: 1.0 to 2.0 mg.

Table 5-2 Supplements for Full-Term Infants

	Vitamin D[a]	Iron[b]	Fluoride[c]
Breastfed infants			
Birth to six months of age	✓		
Six months to one year	✓	✓	✓
Formula-fed infants			
Birth to six months of age			
Six months to one year		✓	✓

[a]Vitamin D supplements are recommended only for as long as breast milk is the infant's major milk.

[b]Infants four to six months of age need additional iron, preferably in the form of iron-fortified cereal for both breastfed and formula-fed infants and iron-fortified infant formula for formula-fed infants.

[c]The Committee on Nutrition of the American Academy of Pediatrics recommends initiating fluoride supplements at six months of age for breastfed infants, formula-fed infants who receive ready-to-use formulas (these are prepared with water low in fluoride), and those who receive formula mixed with water that contains little or no fluoride (less than 0.3 ppm).

Source: Adapted from Committee on Nutrition, American Academy of Pediatrics, Vitamin and mineral supplement needs of normal children in the United States, in *Pediatric Nutrition Handbook*, 3rd ed., ed. L. A. Barness (Elk Grove Village, Ill.: American Academy of Pediatrics, 1993), pp. 34–42; Committee on Nutrition, American Academy of Pediatrics, Fluoride supplementation for children: Interim policy recommendation, *Pediatrics* 95 (1995): 777.

The rapid growth and development during infancy demand a full array of nutrients. When nutrient intakes fall short, growth falters. The next section of this chapter describes how health care professionals use measures of growth to detect nutrient deficiencies.

Nutrition Assessment of Infants

The infant's nutritional health at birth reflects influences experienced during pregnancy, including the adequacy of the mother's nutrition. The main parameters in evaluating the newborn's health are weight, length, and Apgar score. The parameters of interest throughout infancy are those related to growth.

Apgar score: a system of scoring an infant's physical condition right after birth, based on heart rate, respiration rate, color, muscle tone, and responses to stimuli.

Historical Information

A thorough history can reveal medical conditions, eating behaviors, and other factors such as the physical, social, and emotional environments that influence the infant's nutrition status. Environmental factors such as marital strife, financial instability, and an overcrowded or substandard home may lead to failure to thrive (FTT). Comments from caregivers that the infant is "temperamental and difficult to care for" and observations from health care providers that the infant is "irritable and unresponsive" are additional reasons to suspect FTT.

failure to thrive (FTT): inadequate weight gain of infants.

Failure to thrive may be accompanied by medical illness, but most often it reflects a lack of parenting. A fine line separates FTT and child neglect. When health care professionals suspect FTT, the infant requires further evaluation and medical observation. Focal Point 5A discusses FTT further.

A complete diet history for infants includes information about:

- *Type of feeding.* Is the infant breastfed, formula fed, or both? If formula fed, what kind of formula?
- *Quantity and frequency of feeding.* If formula fed, how frequently does the infant feed, and how much does the infant drink at each feeding? If breastfed, how frequently does the infant nurse, and how long does each feeding last?
- *Vitamin and mineral supplements.* Is the infant given supplements? If so, which nutrients are supplemented, and how much of each?
- *Solid food intake.* Is the infant offered solid foods? If so, at what age were solids introduced, which foods does the infant eat, and how often? Are solid foods of the commercial type or home prepared?
- *Feeding behavior.* Does the infant exhibit unusual or abnormal feeding behaviors, such as food aversions?
- *Alternative dietary practices.* Does the infant's family omit any foods or food groups?
- *Food allergies.* Does the infant appear to be allergic to specific foods?

Counseling and further evaluation (such as a detailed dietary analysis) are in order if the assessor suspects that dietary inadequacies or other conditions exist that adversely affect the infant's health.

Anthropometric Data

Genetics is the primary determinant of an individual's growth rate and growth pattern, but nutrition is a substantial factor. Consistent, accurate, and frequent monitoring of growth is important in identifying abnormalities. Growth retardation indicated by measurements below standard for height, weight, or head circumference is an important sign of poor nutrition status. The standards almost universally used for infants and children are the growth charts presented in this and the next chapter. Standard charts compare weight to age, height to age, and weight to height. Although individual growth patterns may vary, in general, an infant's growth curve will stay at about the same percentile throughout the first year. In infants whose growth has been retarded at first, nutrition intervention will ideally induce length and weight gains.

gestational age: the time from the first day of the mother's last menstrual period until birth; also called **fetal age.**

preterm infant: an infant born prior to the 38th week of gestation; also called a **premature** infant.

term infant: an infant born between the 38th and 42nd week of gestation.

post-term infant: an infant born after the 42nd week of gestation.

large for gestational age (LGA): infants whose weight for gestational age falls above the 90th percentile.

appropriate for gestational age (AGA): infants whose weight for gestational age falls between the 10th and 90th percentiles.

Weight at Birth. Birthweight is most often used as an indicator of an infant's probable future health status; it is convenient and easy to determine, and its significance is universally understood. Infants weighing less than 2500 grams (5 1/2 pounds) and 1500 grams (3 1/2 pounds) are defined as low birthweight and very low birthweight, respectively. More precise standards are now available, however, to correlate infants' birthweights with their gestational ages at birth. This is desirable because different birthweights are appropriate for infants of different gestational ages, and because newborn morbidity and mortality correlate both with birthweight and with gestational age.

To evaluate birthweight this way, one must first determine gestational age. Estimates of gestational age are based on precise measurements of fetal growth, including physical characteristics and neurological development. The classification of newborn infants assigns infants to categories based first on gestational age, then on birthweight (see Figure 5-6).[50] The dividing line between preterm and term infants is drawn at 38 weeks; the line between term and post-term infants, at 42 weeks.

Within each gestational age group, three subgroups of infants are classified according to birthweight. Those at or above the 90th percentile in weight for their gestational age category are *large for gestational age*, those between the 10th and 90th percentiles are *appropriate for gestational age*, and those below the 10th percentile are *small for gestational age*. In this manner, nine groups of newborn infants are defined. When all newborns are classified at birth using this system, health care providers can easily recognize those infants who must be watched closely and can predict the types of health problems likely to occur. Infants who are large for gestational age have different problems from those who are small for gestational age. For example, a large-for-gestational-age term or post-term infant would be more likely to be traumatized during delivery, whereas a small-for-gestational-age term or preterm infant would be more likely to suffer from respiratory distress.

The distinction between preterm infants and small-for-gestational-age infants is important in terms of health and development. Preterm infants are born before their gestational development is complete. They may be small, but those who are appropriate in size and weight for gestational age do catch up in growth to their full-term peers. In contrast, small-for-gestational-age infants have experienced fetal growth retardation and do not catch up as well. They may reach only about the 25th percentile for height, and some fail to attain the same mental ability as normal birthweight children.[51] Fetal growth retardation may also have long-term effects on development and immune function.[52]

Although birthweight for gestational age is a useful measure, plain birthweight is often the only one available. In discussions that follow, understand *low birthweight* to refer to an infant with a birthweight below 5 1/2 pounds (2500 grams), unless otherwise indicated.

Infants may lose up to 10 percent of their birthweight during the first few days of life without cause for alarm. Health care providers become concerned when weight loss continues beyond 10 days or the lost weight is not regained to achieve birthweight by three weeks.[53] Such observations indicate failure to thrive.

Length Measurements. Health care professionals use special equipment to measure the length of infants. The assessor lays the barefoot infant on a measuring board that has a fixed headboard and movable footboard attached at right angles to the surface (see Figure 5-7). Often it takes two people to obtain an accurate measurement: one person to hold the infant's head against the headboard and the other to keep the legs straight and to do the measuring. This method provides the most accurate measure possible.

Weight Measurements. Special beam balance and electronic scales are available to weigh infants (see Figure 5-8). Their design allows for in-

small for gestational age (SGA): infants whose weight for gestational age falls below the 10th percentile.

low birthweight: a birthweight of less than 5 1/2 pounds.

Figure 5-6 Newborn Classification System

Newborn infants are classified into one of nine categories based first on gestational age and then on birthweight.

		Gestational age		
		Preterm (<38 weeks)	Term (38–42 weeks)	Post term (>42 weeks)
Birthweight	Small (<10%)	Small for preterm	Small for term	Small for post term
	Appropriate (10–90%)	Appropriate for preterm	Appropriate for term	Appropriate for post term
	Large (>90%)	Large for preterm	Large for term	Large for post term

Know this →

Figure 5-7   Nutrition Assessment of the Infant: How to Measure Length

An infant is measured lying down on a measuring board with a fixed head-board and movable footboard. Note that two people are needed to measure the infant's length.

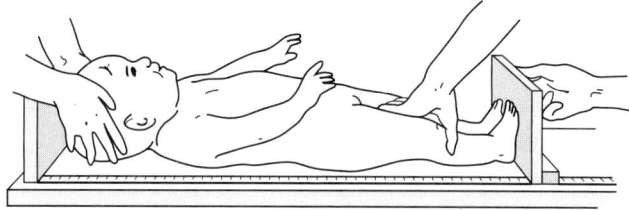

Source: Reprinted with permission of Ross Laboratories, Columbus, OH 43216.

Figure 5-8   Nutrition Assessment of the Infant: How to Measure Weight

Infants sit or lie down on scales that are designed to hold them while they are being weighed.

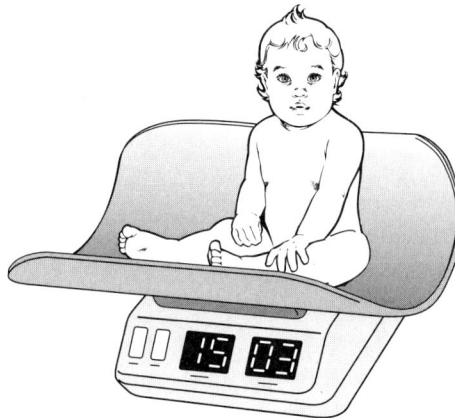

fants to lie or sit on the scales. Infants are weighed without clothing or diapers.

Head Circumference. Assessors may also measure head circumference to confirm that growth is proceeding normally or to help detect protein-energy malnutrition and evaluate the extent of its impact on brain size. The assessor places a nonstretchable tape so as to encircle the largest part of the infant's head: just above the eyebrow ridges, just above the point where the ears attach, and around the occipital prominence at the back of the head. For accuracy in recording, the measurer immediately notes the measure in either inches or centimeters.

Some assessors routinely measure head circumferences, whereas others do so only if they have reason to believe that a child has been severely malnourished. Malnutrition severe enough to reduce head circumference is seldom seen in developed countries, but is not uncommon among the poor in developing countries. The brain grows rapidly during early infancy, and malnourished children with small head circumferences may have fewer brain cells.

It is not uncommon to encounter infants whose head circumference measurements cross percentiles.[54] When researchers plotted head circumference measurements for more than 400 healthy, full-term infants, just over half showed acceleration or deceleration of at least one percentile curve between the ages of 1 month and 12 months. For most of the infants, the change in percentile curve was permanent. The researchers note that when clinicians encounter an infant whose head circumference crosses percentile curves, but whose health and general motor and neurological development appear normal, clinical observation for a certain period, rather than laboratory tests, is in order.

Standards. The growth charts in use today are based on a large sample of children, but the changing ethnicity of the country has prompted concerns about their use.[55] For example, survey data of Mexican-American children from both poor and nonpoor groups show that these children tend to be shorter, heavier, and fatter than other American children.

Concern about reference data used to monitor infant growth focuses on changing feeding methods. The growth charts for infants from birth to 36 months of age are based on data collected in the United States at a time when bottle-feeding was more prevalent than breastfeeding. Research shows that when the growth of breastfed infants is tracked on standard growth curves, they have weight gains similar to formula-fed infants during the first three months of life.[56] Throughout the next nine months, however, breastfed infants gain significantly less weight than formula-fed infants. When the growth of breastfed infants is plotted on these standard charts, they may appear to be faltering, even though they are thriving and healthy. Perhaps growth charts based on breastfed infants are needed.

Growth Charts. To evaluate growth in infants, as long as length is measured lying down, an assessor uses the charts shown in Figures 5-9 (A and B) and 5-10 (A and B). As soon as height can be measured standing, the assessor can switch to the charts presented in Chapter 6. The assessor follows these steps to plot a weight measurement on a percentile graph:

- Select the appropriate chart based on age and sex.
- Locate the infant's age along the horizontal axis on the bottom or top of the chart.
- Locate the infant's weight in pounds or kilograms along the vertical axis on the lower left or right side of the chart.
- Mark the chart where the age and weight lines intersect, and read off the percentile.

Figure 5-9A Nutrition Assessment from Birth to 36 Months: Length and Weight for Age—Girls

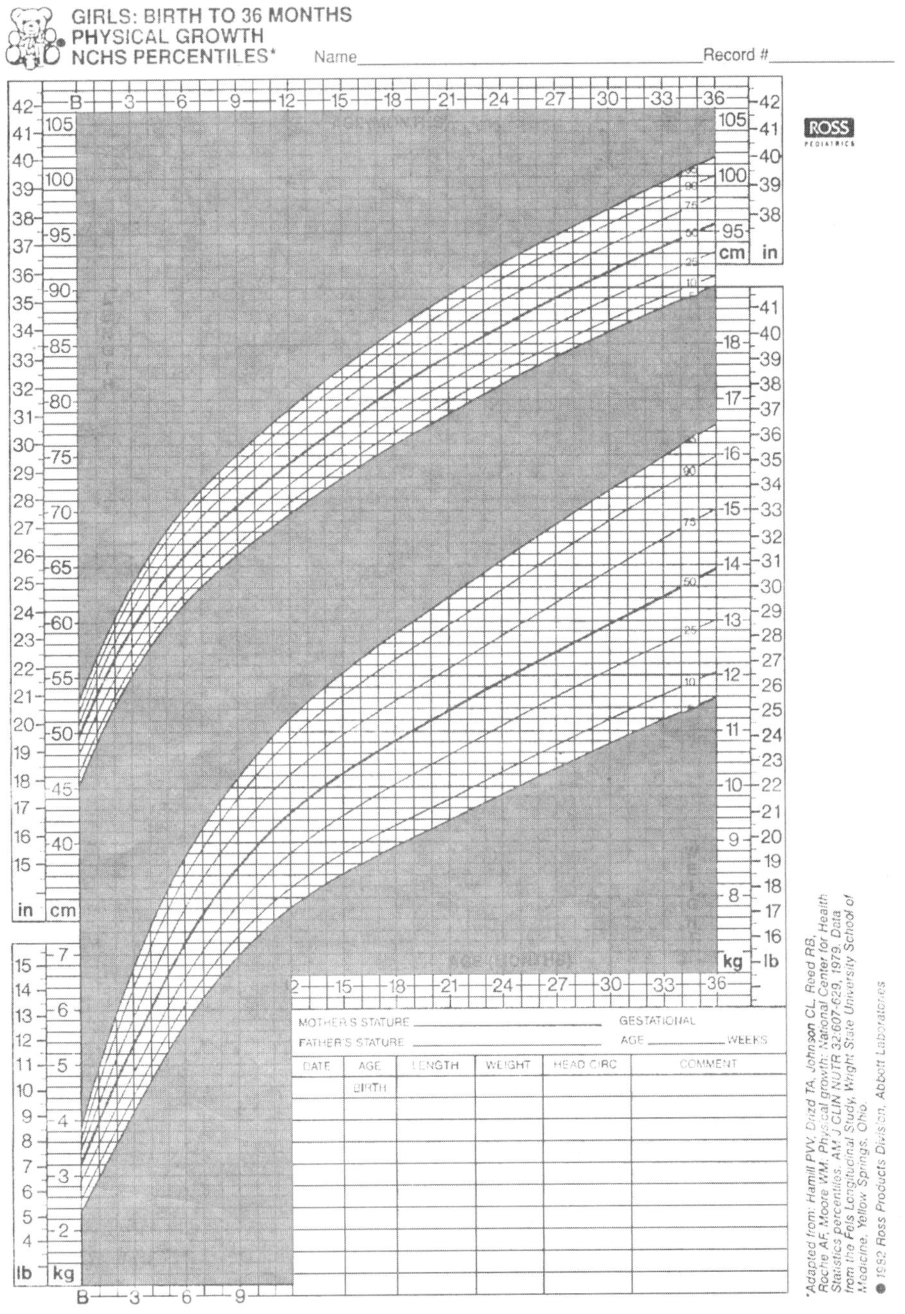

GIRLS: BIRTH TO 36 MONTHS
PHYSICAL GROWTH
NCHS PERCENTILES*

Source: Used with permission of Ross Laboratories, Columbus, OH 43216, from NCHS Growth Charts, © 1982 Ross Laboratories.

Figure 5-9B Nutrition Assessment from Birth to 36 Months: Length and Weight for Age—Boys

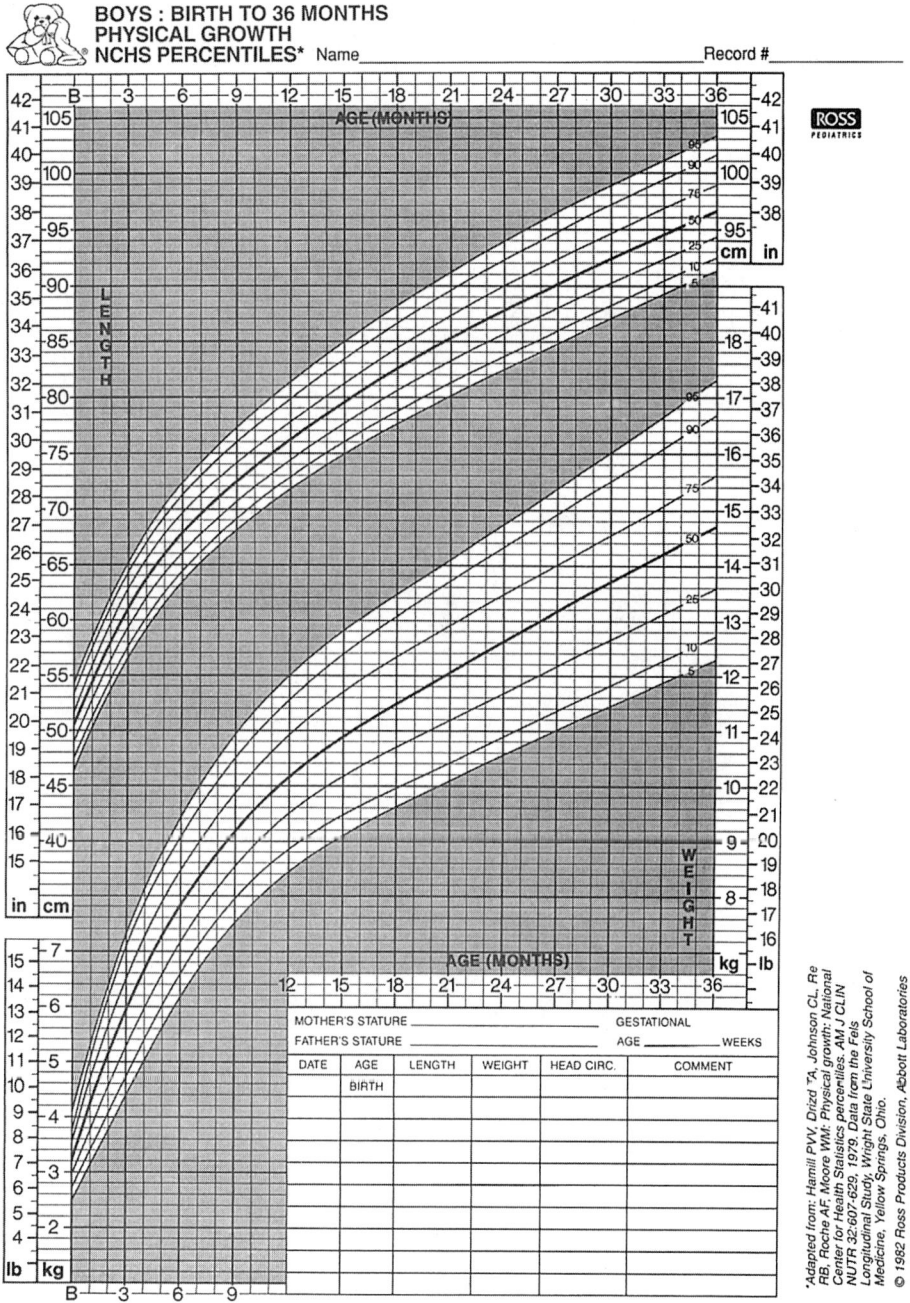

BOYS : BIRTH TO 36 MONTHS
PHYSICAL GROWTH
NCHS PERCENTILES* Name_____ Record #_____

ROSS
PEDIATRICS

MOTHER'S STATURE _____ GESTATIONAL
FATHER'S STATURE _____ AGE _____ WEEKS

DATE	AGE	LENGTH	WEIGHT	HEAD CIRC.	COMMENT
	BIRTH				

*Adapted from: Hamill PVV, Drizd TA, Johnson CL, Re
RB, Roche AF, Moore WM: Physical growth: National
Center for Health Statistics percentiles. AM J CLIN
NUTR 32:607-629, 1979. Data from the Fels
Longitudinal Study, Wright State University School of
Medicine, Yellow Springs, Ohio.

© 1982 Ross Products Division, Abbott Laboratories

Source: Used with permission of Ross Laboratories, Columbus, OH 43216, from NCHS Growth Charts, © 1982 Ross Laboratories.

Figure 5-10A Nutrition Assessment from Birth to 36 Months: Head Circumference for Age and Weight for Length—Girls

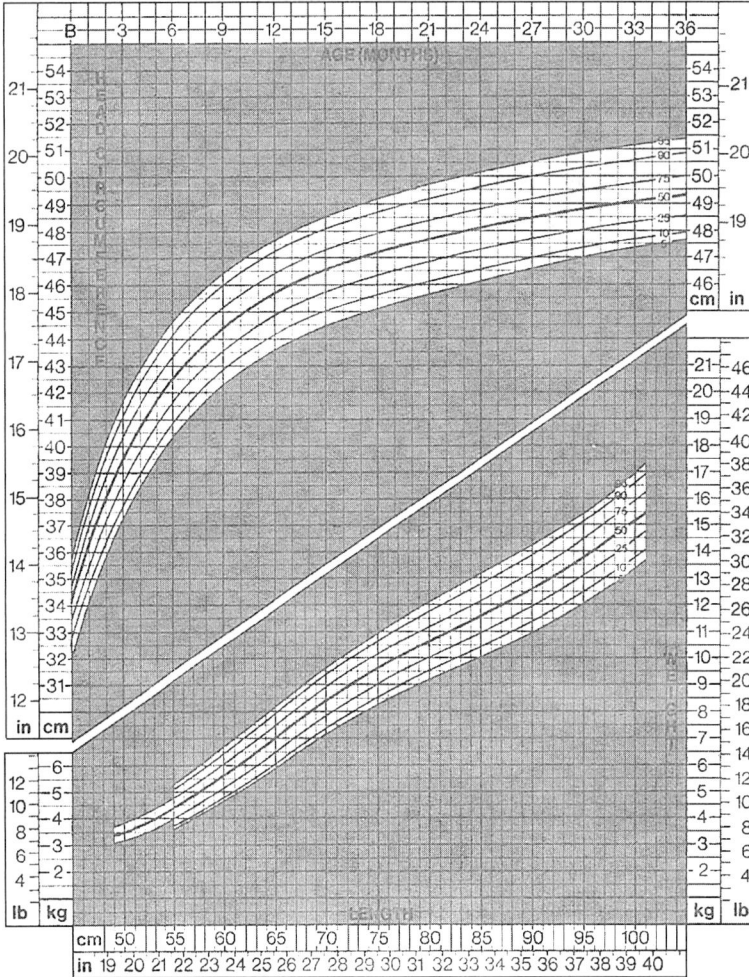

GIRLS: BIRTH TO 36 MONTHS
PHYSICAL GROWTH
NCHS PERCENTILES*   Name_____   Record #_____

Source: Used with permission of Ross Laboratories, Columbus, OH 43216, from NCHS Growth Charts, © 1989 Ross Laboratories.

Figure 5-10B Nutrition Assessment from Birth to 36 Months: Head Circumference for Age and Weight for Length—Boys

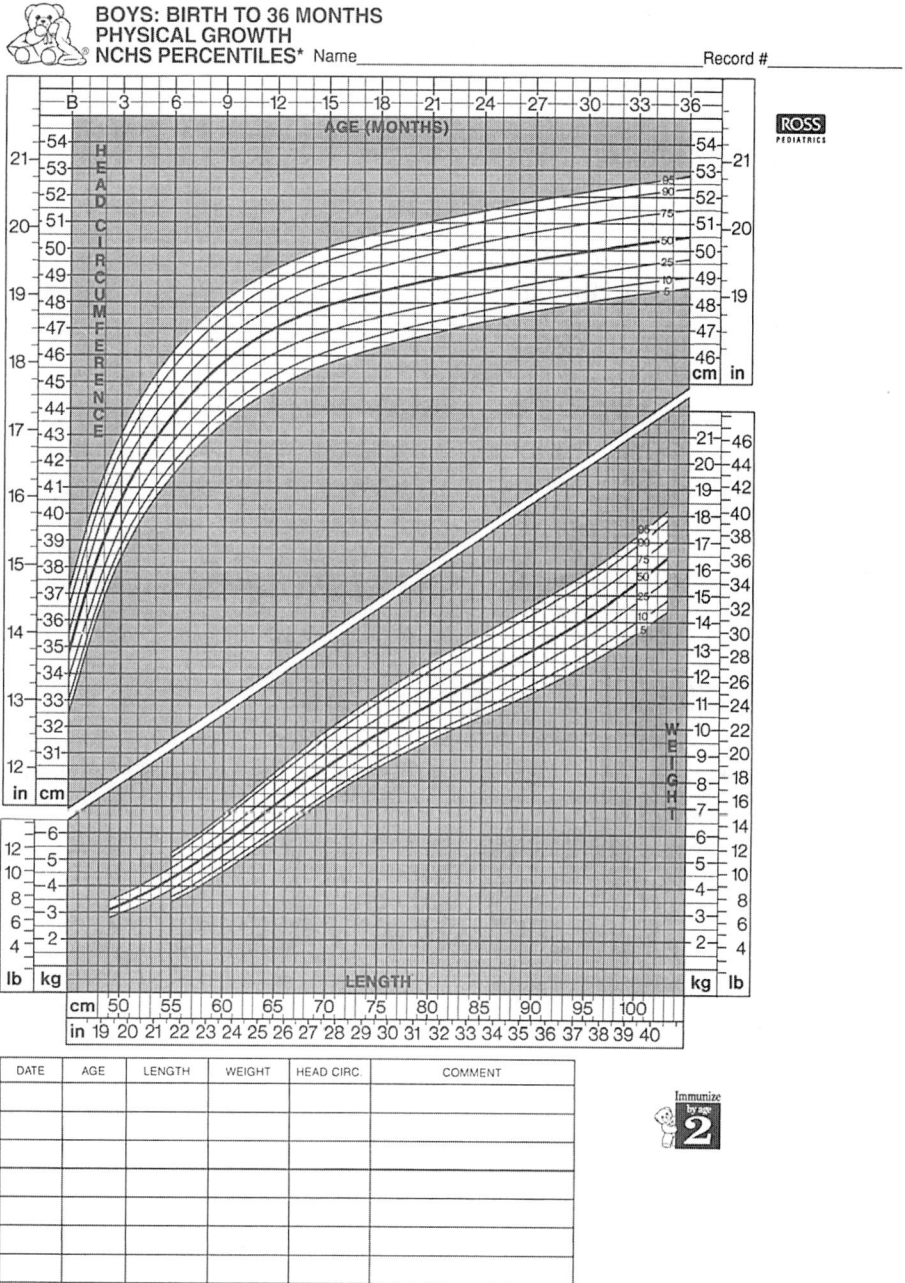

BOYS: BIRTH TO 36 MONTHS
PHYSICAL GROWTH
NCHS PERCENTILES* Name_____ Record #_____

To assess length, height, or head circumference, the assessor follows the same procedure, using the appropriate figure.

With length, weight, and head circumference measures plotted on growth charts, a skilled clinician can begin to interpret the data. Percentile charts divide the measures of a population into 100 equal divisions. Thus half of the population falls at or above the 50th percentile, and half falls below. The use of percentile measures allows for comparisons among people of the same age and sex. For example, a six-month-old female infant whose weight is at the 75th percentile weighs more than 75 percent of the female infants her age. An infant whose weight for length falls below the 5th percentile is defined as FTT, and those below the 10th percentile are suspect for FTT. Such infants require medical evaluation and care.

Physical Examinations and Biochemical Analyses

Clues to an infant's nutrition status can be identified by examining the infant for physical signs of malnutrition. Table 1-8 in Chapter 1 lists some general physical signs of malnutrition valid for all ages, and Table A-5 in Appendix A lists the signs of specific vitamin, mineral, and other imbalances. If the assessor suspects a nutrient deficiency or excess, the next step is to perform the appropriate laboratory tests.

As mentioned earlier, iron deficiency does not usually develop in infants prior to six months of age thanks to the iron they accumulate during gestation, but it becomes the most common nutrient deficiency in infants beyond six months of age. After six months, therefore, assessment of iron status is advisable. At a minimum, laboratory assessment of hemoglobin or hematocrit is suggested. Table 5-3 shows deficient and acceptable values for infants.[57]

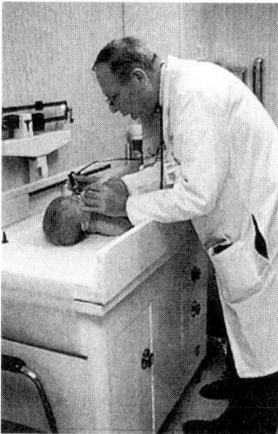

A close physical examination may reveal signs of malnutrition.

Clearly, sound nutrition promotes, and undernutrition impairs, normal growth and development during the first year of life. As the next section shows, common sense in the selection of infant foods and a nurturing, relaxed environment promote an infant's health and well-being.

Table 5-3 Standards for Hemoglobin and Hematocrit for Infants and Young Children

Category	Sex	Hemoglobin (grams per 100 milliliters)[a]		Hematocrit (%)	
		Deficient	Acceptable	Deficient	Acceptable
<0.5 to 4 yr	M-F	<11.0	≥11.0	<0.33	≥0.33

[a]To convert g/100 mL to standard international units (g/L), multiply by 10.

Source: Adapted from Committee on Nutrition, American Academy of Pediatrics, Iron deficiency, in *Pediatric Nutrition Handbook*, 3rd ed., ed. L. A. Barnes (Elk Grove, Ill.: American Academy of Pediatrics, 1993), pp. 227–236.

Feeding the Infant

Infant nutrition and feeding recommendations are much like the infant: constantly changing—in a word, dynamic. Historically, the timing of weaning to cow's milk and of introducing solid foods to the infant's diet has ranged from a few weeks to three years. On the surface, ever-changing infant feeding recommendations and practices are frustrating and confusing for parents and health care providers alike, but their significance goes much deeper. Such changes occur because new knowledge is being gained about the crucial importance of nutrition during infancy. Infancy sets the stage for eating habits that affect nutrition status and health for a lifetime.

All newborns begin life receiving breast milk, formula, or a combination as their only source of food. This continues until solid foods are introduced. Even then, breast milk and formula still contribute significantly to an infant's diet, providing about half of the daily energy intake. Chapter 4 recommended breastfeeding and described breast milk composition in detail; this section focuses on infant formula composition.

Standard Infant Formulas

National and international standards have been established for the nutrient contents of infant formulas. In the United States, the standard developed by the AAP reflects "human milk taken from well-nourished mothers during the first or second month of lactation, when the infant's growth rate is high." Formulas meeting the standard have similar nutrient compositions; small differences are sometimes confusing but usually unimportant.

Throughout the years, infant formula composition and labeling have reflected the knowledge of the time regarding the nutrient needs of infants. In 1941, labeling regulations pertained only to vitamins A, D, C, and thiamin, plus the minerals calcium, phosphorus, and iron. The remaining vitamins and minerals were considered adequate because the formulas of the time contained enough milk to provide them. In 1967, as knowledge about the nutrient needs of infants expanded, the AAP proposed minimum, and in some cases maximum, amounts for vitamins and minerals in formulas. Labeling regulations now specify how nutrient contents and preparation must be listed on labels. The progression of nutrient labeling recommendations for infant formulas is shown in Table 5-4.

To prepare a standard infant formula, manufacturers start with a nonfat cow's milk base or a mixture of nonfat cow's milk and added whey protein (if they are creating a whey-predominant formula, for reasons described later). As mentioned earlier, formula manufacturers replace the fat of cow's milk with vegetable oils. They then add lactose, vitamins, and minerals so that the energy content and nutrient distribution closely resemble those of breast milk. All standard infant formulas in the United States contain lactose as the principal carbohydrate. Fat is

Table 5-4 Labeling Standards and AAP Recommendations for Formulas

1941[a]	1967	1976	Current AAP Recommendations (per 100 kcal)
Protein	Protein	Protein	1.8 to 4.5 g
Fat	Fat	Fat	3.3 to 6.0 g[b]
Carbohydrate	Carbohydrate	Carbohydrate	—
Ash			
Vitamin A	Vitamin A	Vitamin A	75 to 225 μg RE
Vitamin D	Vitamin D	Vitamin D	1.0 to 2.5 μg
	Vitamin E	Vitamin E	0.5 mg α-TE
		Vitamin K	4 μg
Vitamin C	Vitamin C	Vitamin C	8 mg
Thiamin	Thiamin	Thiamin	40 μg
	Riboflavin	Riboflavin	60 μg
	Niacin	Niacin	250 μg
	Vitamin B$_6$	Vitamin B$_6$	35 μg[c]
	Folate	Folate	4 μg
	Pantothenic acid	Pantothenic acid	300 μg
	Vitamin B$_{12}$	Vitamin B$_{12}$	0.15 μg
		Biotin	1.4 μg
		Choline	7 mg
		Inositol	4 mg
Calcium	Calcium	Calcium	60 mg[d]
Phosphorus	Phosphorus	Phosphorus	30 mg[d]
	Magnesium	Magnesium	6 mg
		Sodium	20 to 60 mg (5.8 to 17.5 mEq[e])
		Potassium	80 to 200 mg (13.7 to 34.3 mEq[e])
		Chloride	55 to 150 mg (10.4 to 28.3 mEq[e])
Iron	Iron	Iron	0.15 to 2.5 mg
	Iodine	Iodine	5 to 25 μg
	Copper	Copper	60 μg
		Zinc	0.5 mg
		Manganese	5 μg
		Selenium	3 mg

[a]FDA Regulations—1941 Infant Formula Labeling Requirements, from the *Federal Register*, 1941.

[b]The AAP recommends 300 mg of the essential fatty acid linoleic acid.

[c]The vitamin B$_6$ recommendation provides 15 μg/g protein in formula.

[d]The recommended calcium-to-phosphorus ratio is 1:1 to 2:1.

[e]A milliequivalent (mEq) describes the concentration of electrolytes in a solution.

Source: Adapted from H. P. Sarett, The modern infant formula, in *Infant and Child Feeding*, ed. J. T. Bond (New York: Academic Press, 1981), pp. 99–121; Committee on Nutrition, American Academy of Pediatrics, Recommended nutrient levels of infant formulas, in *Pediatric Nutrition Handbook*, 3rd ed., ed. L. A. Barness (Elk Grove Village, Ill.: American Academy of Pediatrics, 1993), pp. 360–361.

whey protein: the principal protein in human milk, found in the liquid that remains after milk has been coagulated.

the major source of energy, as well as of essential fatty acids, in infant formulas and breast milk. Infant formulas provide 20 kcalories per ounce, as does breast milk.

The proteins of breast milk are whey protein and casein, in a ratio of 80 to 20. In contrast, the whey-to-casein ratio of cow's milk is 20 to 80. Some infant formulas maintain the whey-to-casein ratio of 20 to

80 (casein-predominant formulas), whereas others add whey protein, changing the ratio to 60 to 40 (whey-predominant formulas). Full-term infants grow equally well on either formula. Preterm infants, however, require whey-predominant formulas, which contain an amino acid profile better suited to their metabolic capacities.

Analysis of breast milk reveals that it is a rich source of the amino acid taurine. In 1984, formula makers began to add taurine to infant formulas at a concentration approximating that of breast milk.[58] The retina contains high concentrations of taurine, and although specific functions for taurine in the retina remain to be clarified, a link between taurine availability and the structural and functional integrity of the retina has been demonstrated.[59] Infants receiving taurine-free formulas do not show clinical signs of taurine deficiency, nor are there any known risks associated with its addition. Infants consuming taurine-supplemented formulas have plasma taurine concentrations similar to those of breast-fed infants.

The AAP makes specific recommendations regarding iron. All infant formulas must provide iron in amounts approximately equal to that of breast milk, which averages about 0.5 milligrams per liter. To compensate for the differences in bioavailability between breast milk and formulas, formulas must be fortified with 6 to 12 milligrams iron per liter. Both iron-fortified and nonfortified formulas are available. The AAP recommends, however, that all formula-fed infants receive iron-fortified formulas. Table 5-5 compares the nutrient composition of breast milk, cow's milk, and an infant formula.

Special Infant Formulas

Standard infant formulas are inappropriate for some infants. Special formulas are available, designed to meet the dietary needs of infants with specific conditions such as milk intolerance, prematurity, or inborn errors of metabolism.

Milk Intolerance. Soy formulas are recommended for infants with lactose intolerance. Soy formulas are prepared using soy for the protein source and corn syrup and sucrose instead of lactose. Soy formulas solve the problem of feeding an infant with either a temporary lactase deficiency caused by diarrhea or a congenital lactase deficiency. They are also useful as an alternative to milk-based formulas for vegetarian families.

Some infants with milk allergies are also allergic to soy protein. An infant with multiple food allergies or chronic diarrhea may require a hydrolysate formula. In a hydrolysate formula, the casein is hydrolyzed to amino acids and small peptides to permit easy absorption. Some of these formulas also replace the long-chain triglycerides with medium-chain triglycerides for infants with impaired fat absorption. Hydrolysate formulas are expensive and have an unappealing taste, but for the infant unable to tolerate milk- or soy-based formulas, they are indispensable.

casein: the principal protein in cow's milk, found in coagulated milk curds (*caseus* = cheese).

Soy-based formulas include Prosobee, Nursoy, Isomil, and Soyalac.

Casein hydrolysate formulas include Nutramigen and Pregestimil.

Table 5-5 Comparison of Breast Milk, Cow's Milk, and Infant Formula

Nutrient (per 100 mL)	Breast Milk	Cow's Milk	Infant Formula[a]
Energy-Yielding Nutrients			
Energy (kcal)	64	66	67
Protein (g)	0.9	3.4	1.5
Fat (g)	3.4	3.7	3.7
Carbohydrate (g)	6.6	4.9	7
Minerals			
Sodium (mg)	17	58	18
Potassium (mg)	55	138	71
Chloride (mg)	43	103	43
Calcium (mg)	26	125	51
Phosphorus (mg)	14	96	36
Magnesium (mg)	4	12	5
Iron (mg)	0.05	0.05	1.2[b]
Zinc (mg)	0.2	0.4	0.5
Copper (mg)	0.04	0.01	0.06
Vitamins			
Vitamin A (IU)	190	103	206
Thiamin (μg)	16	44	60
Riboflavin (μg)	36	175	101
Vitamin B_6 (μg)	10	64	41
Niacin (μg)	159	93	777
Pantothenic acid (μg)	198	352	311
Biotin (μg)	1	4	2.3
Folate (μg)	5	5	10
Vitamin B_{12} (μg)	0.04	0.04	0.14
Vitamin C (mg)	4.6	1.2	5.6
Vitamin D (IU)	2.2	3.4	41
Vitamin E (IU)	0.2	0.04	1.7
Vitamin K (μg)	1.5	6.0	5.7
Inositol (mg)	37	17	3
Choline (mg)	6	20	10

[a]Values represent the average for two major commercial products: (1) Similac, Ross Laboratories, and (2) Enfamil, Mead-Johnson Laboratories.

[b]The value represents formulas with iron fortification. The value for unfortified formula is 0.1 milligram.

Source: Adapted with permission from K. J. Motil, Breast-feeding: Public health and clinical overview, in *Pediatric Nutrition*, eds. R. J. Grand, J. L. Sutphen, and W. H. Dietz, Jr. (Stoneham, Mass.: Butterworths, 1987), pp. 251–263. Updated 1997, personal communication.

Preterm Infants. Preterm infants, especially very-low-birthweight infants (less than 1500 grams), have a limited digestive capacity. For this reason, formulas designed for preterm infants contain a mixture of lactose and glucose polymers, and a blend of medium-chain triglycerides and unsaturated long-chain triglycerides. Preterm infants also have greater nutrient needs than full-term infants. For this reason, formulas for preterm infants generally have higher protein, mineral, and vitamin concentrations than standard formulas. They also have an energy density of 24 kcalories per ounce, slightly greater than that of regular infant formulas.

The whey-to-casein ratio in formulas for preterm infants is adjusted to 60 to 40, to approximate that in breast milk. Whey protein is preferred, because it is higher in the amino acid cysteine than casein is. Preterm infants may lack the hepatic enzyme cystathionase, needed to convert the essential amino acid methionine to the nonessential amino acid cysteine.[60] Without this conversion, cysteine becomes a conditionally essential amino acid. Soy formulas are not recommended for preterm infants. Focal Point 5A presents additional information on the nutritional care of preterm infants.

Inborn Errors of Metabolism. Other formulas are available for infants with inborn errors of metabolism who cannot metabolize specific nutrients. These formulas purposely lack one or more nutrients and are called *incomplete* formulas. For this reason, they are not appropriate for other infants. Focal Point 5C describes the special needs of infants with inborn errors of metabolism.

Figure 5-11 illustrates the process of choosing a formula, and Focal Point 4 offers directions for the feeding process itself. Infants born in technologically advanced countries are fortunate to have a diverse array of formulas available to them. At one time, death was the inevitable outcome for infants without access to breast milk. Government regulations for commercially prepared formula ensure protection for formula-fed infants. Mothers can feel confident that infant formulas offer a safe, nutritionally sound alternative to breast milk. At one year of age, cow's milk can replace breast milk or formula in the infant's diet.

Introducing Cow's Milk

The timing of the introduction of whole cow's milk into the infant's diet has been a source of controversy over the years. The AAP currently recommends the introduction of whole cow's milk at 12 months of age.[61]

As mentioned earlier, in some infants, particularly those younger than six months of age, the consumption of whole cow's milk is associated with intestinal bleeding and iron deficiency. Whole cow's milk is also a poor source of iron. Consequently, cow's milk both causes iron loss and fails to replace iron. Furthermore, the bioavailability of iron from infant cereal and other foods is reduced when cow's milk replaces breast milk or iron-fortified formula during the first year. Compared to breast milk or iron-fortified formula, cow's milk is higher in calcium and lower in vitamin C, characteristics that inhibit iron absorption.[62] In short, cow's milk is a poor choice during the first year of life.

Figure 5-11 Choosing a Formula

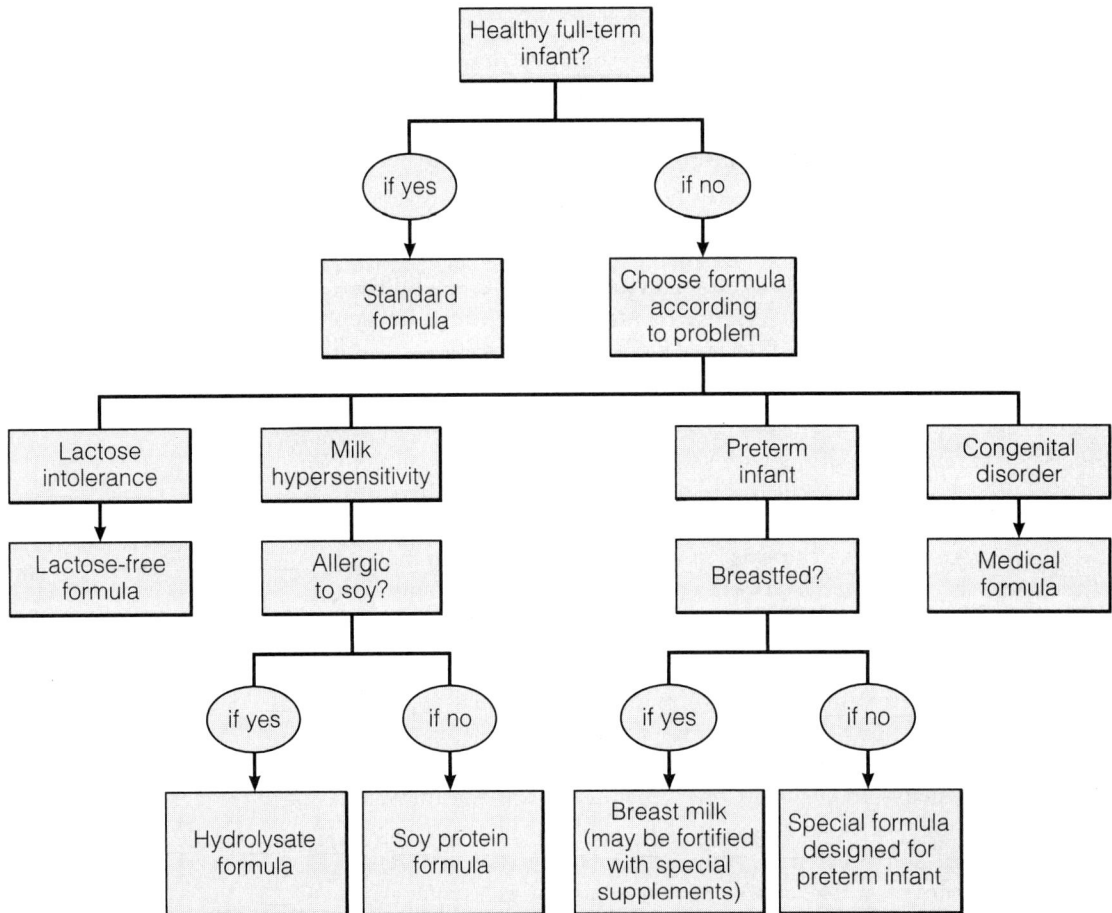

Introducing Solid Foods

The high nutrient needs of infancy must be met by breast milk or formula only at first, and then by a limited diet to which foods are gradually added. Infants only gradually develop the ability to chew, swallow, and digest the wide variety of foods available to adults. The caregiver's selection of appropriate foods at the appropriate stages of development is prerequisite to the infant's optimal growth and health (see Table 5-6).

*The German word **beikost** (BY-cost) is sometimes used to describe any nonmilk food given to an infant.*

When to Begin. Few issues of infant feeding have been debated more than the question of when it is most appropriate to introduce solid foods into the infant's diet. As is true with all aspects of infant feeding, recommendations for the timing of the introduction of solid foods have changed repeatedly over the years. At the beginning of this century, solid foods were not introduced until 10 months of age or older.[63] In the 1950s, the AAP suggested that iron-containing foods be introduced

Table 5-6 Infant Feeding Skills and Recommended Foods

Note: Because each stage of development builds on the previous stage, the foods from an earlier stage continue to be included in all later stages.

Age (months)	Feeding Skill	Appropriate Foods Added to the Diet
0–4	Turns head toward any object that brushes cheek.	Feed breast milk or infant formula.
	Initially swallows using back of tongue; gradually begins to swallow using front of tongue as well.	
	Strong reflex (extrusion) to push food out during first 2 to 3 months.	
4–6	Extrusion reflex diminishes, and the ability to swallow nonliquid foods develops.	Begin iron-fortified cereal mixed with breast milk, formula, or water.
	Indicates desire for food by opening mouth and leaning forward.	Begin pureed vegetables and fruits.
	Indicates satiety or disinterest by turning away and leaning back.	
	Sits erect with support at 6 months.	
	Chewing action begins.	
	Brings hand to mouth.	
	Grasps objects with palm of hand.	
6–8	Able to feed self finger foods.	Begin breads and other cereals.
	Pincher (finger to thumb) grasp developing.	Begin textured vegetables and fruits.
	Begins to drink from cup.	Begin plain, unsweetened fruit juices from cup.
8–10	Begins to hold own bottle.	Begin breads and cereals from table.
	Reaches for and grabs food and spoon.	Begin yogurt.
	Sits unsupported.	Begin pieces of soft, cooked vegetables and fruit from table.
		Gradually begin finely cut meats, fish, casseroles, cheese, eggs, and legumes.
10–12	Begins to master spoon, but still spills some.	Include at least 4 servings of breads and cereals from table, in addition to baby cereal.[a]
		Include at least 2 servings of fruits and 3 servings of vegetables.[a]
		Include 2 servings of meat, fish, poultry, eggs, or legumes.[a]

[a] Serving sizes for infants and young children are smaller than those for an adult. For example, a serving might be 1/2 slice of bread instead of 1, or 1/4 c rice instead of 1/2 c.

Source: Adapted in part from Committee on Nutrition, American Academy of Pediatrics, *Pediatric Nutrition Handbook*, 3rd ed., ed. L. A. Barness (Elk Grove Village, Ill.: American Academy of Pediatrics, 1993), pp. 23–33.

during the third month of life. In the 1970s, most infants received solid foods by six weeks of age. Currently, the AAP recommends the introduction of solid foods between four and six months of age.

The choice of this later time to introduce solid foods accompanied the resurgence of breastfeeding and was probably in the interest of postponing weaning. Formula-fed infants typically receive solid foods earlier than breastfed infants do. Over half of formula-fed infants receive solid foods between two and three months of age, compared with one-fourth of breastfed infants.

In practice, the timing of when to introduce solid foods to infant diets should not depend on whether the infant is breastfed or formula fed, or even on the infant's chronological age. It should depend on the individual infant's nutrient needs and developmental readiness. These factors vary from infant to infant because of differences in growth rates, activity, and environmental conditions.

The main purpose of introducing solid foods to infants is to provide nutrients that are no longer supplied adequately by breast milk or formula alone. The foods chosen must be foods that the infant is developmentally capable of handling both physically and metabolically. Just as it is unrealistic to expect a one-year-old to read and write or ride a bicycle, so it is unrealistic to expect an infant to handle adult foods, or even baby foods, before the physiological capacity exists.

By four to six months of age, not only the nutrient needs but also the energy needs of some infants become too great for breast milk or formula alone to provide. At about four months of age for some infants, and no later than six months of age for all infants, foods other than breast milk or formula should make a significant energy contribution to the diet. It is important, though, that foods not displace breast milk or formula, for they remain the infant's most important sources of nutrients. Foods should be introduced into the diet only as supplemental foods given *in addition to* breast milk or formula, not as replacements for either.

By the end of the first year, infants should be accustomed to eating a variety of foods. As their energy needs increase, their intakes of nonmilk foods will increase as well. Breast milk or formula consumption will decrease gradually. Still, throughout the first year of life, breast milk or formula continues to be the major energy and nutrient source for infants.

What to Introduce First. With continued growth and increasing activity, energy needs become greater and needs for other nutrients intensify as well. Prominent among nutrients demanding attention at this time is iron. Even though the infant is born with ample supplies of iron, by four to six months of age, iron stores have diminished, and additional sources of iron beyond breast milk or formula are needed. Iron-fortified infant cereals are a desirable source of supplemental iron at this time and throughout the first year.

Iron-fortified infant cereal is advantageous as the first solid food for several reasons. The needed iron is provided in a convenient, economical form. Cereals readily mix with fluids, so they are of a consistency easily swallowed by the infant. They provide energy, calcium, phosphorus, and other vitamins and minerals as well as iron, and most infants accept and tolerate them well.

Food Allergies and Intolerances. The timing and type of food introduced may facilitate or prevent the onset of an allergic response. In early infancy, the mucosal barrier of the small intestine is permeable to large molecules. This is advantageous, because it allows maternal antibodies to cross the barrier intact, affording the breastfed infant protection against infection. If, however, inappropriate protein sources, such as those in cow's milk and some solid foods, are fed, undigested proteins may be absorbed and induce allergies. Symptoms of food allergy include nausea, vomiting, abdominal discomfort, respiratory disturbances, and skin rashes.

This chapter focuses on prevention and early detection of allergy; Chapter 6, which deals with children's nutrition, discusses allergies and intolerances in detail. To prevent allergy, and to facilitate its prompt identification should it occur, experts recommend the introduction of single-ingredient foods, one at a time, in small portions, allowing four to five days before introducing the next new food. The gradual introduction of single-ingredient foods permits identification of problem foods. Rice cereal, barley cereal, or other single-ingredient infant cereals are appropriate first foods. Mixed cereals and combination foods can be offered once sensitivity to specific foods has been ruled out.

The introduction of plain cow's milk to an infant's diet should occur after 12 months, for reasons already mentioned (p. 185), and in most infants it should then cause no problems. Adverse reactions may occur in some infants, however, because of either milk protein hypersensitivity (milk allergy) or lactose intolerance. The incidence of milk protein hypersensitivity in infants is estimated to range from 0.5 to as high as 8 percent.[64] The actual incidence is difficult to pinpoint because at present there is no satisfactory, generally accepted way of making the diagnosis. Many infants with milk protein hypersensitivity outgrow it by early childhood.

Other than milk, the foods most often implicated in infants' food allergies include citrus fruits, soy protein, egg whites, and wheat. Some pediatricians recommend delaying the introduction of these foods until around nine months of age to coincide with more complete digestive tract development.

Aside from food allergies, infants may exhibit adverse reactions to foods because of ingredients in them such as bacterial toxins, monosodium glutamate, or the natural laxative in prunes. Adverse reactions may also be caused by digestive tract disorders, such as obstructions or injuries, or inborn errors of metabolism. In some cases it is not possible or necessary to say whether an adverse food reaction is due to allergy or something else, but avoidance of the food, at least for a while, is indicated.

Infants frequently outgrow food allergies and other adverse reactions to foods within the first year. Avoidance of offending foods, therefore, should not have to be permanent unless a physician confirms a food allergy.

Infant Foods

Infant foods should be selected to provide variety, balance, and moderation. Commercial baby foods offer a wide variety of palatable, nutritious foods in a safe and convenient form. Home-made infant foods can be as nutritious as commercially prepared ones, as long as the preparer minimizes nutrient losses during preparation. Ingredients for home-made baby food recipes should be fresh, whole foods without added salt, sugar, or seasonings. Pureed food can be frozen in ice cube trays, providing convenient-sized blocks of food that can be thawed, warmed, and fed to the infant. The preparer should take precautions to guard against food contamination or infection; hands and equipment must be clean.

Commercial infant foods fall into eight groups: cereals, fruits, fruit juices, vegetables, meats, meat and vegetable combinations, yogurts, and desserts. Most are available either in dry form (to which liquid is added) or ready-prepared in jars. In response to consumer concerns, manufacturers have reduced or eliminated salt, sugar, and other additives in commercial infant foods. All infant food ingredients are listed on the label in descending order by weight. In the United States, FDA regulations require that certain nutrients be listed on the label, as shown in Figure 5-12.

The Nutrition Labeling and Education Act of 1990 set forth special rules for the labeling of foods for infants and toddlers.[65] For example, because recommendations to restrict fat do not apply to children under age two, labels on foods for children under two (such as infant meats or cereals), cannot carry information about kcalories from fat or kcalories from saturated fat. In addition, the labels cannot list amounts of saturated fat, polyunsaturated fat, monounsaturated fat, or cholesterol.

Figure 5-12 Infant Food Label

Most fat information on infant food labels is omitted in order to prevent parents from wrongly restricting fat in babies' diets. Parents may unintentionally malnourish their infants by restricting fat because of fears that the infant will become overweight.[66] In fact, infants and young children, because of their rapid growth, need more fat than older children and adults.

Infant Cereals. As mentioned, iron-fortified infant cereal is usually recommended as the first solid food for the infant. Rice cereal is usually the first cereal introduced, because it is the least allergenic. Infant cereals are available as single-grain or mixed varieties. Single-grain cereals are rice, barley, and oatmeal. Mixed cereals include mixed-grain varieties and cereals mixed with fruit. Ready-prepared cereals in jars usually contain fruit, as do the dehydrated varieties. Dry cereals are prepared by adding water, breast milk, or formula to obtain the desired consistency, which should be almost liquid at first. Cereals and other solids should be fed by spoon and offered once a day in small quantities (1 to 2 tablespoons) until the infant becomes accustomed to the food and spoon.

Once the infant is consuming cereals, the caregiver should begin selecting vitamin C-rich foods to accompany them in order to promote optimal iron absorption. Vitamin C greatly enhances the absorption of nonheme iron, which is the common form of iron in infant foods.

Fruits and Fruit Juices. Fruits are often the second food introduced to infants. Commercially prepared infant fruits are fortified with vitamin C and may provide up to half the infant RDA of this vitamin per serving. Most infants like the sweet taste of fruits. Baby food fruits are made from fresh fruits or concentrates. Some of the more tart, acidic, prepared fruits contain small amounts of added sugar, whereas others, such as apples and pears, contain no added sugar. Some infant fruits contain a little modified starch to improve consistency. Modified starches are made from corn, tapioca, potato, or wheat starch treated in such a way as to improve palatability, product stability, and shelf life. The use of modified starches in baby foods has been extensively studied by the FDA and the AAP and is considered safe as practiced.

Fruit juices prepared for infants are also fortified with vitamin C. Fruit juices contain no added sugar and are prepared from pressed whole fruits or concentrates. Several brands of regular fruit juices that are prepared similarly may also be used for infants. All fruit juices should be diluted and introduced in a cup, not a bottle, once the infant is six months of age or older. Some pediatricians recommend waiting until after the infant's first birthday to introduce juice. Juices should be used moderately in the infant diet, so as not to displace other foods. Cases have been reported of children failing to grow and thrive because they were drinking so much juice each day that other, more energy- and nutrient-dense foods were being displaced from their diets.[67]

Vegetables. Commercially prepared baby food vegetables, like all vegetables, are good sources of vitamins and minerals, especially vitamin A and the B vitamins. The introduction of vegetables in infant diets some-times follows that of fruits, but the opposite order may better favor vegetable acceptance. Once an infant is accustomed to the sweet taste of fruits, the tastes of some vegetables may be less appealing than if they had been offered prior to the fruits.

Yogurts. Commercial infant yogurts have a lower sugar content than adult fruit-flavored yogurts and are good sources of calcium and pro-tein. Plain or reduced-sugar varieties of regular yogurt are also accept-able for infants. Yogurt is a popular food for infants and is often intro-duced around eight months of age.

Meats. Meats are generally introduced between eight and ten months of age. Meats for infants are prepared with broth. They are excellent sources of protein, B vitamins, and iron. A suitable choice for the first meat food to offer is a meat-and-vegetable combination, which contains less protein per serving than single-ingredient meats. In most cases, pro-tein from meat sources is not critical to an infant's nutrition, because adequate protein is available from breast milk, formula, and infant ce-real. Meats may be less readily accepted than other foods at first, but this is no cause for concern. As long as the infant is consuming iron-fortified cereal and breast milk or formula daily, the iron and protein contribu-tion of meats is dispensable.

Feeding Infants in Vegetarian Families. The young infant is a lacto-vegetarian. As long as the infant has access to sunlight as a source of vi-tamin D and to sufficient quantities of breast milk from a mother who eats an adequate diet, the infant will thrive during the early months. The same is true of infants fed standard or soy-based infant formulas.

Infants beyond about 4 months of age present a greater challenge in terms of meeting nutrient needs by way of vegetarian and, especially, vegan diets. Continued breastfeeding or formula feeding is recom-mended, but supplementary feedings are necessary to ensure adequate energy and iron intakes. The risks of poor nutrition status in infants in-crease with weaning and reliance on table foods. Infants who receive a well-balanced vegetarian diet that includes milk products and a va-riety of other foods can easily meet their nutritional requirements for growth. This is not always true for vegan infants. Researchers studying growth differences between vegan and nonvegetarian infants and chil-dren find consistent, significant depressions in growth of vegan infants around the time of transition from breast milk to solid foods.[68] Severely restrictive vegan diets pose the greatest threats to infants' health and nutrition status. Parents or caregivers who choose to feed their infants vegan diets should consult with their pediatrician.

Protein-energy malnutrition and deficiencies of vitamin D, vita-min B_{12}, iron, and calcium have been reported in infants fed vegan diets.[69] Vegan diets that are high in fiber, other complex carbohy-drates, and water have a low kcaloric density. Infants in families who eat such diets often receive gruel mixtures of cereal grains when they are

vegan (VAY-gun or VEJ-an) **diets:** diets from which all animal-derived foods (meat, poultry, fish, eggs, and dairy products) are excluded.

weaned from the breast. Cereals absorb significant amounts of water when cooked, sometimes increasing their volume three to four times. The stomach capacity of infants is limited; when fed gruel mixtures of low kcaloric density, infants cannot consume enough volume to meet their energy needs. This problem can be partially alleviated by providing more nut butters, legumes, dried fruit spreads, and mashed avocado, while limiting the infant's intake of foods with low kcaloric density such as vegetables and gruels. Calcium and vitamin D-fortified soy milk, vitamin B_{12}-fortified foods or supplements, and the inclusion of vitamin C-containing foods at meals to enhance iron absorption will alleviate other nutrient deficiencies in vegan infant diets.

Infants and young children in vegetarian families should be given iron-fortified infant cereals well into the second year of life. Legumes and whole-grain foods can be added to their diets in place of meat.

Foods to Omit. Concentrated sweets, including baby food "desserts," have no place in an infant's diet. They convey no nutrients to support growth, and the extra food energy can promote obesity. Canned vegetables are also inappropriate for infants, as they often contain too much sodium. Honey and corn syrup should never be fed to infants because of the risk of botulism.* Infants and even young children cannot safely chew and swallow popcorn, whole grapes, whole beans, hot dog slices, hard candies, and nuts; they can easily choke on these foods, a risk not worth taking.

Foods at One Year. At one year of age, whole cow's milk becomes the primary source of most of the nutrients an infant needs; 2 to 3 1/2 cups a day meets those needs sufficiently. More milk than this displaces foods necessary to provide iron and can lead to milk anemia. Children one to two years old should drink whole milk, not reduced fat, low-fat, or nonfat milk. If powdered milk is used, it should be one of the fat-containing varieties. Other foods—meats, iron-fortified cereals, enriched or whole-grain breads, fruits, and vegetables—should be supplied in variety and in amounts sufficient to round out total energy needs. Ideally, a one-year-old will sit at the table, eat many of the same foods everyone else eats, and drink liquids from a cup, not a bottle. Table 5-7 shows a meal plan that meets a one-year-old's requirements.

Activity

Infants normally are spontaneously active; they need no special programs or equipment. Caregivers can provide opportunities for normal activity by giving them freedom of movement and the stimulation of

botulism (BOT-chew-lism): an often fatal food-borne illness caused by the ingestion of foods containing a toxin produced by bacteria that grow without oxygen.

milk anemia: iron deficiency anemia that develops when an excessive milk intake displaces iron-rich foods from the diet.

---

*In infants, but not in older individuals, ingestion of *Clostridium botulinum* spores can cause illness when the spores germinate in the intestine and produce toxin, which is absorbed. Symptoms include poor feeding, constipation, loss of tension in the arteries and muscles, weakness, and respiratory compromise. Infant botulism has been implicated in 5 percent of cases of sudden infant death syndrome (SIDS).

Ideally, a one-year-old eats many of the same foods as the rest of the family.

Table 5-7   Meal Plan for a One-Year-Old

Breakfast	*Afternoon Snack*
1 c whole milk	1/2 c whole milk
3 tbs cereal	Teething crackers
2 to 3 tbs fruit[a]	1 tbs peanut butter
Teething crackers	*Dinner*
Morning Snack	1 c whole milk
1/2 c whole milk	1 egg
1 to 2 tbs fruit[a]	2 tbs cereal or potato
Teething crackers	2 to 3 tbs vegetables[b]
Lunch	2 to 3 tbs fruit[a]
1 c whole milk	
2 to 3 tbs vegetables[b]	
2 tbs chopped meat or well-cooked, mashed legumes	

[a]Include citrus fruits, melons, and berries.
[b]Include dark green, leafy and deep yellow vegetables.

play. (Infants benefit, too, from the age-old tradition of making sure they have "plenty of sunshine and fresh air.") An active infant is on the way to becoming an active child; activity defends the infant from the problem of obesity that creeps into many people's lives as early as young childhood.

Table 5-8   Mealtimes with a One-Year-Old

These recommendations reflect a spirit of tolerance that serves the best interest of the child emotionally as well as physically.

- Discourage unacceptable behavior, such as standing at the table or throwing food, by removing the child from the table to wait until later to eat. Be consistent and firm, not punitive. The child will soon learn to sit and eat.

- Let the child explore and enjoy food, even if this means eating with fingers for a while. Use of the spoon will come in time.

- Don't force food on children. Rejecting new foods is normal; acceptance is more likely as infants and children become familiar with new foods through repeated opportunities to taste them.

- Provide nutritious foods, and let the child choose which ones and how much they will eat. Gradually they will acquire a taste for different foods.

- Limit sweets. Infants and young children have little room in their daily energy allowance for empty-kcalorie foods. Do not use sweets as a reward for eating meals.

- Don't turn the dining table into a battleground. Make mealtimes enjoyable. Teach healthy food choices and eating habits in a pleasant environment.

Feeding an infant appropriately supports sound nutrition and health, and eating habits acquired during infancy and childhood influence the overall food attitudes of the individual throughout life. The nurturing of an infant, however, involves more than nutrition. In light of infants' developmental and nutrient needs, and in the face of their often contrary and willful behavior, a few feeding guidelines may be helpful (see Table 5-8). Those who care for infants are responsible for providing not only nutritious milk, foods, and water, but also a safe, loving, secure environment in which the infants may grow and develop.

CHAPTER FIVE NOTES

1. The gastrointestinal tract: Development and nutrition, in *Dynamics of Infant Physiology and Nutrition*, ed. L. J. Filer (Bloomfield, N.J.: Health Learning Systems, 1982), pp. 1–16.

2. Committee on Nutrition, American Academy of Pediatrics, *Pediatric Nutrition Handbook*, 3rd ed., ed. L. A. Barness (Elk Grove Village, Ill.: American Academy of Pediatrics, 1993), pp. 23–33.

3. J. A. Olson, Recommended dietary intakes (RDI) of vitamin A in humans, *American Journal of Clinical Nutrition* 45 (1987): 704–716.

4. P. S. W. Davies, Energy requirements and energy expenditure in infancy, *European Journal of Clinical Nutrition* 46 (1992): S29–S35.

5. Committee on Dietary Allowances, Food and Nutrition Board, *Recommended Dietary Allowances*, 10th ed. (Washington, D.C.: National Academy Press, 1989), p. 35.

6. R. G. Whitehead and coauthors, A critical analysis of measured food energy intakes during infancy and early childhood in comparison with current international recommendations, *Journal of Human Nutrition* 35 (1981): 339–348; S. J. Fomon and E. F. Bell, Energy, in *Nutrition of Normal Infants*, ed.

S. J. Fomon (St. Louis: Mosby, 1993), pp. 103–120.

7. R. G. Whitehead and A. A. Paul, Human lactation, infant feeding and growth: Secular trends, in *Nutritional Needs and Assessment of Normal Growth*, eds. M. Gracey and F. Falkner (New York: Raven Press, 1985), pp. 85–122.

8. M. J. Heinig and coauthors, Energy and protein intakes of breast-fed and formula-fed infants during the first year of life and their association with growth velocity: The DARLING Study, *American Journal of Clinical Nutrition* 58 (1993): 152–161.

9. Committee on Dietary Allowances, 1989, pp. 56–58.

10. W. C. Heird, Nutritional requirements during infancy and childhood, in *Modern Nutrition in Health and Disease*, 8th ed., eds. M. E. Shils, J. A. Olson, and M. Shike (Philadelphia: Lea & Febiger, 1994), pp. 740–758; G. E. Gaull, Taurine in pediatric nutrition: Review and update, *Pediatrics* 83 (1989): 433–442.

11. Committee on Dietary Allowances, 1989, pp. 52–77.

12. Committee on Dietary Allowances, 1989, pp. 62–64.

13. C. Garza, N. F. Butte, and A. S. Goldman, Human milk and infant formula, in *Textbook of Pediatric Nutrition*, 2nd ed., eds. R. M. Suskind and L. Lewinter-Suskind (New York: Raven Press, 1993), pp. 33–42.

14. A. C. Goedhart and J. C. Bindels, The composition of human milk as a model for the design of infant formulas: Recent findings and possible applications, *Nutrition Research Reviews* 7 (1994): 1–23.

15. J. A. Nettleton, Are n-3 fatty acids essential nutrients for fetal and infant development? *Journal of the American Dietetic Association* 93 (1993): 58–64.

16. Nettleton, 1993.

17. J. Farquharson and coauthors, Infant cerebral cortex phospholipid fatty acid composition and diet, *Lancet* 340 (1992): 810–813; M. Makrides and coauthors, Fatty acid composition of brain, retina, and erythrocytes in breast- and formula-fed infants, *American Journal of Clinical Nutrition* 60 (1994): 189–194.

18. M. Makrides and coauthors, Are long-chain polyunsaturated fatty acids essential nutrients in infancy? *Lancet* 345 (1995): 1463–1468.

19. M. A. Crawford, The role of essential fatty acids in neural development: Implications for perinatal nutrition, *American Journal of Clinical Nutrition* 57 (1993): 703S–710S.

20. Committee on Nutrition, 1993, pp. 23–33.

21. Olson, 1987.

22. D. A. Bender, Vitamin D, in *Nutritional Biochemistry of the Vitamins* (New York: Cambridge University Press, 1992), pp. 51–86.

23. Committee on Nutrition, 1993, pp. 34–42.

24. S. J. Fomon and E. F. Bell, Vitamin E, in *Nutrition of Normal Infants*, ed. S. J. Fomon (St. Louis: Mosby, 1993), pp. 339–347.

25. J. A. Olson, Recommended dietary intakes (RDI) of vitamin K in humans, *American Journal of Clinical Nutrition*, 45 (1987): 687–692.

26. V. Herbert, Recommended dietary intakes (RDI) of folate in humans, *American Journal of Clinical Nutrition* 45 (1987): 661–670.

27. Committee on Nutrition, 1993, pp. 362–364.

28. H. L. Greene, Disorders of the water-soluble vitamin B-complex and Vitamin C, in *Textbook of Pediatric Nutrition*, 2nd ed., eds. R. M. Suskind and L. Lewinter-Suskind (New York: Raven Press, 1993), pp. 73–89.

29. Committee on Dietary Allowances, 1989, p. 180.

30. Committee on Dietary Allowances, 1989, p. 180.

31. L. S. Hillman, Mineral and vitamin D adequacy in infants fed human milk or formula between 6 and 12 months of age, *Journal of Pediatrics* (supplement) 117 (1990): S134–S142.

32. J. B. Anderson and C. Barrett, Dietary phosphorus: The benefits and problems, *Nutrition Today*, April 1994, pp. 29–34.

33. Committee on Dietary Allowances, 1989, p. 254.

34. Garza, Butte, and Goldman, 1993.

35. J. Raloff, Umbilical clamping affects anemia risk, *Science News* 149 (1996): 263; R. Grajeda, R. Perez-Escamilla, and K. Dewey, Delayed clamping of the umbilical cord improves hematologic status of Guatemalan infants at 2 mo of age, *American Journal of Clinical Nutrition* 65 (1997): 425–431.

36. F. R. Moya, Nutritional requirements of the term newborn, *Textbook of Pediatric Nutrition*, 2nd ed., eds. R. M. Suskind and L. Lewinter-Suskind (New York: Raven Press, 1993), pp. 9–22.

37. P. R. Dallman, Review of iron metabolism, in *Dietary Iron: Birth to Two Years*, ed. L. J. Filer (New York: Raven Press, 1989), pp. 1–11.

38. E. E. Ziegler, Intestinal blood loss by normal infants fed cow's milk, in *Dietary Iron: Birth to Two Years*, ed. L. J. Filer (New York: Raven Press, 1989), pp. 75–80.

39. M. Lane and C. L. Johnson, Prevalence of iron deficiency, in *Iron Nutrition Revisited—Infancy, Childhood, Adolescence*, Report of the 82nd Ross Conference on Pediatric Research, eds. F. A. Oski and H. A. Pearson (Columbus, Ohio: Ross Laboratories, 1981), pp. 31–46; M. F. Picciano and R. H. Deering, The influence of feeding regimens on iron status during infancy, *American Journal of Clinical Nutrition* 33 (1980): 746–753.

40. R. Yip, The changing characteristics of childhood iron nutritional status in the United States, in *Dietary Iron: Birth to Two Years*, ed. L. J. Filer (New York: Raven Press, 1989), pp. 37–56; E. E. Ziegler and S. J. Fomon, Strategies for the prevention of iron deficiency: Iron in infant formulas and baby foods, *Nutrition Reviews* 54 (1996): 348–354.

41. F. A. Oski, Iron deficiency in infancy and childhood, *New England Journal of Medicine* 329 (1993): 190–193.

42. N. F. Krebs and K. M. Hambridge, Zinc requirements and zinc intakes of breast-fed infants, *American Journal of Clinical Nutrition* 43 (1986): 288–292; J. A. Milner, Trace minerals in the nutrition of children, *Journal of Pediatrics* 117 (1990): S147–S155.

43. Subcommittee on Nutrition during Lactation, Food and Nutrition Board, Institute of Medicine, National Academy of Sciences, Milk composition, in *Nutrition during Lactation* (Washington, D.C.: National Academy Press, 1991), pp. 113–152.

44. Krebs and Hambridge, 1986.

45. P. A. Walravens and coauthors, Zinc supplements in breastfed infants, *Lancet* 340 (1992): 683–685.

46. Zinc, copper, and manganese, in *Nutrition of Normal Infants*, ed. S. J. Fomon (St. Louis: Mosby, 1993), pp. 261–280.

47. Moya, 1993.

48. Committee on Nutrition, 1993, pp. 34–42.

49. American Academy of Pediatrics, Controversies concerning vitamin K and the newborn, *Pediatrics* 91 (1993): 1001–1003.

50. F. C. Battaglia and L. O. Lubchenco, A practical classification of newborn infants by birthweight and gestational age, *Journal of Pediatrics* 71 (1967): 159.

51. J. M. Tanner, Growth before birth, in *Foetus into Man: Physical Growth from Conception to Maturity* (London: Open Books Publishing, 1978), p. 46; M. Hack and coauthors, Effect of very low birth weight and subnormal head size on cognitive abilities at school age, *New England Journal of Medicine* 325 (1991): 231–237.

52. R. K. Chandra, Concentrations and production of IgG subclasses in preterm and small-for-gestation low birth weight infants, *Monographs in Allergy* 23 (1988): 156–159.

53. Committee on Nutrition, 1993, pp. 1–10.

54. M. Jaffe and coauthors, Variability in head circumference growth rate during the first 2 years of life, *Pediatrics* 90 (1992): 190–192.

55. B. J. Scott, H. Artman, and S. T. St. Jeor, Growth assessment in children: A review, *Topics in Clinical Nutrition*, December 1992, pp. 5–31.

56. K. G. Dewey and coauthors, Growth of breast-fed and formula-fed infants from 0 to 18 months: The DARLING Study, *Pediatrics* 89 (1992): 1035–1041; K. G. Dewey and coauthors, Growth of breast-fed infants deviates from current reference data: A pooled analysis of U.S., Canadian, and European data sets, *Pediatrics* 96 (1995): 495–503.

57. Committee on Nutrition, 1993, pp. 227–236.

58. T. A. Picone, Taurine update: Metabolism and function, *Nutrition Today*, August 1987, pp. 16–20.

59. G. E. Gaull, Taurine in pediatric nutrition: Review and update, *Pediatrics* 83 (1989): 433–442.

60. O. G. Brooke, Nutritional requirements of low and very low birthweight infants, *Annual Review of Nutrition* 7 (1987): 91–116.

61. Committee on Nutrition, American Academy of Pediatrics, The use of whole cow's milk in infancy, *Pediatrics* 89 (1992): 1105–1109.

62. Committee on Nutrition, 1992.

63. S. J. Fomon, History, in *Normal Nutrition of Infants*, ed. S. J. Fomon (St. Louis: Mosby, 1993), pp. 6–14.

64. C. W. Lo and R. E. Kleinman, Infant formula, past and future: Opportunities for improvement, *American Journal of Clinical Nutrition* 63 (1996): 646S–650S.

65. P. Kurtzweil, Labeling rules for young children's food, *FDA Consumer*, March 1995, pp. 14–18.

66. J. K. Jarvis and G. D. Miller, Fat in infant diets, *Nutrition Today* 31 (1996): 182–190.

67. M. M. Smith and F. Lifshitz, Excess fruit juice consumption as a contributing factor in nonorganic failure to thrive, *Pediatrics* 93 (1994): 438–443.

68. P. C. Dagnelie and W. A. van Staveren, Macrobiotic nutrition and child health: Results of a population-based, mixed-longitudinal cohort study in the Netherlands, *American Journal of Clinical Nutrition* 59 (1994): 1187S–1196S.

69. Dagnelie and van Staveren, 1994.

Nutrition for High-Risk Infants

Each stage of life presents unique nutrition concerns, challenges, and priorities. The nutrition concerns and challenges of infancy include premature and growth-retarded newborns, protein-energy malnutrition (PEM), and failure to thrive (FTT). The accompanying glossary defines terms related to these concerns. The priority in each situation is nutrition intervention that successfully restores growth, nutrition status, and health.

Preterm Infants

The terms *preterm* and *premature* were introduced on p. 172, and are used interchangeably to refer to a shortened gestation period. They imply incomplete fetal development, or immaturity, of many body systems. Preterm infants face physical independence before the growth of some of their organs and body tissues is complete. The rate of weight gain in the fetus is greater during the last trimester of gestation than at any other time in the life span. Therefore, a preterm infant is most often a low-birthweight infant as well. With a premature birth, the infant is deprived of the nutritional support of the placenta during a time of maximal growth.

Nutrient Needs of Preterm Infants

The last trimester of gestation is also a time of building nutrient stores. Being born with limited nutrient stores intensifies the precarious situation for the infant. Further compromising the nutrition status of preterm infants is their physical and metabolic immaturity. Nutrient absorption, especially that of fat and calcium, from an immature gastrointestinal tract is impaired.[1] Immature renal function makes high plasma concentrations of certain amino acids likely, which in turn may damage the immature brain and liver. In short, preterm, low-birthweight infants are candidates for nutrient imbalances. For these reasons, premature infants require special dietary and medical attention. Table FP5A-1 summarizes risk factors for nutrient deficiencies in premature infants.

Few guidelines are available concerning preterm infant nutrient requirements. Authorities on infant nutrition disagree as to how fast the

GLOSSARY

acute PEM: protein-energy malnutrition caused by recent severe food restriction; characterized in children by thinness for height (wasting).

catch-up growth: the acceleration in growth that occurs when a period of growth retardation ends and favorable conditions are restored. It is a self-correcting response that, at best, restores the individual to his or her original growth pattern.

chronic PEM: protein-energy malnutrition caused by long-term food deprivation; characterized in children by short height for age (stunting).

hemolytic anemia: anemia characterized by breakage of the red blood cells, with resultant low hemoglobin levels and an increased production of immature red blood cells (*heme* = blood; *lysis* = to break).

osteopenia: a metabolic bone disease common in preterm infants; also called **rickets of prematurity.**

parenteral nutrition: the delivery of nutrients through a vein, bypassing the intestines. In contrast, **enteral** refers to delivering nutrients into the intestines (*para* = opposite; *enteron* = intestine).

protein-energy malnutrition (PEM), also called **protein-kcalorie malnutrition (PCM):** a deficiency of both protein and energy; the world's most widespread malnutrition problem.

Table FP5A-1 Risk Factors for Nutrient Deficiencies in Premature and Low-Birthweight Infants

Reduced nutrient stores: Premature infants are born before accretion of some nutrients (glycogen, fat, protein, fat-soluble vitamins, calcium, phosphorus, magnesium, and trace minerals) is complete.

Rapid growth rate: The accelerated growth rate needed to achieve "catch-up" growth intensifies energy and nutrient needs.

Impaired nutrient absorption: Insufficient amounts of bile salts and fat-digesting enzymes such as pancreatic lipase inhibit absorption of fat and fat-soluble vitamins.

Limited excretory function and immature renal function: Premature infants have a limited ability to concentrate and dilute urine, eliminate wastes, and handle a solute load—characteristics that impair fluid and acid–base balance, limit total protein intake, and favor mineral loss.

Vulnerability to intestinal dysfunction, respiratory distress, and infection: These and other afflictions increase energy expenditure and limit the ability to feed.

Low-birthweight babies need special care and nourishment.

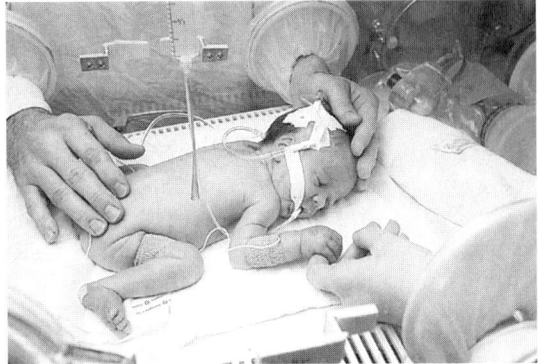

do not respond to the administration of a water-miscible form of vitamin E. They may require supplementation by way of parenteral nutrition until oral feeding is possible.

Folate. Premature, low-birthweight infants have small, readily depleted reserves of folate. At birth, plasma folate in both premature and full-term infants is higher than in adults, but only temporarily. Plasma folate concentrations decline in all infants, but in premature infants the decline is more rapid. Infants with the lowest birthweights experience the greatest declines. A maintenance dose of 50 micrograms per day is recommended to prevent anemia.[4] Premature infants may also experience deficiencies of other B vitamins, because they consume small quantities of breast milk or formula; they may require supplementation.

preterm infant should grow and what the body composition should be. Some authorities advocate using the intrauterine growth rate of full-term infants, without metabolic complications, as the standard for determining the nutrient needs of preterm infants.[2]

What *is* known about low-birthweight infants, and especially very-low-birthweight infants, is that nutrient deficiencies appear during the first days of life. Deficiencies of the fat-soluble vitamins, calcium, iron, and zinc are common.[3]

Vitamin E. Vitamin E deficiency in premature infants is associated with hemolytic anemia. For the premature infant, the attainment of adequate vitamin E status is difficult. Many of these infants are so small that ingesting adequate breast milk or formula orally is impossible at first. In addition, their immature digestive systems do not absorb fat efficiently, and some also

Iron. Low-birthweight infants have a greater requirement for dietary iron than do full-term normal-weight infants.[5] Low-birthweight infants have less stored iron on a weight basis to begin with, and their more rapid growth rate exhausts their iron stores sooner. For these reasons, breast-fed low-birthweight infants are given ferrous sulfate drops at a dose of 2 milligrams per kilogram body weight, not to exceed 15 milligrams per day, beginning at one month of age.[6] Preterm infants who receive iron-fortified formula do not require supplementation.

Calcium. Infants who are born 8 to 10 weeks before term have acquired only about 30 percent as much calcium as full-term infants, so their calcium requirements are high.[7] They miss out on the normal mineralization of bone that takes place during the last trimester of gestation. Their inadequate bone mineralization may result in the metabolic bone disease referred to as osteopenia, or the rickets of prematurity. The probability that this condition will occur varies directly with the infant's weight: The smaller the infant, the greater the risk of bone disease.

Feeding Preterm Infants

Disagreement abounds regarding the best method of feeding preterm infants. Once the infant can accept feedings orally, the question arises, which type of feeding best meets the preterm infant's special nutrition needs? Researchers have compared three sources of nutrients: breast milk from mothers of preterm infants, breast milk from mothers of term infants, and formulas designed for preterm infants. They find that preterm breast milk and special formulas support infant growth better than full-term breast milk, even though preterm infants tolerate full-term milk well.

Preterm Breast Milk. A preterm mother's milk is well suited to meet a preterm infant's needs. It differs both in some nutrient concentrations and in milk volume from that of term mothers. During early lactation, preterm milk contains higher concentrations of protein and is lower in volume than term milk. The low milk volume is advantageous because preterm infants consume small quantities of milk per feeding, and the higher protein concentration allows for better growth. Controversy surrounds the question of whether the high protein concentrations reflect a small milk volume or a premature stage of lactation. The physical and hormonal activities of the reproductive cycle and of mammary gland development have been prematurely halted. The composition of preterm milk closely resembles that of colostrum, the earliest secretion of lactation. In either case, most authorities agree that preterm breast milk best meets the specific needs of a preterm infant.

A preterm infant's own mother's milk may positively influence intellectual development. One group of researchers noted that the greater proportion of mother's milk the infant received, the higher the IQ at 7 1/2 years of age.[8] These findings offer one more reason why health care workers should actively encourage mothers of premature infants to provide at least some of their own breast milk to their infants.

Supplementation. Theoretical estimates of the requirements of preterm infants suggest that preterm breast milk may be an inadequate source of calcium and phosphorus.[9] The combination of preterm breast milk fortified with a preterm supplement supports growth at a rate that approximates the growth rate that would have occurred in utero. In many instances, these supplements of nutrients specifically designed for preterm infants are added to the mother's expressed breast milk and fed to the infant from a bottle.

The compromised nutrition status of premature and low-birthweight infants requires skillful monitoring and care. The goal is to provide optimal nutrition that corrects deficiencies and promotes growth and development. The appropriate nutrition intervention for these infants depends on several important factors—birthweight, gestational age, and the infant's health at birth. Nutrient needs and feeding methods will vary as the infant matures.

Protein-Energy Malnutrition

When people are deprived of protein, energy, or both, the result is protein-energy malnutrition (PEM). PEM most often strikes in infancy and early childhood. It is the most widespread form of malnutrition in the world today. Most of the 33,000 children who die each day are malnourished and suffer from infectious diseases.[10]

Early malnutrition can retard both mental development and physical growth. One question of critical importance asks whether the effects of malnutrition in infancy or early childhood leave a permanent mark or can be rectified by subsequent good nutrition. The answer may be either, depending on the timing, severity, and duration of the deprivation.

Physical Growth

Physical growth may in some cases be less vulnerable than mental development to nutritional insult. Except for those who are severely malnourished at the earliest ages, children have an amazing ability to return to their predicted growth patterns with adequate nutrition—that is, they can catch up. In fact, they show astonishingly rapid rates of catch-up growth. The rate of catch-up growth is influenced by the age and stage of development at which the nutrition insult occurs, the nature of the initial deficit, and the composition of the rehabilitation diet.[11] For example, children who have suffered from a re-

cent (several weeks) shortage of food (acute PEM) may be thin for their height (wasted), but not stunted (short height for their age). These children can achieve remarkable rates (up to 20 times normal for chronological age) of catch-up growth in weight. In contrast, children who have suffered from long-term (years) food deprivation (chronic PEM) tend to catch up much more slowly. Nevertheless, catch-up velocity in height can reach four times the normal velocity for chronological age (see Figure FP5A-1). The extent of catch-up growth depends on the nutrient and growth status prior to the onset of PEM.[12]

Mental Development

Ascertaining the impact infant malnutrition has on mental development and intellect is difficult, if not impossible. Nutrition is only one of many interrelated factors that influence mental development and intellectual performance,

Figure FP5A-1 Catch-Up Growth in a Child Following Two Periods of Starvation

The graph shows that growth in height nearly ceased during periods of starvation and that growth accelerated markedly as soon as the child began eating adequate food.

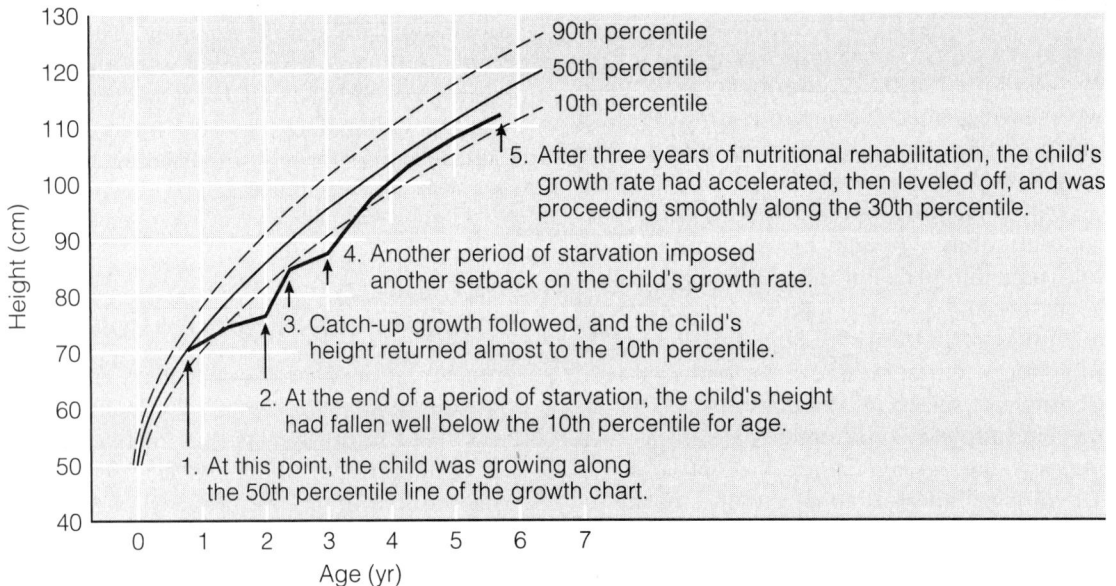

Source: Adapted from A. Prader, J. M. Tanner, and G. A. von Harnack, Catch-up growth, following illness or starvation: An example of developmental canalization in man, *Journal of Pediatrics* 62 (1963): 646–659.

and these characteristics are less easily measured than growth. The importance of understanding these relationships, however, merits perseverance.

Researchers have examined the effects of severe malnutrition during infancy on intellectual functions as well as on subsequent physical growth.[13] One study based on a 15-year follow-up of 20 severely malnourished infants found evidence of gross, irreversible intellectual damage. Despite improved nutrition, the damage appeared to be permanent. Over the years, with improved nutrition, the difference in mean height between the malnourished and well-nourished children diminished, but the difference in mean head circumference increased. Evidently, a reduced head circumference reflects suboptimal brain growth and tends to be a permanent effect of early severe malnutrition.

Another group of researchers studied children between the ages of 5 and 11 who had suffered from severe PEM during the first year of life.[14] They found that these previously malnourished children scored lower on IQ tests and in eight out of nine school subjects compared with children with similar social backgrounds but without a history of malnutrition. Compared with their peers, the previously malnourished children were less emotionally stable and less well behaved in the classroom.

Socioeconomic Factors. Economic and social environment can modify the effects of infant malnutrition. The intellectual capacities of young boys who had suffered severe malnutrition (protein deprivation, energy deprivation, or a combination of the two) during their first two years of life were compared with those of similar boys who had never been malnourished. The malnourished boys came from differing social and economic backgrounds, and the effects of those backgrounds were compared. When malnutrition was surrounded by an unfavorable social and economic environment, it resulted in intellectual impairment.[15] The effect of malnutrition on intellectual development was negligible, however, when the malnutrition occurred in a generally favorable social and economic environment.

In both developing and developed nations, low birthweight and growth retardation are most prevalent among those in the lowest income groups.[16] The living conditions of children influence the extent to which they might recover from early malnutrition. The likelihood of an impaired developmental outcome from early malnutrition is greatest among those in the lowest income groups, where health care and educational opportunities are fewest. Thus, the nature of the malnutrition episode as well as the nature of the environment in which children grow determines the developmental outcome.

Critical Periods. Child nutrition experts attending a symposium to study the effects of early malnutrition on mental development made the following conclusions after extensively reviewing five large studies:[17]

■ Within 40 months following birth, nutritional rehabilitation alone, in malnourished children, is effective.

■ Psychosocial intervention, however, is still markedly effective past this 40-month growth period.

Because each body system has a particular schedule of growth, the time of deprivation has different effects on each system. The growth of the brain illustrates this concept. The timing of brain cell replication varies among different species, but in general, it occurs most rapidly just prior to, or immediately following, birth. In human beings, brain cell number rises rapidly until birth, continues to rise until two years of age, and then rises more slowly until maturity.[18] The brain stops its rapid cell replication earlier than other organs do. More than 70 percent of total adult brain weight is achieved at approximately two years of age.

In one study of rats, malnutrition during the period of active brain cell division reduced the total brain cell number finally achieved.[19] Based on this information, researchers have speculated that if the response to malnutrition in the human brain is similar to that in the rat brain, then the critical postnatal period of cell division would be the first six months of life. The researchers then compared well-nourished infants who died in accidents at various times within the first year of life with infants who died of severe malnutrition at the same times. They examined both brain protein, which reflects the total mass of cells in the brain, and brain DNA, which reflects the number of cells. (The ratio of protein to DNA reflects cell size.) Brain cell sizes were normal in both groups, but cell numbers were significantly reduced in the brains of the malnourished infants.[20] Severe early malnutrition can, in rare cases, curtail the normal increase in brain cell number in human beings.

Failure to Thrive

As you read the following section, keep in mind that although PEM and FTT are described separately, they can overlap. In both cases, the infants are malnourished and appear lethargic, but their environments may differ.

The family of an infant with PEM lacks adequate food, income, housing, sanitation, health care, and education. In short, the family lives in poverty. The infant usually suffers an infection and falls into a vicious malnutrition cycle.

The family of an FTT infant lives with financial instability, in overcrowded or substandard housing, and with emotional discord. Often the father is absent and the mother feels overwhelmed. Described as "sickly," the infant may not even have a medical disease. Such an infant simply does not receive enough energy either because food is limited, or because the infant refuses to eat or vomits on eating. In short, the family lives in strife and the infant lacks nourishment—both nutrients from food and love from parents.

At one time, if no medical cause could be found, the growth delay of FTT was attributed solely to problems in the social environment. Three decades ago research proved that in all children with FTT, nutrition has been inadequate.[21]

Growth failure of FTT infants is sometimes caused by a medical disorder, but most often, growth failure can be attributed to such factors as feeding and behavior problems and interaction difficulties between mother and child. For example, FTT infants may respond poorly to food or refuse to eat. Mothers of FTT infants are less affectionate and less verbal with their infants than mothers of healthy infants. The FTT infant's needs are less often fulfilled, including the need for food. Treatment of FTT usually includes medical, nutritional, and social intervention. As with PEM, the earlier FTT is treated, the greater the chance of long-term success.

Without a full gestation, a full array of nutrients, or full parental attention, preterm infants and those with PEM and FTT face life-threatening problems. Fortunately, early identification and nutrition intervention can enhance their survival and long-term physical growth and mental development.

FOCAL POINT 5A NOTES

1. O. G. Brooke, Nutritional requirements of low and very low birthweight infants, *Annual Review of Nutrition* 7 (1987): 91–116.
2. D. A. Anderson, Nutrition for premature infants, in *Handbook of Pediatric Nutrition*, eds. P. M. Queen and C. E. Lang (Gaithersburg, Md.: Aspen Publishers, 1993), pp. 83–106.
3. D. A. Clark, Nutritional requirements of the premature and small for gestational age infant, in *Textbook of Pediatric Nutrition*, 2nd ed., eds. R. M. Suskind and L. Lewinter-Suskind (New York: Raven Press, 1993), pp. 23–31.

4. Brooke, 1987; P. R. Dallman, Nutritional anemias in childhood: Iron, folate, and vitamin B_{12}, in *Textbook of Pediatric Nutrition*, 2nd ed., eds. R. M. Suskind and L. Lewinter-Suskind (New York: Raven Press, 1993), pp. 91–105.
5. Dallman, 1993.
6. Committee on Nutrition, American Academy of Pediatrics, *Pediatric Nutrition Handbook*, 3rd ed., ed. L. A. Barness (Elk Grove Village, Ill.: American Academy of Pediatrics, 1993), pp. 34–42.
7. Clark, 1993.
8. A. Lucas, Influence of neonatal nutrition on

long-term outcome, in *Nutrition of the Low Birthweight Infant*, eds. B. L. Salle and P. R. Swyer (New York: Vevey/Raven Press, 1993), pp. 183–196.

9. Committee on Nutrition, 1993, pp. 115–124.

10. D. G. Schroeder and R. Martorell, Enhancing child survival by preventing malnutrition, *American Journal of Clinical Nutrition* 65 (1997): 1080–1081.

11. B. J. Scott, H. Artman, and S. T. St. Jeor, Growth assessment in children: A review, *Topics in Clinical Nutrition*, December 1992, pp. 5–31.

12. T. O. Scholl and coauthors, A prospective study of the effects of clinically severe protein-energy malnutrition on growth, *Acta Paediatrics Scandinavia* 69 (1980): 331–335.

13. M. B. Stoch and P. M. Smythe, 15-year developmental study on effects of severe undernutrition during infancy on subsequent physical growth and intellectual functioning, *Archives of Disease in Childhood* 51 (1976): 327–336.

14. J. R. Galler and coauthors, The influence of early malnutrition on subsequent behavioral development. I. Degree of impairment in intellectual performance, *Journal of the American Academy of Child Psychiatry* 22 (1983): 8–15, as cited in L. M. Roeder, The social influence: Diet, child care and intellectual performance, in *Malnutrition and the Infant Brain: Proceedings of an International Symposium*, eds. N. M. van Gelder, R. F. Butter-

worth, and B. D. Drujan (New York: Wiley-Liss, 1990), pp. 253–265.

15. S. A. Richardson, The relation of severe malnutrition in infancy to the intelligence of school children with differing life histories, *Pediatric Research* 10 (1976): 57–61.

16. R. Buzina and coauthors, Workshop on functional significance of mild-to-moderate malnutrition, *American Journal of Clinical Nutrition* 50 (1989): 172–176.

17. N. M. van Gelder, Introduction, in *Malnutrition and the Infant Brain: Proceedings of an International Symposium*, eds. N. M. van Gelder, R. F. Butterworth, and B. D. Drujan (New York: Wiley-Liss, 1990), p. xvii.

18. E. M. Widdowson, Nutrition and cell and organ growth, in *Modern Nutrition in Health and Disease*, 8th ed., eds. M. E. Shils, J. A. Olson, and M. Shike (Philadelphia: Lea & Febiger, 1994), pp. 728–739.

19. M. Winick and A. Noble, Cellular response in rats during malnutrition at various ages, *Journal of Nutrition* 89 (1966): 300–306.

20. M. Winick and P. Rosso, The effect of severe early malnutrition on cellular growth of human brain, *Pediatric Research* 3 (1969): 181–184.

21. C. Whitten, M. Pettit, and J. Fischoff, Evidence that growth failure from maternal deprivation is secondary to undereating, *Journal of the American Medical Association* 209 (1969): 1675–1679.

Iron Deficiency and Infant Behavior

This Focal Point examines some of the recent research findings on iron deficiency and infant behavior. Many studies in the last two decades have documented the relationship between iron deficiency and poor performance on behavioral tests. Almost without exception, iron-deficient anemic infants perform poorly on tests of mental and motor development.[1] In addition, iron-deficient anemic infants are often unhappy, tense, fearful, withdrawn, and less active than their iron-sufficient peers.

Despite extensive research on this topic, the question of whether iron deficiency *causes* poor behavioral performance continues to be debated. Several research problems fuel the debate on this topic. For example, difficulties in identifying and excluding confounding factors that affect mental development and weaknesses in experimental design impose limits on interpretation of results. In addition, researchers have yet to agree on the exact criteria by which to define iron deficiency, often using different biochemical indexes and cutoff points. Although several biochemical tests are available to evaluate iron status, the lack of one perfect test to diagnose iron deficiency complicates its detection and confounds study results. Researchers relying on only one measure of iron status risk error in classifying individuals being studied.[2]

A point of clarification is also needed. Iron deficiency and anemia are not one and the same, although they often go hand in hand. Infants may be iron deficient without being anemic. An equally important concept is that not all anemia is caused by iron deficiency.[3] Because of the strong association between iron deficiency and anemia, however, these terms are sometimes used interchangeably. The term *iron deficiency* generally refers to impairment of the production of essential iron compounds such as hemoglobin, because of a lack of iron; the body's iron stores are low.[4] The term *iron deficiency anemia* refers to the hematologic state resulting from a severe deficiency; the body's iron stores are severely depleted.

Infant research entails all the many problems associated with adult research, as well as specific problems of its own. Researchers encounter difficulty in selecting appropriate measurements of behaviors and intelligence, because infant responses are limited. Language as a component of intelligence is easily measured later, in an older child, but not in an infant. Despite such shortcomings, tests such as the Bayley Infant Scale of Mental and Motor Development are available to researchers who study infant behavior. The Bayley scale attempts to establish norms for certain behaviors that emerge during infancy, such as reaching for toys or responding to voices, and to assign levels of mental development based on these standards.

Research on the effects of iron deficiency on infant behavior has proceeded despite these pitfalls. In general, the results suggest that iron deficiency anemia impairs behavioral development; iron-deficient anemic infants score lower on the Bayley test than those who are iron sufficient.[5] Evidence that iron deficiency *without* anemia impairs infant development, however, is lacking.[6]

In one of these studies, infants who had been anemic for more than three months scored lower on the Bayley test than those who had been anemic for less than three months.[7] After both short-term (10 days) and long-term (three months) iron therapy, even when the anemia had been reversed, test scores of infants who had been anemic for longer than three months did not improve.[8] These results suggest that when iron deficiency progresses to anemia at a critical period of brain growth, permanent abnormalities may occur.

Additional research has been conducted to determine whether infants with iron deficiency anemia have long-term developmental impairments.[9] Researchers who had documented and corrected the iron status of a group of infants initiated a follow-up evaluation of the children five years later. At five years of age, all the children had excellent iron status and growth, but those who had had moderate iron deficiency anemia (hemoglobin ≤10.0 g/dL) as infants still tested lower in mental and motor functioning compared with the other children. The authors concluded that children who have iron deficiency anemia in infancy are more likely to have long-term developmental impairments.

The findings of studies such as those just described support the theory that iron deficiency anemia during infancy can have irreversible, adverse effects on development. The findings of a well-controlled study of Indonesian infants, however, suggest that the developmental effects of iron deficiency anemia can be completely reversed by appropriate iron therapy if the anemia is of short duration.[10] Questions about the role of timing, severity, and duration of iron deficiency on reversal of developmental effects remain to be answered.

Now, how to account for the effects of iron deficiency on behavior? Of the many possible mechanisms by which iron deficiency anemia can delay development, reduction of iron in the brain is the one that most research focuses on.[11] The brain contains significant quantities of iron, much of it in iron-dependent enzymes that participate in the synthesis and catabolism of neurotransmitters such as norepinephrine and serotonin.[12] The breakdown of these neurotransmitters to their excretion products requires the iron-dependent enzyme monoamine oxidase. When monoamine oxidase activity is depressed, the rate of breakdown of these neurotransmitters is reduced. In fact, iron-deficient children excrete excessive amounts of norepinephrine in their urine, an abnormality unique to the anemia of iron deficiency and reversible with iron therapy.[13] Researchers speculate that the behavioral abnormalities seen in iron-deficient children may be secondary to abnormal amounts of neurotransmitters in the central nervous system caused by the reduced activity of monoamine oxidase. The precise mechanism by which iron deficiency anemia affects mental and motor function in infants, however, remains to be determined.

In summary:

- Iron deficiency anemia is a risk factor for delayed mental and motor development during infancy.

- The developmental effects of iron deficiency anemia during infancy may be reversible; the reversibility of the effects depends on timing (age of infant), severity, and duration of the anemia.

- The precise mechanism by which iron deficiency anemia impairs development remains to be clarified.

Despite some uncertainty, the findings of studies on infants and iron deficiency anemia offer convincing support for making programs and interventions aimed at the prevention of iron deficiency a priority.

FOCAL POINT 5B NOTES

1. E. Pollitt, Iron deficiency and cognitive function, *Annual Review of Nutrition* 13 (1993): 521–537.
2. T. Walter, Infancy: Mental and motor development, *American Journal of Clinical Nutrition* 50 (1989): 655S–661S.
3. R. Yip, Iron nutritional status defined, in *Dietary Iron: Birth to Two Years*, ed. L. J. Filer (New York: Raven Press, 1989), pp. 19–36.
4. P. R. Dallman, Review of iron metabolism, in *Dietary Iron: Birth to Two Years*, ed. L. J. Filer (New York: Raven Press, 1989), pp. 1–11.
5. T. Walter and coauthors, Iron deficiency anemia: Adverse effects on infant psychomotor development, *Pediatrics* 84 (1989): 7–17; B. Lozoff and coauthors, Iron deficiency anemia and iron therapy effects on infant developmental test performance, *Pediatrics* 79 (1987): 981–995; B. Lozoff, E. Jimenez, and A. W. Wolf, Long-term developmental outcome of infants with iron deficiency, *New England Journal of Medicine* 325 (1991): 687–694.
6. Pollitt, 1993.
7. Walter and coauthors, 1989.
8. Walter, 1989.
9. Lozoff, Jiminez, and Wolf, 1991.

10. P. Idjradinata and E. Pollitt, Reversal of developmental delays in iron-deficient anemic infants treated with iron, *Lancet* 341 (1993): 1–4.

11. Pollitt, 1993.

12. M. B. H. Youdim, Neuropharmacological and neurobiochemical aspects of iron deficiency, in *Brain, Behaviour, and Iron in the Infant Diet*, ed. J. Dobbing (London: Springer-Verlag, 1990), pp. 83–106.

13. F. A. Oski and coauthors, Effect of iron therapy on behavior performance in nonanemic, iron-deficient infants, *Pediatrics* 71 (1983): 877–880, as cited in J. L. Beard, J. R. Connor, and B. C. Jones, Iron in the brain, *Nutrition Reviews* 51 (1993): 157–170.

Inborn Errors of Metabolism

Focal Point 2A lays the foundation for a closer look at the ramifications of a genetic error in protein synthesis. When enzymes are not synthesized in sufficient quantities or have abnormal structures, metabolic reactions and transport of nutrients cannot proceed. If an enzyme that converts compound A to compound B is missing or malfunctioning in the metabolic pathway, then compound A accumulates and compound B becomes deficient. Both the excess of compound A and the lack of compound B can lead to a variety of problems and, in many cases, death. Furthermore, more compound A and less compound B become available for use in other metabolic pathways. These consequences, in turn, create excesses and deficiencies of other metabolites that present another array of problems. The diseases that result from these "inherited biochemical blocks in normal metabolic pathways" are known as "inborn errors" of metabolism.[1] The glossary on p. 210 defines related terms.

In some cases, inborn errors have severe consequences, causing physical and mental retardation. Without proper diagnosis and treatment, they can be lethal. As is true of most medical disorders, the earlier the diagnosis and treatment, the better the prognosis.

One goal of medical research and practice is to detect genetic defects before they cause harm. Recent advances in medical technology allow clinicians to study the developing fetus and identify abnormal conditions during gestation. One technique, amniocentesis, analyzes amniotic fluid that has been removed from the amniotic sac by a needle or syringe. This technique has proven most valuable in identifying more than 70 different inborn errors of metabolism. Analysis of the amniotic fluid identifies specific enzymes and measures their concentrations, revealing abnormalities. A genetic technique called restriction enzyme analysis allows scientists to locate specific genes. Such a technique holds promise of actually correcting a genetic defect in the future. For now, prenatal diagnosis of genetic diseases allows nutritional intervention during gestation.

Nutrition plays a primary role in treating many inborn errors of metabolism. With an understanding of the biochemical pathway involved, a clinician can manipulate the diet to compensate for excesses and inadequacies. Dietary management of genetic diseases involves restricting dietary precursors that occur prior to the error in the metabolic pathway, administering pharmacological doses of specified nutrients, or replacing needed products. The goal of therapy is to:

- Prevent the toxic accumulation of metabolites.
- Replace essential nutrients that are deficient as a result of the defective metabolic pathway.
- Provide a diet that supports normal growth, development, and maintenance.

Meeting these three objectives is a major challenge that was unattainable until earlier this century. Increased knowledge about the body's many biochemical pathways, coupled with current technology that can synthesize formulas of specific nutrient compositions, has greatly enhanced the treatment of inborn errors.

The number of possible inborn errors is limited only by the number of possible gene mutations, for genes carry the codes to synthesize enzymes in the body. Consider that a small bacterial cell carries genes for at least 1000 different enzymes and that mutations are possible in all these genes. Human cells have 1000 times as many proteins of major importance, and in theory, any of them can appear in different mutant forms. Not only can the genes for enzymes be affected, but also those for other proteins, such as the "pumps" that move nutrients into and out of cells, the carriers that transport them in the blood, and the structural proteins of cell membranes and connective tissue.

GLOSSARY

galactosemia: an inborn error of metabolism in which galactose cannot be metabolized normally to compounds the body can handle and an alternative metabolite accumulates in the tissues, causing damage.

inborn error of metabolism: an inherited flaw evident as a metabolic disorder or disease present from birth.

phenylketonuria (PKU): an inborn error of metabolism in which phenylalanine, an essential amino acid, cannot be converted to tyrosine. Alternative metabolites of phenylalanine (phenylketones) accumulate in the tissues, causing damage, and overflow into the urine.

This discussion provides a look at a few inborn errors that respond to nutrition therapy, focusing primarily on the most common one—phenylketonuria, one of several inborn errors that affect amino acid metabolism. Two inborn errors of carbohydrate metabolism, galactosemia and glycogen storage disease, are also discussed.

Classic Phenylketonuria

Phenylketonuria (PKU) affects approximately 1 out of every 12,000 newborns in the United States each year. The ability to detect and treat PKU has saved and significantly improved the lives of many people.

Classic PKU results from a deficiency of the enzyme phenylalanine hydroxylase. This enzyme hydroxylates the essential amino acid phenylalanine, converting it to another amino acid, tyrosine (see Figure FP5C-1). Without the phenylalanine hydroxylase enzyme, abnormally high concentrations of phenylalanine and other related metabolites (phenylketones) accumulate and damage the developing nervous system. Simultaneously, the body cannot make tyrosine or other metabolites (such as the neurotransmitter epinephrine) that normally derive from tyrosine. Under these conditions, tyrosine becomes an essential amino acid; that is, the body cannot make it, and therefore the diet must supply it. To ensure that blood phenylalanine and tyrosine concentrations remain within an acceptable range, children with PKU receive blood tests periodically.

PKU is a hidden disease that is detected by means of routine screening of the blood at birth. Diagnosis and treatment beginning in the first few days of life can prevent the devastating effects of PKU. For these reasons, all newborns in the United States are tested for PKU.[2] Tests must be conducted after infants have consumed several meals containing protein (usually after 24 hours and before seven days). Before the 1960s, when screening became routine, an infant with PKU would suffer the dire consequences of uncorrected high phenylalanine concentrations. At first, the only signs are a skin rash and light skin pigmentation. Between three and six months, signs of developmental delay begin to appear. The infant becomes irritable and unable to sleep restfully. Seizures are common. By one year, irreversible brain damage is clearly evident, and the child will score poorly on developmental and intellectual tests.

Nutrition Therapy for PKU

The effect of nutrition intervention in PKU is remarkable. In almost every case, dietary management can prevent the devastating array of symptoms described. Essentially, the diet restricts phenylalanine and supplements tyrosine to maintain blood concentrations of both amino acids within a safe range. As most dietitians can attest, this is more easily said than done.

Because phenylalanine is an essential amino

Figure FP5C-1 The Biochemical Pathway in PKU

Normal

Normally, the amino acid phenylalanine follows two pathways, one in the liver, the other in the kidneys. In the liver, the enzyme phenylalanine hydroxylase adds a hydroxyl group (OH) to produce the amino acid tyrosine. Tyrosine, in turn, produces melanin, the pigmented compound found in skin and brain cells; the neurotransmitters epinephrine and norepinephrine; and the hormone thyroxin. In the kidneys, enzymes convert phenylalanine to by-products that are excreted.

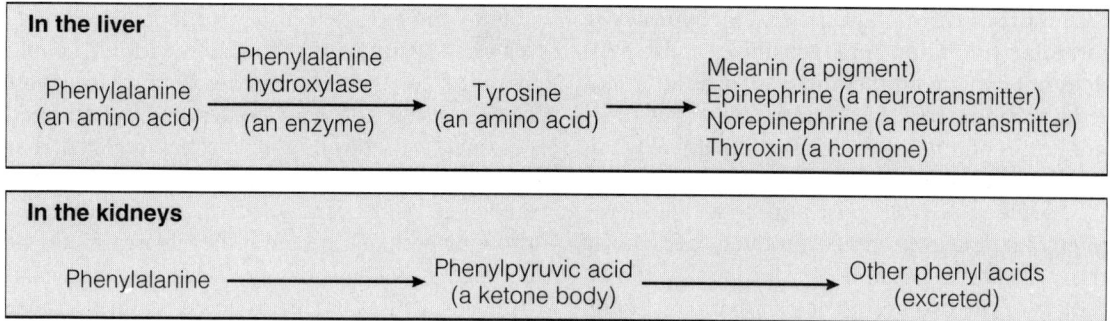

In the liver

Phenylalanine (an amino acid) →[Phenylalanine hydroxylase (an enzyme)]→ Tyrosine (an amino acid) →
- Melanin (a pigment)
- Epinephrine (a neurotransmitter)
- Norepinephrine (a neurotransmitter)
- Thyroxin (a hormone)

In the kidneys

Phenylalanine ———→ Phenylpyruvic acid (a ketone body) ———→ Other phenyl acids (excreted)

In PKU

Individuals with PKU lack the liver enzyme phenylalanine hydroxylase, impairing conversion of phenylalanine to tyrosine. Phenylalanine accumulates in the liver and blood, reaching the kidneys in abnormally high concentrations. In the kidneys, an aminotransferase enzyme converts phenylalanine to the ketone body phenylpyruvic acid, which spills into the urine — thus the name phenylketonuria.

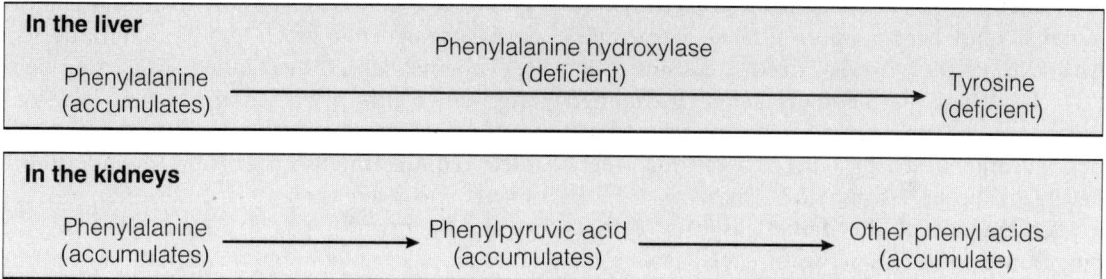

In the liver

Phenylalanine (accumulates) —[Phenylalanine hydroxylase (deficient)]→ Tyrosine (deficient)

In the kidneys

Phenylalanine (accumulates) ———→ Phenylpyruvic acid (accumulates) ———→ Other phenyl acids (accumulate)

acid, the diet cannot exclude it completely. If phenylalanine intake is too low, children suffer bone, skin, and blood disorders, and growth and mental retardation. Therefore, the diet must strike a perfect balance, providing enough phenylalanine to support normal growth and health, but not so much as to cause harm. The problem is not that children with PKU require less phenylalanine than other children, but that they cannot handle excesses without detrimental effects. With a controlled phenylalanine intake, children with PKU can lead normal, happy lives.

To control phenylalanine intake requires strict dietary management that was impossible prior to 1958, when a special low-phenylalanine formula became commercially available. This type of formula is now the primary source of energy and protein for children with PKU. Their diet excludes high-protein foods such as meat, fish, poultry, cheese, eggs, milk, nuts, and dried beans and peas. Also excluded are commercial breads and pastries made from regular flour, which has a high protein and phenylalanine content. Basically, the diet allows foods that contain some phenylalanine, such as fruits, vegetables, and cereals, and those that contain none, such as fats, sugar, jellies, and some candies. Clearly, it is impossible to create such a diet using only whole, natural foods, and children who depend primarily on a formula for their nourishment risk multiple trace mineral deficiencies.[3] Health care professionals monitor trace mineral status and supplement as needed.

Infants receive a special casein hydrolysate formula with a low phenylalanine content. It does not contain all the phenylalanine an infant requires, so parents supplement it with measured quantities of milk, rice cereal, and baby foods as the infant develops.

People with PKU must also be aware of the phenylalanine content in products containing the sweetener aspartame. Aspartame is a combination of the two amino acids aspartic acid and phenylalanine and therefore contributes phenylalanine to the diet. Sold under the trade name NutraSweet, aspartame is an ingredient in many foods and beverages such as powdered drink mixes, instant puddings, gelatin desserts, breakfast cereals, chewing gums, and the tabletop sweetener Equal. Products sweetened with aspartame must bear a warning label for people with PKU. People with PKU need to consult with their physicians or dietitians before including aspartame in their diets. For adolescents on phenylalanine-restricted diets, occasional diet beverages appear to cause no harm.[4]

Perhaps one of the hardest aspects of this diet is the children's sense of social isolation. From birth, children with PKU require a "special diet" and cannot eat the foods that other children are eating. Some commercially available low-protein products, such as cookies, contain very little, if any, phenylalanine and allow children to share treats with others. Teachers, friends, and family members must understand that they cannot offer food to children with PKU without permission from the children's parents. Until the children are old enough to understand the dietary restrictions, parents must teach them to ask before eating any food. Parents who have learned positive and creative problem-solving skills can effectively resolve situations involving dietary decisions. Consequently, their children are more likely to eat appropriate foods and maintain phenylalanine levels within normal ranges than children of parents without such skills.[5]

During the early years of central nervous system development, prompt nutrition intervention is critical to preventing irreversible mental retardation in the young PKU child. Less clear is the length of time the nervous system is vulnerable to the PKU defect. Until the late 1970s, researchers assumed that the child with PKU could abandon the special diet after the first few years of life when the central nervous system had completed its development. They realized that with a regular diet, phenylalanine and associated metabolite concentrations would rise, but thought perhaps these high levels would not be damaging. Unfortunately, elevated phenylalanine concentrations in the older child do cause problems such as short attention span, poor short-term memory, and poor eye-to-hand coordination, although the damage is less severe than at an earlier age. In general, children with PKU who have discontinued their controlled diet experience problems in school performance, mood, and behavior. For these reasons, clinicians now encourage children to continue the low-phenylalanine diet indefinitely. Convincing adolescents to return to a phenylalanine-restricted diet after several years of unrestricted diets requires intense education and reinforcement, and even then it is quite often unsuccessful. Reinstituting a controlled diet, however, does improve blood phenylalanine concentrations, behavior, and IQ scores.

Therapy for inborn errors goes beyond nutrition to include psychological counseling for the people who are affected and their families. A genetic disorder is a lifelong problem that affects the entire family. All family members are at high risk for being carriers, and they inevitably become involved in the care and management of the person with the inborn error. Therefore, families must learn how to handle the impact such a diagnosis has on their relationships.

Classic PKU is not the only inborn error to raise blood concentrations of phenylalanine. Malignant hyperphenylalaninemia has the same result, but because the error occurs elsewhere in the metabolic pathway, it does not respond to phenylalanine restriction alone. In this case, treatment involves replacing missing cofactors and neurotransmitters instead of restricting the precursor in the pathway.

Other Inborn Errors

To this point, the discussion has focused on PKU, an example of a defect in amino acid metabolism. The following paragraphs describe two inborn errors of carbohydrate metabolism—galactosemia and glycogen storage disease.

Galactosemia

Galactosemia is an inborn error of carbohydrate metabolism in which the body cannot use the monosaccharide galactose. Three enzymes are required for the conversion of galactose to glucose; in galactosemia, at least one of those enzymes (most commonly galactose-1-phosphate uridyl transferase) is missing or defective. The unmetabolized product (galactose-1-phosphate) accumulates and follows an alternative pathway to form an abnormal product that causes growth failure, liver enlargement, kidney failure, and other neurological abnormalities that lead to coma and death. Early introduction of a galactose-restricted diet prevents or minimizes most of these symptoms. However, it may not prevent ovarian damage, some visual and speech problems, or other neurological abnormalities.

Dietary adjustment in galactosemia is simpler than in PKU for a couple of reasons. First, galactose is not an essential nutrient, as is phenylalanine. The PKU diet is a balancing act between providing enough phenylalanine for normal growth and development on the one hand and assuring that not enough is left over to be toxic on the other. The galactosemia diet needs only to exclude galactose. Second, because galactose occurs only in lactose (the sugar in milk), treatment depends chiefly on the careful restriction of milk and milk products. This is not to say that the diet is easy to follow; many commercially prepared products contain milk. Still, milk is less widespread in the diet than the amino acid phenylalanine, which appears in all protein foods.

Glycogen Storage Diseases

Other genetic diseases affecting carbohydrate metabolism include the glycogen storage diseases. At least 10 distinct abnormalities of glycogen storage have been identified. These diseases are characterized by glycogen abnormalities, including quantity, location, or structure.

The most common and severe of the glycogen diseases is caused by a deficiency of the enzyme glucose-6-phosphatase. This enzyme converts glucose-6-phosphate to glucose in the gluconeogenesis pathways and glycogenolysis pathways. The body's inability to handle glucose-6-phosphate results in hypoglycemia, glycogen accumulation, and their associated metabolic consequences. Dietary management of this type of glycogen storage disease involves constant provision of carbohydrate (glucose). To accomplish this, children eat frequent (at least every three hours) high-carbohydrate meals (60 to 70 percent of total kcalories) composed primarily of starch foods. To maintain glucose concentrations during sleep, children receive a special glucose polymer formula by tube feeding.

Another type of glycogen storage disease results from a deficiency of the glycogen-debranching enzyme amylo-1,6-glucosidase, which inhibits glycogen breakdown at the branch points and results in short-branched glycogen. This fairly harmless disorder can usually be managed by diet. In contrast, another type of glycogen storage disease is due to a deficiency of the branching enzyme, which results in abnormally long glycogen chains with few branch points. This form of glycogen is insoluble and accumulates in many of the body's tissues, including the liver, causing liver enlargement. Liver disorders usually progress to cirrhosis and death by age four. Liver transplantation is an effective treatment that seems to correct enzyme activity not only in the liver, but in other tissues as well.[6]

As scientific understanding of human genetics and biochemistry increases, more and more inborn errors affecting enzyme function are being recognized. Understanding proteins and their roles as enzymes in metabolism makes it possible to compensate for these defects of metabolism that otherwise would destroy the quality of life. Diet cannot always be tailored to prevent the defects of inborn errors, but in many such diseases diet can make a dramatic difference in people's lives.

FOCAL POINT 5C NOTES

1. A. E. Garrod, Inborn errors of metabolism (Croonian lectures), *Lancet* 2 (1908), as cited by M. A. Wallen and S. Packman, Nutrition and inborn errors of metabolism, in *Nutrition Update*, vol. 2, eds. J. Weinger and G. M. Briggs (New York: Wiley, 1985), pp. 53–75.

2. Committee on Genetics, American Academy of Pediatrics, Issues in newborn screening, *Pediatrics* 89 (1992): 345–349.

3. C. Reilly and coauthors, Trace element nutrition status and dietary intake of children with phenylketonuria, *American Journal of Clinical Nutrition* 52 (1990): 159–165; S. Stepnick-Gropper and coauthors, Trace element status of children with PKU and normal children, *Journal of the American Dietetic Association* 88 (1988): 459–465.

4. L. C. Wolf-Novak and coauthors, Aspartame ingestion with and without carbohydrate in phenylketonuric and normal subjects: Effect on plasma concentrations of amino acids, glucose, and insulin, *Metabolism* 39 (1990): 391–396.

5. A. M. B. Fehrenbach and L. Peterson, Parental problem-solving skills, stress, and dietary compliance in phenylketonuria, *Journal of Consulting and Clinical Psychology* 57 (1989): 237–241.

6. R. S. Selby and coauthors, Liver transplantation for Type IV glycogen storage disease, *New England Journal of Medicine* 324 (1991): 39–42.

Nutrition during Childhood

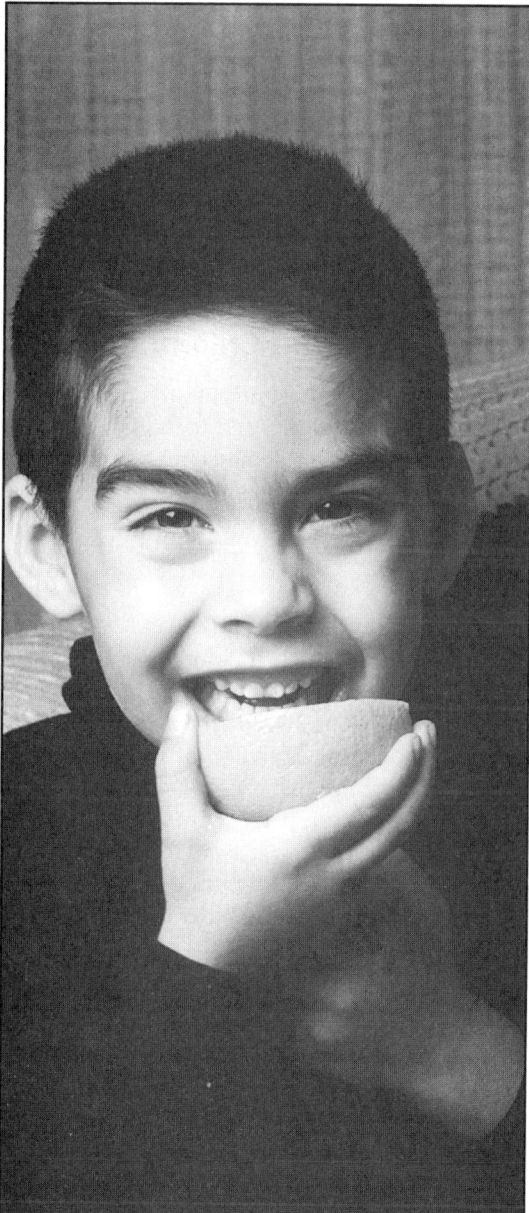

utrient needs change throughout the growing years, depending on genetics, rates of growth, activity, and many other factors. Nutrient needs also vary from individual to individual, but generalizations are possible and useful. Sound nutrition throughout childhood promotes normal growth and development, facilitates academic and physical performance, and may help prevent the development of obesity, heart disease, cancer, and other degenerative diseases in adulthood. As children enter the teen years, a nutrient foundation laid by years of eating nutritious foods prepares them to meet the upcoming demands of rapid growth.

Delivering nutrients in the form of meals and snacks that are nutritious and delicious to children is a challenge for caregivers. Children develop likes and dislikes without regard to their nutrient needs; they are easily influenced by peers, the media, and their taste buds, although it is a food's nutritional quality that has the greatest impact on a child's health. Childhood is a time of developing food habits that will be carried into the future. This chapter focuses on nutrition for toddlers, preschool, and grade-school-age children.

Growth and Development during Childhood

After age one, growth rate slows, but the body continues to change dramatically. Many developmental changes take place to transform the one-year-old child into a child ready to enter adolescence. As with infants, growth is the parameter always monitored as an index of the nutrition status of children.

Growth

2 1/2 inches = approximately 5 cm.

5 lb = 2.2 kg.

After having grown all of 10 inches during the first dramatic year of life, a child grows approximately 5 inches in height between the ages of one and two. Thereafter, the rate slows to about 2 1/2 inches per year until adolescence, when this steady rate of increase in height rises abruptly and markedly. Weight gain settles into a similar pattern, averaging out as an increase of approximately 5 to 6 pounds per year until the onset of adolescence.

The lengthening of bones and development of muscles are reflected outwardly as the obvious signs of a child's growth. Other tissues such as connective tissues, teeth, body fat, skin, and the nervous system are also growing.

Development

Increases in height and weight are only two of the many changes growing children experience. At age one, children can stand alone and are beginning to toddle; by two, they can walk and are learning to run; by three, they can jump and are climbing with confidence. Bones and

The body shape of a one-year-old (left) changes dramatically by age two (right). The two-year-old has lost much of the baby fat; the muscles (especially in the back, buttocks, and legs) have firmed and strengthened, and the leg bones have lengthened.

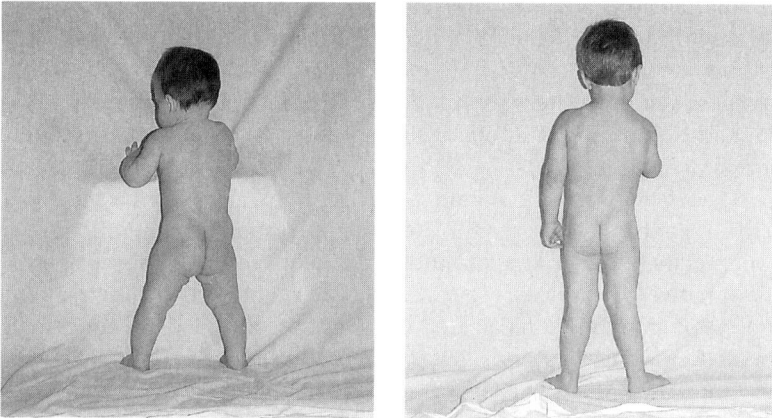

muscles increase in both mass and density to make these new accomplishments possible.

Growth slows as children enter the second year of life, but development continues to progress rapidly. The brain and the rest of the central nervous system mature at a tremendous pace, as evidenced by increasing muscle control, coordination, and the ability to perform new skills. By age two, most of the primary teeth have erupted, and control of the jaw muscles is voluntary. During the second year of life, children can handle many more types of foods than they could at age one.

At the same time, children who as infants were willing to taste anything and everything, whether it was food or not, become assertive and selective about what they will eat or drink and how they will do it. This behavior reflects children's psychological development. Two-year-olds would rather show their independence than eat. They constantly strive to prove that they are separate people from their parents. One way of doing this is by saying no a lot. Yet toddlers need to have limits. For parents or caregivers of toddlers, the task is to allow for independence and curiosity, but to avoid giving in to unreasonable demands by letting the child know that limits exist. This is true with food as well as with other aspects of the toddler's life. A later section offers suggestions for avoiding power struggles over food.

Energy and Nutrient Needs during Childhood

Steady growth during childhood necessitates a gradual increase in intakes of most nutrients. Dietary recommendations cluster children into age groupings that reflect similarities in growth rate, biological changes, and hormone status (review Table 1-2 on p. 10).

Ideally, children accumulate stores of nutrients before adolescence. Then, when they take off on the adolescent growth spurt and their nutrient intakes cannot keep pace with the demands of rapid growth, they can draw on the nutrient stores accumulated earlier. This is especially true of calcium. An adequate calcium intake promotes optimal bone density; the denser the bones are in childhood, the better prepared they will be to support teen growth and still withstand the inevitable bone losses of later life.[1] The way children eat therefore influences their nutritional health during childhood, during their teen years, and for the rest of their lives.

Besides helping to maintain a healthy body, most of the nutrients actively contribute to growth. Consider, for example, the growth of a bone. Fetal bones develop into those of an adult through a series of well-coordinated, dynamic processes. Bones are continually undergoing a process of growth and repair that involves the loss of existing bone and the deposition of new bone. Vitamin C helps form the collagen matrix on which the minerals are deposited; and vitamin D, calcium, phosphorus, magnesium, and fluoride are required to mineralize the bone. Many nutrients perform roles during growth in a variety of body systems. For example, vitamin A supports the growth of skin and other epithelial tissues such as the lining of the digestive tract and the lining of the respiratory system, promotes normal vision, and supports the immune system.

epithelial tissues: the layers of the body that serve as selective barriers between the body's interior and the environment.

Energy and Energy Nutrients

Children's appetites begin to diminish around one year, consistent with the slowing of growth. Thereafter, children spontaneously vary their food intakes to coincide with their growth patterns; they demand more food during periods of rapid growth than during slow growth. Sometimes they seem insatiable; other times they seem to live on air and water.

Children's energy intakes also vary widely from meal to meal. Even so, their total daily intakes remain remarkably constant.[2] If children eat less at one meal, they typically eat more at the next, and vice versa. Overweight children are an exception: They do not always adjust their energy intakes appropriately and may eat in response to external cues such as television commercials, disregarding appetite regulation signals.

Energy RDA during childhood:
  1300 kcal/day (1 to 3 yr).
  1800 kcal/day (4 to 6 yr).
  2000 kcal/day (7 to 10 yr).

Canadian RNI during childhood:
  1100 kcal/day (1 yr).
  1300 kcal/day (2 to 3 yr).
  1800 kcal/day (4 to 6 yr).
  1900 kcal/day (7 to 9 yr, female).
  2200 kcal/day (7 to 9 yr, male).

Energy. Individual children's energy needs vary widely, depending on their growth and physical activity. A one-year-old child needs about 1000 kcalories a day; a three-year-old needs only 300 kcalories more. By age 10, a child needs about 2000 kcalories a day. Total energy needs increase slightly with age, but energy needs per kilogram of body weight actually decline gradually.

Protein RDA during childhood:
  16 g/day (1 to 3 yr).
  24 g/day (4 to 6 yr).
  28 g/day (7 to 10 yr).
Canadian RNI during childhood:
  13 g/day (1 yr).
  16 g/day (2 to 3 yr).
  19 g/day 4 to 6 yr).
  26 g/day (7 to 9 yr).

Protein. Like energy needs, total protein needs increase slightly with age, but when the child's body weight is considered, the protein requirement actually declines gradually. The estimation of protein needs considers the requirements for maintaining nitrogen balance, the quality of protein consumed, and the added needs of growth.

Vitamins and Minerals

The vitamin and mineral needs of children increase with their ages (see inside front cover). A balanced diet of nutritious foods can easily meet children's needs for these nutrients, with the notable exception of iron. Iron deficiency anemia is a major problem worldwide, as well as being the most prevalent nutrient deficiency among U.S. and Canadian children. Children's typically low iron intakes leave many of them in marginal iron status. Reducing iron deficiency among young children is one of the foremost health priorities in the United States.[3] Internationally, the World Health Organization is collaborating with a United Nations subcommittee on nutrition to develop a 10-year plan for eliminating iron deficiency.[4]

✶ Healthy People 2000

Reduce iron deficiency to less than 3% among children aged one through four years.

To help prevent iron deficiency, children's foods must deliver approximately 10 milligrams of iron per day. To achieve this goal, snacks and meals should include the iron-rich foods listed in Table 6-1. Milk should be limited to three or four cups a day, so that it will not displace lean meats, fish, poultry, eggs, legumes, and whole-grain or enriched breads and cereals.

Nationwide surveys of children's diets show that the average iron intakes of most age groups have improved over the past 20 years. As many as 20 percent of children under five years of age, however, still receive less than 50 percent of their iron RDA.[5] Considering that milk, which contains little iron, has a prominent place in most younger children's diets, the low iron intakes of this group are not surprising.

The body's iron status reflects its ability to conserve and recycle iron once it has been absorbed. Thus iron status depends on more than dietary intake alone, but diet is more critical to iron balance in children than it is in adults. In adult males, about 95 percent of the required iron is recycled, and only 5 percent need come from the diet. By comparison, the percentages for children are 70 and 30, respectively.

Iron RDA during childhood: 10 mg/day (1 to 10 yr).
Canadian RNI during childhood:
6 mg/day (1 to 3 yr).
8 mg/day (4 to 9 yr).

Nutrient Supplements for Children

Many parents provide their children with vitamin and mineral supplements, relying on them as insurance against possible nutrient deficiencies. Supplements are reasonably inexpensive and available without a

Table 6-1 Iron-Rich Foods Children Like[a]

Breads, Cereals, and Grains

Canned macaroni (1/2 c)
Canned spaghetti (1/2 C)
Cream of wheat (1/4 c)
Fortified dry cereals (1 oz)[b]
Noodles, rice, or barley (1/2 c)
Tortillas (1 flour, 2 corn)
Whole-wheat, enriched, or fortified bread (1 slice)
Bran muffins

Vegetables

Baked flavored potato skins (1/2 skin)
Cooked mushrooms (1/2 c)
Cooked mung bean sprouts or snow peas (1/2 c)
Green peas (1/2 c)
Mixed vegetable juice (1 c)

Fruits

Apple juice (1 c)
Canned plums (3 plums)
Cooked dried apricots (1/2 c)
Dried peaches (4 halves)
Raisins (1 tbs)

Meats and Legumes

Bean dip (1/4 c)
Canned pork and beans (1/3 c)
Mild chili or other bean/meat dishes (1/4 c) such as burritos
Liverwurst (1/2 oz)
Meat casseroles (1/2 c)
Peanut butter and jelly sandwich (1/2 sandwich)
Lean roast beef or cooked ground beef (1 oz)
Sloppy joes (1/2 sandwich)

---

[a]Each serving contains at least 1 milligram iron, or one-tenth of a child's RDA for iron. Bioavailability varies. Vitamin C-rich foods included with these foods increase iron absorption.

[b]Some fortified breakfast cereals contain more than 10 milligrams iron per half-cup serving (read the labels).

prescription. Children's supplements typically include most of the vitamins. They may or may not provide iron, zinc, calcium, or other minerals as well.

For most children, routine supplementation is unnecessary. Diets of healthy, well-fed children generally supply enough nutrients; supplements do not improve their nutrition status. Particular groups of children, however, may benefit from supplementation.[6] These groups include:

- Children who suffer from malnutrition.
- Children with anorexia or poor eating habits.
- Children adhering to restricted diets, such as total vegetarian diets.
- Children with chronic diseases, such as cystic fibrosis.

Unfortunately, malnourished children, who most need more food or supplements to provide the essential nutrients they lack, are the least likely to have access to them. For children who do take supplements, any multivitamin-mineral product that provides a complete array of vitamins and minerals at approximately the RDA levels is appropriate.

Little effort is required to convince a child to take a vitamin supplement. Fruit-flavored, chewable vitamins shaped like cartoon characters entice young children to accept them. These cute, flavorful tablets also have the potential to cause poisoning in children who take large doses of them. Of most concern are the supplements that contain iron.

Dangers of Iron Supplements. Iron-containing supplements are the leading cause of poisoning deaths for children under six years of age in the United States.[7] Food and Drug Administration (FDA) regulations require supplements with 30 or more milligrams of iron to carry label warnings about the potential danger of iron toxicity in children. The regulation also requires that such supplements be packaged in "non-reusable unit-dose packaging" such as blister packs.[8] A mild overdose of iron-containing vitamins causes nausea, vomiting, diarrhea, and bleeding in the GI tract. More severe overdoses result in shock, liver damage, coma, and, in some cases, death. The ingestion of 30 milligrams of iron per kilogram of body weight is toxic, and if it occurs, vomiting should be induced immediately.

Benefits of Fluoride Supplements. In regions without water fluoridation, fluoride supplements are recommended for all children. Fluoride makes the bones stronger and the teeth more resistant to decay. The recommended fluoride dosage varies from 0.25 to 1.0 milligram per day, depending on the child's age and the fluoride concentration of the local water.*[9]

*The National Research Council of the National Academy of Sciences recommends fluoridation of drinking water to raise the concentration to about 1 part fluoride per 1 million parts water, an amount that offers the greatest protection against dental caries at virtually no risk of toxicity.

The next section describes the nutrition assessment tools that can alert health care professionals to the signs of malnutrition in children. Once alerted, health care professionals can act promptly to prevent or reverse malnutrition.

Nutrition Assessment of Children

Nutrition assessment of children relies on the same four approaches as are used for infants and adults; historical information, anthropometric data, physical examinations, and biochemical analyses. A fifth approach—observation of behavior—is also useful in assessing children's nutrition status and is included in the discussion on physical examinations.

Historical Information

A child's nutrition status often correlates with the economic status of the family. The socioeconomic history, including the extent of participation in food assistance programs, can reveal useful information regarding the child's food intake. Questions similar to those listed in Form A-1 in Appendix A ascertain information about a child's nutrition status.

Diet history information can provide clues to children's iron and calcium status, two nutrients that typically fall short in unbalanced diets. If the assessor suspects that one or both of these nutrients is lacking in a child's diet, nutrition counseling for the parents and further evaluation of the child are appropriate.

Young children often snack between meals, so it is important to include such information in the diet history. Snacks can have a major influence on a child's nutrient intake.

As discussed in a later section of this chapter, overweight children often become overweight adults. Diet history information can identify eating habits that promote obesity. At the other extreme, the diet history can help identify children with inadequate energy intakes—whether due to poverty or to parent-imposed food restrictions that undermine the child's health.

Anthropometric Data

Anthropometric measures on children can throw light not only on their growth and development but also on the probability of their becoming obese—a likelihood in our society. Often the best clinical assessment of obesity is the trained eye. It is quite likely that children who are visually identified as obese are also obese by anthropometric criteria. When visual assessment is questionable, both weight-for-height and fatfold measurements should be taken to confirm the diagnosis.

Weight–Height Measures. As soon as they can stand upright, children can be weighed on the same beam balance scale used for adults. Assessors can also measure children's standing height, but they must use care in doing so. It is especially important never to use the flimsy rod on a standard scale when measuring children's height, for the errors that arise in doing so are greater in proportion to a child's short height than to an adult's. The best way to measure height is with the child's back against a flat wall to which a measuring board or nonstretchable measuring tape or stick is attached. The child stands erect, without shoes, with heels together. The child's line of sight should be horizontal, with the heels, buttocks, shoulders, and head touching the wall. The assessor uses a block, book, or other inflexible object to ensure that the top of the head is measured at an exact right angle to the wall. The assessor carefully checks the height measurement and immediately records the result in either inches or centimeters. Such a practice prevents misplacing or forgetting the measurement. To evaluate growth in children, an assessor uses the standard growth charts shown in Figures 6-1 (A and B) and 6-2 (A and B). Weight and height measurements are taken at regular intervals and compared both with these population standards and with previous measures of the child.

Standing "at attention" allows for an accurate height measurement.

Fatfold Measures. In contrast to weight and height measures, fatfold measures provide a direct measure of body fat.[10] Some researchers have developed new population-specific fatfold equations for children and adolescents. These new equations, from which estimates of percent body fat can be derived, take into account the differences in body chemistry between children and adults. The equations that are now used assume a constant adult-derived body density that may overpredict body fatness when applied to children.[11] The new fatfold equations look promising, but need further research.

PEM. Protein-energy malnutrition (PEM) is detectable by taking anthropometric measures.[12] Height for age below the 85th percentile on standard growth charts indicates *stunted* growth of a degree consistent with severe PEM. Weight for age at the 60th percentile or below indicates *wasting* consistent with severe PEM. Physical examination and biochemical testing can help confirm PEM, so that appropriate therapy can be instituted.

Reminder: **Stunted** *growth reflects long-term food deprivation, whereas* **wasting** *reflects recent short-term malnutrition.*

Physical Examinations

The effects of malnutrition during childhood are not limited to growth impairment; they are diverse and numerous and include behavioral as well as physical effects. If a child looks unhealthy and acts abnormally, consider that the cause *may* be malnutrition. This advice may sound obvious, but surprisingly, parents and medical practitioners often overlook the possibility that malnutrition may account for abnormalities of appearance and behavior. No physical or behavioral symptom alone is

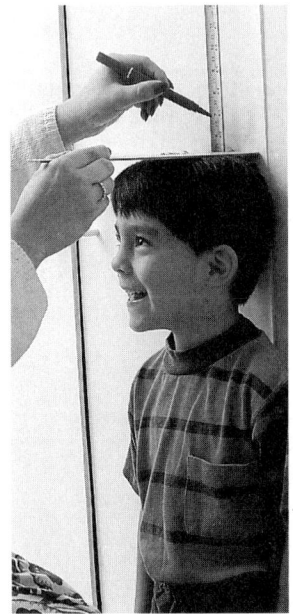

Figure 6-1A Nutrition Assessment from 2 to 18 Years: Height and Weight for Age—Girls

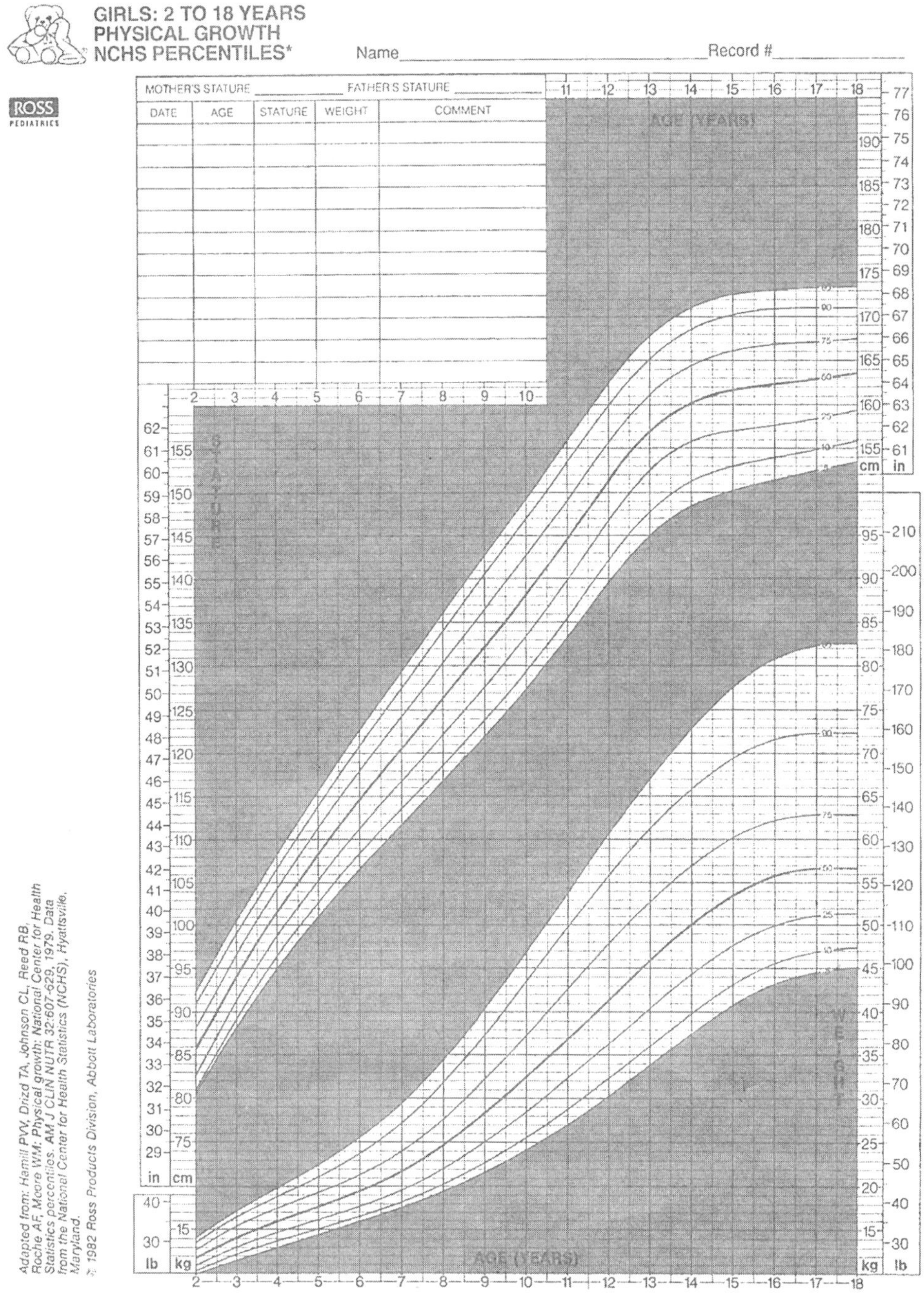

Figure 6-1B Nutrition Assessment from 2 to 18 Years: Height and Weight for Age—Boys

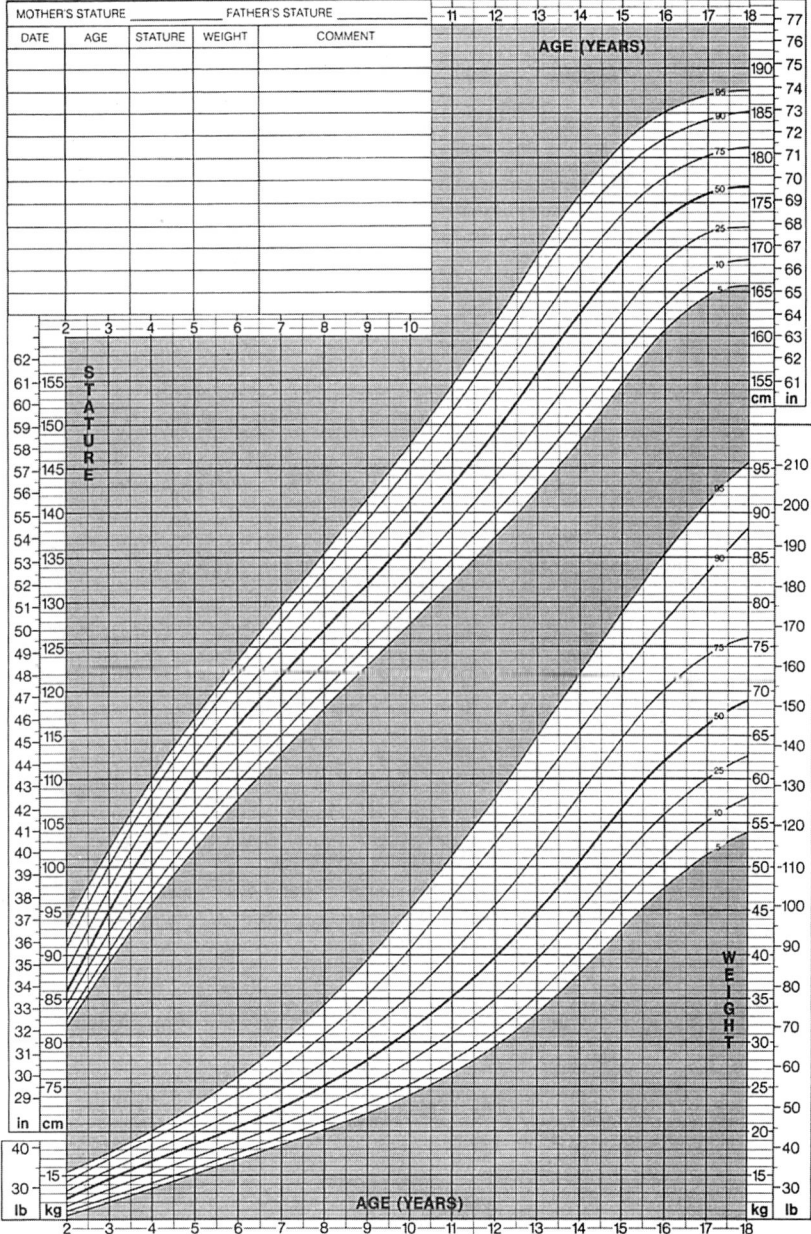

BOYS: 2 TO 18 YEARS
PHYSICAL GROWTH
NCHS PERCENTILES*

Source: Used with permission of Ross Laboratories, Columbus, OH 43216, from NCHS Growth Charts, © 1982 Ross Laboratories.

Figure 6-2A Nutrition Assessment during Prepubescence: Weight for Height—Girls

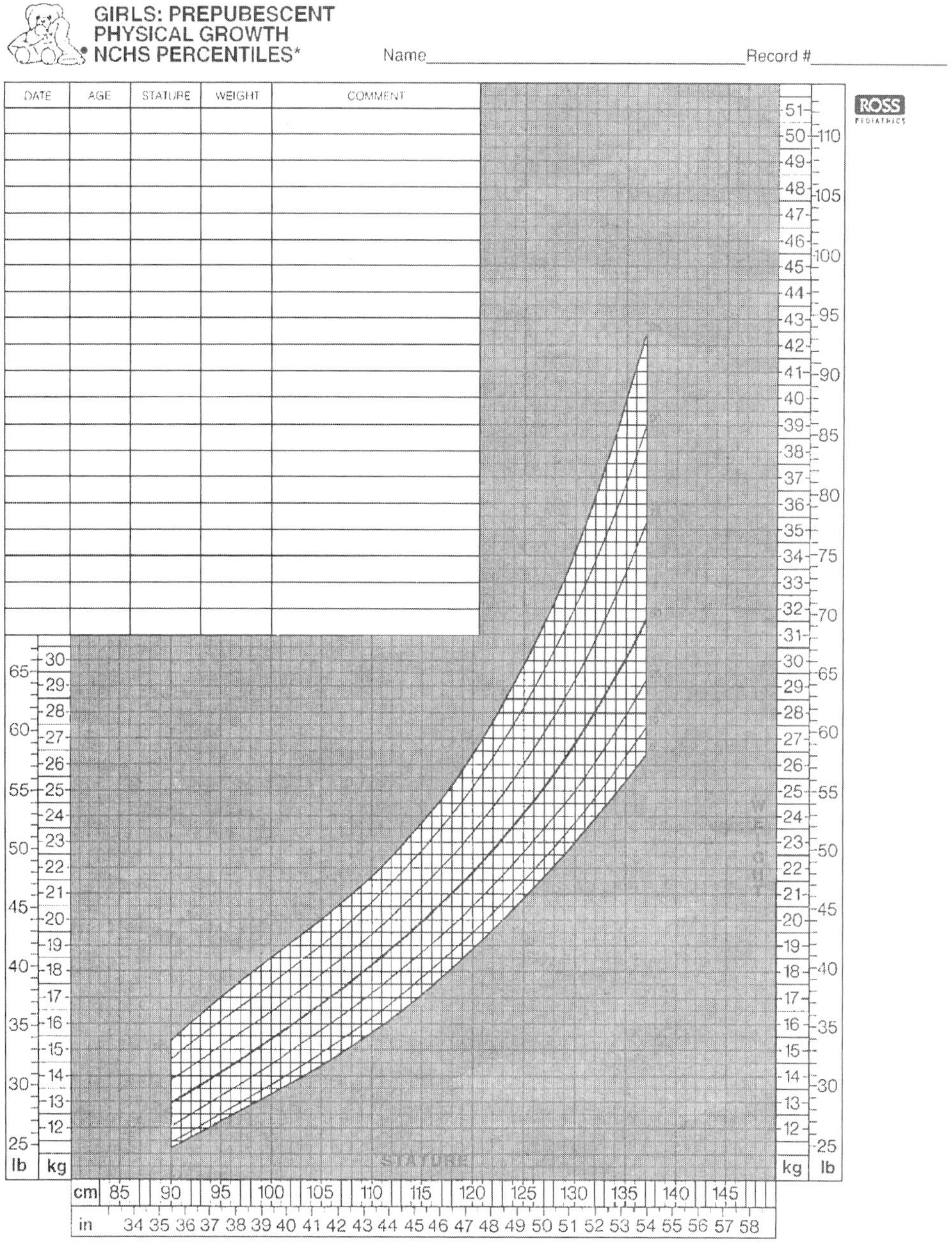

Figure 6-2B Nutrition Assessment during Prepubescence: Weight for Height—Boys

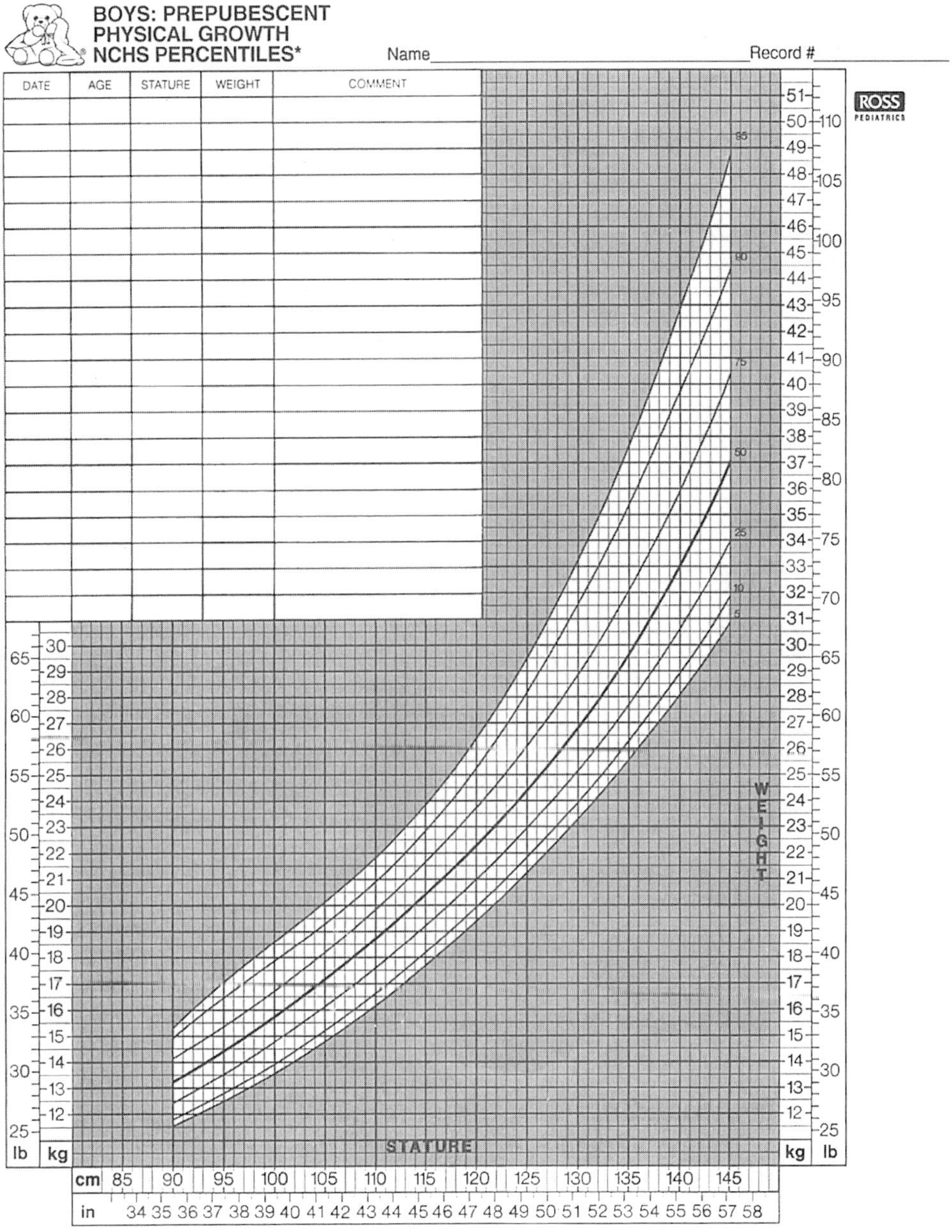

Source: Used with permission of Ross Laboratories, Columbus, OH 43216, from NCHS Growth Charts, © 1982 Ross Laboratories.

Table 6-2 Physical Signs of Health and Malnutrition

Body System	Healthy	Malnourished	Nutrients of Concern
Hair	Shiny, firm in the scalp	Dull, brittle, dry, loose; falls out	Protein, energy
Eyes	Bright, clear pink membranes; adjust easily to light	Pale membranes; spots; redness; adjust slowly to darkness	Vitamin A, the B vitamins, zinc, and iron
Teeth and gums	No pain or caries, gums firm, teeth bright	Missing, discolored, decayed teeth; gums bleed easily and are swollen and spongy	Minerals and vitamin C
Face	Clear complexion without dryness or scaliness	Off-color, scaly, flaky, cracked skin	Protein, energy, vitamin A, and iron
Glands	No lumps	Swollen at front of neck, cheeks	Protein, energy, and iodine
Tongue	Red, bumpy, rough	Sore, smooth, purplish, swollen	B vitamins
Skin	Smooth, firm, good color	Dry, rough, spotty; "sandpaper" feel or sores; lack of fat under skin	Protein, energy, essential fatty acids, vitamin A, the B vitamins, and vitamin C
Nails	Firm, pink	Spoon-shaped, brittle, ridged	Iron
Internal systems	Regular heart rhythm, heart rate, and blood pressure; no impairment of digestive function, reflexes, or mental status	Abnormal heart rate, heart rhythm, or blood pressure; enlarged liver, spleen; abnormal digestion; burning, tingling of hands, feet; loss of balance, coordination; mental confusion, irritability, fatigue	Protein, energy, and minerals
Muscles and bones	Muscle tone; posture, long bone development appropriate for age	"Wasted" appearance of muscles; swollen bumps on skull or ends of bones; small bumps on ribs; bowed legs or knock-knees	Protein, energy, and vitamin D

diagnostic of a particular nutrient deficiency, but it does suggest the need for further tests. Whereas the signs of malnutrition can be mistaken for those of disease, neglect, or other causes, the signs of good health and sound nutrition are unmistakable, as Table 6-2 shows.

Biochemical Analyses

The biochemical tests used for adults are also suitable for children, although in some cases the standards are different. Some biochemical tests are part of routine health care evaluation, whereas others are per-

Healthy, well-nourished children are alert in the classroom and energetic at play.

formed only when history information or other assessment measures suggest nutritional risk. Table A-6 in Appendix A lists biochemical tests useful for assessing nutrition status.

Nutrition-Related Concerns during Childhood

Most children in the United States and Canada are well nourished. Their average energy intakes are sufficient to support normal growth, and their average nutrient intakes, except for iron, meet or exceed recommendations. Malnutrition does appear, however, in certain circumstances. Low-income children, for example, may be hungry and malnourished. Adverse reactions to foods may lead to malnutrition when parents eliminate nutrient-rich foods from their child's diet in an effort to treat a food allergy. In some cases, malnutrition is mistakenly blamed for a child's hyperactive behavior when in fact other factors are responsible. These nutrition-related concerns are featured in this section.

Hunger and Malnutrition in Children

An estimated 11 million U.S. children under age 12 are hungry. When hunger is chronic, children become malnourished. Worldwide, malnutrition takes a devastating toll on children, contributing to nearly half of the deaths of children under four years old. Vitamin A deficiency afflicts more than 5 million children worldwide, leading to blindness, stunted growth, and infections. Zinc deficiency also retards growth and typically accompanies protein-energy malnutrition and vitamin A deficiency.

A root cause of hunger in the United States is poverty. Homeless people are much more likely to be hungry than people who are housed, and children are the fastest growing segment of the homeless population. Homeless children from large families and from families headed by single mothers are especially vulnerable to hunger and its adverse consequences. Fortunately, federal programs that fund school breakfast and lunch ensure that most impoverished children receive at least two reasonably nutritious meals a day. Focal Point 5A describes the effects of malnutrition on physical growth and mental development. This section focuses on the behavioral consequences of hunger and malnutrition in children.

🏃 Healthy People 2000

Reduce growth retardation among low-income children aged five years and younger to less than 10%.

Hunger and Behavior. Even when hunger is temporary, as when a child misses one meal, behavior and academic performance may be affected. Children who eat nutritious breakfasts function better than their peers who do not. Young children who participate in the federally funded School Breakfast Program improve their scores on achievement tests and are tardy or absent significantly less often than children who qualify for the program but do not participate. Without breakfast, even healthy children perform poorly in tasks requiring concentration, their attention spans are shorter, and they score lower on IQ tests than their well-fed peers; malnourished children are particularly vulnerable. Common sense dictates that it is unreasonable to expect anyone to learn and perform work when no fuel has been provided. By late morning, discomfort from hunger may become distracting even if a child has eaten breakfast.

The problem children face when attempting morning schoolwork on an empty stomach appears to be at least partly due to low blood glucose. The average child up to age 10 or so needs to eat about every four hours to maintain a blood glucose concentration high enough to support the activity of the brain and the rest of the nervous system. A child's brain is as big as an adult's, and the brain is the body's chief glucose consumer. A child's liver is much smaller than an adult's, however, and the liver is responsible for storing glucose as glycogen and releasing it into the blood as needed. A child's liver can store only about a four-hour supply of glycogen—hence the need to eat fairly often. Teachers

The brain uses about three times as much glucose per day as the rest of the body.

aware of the late-morning slump in their classrooms wisely request that midmorning snacks be provided; snacks improve classroom performance all the way to lunchtime. For the child who hasn't had breakfast, the morning's lessons may be lost altogether.

Eating breakfast also helps children to meet their nutrient needs each day. Children who skip breakfast typically do not make up the deficits at later meals—they simply have lower intakes of energy, vitamins, and minerals than those who eat breakfast.[13]

Iron Deficiency and Behavior. Iron deficiency has well-known and widespread effects on children's behavior. In addition to carrying oxygen in the blood, iron transports oxygen within cells, which use it to help produce energy. Iron is also used to make neurotransmitters—most notably, those that regulate the ability to pay attention, which is crucial to learning. Consequently, an iron deficiency not only causes an energy crisis but also directly affects mood, attention span, and learning ability.

Iron deficiency is usually diagnosed by a deficit of iron in the *blood*, after the deficiency has progressed all the way to anemia. A child's *brain*, however, is sensitive to low iron concentrations long before the blood effects appear. Research has shown that iron deficiency lowers the "motivation to persist in intellectually challenging tasks," shortens the attention span, and impairs overall intellectual performance. Anemic children perform less well on tests and are more disruptive than their nonanemic classmates. Focal Point 5B explores the relationships between iron deficiency and behavior in greater detail.

Other Nutrient Deficiencies and Behavior. Iron is not the only nutrient whose deficiency can affect behavior. A child with any of several nutrient deficiencies may be irritable, aggressive, disagreeable, or sad and withdrawn (see Table 6-3). Such a child may be labeled "hyperactive," "depressed," or "unlikable," when in fact these traits may arise from simple, even marginal, malnutrition. In any such case, inspection of the child's diet by a registered dietitian or other qualified health care professional is clearly in order. Should suspicion of dietary inadequacies be raised, no matter what other causes may be implicated, the people responsible for feeding the child should take steps to correct those inadequacies promptly.

Adverse Reactions to Foods

Adverse reactions to foods can threaten nutritional health to varying extents, depending on the severity and duration of the reactions and the foods they involve. Temporary reactions may lead to permanent avoidance of foods; permanent reactions, if not detected and treated, can cause chronic illness.

Food Intolerances. Not all adverse reactions to foods are food allergies, although even physicians may describe them as such. Signs of adverse reactions to foods include stomachaches, headaches, pain, rapid pulse

adverse reactions: unusual responses to food (including intolerances and allergies).

food intolerances: adverse reactions to foods that do not involve the immune system.

Table 6-3 Behavioral Symptoms of Nutrient Deficiencies

- *Protein-energy deficiency*—apathy, fretfulness, lack of energy, and lack of interest in food.
- *Thiamin deficiency*—confusion, uncoordinated movements, depressed appetite, irritability, insomnia, fatigue, and general misery.
- *Riboflavin deficiency*—depression, hysteria, psychopathic behavior, lethargy, and hypochondria evident before deficiency can be detected by clinical symptoms.
- *Niacin deficiency*—irritability, agitated depression, headaches, sleeplessness, memory loss, emotional instability (early signs of pellagra onset), and mental confusion progressing to psychosis or delirium.
- *Vitamin B$_6$ deficiency*—irritability, insomnia, weakness, mental depression, abnormal brainwave pattern, convulsions, the mental symptoms of anemia,[a] fatigue, and headache.
- *Folate deficiency*—the mental symptoms of anemia,[a] tiredness, weakness, forgetfulness, mild depression, abnormal nerve function, headache, disorientation, confusion, and inability to perform simple calculations.
- *Vitamin B$_{12}$ deficiency*—degeneration of peripheral nervous system and anemia.
- *Vitamin C deficiency*—hysteria, depression, listlessness, lassitude, weakness, aversion to work, hypochondria, social introversion, possible iron deficiency anemia, and fatigue.
- *Vitamin A deficiency*—anemia.
- *Iron deficiency*—fatigue, weakness, headaches, pallor, listlessness, and the mental symptoms of anemia.[a]
- *Magnesium deficiency*—apathy, personality changes, and hyper-irritability.
- *Copper deficiency*—iron deficiency anemia.
- *Zinc deficiency*—poor appetite, failure to grow, iron deficiency anemia, irritability, emotional disorders, and mental lethargy.

[a]The mental symptoms of anemia can include any or all of the following: lack of appetite, apathy, listlessness, clumsiness, conduct disturbances, shortened attention span, hyperactivity, irritability, learning disorders (vocabulary, perception), lowered IQ, low scores on latency and associative reactions, reduced physical work capacity, and repetitive hand and foot movements. These symptoms are not caused by anemia itself but by iron deficiency in the brain. Children with much more severe anemias from other causes, such as sickle cell anemia and thalassemia, show no reduction in IQ when compared with children without anemia.

rate, nausea, wheezing, hives, bronchial irritation, coughs, and other such discomforts. Among the causes may be reactions to chemicals in foods, such as the flavor enhancer monosodium glutamate (MSG), the natural laxative in prunes, and the mineral sulfur; digestive disorders, such as obstructions or injuries; enzyme deficiencies, such as lactose intolerance; and even psychological aversions. These reactions involve symptoms but no antibody production. Therefore, they are food intolerances, not allergies.[14]

Pesticides on produce may also cause adverse reactions. Pesticides may linger in the foods to which they were applied in the field.[15] Health risks from pesticide exposure are probably small for healthy adults, but children may be vulnerable to some types of pesticide poisoning.[16] The FDA, EPA, and USDA are all considering proposals to revise food safety and pesticide laws to protect infants and children from pesticide risks.[17] When setting tolerance levels, the agencies will first identify foods that children commonly eat in large amounts and then consider the effects of pesticide exposure during developmental stages.[18]

Food Allergies. A true food allergy occurs when a whole food protein or other large molecule enters the body and elicits an immunologic response. (Proteins and other large molecules of food are normally dismantled in the digestive tract to smaller ones that are absorbed without such a reaction.) The body's immune system reacts to these large food molecules as it does to other antigens—by producing antibodies, histamines, or other defensive agents.

Allergies may have one or two components. They always involve antibodies; they may or may not involve symptoms. This means that allergies can be diagnosed only by testing for antibodies. Even symptoms exactly like those of an allergy may not be caused by one.

Allergic reactions to food may be immediate or delayed. In both cases, the antigen interacts immediately with the immune system, but the timing of symptoms varies from minutes to 24 hours after consumption of the antigen. Identifying the food that causes an immediate allergic reaction is fairly easy, because the symptoms appear within minutes after eating the food. Identifying the food that causes a delayed reaction is more difficult, because the symptoms may not appear until a day later. By this time, many other foods will have been eaten, complicating the picture.

Almost 75 percent of adverse reactions are caused by three foods—eggs, peanuts, and milk.[19] Allergic reactions to single foods are common. Reactions to multiple foods are the exception, not the rule.

Identifying a true food allergy requires a thorough health history, physical examination, and diagnostic tests to eliminate other diseases.[20] Skin pricks with food extracts are one of the most common tests for food allergies, even though the high incidence of false positive results can complicate diagnosis. Physicians also conduct dietary trials that first eliminate the offending food and then reintroduce it in small quantities to substantiate that reactions occur only when that particular food is eaten.[21] Once a food allergy has been diagnosed, therapy requires strict elimination of the offending food.

Food allergies are most common during the first few years of life, but then children typically outgrow (become tolerant to) their hypersensitivity. Between 2 and 8 percent of young children are allergic to certain foods, whereas only 2 percent of adults have food allergies.[22] Developing tolerance is most likely if the offending food can be identified and eliminated from the diet for at least a year or two.[23]

food allergies: adverse reactions to foods that involve an immune response; also called **food hypersensitivity reactions.**

antigens: substances that elicit the formation of antibodies or an inflammation reaction from the immune system. A bacterium, a virus, a toxin, and a protein in food that causes allergy are all examples of antigens.

histamines (HISS-tah-means, or HISS-tah-mens): substances produced by cells of the immune system as part of a local immune reaction to an antigen, causing inflammation.

A person who produces antibodies without *having any symptoms has an **asymptomatic allergy**; a person who produces antibodies and has symptoms has a **symptomatic allergy**.*

Eggs, peanuts, and milk are most likely to induce symptoms in people with food allergy.

When parents stop serving a suspected food to their child, they risk the child's suffering nutrient deficiencies. They should be sure to include other foods that offer the same nutrients as the omitted food. Children with allergies, like all children, need all their nutrients.

Hyperactivity and "Hyper" Behavior

Because malnutrition can impair children's behavior in many ways, people look to food habits for explanations of hyperactivity. Hyperactivity is not caused by a poor diet, but a poor diet may be part of a cluster of factors seen in a hyperactive child's life.

tension–fatigue syndrome: apparent hyperactivity caused in a child by the combination of lack of sleep, overstimulation, and anxiety.

Tension–Fatigue Syndrome. Children can become excitable, rambunctious, and unruly out of a desire for attention, lack of sleep, overstimulation, too much television, or a lack of physical activity. Together, these factors produce the tension–fatigue syndrome, which suggests that more consistent care is needed. It helps most to insist on regular hours of sleep, regular mealtimes, and regular outdoor activity.

hyperactivity: excessive activity that may cause behavior and learning problems. Behavior problems include impulsiveness and restlessness. Learning problems reflect a short attention span. Professionals call this syndrome **attention deficit hyperactivity disorder (ADHD).**

Hyperactivity Disorder. Hyperactivity affects behavior and learning in about 5 percent of young school-age children. Left untreated, hyperactivity can interfere with a child's social development and ability to learn. Treatment focuses on relieving the symptoms and controlling the associated problems; the cause of hyperactivity disorder is unknown, so there is no cure.

Physicians often manage hyperactivity through behavior modification, special educational techniques, psychological counseling, and drug therapy. The drugs most commonly prescribed are stimulants. Stimulants help many hyperactive children to focus their attention, which improves their behaviors and social interactions with others. The effects are often immediate, but diminish within a few hours, making multiple daily doses necessary. Many parents and some clinicians have expressed concerns about such treatment because of its ineffectiveness in some children, potential adverse side effects, and emotional implications.

Many parents mistakenly believe a solution may lie in manipulating the diet—most commonly, by eliminating sugar or food additives. This notion is appealing because diet is one area of a child's life over which parents feel they can exert some control. If problems might be solved by adding carrots or eliminating candy, then parents are eager to give diet advice a try. Although nutrition should be considered whenever a person's health is less than optimal, it is unwise to jump at appealing solutions that are unfounded. Several studies have found no convincing evidence that sugar causes hyperactivity or worsens behavior.[24] Recommendations to restrict sugar in children's diets to prevent or treat behavior problems are not supported by research findings. Sugar can influence children's behavior only by displacing nutritious foods and contributing to nutrient deficiencies.

Caffeine and Behavior. Caffeine is often overlooked as a source of "hyper" behavior in children, but it is a matter of some concern to pediatricians. A 12-ounce soft drink may contain as much as 50 milligrams caffeine; in the body of a 60-pound child, two or more such beverages are equivalent to the caffeine in 8 cups of coffee for a 175-pound adult. Children who are troubled by sleeplessness, restlessness, and irregular heartbeats may need to limit their caffeine consumption. Children not accustomed to caffeine who are given doses equivalent to about two cola beverages a day become noticeably inattentive and restless. As long as children are surrounded by attractive temptations such as cola beverages, adults must intervene until the children learn to control consumption themselves. Table 6-4 lists the caffeine contents of foods, beverages, and medicines.

Hunger, allergic reactions, and hyperactivity can all adversely affect a child's nutrition status and health. Fortunately, each of these problems has solutions. They may not be easy solutions, but at least we have a reasonably good understanding of the problems and ways to correct them. Given the resources, success is achievable. Such is not the case with the most pervasive health problem for children in the United States—obesity.

Childhood Obesity

If you were to ask two experts in the field of obesity, "What causes obesity?" you would quite likely receive two different, but correct, answers. One might authoritatively respond, "Excess body fat accumulates when people consistently take in more food energy than they spend. They do this because of a multitude of interrelated causes, including genetic, environmental, behavioral, and metabolic factors." The other expert might simply respond, "We don't know." Without a clear understanding of the causes, solutions have been elusive. Yet the problems have been overwhelming and demand attention.

An estimated 25 percent of the children in the United States, ages 6 to 11, and 3 to 8 percent of those ages 2 to 5, are overweight.[25] Most disturbing is the realization that the incidence has increased dramatically over the past three decades. Children in the United States are becoming fatter.

Effects of Obesity

The single most important problem for obese children is the potential of becoming obese adults with all the social, economic, and medical ramifications that obesity entails. They have other problems, though, arising from differences in their growth, physical health, and psychological development.

Growth. Obese children develop a characteristic set of physical traits. They typically begin puberty earlier and so grow taller than their peers at first, but then stop growing at a shorter height. They develop greater

Table 6-4 Caffeine Content of Beverages, Foods, and Over-the-Counter Drugs

Beverages and Foods	Average (mg)	Range (mg)
Coffee (5-oz cup)		
Brewed, drip method	130	110–150
Brewed, percolator	94	64–124
Instant	74	40–108
Decaffeinated, brewed or instant	3	1–5
Tea (5-oz cup)		
Brewed, major U.S. brand	40	20–90
Brewed, imported brands	60	25–110
Instant	30	25–50
Iced (12-oz can)	70	67–76
Soft drinks (12-oz can)		
Dr Pepper, Mr. Pibb		40–43
Colas and cherry cola		
Regular or diet		30–49
Caffeine-free		0–trace
Jolt		72
Mountain Dew, Mello Yello, Surge		50–55
Fresca, Hires Root Beer, 7-Up, Sprite, Squirt	0	
Water Joe (17-oz bottle)	60	
Cocoa beverage (5-oz cup)	4	2–20
Chocolate milk beverage (8 oz)	5	2–7
Milk chocolate candy (1 oz)	6	1–15
Dark chocolate, semisweet (1 oz)	20	5–35
Baker's chocolate (1 oz)	26	
Chocolate flavored syrup (1 oz)	4	
Drugs[a]		
Cold remedies (standard dose)		
Dristan	0	
Coryban-D, Triaminicin	30	
Diuretics (standard dose)		
Aqua-ban, Permathene H_2Off	200	
Pre-Mens Forte	100	
Pain relievers (standard dose)		
Excedrin	130	
Midol, Anacin	65	
Aspirin, plain (any brand)	0	
Stimulants		
Caffedrin, NoDoz, Vivarin	200	
Weight control aids (daily dose)		
Prolamine	280	
Dexatrim, Dietac	200	

Note: A pharmacologically active dose of caffeine is defined as 200 milligrams.

[a]Because products change, contact the manufacturer for an update on products you use regularly.

bone and muscle mass, possibly because their skeletons respond to the demand of having to carry more weight—both fat and fat-free weight. Consequently, they appear "stocky" even when they lose their excess fat. Obese children have a faster metabolic rate, apparently due to their abundant lean body mass.

A relationship between obesity and maturation is also evident, although not clearly understood. Early-maturing females are both shorter and fatter than their peers. By age 30, they have accumulated 30 percent more fat. Such evidence of the inverse relationship between age of menarche and fatness in women is apparent throughout their lives.

Physical Health. Like obese adults, obese children display a blood lipid profile indicative that atherosclerosis is beginning to develop: high levels of total cholesterol, triglycerides, low-density lipoprotein (LDL) cholesterol, and very-low-density lipoprotein (VLDL) cholesterol. Obese children also tend to have high blood pressure; in fact, obesity is the leading cause of pediatric hypertension. These relationships between childhood obesity and cardiovascular disease risks are discussed fully in Focal Point 6C.

Psychological Development. Obesity often causes psychological problems. Because people frequently judge others on appearance more than on character, obese children are often victims of prejudice. Many suffer discrimination by adults and rejection by their peers. They are teased and called names. They often have poor self-images, a sense of failure, and a passive approach to life. Television shows, which are a major influence in children's lives, frequently portray the fat person as the bumbling misfit.

The meaning body shape has for children is illustrated by research in which children have been asked to assign adjectives ascribing behavior and personality traits to silhouettes representing various body types. Their responses reveal that they associate a common stereotype with each body image. They commonly describe muscular body types favorably; slender body types unfavorably *personally*; and bulky body types unfavorably *socially*. These stereotypes are not typically related to the body type of the child assigning the description; overweight children themselves share the negative view of fat children. The general consensus is that obese children are children that others "did not like so well." No doubt, childhood obesity creates heartaches. Children are often reasonably accurate in their perceptions of their own body shape, and they clearly prefer to look like the muscular, athletic image. In general, children believe thin children have more friends and are better looking and smarter than fat children.

Personality problems may also arise from the other physical changes that accompany obesity. As mentioned, puberty comes earlier. What effect a shortened childhood, or being one of the first in the class to begin the transition, has on a child is difficult to determine. The teen years are a time of finding a place in the social system. What you look like, who your friends are, and what activities you are involved in all seem overwhelmingly important. These are not easy years for most children. Because obesity can cause psychological distress, weight loss treatment must include ways to cope with these problems as well.

Prevention and Treatment of Obesity

Medical science has worked wonders in preventing or curing many of even the most serious childhood diseases, but obesity remains a challenge. Once excess fat has been stored, it is stubbornly difficult to remove. In light of all this, parents are encouraged to make major efforts to prevent childhood obesity.

Their efforts, however, must be balanced. Both parents and professionals need to realize that demanding that children conform to the image presented by our society creates pressures. Although not always easy to accomplish, a relaxed, nonjudgmental attitude toward the child, whatever the child's weight, will assure the best emotional growth potential.

Because many treatment programs have a high rate of failure, it is important to ask if the consequences of failing at a weight loss attempt are worse than being overweight. Other questions to consider include, When is intervention most appropriate? Will the problem resolve itself without intervention? What are the child's needs and wants? Before implementing a treatment program, a health care provider must assess the medical, social, and economic risks of the child's obesity. These risks must be balanced against the potential benefits of success and the possible harm of failure.

Treatment must consider the many aspects of the problem and possible solutions. An integrated approach is recommended, involving diet, physical activity, psychological support, and behavioral changes.

Diet. The initial goal for obese children is to reduce the rate of weight gain. Continued growth will then accomplish the desired change in weight for height. Treatment should begin with this conservative approach before other more drastic measures are taken.

Whether the goal is to treat or prevent obesity, the following strategies may be helpful:

■ Serve family meals that reflect kcalorie control both in the foods offered and in the ways foods have been prepared.

■ Encourage children to eat slowly, to pause and enjoy their table companions, and to stop eating when they are full.

■ Teach them how to select low-fat snacks and to serve themselves appropriate portions at meals.

■ Never force children to clean their plates.

Physical Activity. Underactivity is probably the single most important contributor to obesity. In turn, television watching may contribute most to physical inactivity.[26] Watching television contributes to obesity

Television watching influences children's eating habits and activity patterns.

in several ways. First, television viewing requires little energy beyond the resting metabolic rate. Second, it replaces time spent in more vigorous activities. Third, watching television correlates with between-meal snacking, eating the high-kcalorie, high-fat foods most heavily advertised, and influencing family food purchases. Nonnutritious foods and beverages appear not only in commercials, but also within the television programs themselves. People, especially children, may forget that eating and drinking these foods will bring about weight gain when they see television stars indulging in such behavior and remaining thin.

Children who watch the most television have the greatest prevalence of obesity: Obesity increases by 2 percent for each additional hour of television viewed per day.[27] The relationship between television and obesity remains strong when variables such as prior obesity and socioeconomic class are considered.

The many benefits of physical activity are well known, but often are not incentive enough to motivate overweight people, especially children. Ideally, parents will encourage daily physical activity to promote strong skeletal, muscular, and cardiovascular development and instill in their children the desire to be physically active throughout life. Physical activity is a natural and lifelong behavior of healthy living.[28] It can be as simple as riding a bike, playing tag, jumping rope, or doing chores. It need not be an organized sport; it just needs to be some activity on a regular basis.

TV fosters obesity because it:
- *Requires little energy beyond basal metabolism.*
- *Replaces vigorous activities.*
- *Encourages snacking.*
- *Promotes a sedentary lifestyle.*

Psychological Support. Programs that involve parents in treatment report greater weight losses than programs in which parents are not involved. Because obesity in parents and children tends to be positively correlated, both benefit from a weight loss program. Parental attitudes about food greatly influence their children's eating behavior, and so it is important that the influence be positive. Otherwise, eating problems may become exacerbated. Unaware that they are teaching their children, parents pass on lessons at the dinner table, on television trays, and in drive-through restaurants. Parents who have failed in the battle against being overweight themselves may model for their children the behaviors that have led to their failure—eating too much, dieting inappropriately, exercising too little. Those who have fears about their child becoming overweight may inadvertently teach their children fears that are unproductive.

Some childhood obesity treatment programs provide specific instructions for parents, whereas others allow passive observation. In general, the parent's role is to:

■ Provide a nutritionally adequate diet.
■ Serve food under pleasant conditions.
■ Allow the child to eat according to need.
■ Allow the child to eat without being nagged or cajoled.
■ Encourage the child to enjoy nutritious foods.
■ Encourage the child to be physically active.

The school can also play a supportive role in the treatment of childhood obesity by way of the foods served, the physical activity program, and the classroom teaching. The school cafeteria can offer lunches that meet both the nutrient and energy needs of children, without excess fat or kcalories. In physical education class, the instructor can instill motivation and emphasize lifelong activities; physical activities that can be done alone and at any age with minimal equipment are most valuable throughout life. In the classroom, nutrition education can include lessons on:

■ Body fatness.
■ Relative kcalorie value of foods.
■ Relative kcalorie value of activities.
■ Nutrient density.
■ Energy balance.
■ Benefits of physical activity.
■ Food sources and variety.
■ Cooking.
■ Learning to enjoy the tastes of new foods.

The public school system offers many opportunities to attack the problem of obesity. It can reach a large number of children, the treatment can be continuous, and it can be offered at minimal cost. The setting is also conducive to nutrition education, an important facet of treatment. Perhaps most importantly, a broad base of social support can be established. School personnel, family, and peers can offer an over-weight child the reinforcement needed to continue efforts in a weight loss program.

Behavioral Changes. Behavior modification techniques have shown modest success when used to control obesity in children. In contrast to traditional weight loss programs that focus on *what* to eat, behavioral programs focus also on *how* to eat. These techniques involve changing learned habits that lead a child to eat excessively.

Obesity is prevalent in our society. Its far-reaching effects lend urgency to the need to find a remedy. Because treatment of obesity is frequently unsuccessful, it is most important to prevent its onset. In teaching children how to maintain appropriate body weight, it is important to use sensitivity. Children can easily get the impression that their worth is tied to their body weight. Some parents fail to realize that society's ideal of slimness can be perilously close to starvation, and that a child encouraged to "diet" cannot obtain the energy and nutrients required for normal growth and development. Even healthy children without diagnosable eating disorders have been observed to limit their growth through "dieting."[29] Weight control in truly overweight children should be overseen by a health care professional to ensure a healthy eating plan that will not compromise growth.

Food Choices and Fitness during Childhood

To provide all the needed nutrients, a child's meals and snacks should include a variety of foods from each food group—in amounts suited to the child's appetite and needs. Serving sizes increase with age. A portion of meat, grains, fruits, or vegetables for children is loosely defined as 1 tablespoon for each year. Thus, at four years of age, a portion is about 4 tablespoons, or 1/4 cup. This rule of thumb applies until children reach age 12. Table 6-5 offers a daily food pattern for children.

Parents and other caregivers must balance food choices skillfully to ensure adequate nutrition. Some food intakes that seem fairly nutritious are not, as the following comparison demonstrates. Consider the sample menus A and B shown in Figure 6-3. They were designed for two fairly typical five-year-old children. Note that the planners of both sets of menus were conscientious: Both menus provide adequate kcalories; three meals plus snacks; and foods from all five groups. But as Figure 6-4 on p. 244 shows, sample menu A provided adequate amounts of all the recommended nutrients, whereas sample menu B fell short.

Inspection of the two sample menus as compared with the recommended food pattern in Table 6-5 shows that although all food groups were represented, not all were represented in sufficient amounts or by nutrient-dense choices. Consequently, the choices on menu B fell significantly short in fiber, vitamin A, pantothenic acid, calcium, iron, and

Table 6-5 Children's Daily Food Patterns for Good Nutrition

Food Group	Servings per Day	Average Size of Serving		
		1 to 3 years	4 to 6 years	7 to 12 years
Bread and cereals (whole grain or enriched)[a]	6 or more	1/2 slice	1 slice	1 to 2 slices
Vegetables[b]	3 or more	2–4 tbs or 1/2 c juice	1/4–1/2 c or 1/2 c juice	1/2–3/4 c or 1/2 c juice
Fruits[b]	2 or more	2–4 tbs or 1/2 c juice	1/4–1/2 c or 1/2 c juice	1/2–3/4 c or 1/2 c juice
Meat and meat alternates[c]	2 or more	1–2 oz	1–2 oz	2–3 oz
Milk and milk products[d]	3 to 4	1/2–3/4 c	3/4 c	3/4–1 c

[a] 1 slice bread = 3/4 c dry cereal, 1/2 c cooked cereal, 1/2 c potato, rice, or noodles.

[b] Vitamin C source (citrus fruits, berries, tomatoes, broccoli, cabbage, cantaloupe) daily; vitamin A source (spinach, carrots, squash, tomato, cantaloupe) 3 to 4 times weekly. To help meet iron needs, include 1 cup of dark green vegetables daily.

[c] 1 oz meat, fish, poultry = 1 egg, 1 frankfurter, 2 tbs peanut butter, 1/2 c cooked legumes.

[d] 1/2 c milk = 1/2 c cottage cheese, pudding, yogurt; 3/4 oz cheese; 2 tbs dried milk. Children who do not use milk or milk products should use soy milk fortified with calcium, vitamin D, and vitamin B_{12}.

Source: Adapted from P. M. Queen and R. R. Henry, Growth and nutrient requirements of children, in *Pediatric Nutrition,* eds. R. J. Grand, J. L. Sutphen, and W. H. Dietz, Jr. (Boston: Butterworths, 1987), p. 347.

zinc. In both quality and quantity, fruits and vegetables on menu B were lacking, consisting of an apple, applesauce, and french fries. This child needed at least two more servings of vegetables according to Table 6-5, and better choices would have raised the child's low fiber and vitamin A intakes. Milk and milk products were also underrepresented, accounting for the child's low calcium intake that day. A glass or two of milk instead of the lemonade or grape soda would have raised the calcium intake. The meat group and the bread and cereal group were inadequately represented as well; as a consequence, iron and zinc were lacking for the day. A nutrient-rich serving of whole-grain cereal or bread in place of the doughnut at breakfast and an additional serving of grains and legumes or meat at dinner would have boosted the intakes of these nutrients considerably. All of these additional foods would contribute enough pantothenic acid to increase its intake as well.

For a similar energy intake, the foods in menu A provided more nutrients and offered a greater variety of foods. They also did a lot more than the foods in menu B to promote healthful eating habits. Clearly, a little extra effort and planning can make a big difference in a child's day-to-day nutrient intake.

Consider that the child represented by menu B received three meals and a snack on this day. Imagine how much worse the picture is when a child skips breakfast or when more sugary foods take the place of some of the nourishing ones. Chances are, the nutrients missed from a skipped breakfast will not be "made up" at lunch and dinner, but will be completely left out that day.

Figure 6-3 Sample Menus A and B

Sample Menu A

Sample Menu B

Breakfast:	**Breakfast:**
1/2 c cereal with 1/2 c 2% milk	1 glazed doughnut
1/2 banana	1 c 2% milk
3/4 c orange juice	
Snack:	
2 oz raisin–sunflower seed mixture	
3/4 c tomato juice	
Lunch:	**Lunch:**
Peanut butter and jelly sandwich on whole-wheat bread	Peanut butter and jelly sandwich on white bread
3/4 c 2% milk	1 c lemonade
1/2 carrot cut into sticks	1 medium apple
1 medium apple	1 small bag potato chips
Snack:	**Snack:**
6 oz fruit yogurt	6 oz fruit yogurt
Dinner:	**Dinner:**
1 baked chicken drumstick	1 baked chicken drumstick
1/2 c macaroni and cheese	10 large french fries with 1 tbs catsup
1/2 c apple juice	1/2 c applesauce
1/4 c broccoli with 1 tsp margarine	8 oz grape soda
Snack:	**Snack:**
1 peanut butter cookie	1 oz jelly beans

To ensure that children have healthy appetites and plenty of room for nutritious foods when they are hungry, parents and teachers must limit access to candy, cola, and other concentrated sweets. If such foods are permitted in large quantities, the only possible outcomes are nutrient deficiencies, obesity, or both. The preference for sweets is innate; most children do not naturally select nutritious foods on the basis of taste. In one study, when children were allowed to create meals freely from a variety of foods, they selected foods that provided 25 percent of the kcalories from sugar.[30] When their parents were watching, or even when they thought their parents were watching, the children improved their selections. Overweight children, especially, need help in selecting nutrient-dense foods that will meet their nutrient needs within their energy allowances.

Sweets need not be banned altogether. Children who are exceptionally active can enjoy high-kcalorie foods such as ice cream or pudding from the milk group or pancakes from the bread group. As for sedentary children, they need to become more active, and then they, too, can enjoy some of these foods without unhealthy weight gain.

244 ■ *Chapter Six*

Figure 6-4 Nutrient Comparisons of Menus A and B

Sample menu A meets a child's nutrient needs much better than sample menu B.

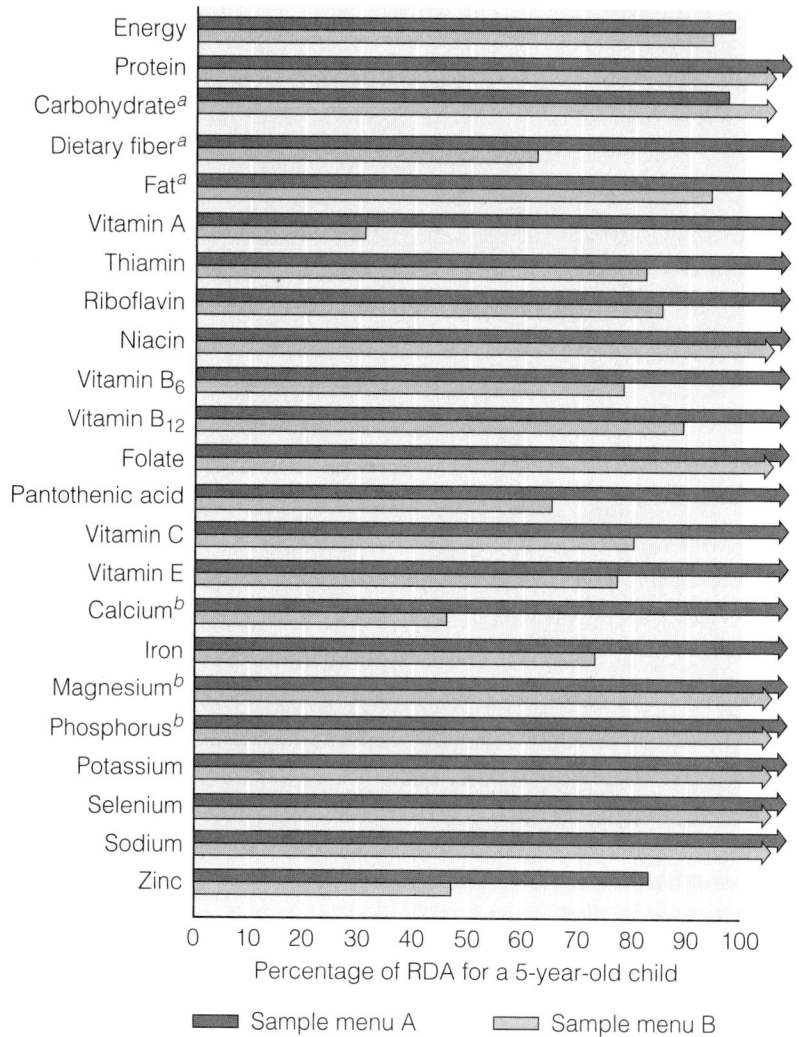

Percentage of RDA for a 5-year-old child

■ Sample menu A   □ Sample menu B

[a]Because there are no RDA for these nutrients, recommendations were based on 57 percent kcalories from carbohydrate, 30 percent kcalories from fat, and 18 grams of fiber.

[b]Values reflect DRI.

Mealtimes at Home

Feeding children requires not only providing a variety of nutritious foods but also nurturing the children's self-esteem and well-being. Parents face a number of challenges in preparing meals that appeal to their children's tastes as well as providing needed nutrients. Because the interactions between parents and children regarding food intake can set the stage for lifelong attitudes and habits, a child's preferences should be treated with respect, even when nutrient needs must take precedence.

Avoiding Power Struggles. It is not surprising that problems over food often arise during the second or third year, when children begin asserting their independence. Many of these problems stem from the conflict between children's developmental stages and capabilities and parents who, in attempting to do what they think is best for their children, try to control every aspect of eating. Such conflicts can disrupt children's abilities to regulate their own food intakes or to determine their own likes and dislikes. For example, many people share the misconception that children must be persuaded or coerced to try new foods. In fact, the opposite is true. When children are forced to try new foods, even by way of rewards, they are less likely to try those foods again than are children who are left to decide for themselves. The parent is responsible for *what* the child is offered to eat, but the child is responsible for *how much* and even *whether* to eat.[31]

When introducing new foods at the table, parents are advised to offer them one at a time and only in small amounts at first. The more often a food is presented to a young child, the more likely the child will accept that food. Whenever possible, offer the new food at the beginning of the meal, when the child is hungry, and allow the child to make the decision to accept or reject it. Never make an issue of food acceptance. A power struggle almost invariably sets a firm pattern of resistance and permanently closes the child's mind.

Honoring Children's Preferences. Researchers attempting to explain children's food preferences encounter contradictions. Children say they like colorful foods, yet most often reject green and yellow vegetables while favoring brown peanut butter and white potatoes, apple wedges, and bread. They do like raw vegetables better than cooked ones, though, so it is wise to offer vegetables that are raw or slightly undercooked. Foods should be warm, not hot, because a child's mouth is much more sensitive than an adult's. The flavor should be mild (a child has more taste buds), and smooth foods such as mashed potatoes or split-pea soup should have no lumps (a child wonders, with some disgust, what the lumps might be). Vegetables should be served separately and be easy to eat.

Young children like to eat at little tables and to be served small portions of food. They also love to eat with other children and have been observed to stay at the table longer and eat much more when in the company of their peers. Parents who serve food in a relaxed and casual manner, without anxiety, provide an environment that enhances the child's enjoyment of food.

Little children like little tables and little portions.

Preventing Choking. Parents must always be alert to the dangers of choking. A choking child is silent, and an adult should always be present whenever a child is eating. Make sure the child sits when eating; choking is more likely when a child is running or falling. Round foods such as grapes, nuts, hard candies, and hot dog pieces are difficult to control in a mouth with few teeth, and they can easily become lodged in the small opening of a child's trachea. Other potentially dangerous foods include tough meat, popcorn, and chips.

Playing First. Children may be more relaxed and attentive at mealtime if outdoor play or other fun activities are scheduled before, rather than immediately after, mealtime. A number of schools have discovered that children eat a better lunch if it is served after, rather than before, recess. Otherwise, children "hurry up and eat" so that they can go and play.

Children enjoy eating the foods they help to prepare.

Allowing Child Participation. Allowing children to help plan and prepare the family's meals provides enjoyable learning experiences and encourages children to eat the foods they have prepared. Vegetables are pretty, especially when fresh, and provide opportunities for children to learn about color, growing vegetables and their seeds, and shapes and textures—all of which are fascinating to young children. Measuring, stirring, decorating, and arranging foods are skills that even a very small child can practice with enjoyment and pride.

Providing Snacks. Parents may find that their children snack so much that they aren't hungry at mealtimes. Instead of teaching children *not* to snack, teach them *how* to snack. Provide snacks that are as nutritious as the foods served at mealtime. Snacks can even be mealtime foods that are served individually over time, instead of all at once on one plate. When providing snacks to children, think of the food groups and offer such snacks as pieces of cheese, tangerine slices, carrot sticks, and peanut butter on whole-wheat crackers (see Table 6-6). Snacks that are easy to prepare should be readily available to children, especially if they arrive home after school before their parents.

Planning Vegetarian Meals. Parents or caregivers who plan vegetarian diets for children must be aware of the nutrition challenges such diets present. Well-planned vegetarian diets, especially those that include eggs, milk, and milk products, can easily provide adequate nutrient intakes for growing children. Vegetarian diets that exclude these foods (vegan diets) may be low in energy, vitamin B_{12}, vitamin D, and calcium. Caregivers can use the daily food patterns presented in Table 6-5 on p. 242, paying particular attention to the notes at the bottom of the table, to plan nutritionally sound vegetarian diets for children.

Caring for Sick Children. Children sometimes lose their appetites when they are sick with colds or flu and sometimes for no apparent reason at all. On a short-term basis, this is usually nothing to worry about. As children who are ill recover, so will their appetites. The child who is sick should be encouraged to drink plenty of fluids. Focal Point 6B gives additional suggestions on caring for the sick child.

Preventing Dental Caries. Children frequently snack on sticky, sugary foods that stay on the teeth and provide an ideal environment for the

Table 6-6 Healthful Snack Ideas—Think Food Groups, Alone and in Combination

Selecting two or more foods from different food groups adds variety and nutrient balance to snacks. The combinations are endless, so be creative.

Grain Products

Grain products are filling snacks, especially when combined with other foods:

- Cereal with fruit and milk.
- Crackers and cheese.
- Wheat toast with peanut butter.
- Popcorn with grated cheese.
- Oatmeal raisin cookies with milk.

Vegetables

Cut-up fresh, raw vegetables make great snacks alone or in combination with foods from other food groups:

- Celery with peanut butter.
- Broccoli, cauliflower, and carrot sticks with a flavored cottage cheese dip.

Fruits

Fruits are delicious snacks and can be eaten alone—fresh, dried, or juiced—or combined with other foods:

- Apples and cheese.
- Bananas and peanut butter.
- Peaches with yogurt.
- Raisins mixed with sunflower seeds or nuts.

Meats and Meat Alternates

Meat and meat alternates add protein to snacks:

- Refried beans with nachos and cheese.
- Tuna on crackers.
- Luncheon meat on wheat bread.

Milk and Milk Products

Milk can be used as a beverage with any snack, and many other milk products, such as yogurt and cheese, can be eaten alone or with other foods as listed above.

growth of bacteria that cause dental caries. Teach children to brush and floss after meals, to brush or rinse after eating snacks, to avoid sticky foods, and to select crisp or fibrous foods frequently.

Serving as Role Models. In an effort to practice these many tips, parents may overlook perhaps the single most important influence on their children's food habits—themselves. Parents who don't eat carrots shouldn't be surprised when their children refuse to eat carrots. Likewise, parents who dislike the smell of brussels sprouts may not be able to persuade children to try them. Children learn much through imitation. Parents and older siblings set an irresistible example by enjoying nutritious foods.

While serving and enjoying food, caregivers can promote physical and emotional growth at every stage of a child's life. They can help their children to develop both a positive self-concept and a positive attitude toward food. If the beginnings are right, children will grow without the conflicts and confusions over food that can lead to nutrition and health problems.

Nutrition at School

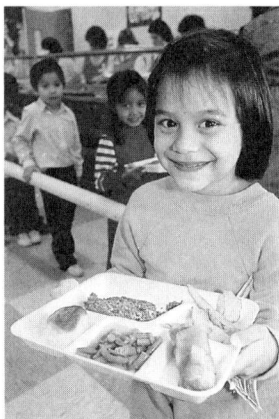

School lunches provide children with nourishment at little or no charge.

While parents are doing what they can to establish good eating habits in their children at home, others are preparing and serving foods to their children at day care centers and schools. The U.S. government funds several programs to provide nutritious, high-quality meals for children at school. Both the School Breakfast Program and the National School Lunch Program provide meals at a reasonable cost to children from families with the financial means to pay. Meals are available free or at reduced cost to children from low-income families. In addition, children begin to learn about food and nutrition in the classroom. Meeting the nutrition and education needs of children is critical to supporting their healthy growth and development.[32]

🏃 Healthy People 2000

Increase to at least 75% the proportion of the nation's schools that provide nutrition education from preschool through grade 12, preferably as part of quality school health education.

School Breakfast. The School Breakfast Program is available in slightly more than half of the nation's schools, and about 5 million children participate in it. The school breakfast must contain at a minimum the following:

- One serving of fluid milk.
- One serving of fruit or vegetable or full-strength juice.

■ Two servings of bread or bread alternates, two servings of meat or meat alternates, or one of each.

Surveys show that the majority of children who eat school breakfasts are from low-income families. As research results continue to emphasize the positive impact breakfast has on school performance and health, campaigns to expand school breakfast programs are under way.

School Lunch. Nearly 25 million children receive lunches through the National School Lunch Program—half of them free or at a reduced price. School lunches are designed to provide at least a third of the RDA for each nutrient and must include specified numbers of servings of milk, protein-rich food (meat, poultry, fish, cheese, eggs, legumes, or peanut butter), vegetables, fruits, and breads or other grain foods. Table 6-7 shows school lunch patterns for children of different ages.

⋆ Healthy People 2000

Increase to at least 90% the proportion of school lunch and breakfast services and increase to at least 50% the proportion of child care foodservices with menus that are consistent with the nutrition principles in the *Dietary Guidelines for Americans*.

Table 6-7 School Lunch Patterns for Different Ages

Food Group	Preschool (Age)		Grade School through High School (Grade)		
	1 to 2	*3 to 4*	*K to 3*	*4 to 6*	*7 to 12*
Meat or meat alternate					
1 serving:					
Lean meat, poultry, or fish	1 oz	1 1/2 oz	1 1/2 oz	2 oz	3 oz
Cheese	1 oz	1 1/2 oz	1 1/2 oz	2 oz	3 oz
Large egg(s)	1	1 1/2	1 1/2	2	3
Cooked dry beans or peas	1/2 c	3/4 c	3/4 c	1 c	1 1/2 c
Peanut butter	2 tbs	3 tbs	3 tbs	4 tbs	6 tbs
Vegetable and/or fruit					
2 or more servings, both to total	1/2 c	1/2 c	1/2 c	3/4 c	3/4 c
Bread or bread alternate					
Servings	5 per week	8 per week	8 per week	8 per week	10 per week
Milk					
1 serving of fluid milk	3/4 c	3/4 c	1 c	1 c	1 c

Table 6-8 Lunchbox Logic

To prevent food spoilage, and to maintain the freshness and appeal of fruits, vegetables, juices, and dairy foods, use resealable containers (or less environmentally friendly disposable packages) and pack a small "blue ice" block in lunchboxes. Remember that leftovers from the evening meal make great lunches the next day.

The following lists offer nutritious food ideas from each of the five food groups. Look to Table 6-6 for nutritious snack ideas.

Breads, Cereals, Rice, and Pasta Group

Whole-grain breads, muffins, crackers, and cereals

Rice or popcorn cakes	Bagels (mini size are ideal)
Rice	Whole-wheat pretzels
Pasta	Popcorn*
Bread sticks	

Vegetable Group

Raw or cooked vegetables (cut into easy-to-handle pieces)

Vegetable juices, such as V-8 or tomato juice

Fruit Group

Fresh fruits (cut into easy-to-handle pieces)

Canned fruits, packed in their own juices

Fruit juices

Dried fruits, such as raisins and apricots

Milk, Yogurt, and Cheese Group

2% low-fat milk	Cottage cheese
Yogurt	Cheese (cubes, slices, or sticks)

Meat, Poultry, Fish, Dry Beans, Eggs and Nuts Group

Bean spreads	Hard-boiled or deviled eggs
Nut mixes*	Egg, tuna, chicken, or turkey salad
Peanut butter	

Cooked fish, shrimp, chicken, or other meat (cut into bite-size pieces)

*Popcorn and nuts are inappropriate for children under the age of four, because of the danger of choking.

Lunch Box Logic. Some children bring their own lunch to school on one or more days each week. Packed lunches from home, like school lunches, should provide about one-third of the child's nutrients for the day. When packing children's lunches, keep the five food groups in mind, and include at least the following:

- 2 servings of breads or grains.
- 2 servings of vegetables or fruits.
- 1 serving of milk.
- 1 serving of meat or other protein-rich alternate.

Allow children to help pack their own lunches so they will choose foods they like to eat. Table 6-8 offers suggestions for nutritious lunches from home.

Healthful food choices and regular physical activity both promote growth and help prevent the degenerative diseases of later life. They also prepare a child for the teen years, the focus of the next chapter.

CHAPTER SIX NOTES

1. V. Matkovic and J. Z. Ilich, Calcium requirements for growth: Are current recommendations adequate? *Nutrition Reviews* 51 (1993): 171–180.

2. L. L. Birch and coauthors, Effects of a non-energy fat substitute on children's energy and macronutrient intake, *American Journal of Clinical Nutrition* 58 (1993): 326–333; S. Shea and coauthors, Variability and self-regulation of energy intake in young children in their everyday environment, *Pediatrics* 90 (1992): 542–546; L. L. Birch and coauthors, The variability of young children's energy intake, *New England Journal of Medicine* 324 (1991): 232–235.

3. P. L. Splett and M. Story, Child nutrition: Objectives for the decade, *Journal of the American Dietetic Association* 91 (1991): 665–668.

4. N. S. Scrimshaw, Iron deficiency, *Scientific American*, October 1991, pp. 46–52.

5. E. Kennedy and J. Goldberg, What are American children eating? Implications for public policy, *Nutrition Reviews* 53 (1995): 111–126.

6. Committee on Nutrition, American Academy of Pediatrics, *Pediatric Nutrition Handbook*, 3rd ed., ed. L. A. Barness (Elk Grove Village, Ill.: American Academy of Pediatrics, 1993), pp. 34–42.

7. A. T. Hingley, Preventing childhood poison-ing, *FDA Consumer*, March 1996, pp. 7–11.

8. Department of Health and Human Services, Food and Drug Administration, Iron-containing supplements and drugs: Label warning statements and unit dose packaging requirements; Final rule, *Federal Register*, January 15, 1997.

9. Committee on Nutrition, 1993, p. 167–178.

10. W. H. Dietz, Childhood obesity, in *Child Health, Nutrition, and Physical Activity*, eds. L. W. Y. Cheung and J. B. Richmond (Champaign, Ill.: Human Kinetics, 1995), pp. 155–169.

11. K. F. Janz and coauthors, Cross-validation of the Slaughter skinfold equations for children and adolescents, *Medicine and Science in Sports and Exercise* 25 (1993): 1070–1076.

12. S. Krug-Wispé, Nutritional assessment, in *Handbook of Pediatric Nutrition*, eds. P. M. Queen and C. E. Lang (Gaithersburg, Md.: Aspen Publishers, 1993), pp. 26–82.

13. T. A. Nicklas and coauthors, Breakfast consumption affects adequacy of total daily intake in children, *Journal of the American Dietetic Association* 93 (1993): 886–891.

14. H. A. Sampson and D. D. Metcalfe, Food allergies, *Journal of the American Medical Association* 268 (1992): 2840–2844.

15. C. F. Chaisson, B. Petersen, and J. S. Doug-

lass, *Pesticides in Foods: A Guide for Professionals* (Chicago: American Dietetic Association, 1991), pp. 2–3.

16. National Academy of Sciences Committee, as quoted by J. Raloff and D. Pendick, Pesticides in produce may threaten kids, *Science News* 144 (1993): 4–5.

17. Three agencies propose pesticide reforms, *FDA Consumer*, January–February 1994, p. 3.

18. C. Marwick, Pesticides pose concern about children's diet, *Journal of the American Medical Association* 270 (1993): 802, 805.

19. S. A. Bock and F. M. Atkins, Patterns of food hypersensitivity during sixteen years of double-blind, placebo-controlled food challenges, *Journal of Pediatrics* 117 (1990): 561–567.

20. Sampson and Metcalfe, 1992.

21. V. L. Olejer, Food hypersensitivities, in *Handbook of Pediatric Nutrition*, eds. P. M. Queen and C. E. Lang (Gaithersburg, Md.: Aspen Publishers, 1993): 206–231.

22. A. T. Hingley, Food allergies: When eating is risky, *FDA Consumer*, December 1993, pp. 27–31.

23. Sampson and Metcalfe, 1992.

24. J. W. White and M. Wolraich, Effect of sugar on behavior and mental performance, *American Journal of Clinical Nutrition* (supplement) 62 (1995): 242–249; M. L. Wolraich and coauthors, Effects of diets high in sucrose or aspartame on the behavior and cognitive performance of children, *New England Journal of Medicine* 330 (1994): 301–307; E. H. Wender and M. V. Solanto, Effects of sugar on aggressive and inattentive behavior in children with attention deficit disorder with hyperactivity and normal children, *Pediatrics* 88 (1991): 960–966.

25. Dietz, 1995; C. L. Ogden and coauthors, Prevalence of overweight among preschool children in the United States, 1971 through 1994, *Pediatrics*, 1997, URL: http// www.pediatrics.org/cgi/content/full/99/4/e1.

26. S. L. Gortmaker, W. H. Dietz, Jr., and L. W. Y. Cheung, Inactivity, diet, and the fattening of America, *Journal of the American Dietetic Association* 90 (1990): 1247–1255; E. Obarzanek and coauthors, Energy intake and physical activity in relation to indexes of body fat: The National Heart, Lung, and Blood Institute Growth and Health study, *American Journal of Clinical Nutrition* 60 (1994): 15–22.

27. W. H. Dietz, Jr., and S. L. Gortmaker, Do we fatten our children at the television set? Obesity and television viewing in children and adolescents, *Pediatrics* 75 (1985): 807–812.

28. Committee on Sports Medicine and Fitness, Fitness, activity, and sports participation in the preschool child, *Pediatrics* 90 (1992): 1002–1004.

29. F. Lifshitz and N. Moses, Nutritional dwarfing: Growth, dieting, and fear of obesity, *Journal of the American College of Nutrition* 7 (1988): 367–376.

30. R. E. Klesges and coauthors, Parental influence on food selection in young children and its relationships to childhood obesity, *American Journal of Clinical Nutrition* 53 (1991): 859–864.

31. E. Satter, *How to Get Your Kid to Eat . . . But Not Too Much* (Palo Alto, Calif.: Bull Publishing, 1987), pp. 13–28.

32. Position of The American Dietetic Association: Nutrition standards for child care programs, *Journal of the American Dietetic Association* 94 (1994): 323.

Lead Toxicity

Old, lead-based paint threatens the health of an exploring child.

At nine months, Joey crawled about exploring the world around him—touching and tasting everything, as all infants do. He chewed on table legs, toys, the spindles of railings with flaky paint—whatever was in his reach. He eagerly drank his morning bottle of formula, which his mother prepared with the first water drawn from the tap. Not until he was two did he begin to toddle about while his parents watched proudly. At four, he amused his parents when he'd chase after balls tossed his way, but he couldn't catch them. By age five, he was a cautious, quiet preschooler who clung tightly to stair railings with both hands as he slowly climbed up or down.

Joey was late in walking, small for his age, seldom played as vigorously as other children, and was prone to small health disturbances such as diarrhea, irritability, and lethargy. His kindergarten teacher reported that Joey had some difficulty hearing and that his progress was slower than expected. While his health quietly deteriorated, his parents thought these subtle symptoms were within the range of normal vari-

ations seen in children. Finally, a pediatrician detected lead toxicity and started treating Joey with lead-scavenging drugs.[*1] Joey is now growing normally and playing vigorously, although he still has minor learning disabilities. His physician expects the deficits in brain function to persist into adulthood.[2]

For children like Joey, the diagnosis often comes too late. Even one year of lead exposure can permanently injure the brain and other parts of the nervous system and psychological functioning. Furthermore, the effects occur with even low exposure.[3] The lead poisoning threshold—the amount of lead in the blood recognized to cause harm—is now known to be 10 micrograms per 100 milliliters of blood; earlier it was thought to be 25.[4] Health agencies point to lead poisoning as the most serious environmental threat to young children.[5] The FDA has proposed reducing the acceptable level of lead in foods 10-fold—from its 1958 limit of 10 parts per million to 0.5 to 1.0 parts per million.[6]

This Focal Point first describes how lead disrupts body processes and impairs nutrition status and then points out sources of lead in the environment. Perhaps, with awareness, we can make changes that will safeguard the health of our children.

Lead in the Body

Minerals serve the body in many ways: maintaining fluid and electrolyte balance, providing structural support to the bones, transporting oxygen, and assisting enzymes. In contrast to those minerals that the body requires, lead impairs the body's growth, work capacity, and general health.

Like other minerals, lead is indestructible; the body cannot change its chemistry. Chemically similar to nutrient minerals like iron, calcium, and zinc (they are all cations carrying two

*The majority of children with high blood lead levels (45 μg/dL) are treated using a process called chelation—using drugs (most often succimer or calcium disodium EDTA) that bind to lead in the blood and carry it out in the urine.

Figure FP6A-1   Lead Displaces Iron

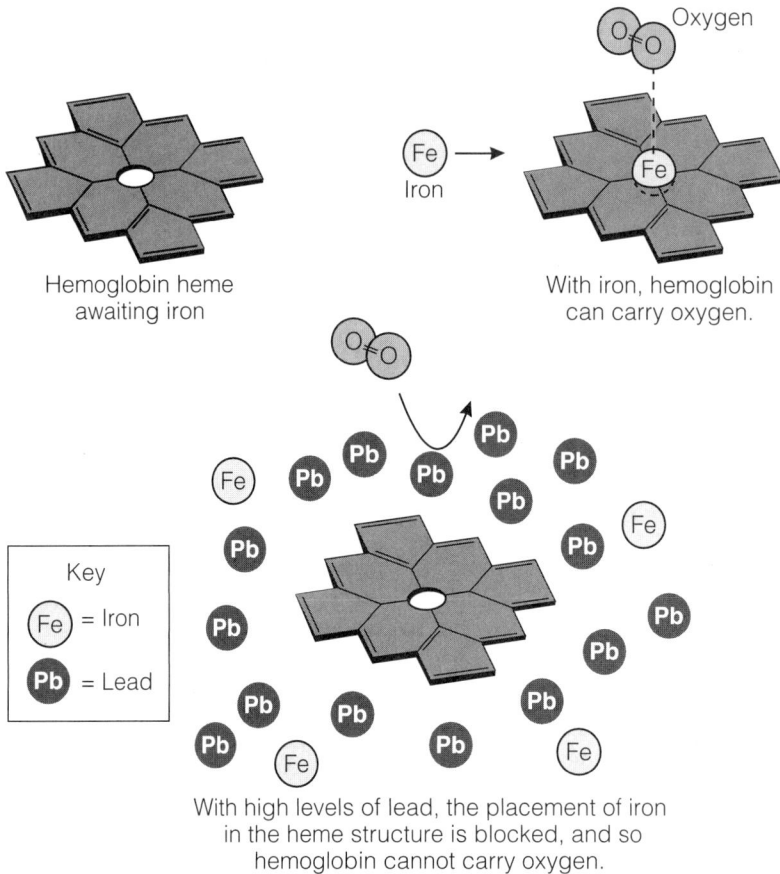

Hemoglobin heme
awaiting iron

With iron, hemoglobin
can carry oxygen.

Oxygen

Key

Fe = Iron

Pb = Lead

With high levels of lead, the placement of iron
in the heme structure is blocked, and so
hemoglobin cannot carry oxygen.

positive charges), lead displaces them from some of the metabolic sites they normally occupy, but is then unable to perform their roles. For example, lead interferes with the enzymes that facilitate heme formation (see Figure FP6A-1). Lead damages many body systems, particularly the vulnerable nervous system, kidneys, and bone marrow. It impairs such normal activities as growth by interfering with hormone activity.[7] The greater the exposure, the more damaging the effects. Table FP6A-1 lists symptoms of lead toxicity.

Lead and Growing Children

The body readily absorbs lead, especially during times of rapid growth. Thereafter, it hoards lead possessively—just as it does iron. During preg-

nancy, lead moves across the placenta, inflicting severe damage on the developing fetal nervous system. Infants and young children absorb five to ten times as much lead as adults do. One out of every six children from six months to five years old and one out of every nine fetuses are exposed to harmful doses of lead. Lead toxicity is most prevalent among children under six—as many as 1.7 million children may have blood lead concentrations high enough to cause mental, behavioral, and other health problems.[8]

Lead poisoning in infants most often comes from infant formula made with contaminated water.[9] The water, in turn, receives its lead burden from lead-soldered plumbing. The first water drawn from the tap each day is highest in lead—therefore, a person living in a house with old, lead-soldered plumbing should let the water run

Table FP6A-1 Symptoms of Lead Toxicity

In children

- Learning disabilities
- Low IQ
- Behavior problems
- Slow growth
- Iron deficiency anemia
- Nervous system disorders
- Impaired concentration
- Reduced short-term memory
- Slow reaction time
- Seizures
- Impaired hearing
- Poor coordination

In adults

- Hypertension
- Kidney failure
- Reproductive complications

a few minutes before drinking or using it to prepare formula or food.

Lead intoxication in young children comes from their own behaviors and activities—putting their hands in their mouths, playing in dirt, and eating nonfood items (see Figure FP6A-2). The toddler years see a marked rise in blood lead. Tragically, a child's neuromuscular system is maturing at precisely the same time. No wonder children with high blood lead experience impairment of balance, motor development, and the relaying of nerve messages to and from the brain. Among children two and three years old, those with the highest blood lead suffer the greatest developmental delays at age four. Parents and teachers of older children with high bone lead levels report antisocial and delinquent behavior among these children.[10] Researchers studying children's development and behavior must now consider the possibility that lead poisoning may affect their results.

Excess lead in the blood also deranges the structure of red blood cell membranes, making them leaky and fragile. Lead interacts with white blood cells, too, impairing their ability to fight infection, and it binds to antibodies, thereby impairing the body's resistance to disease. In short, lead's interactions with a variety of substances in the body have profound adverse effects.[11]

The Malnutrition–Lead Connection

Nutrient deficiencies enhance lead toxicity.[12] Among children, those who are malnourished are most vulnerable to lead poisoning. Children absorb more lead if their stomachs are empty, if they have low calcium or zinc intakes, and, of greatest concern because it is so common, if they have iron deficiencies.

Lead poisoning can cause iron deficiency, and iron deficiency weakens the body's defenses against lead absorption. In fact, the interactions between lead poisoning and iron deficiency are so strong that lead poisoning is considered an adverse consequence of iron deficiency. A child with adequate iron stores is not immune, but a child with iron deficiency anemia is three times as likely to have high blood lead. Common to both iron deficiency and lead poisoning is a low socioeconomic background. Another common factor is pica—a craving for nonfood items. Many children with lead poisoning eat dirt or newspapers, two common sources of lead.

The anemia brought on by lead poisoning may be mistaken for a simple iron deficiency and therefore may be incorrectly treated. Like iron deficiency, mild lead toxicity has nonspecific effects, including diarrhea, irritability, reduced ability of the blood to carry oxygen, and fatigue. The symptoms are not reversible by adding iron to the diet; exposure to lead must stop. With further exposure, the signs become more pronounced: Children lose cognitive, verbal, and perceptual abilities and develop learning disabilities and behavior problems. Still more severe lead toxicity can cause irreversible nerve damage, paralysis, mental retardation, and death.

Just as iron deficiency enhances lead uptake, an inadequate calcium intake enhances lead absorption and retention. Zinc deficiency also enhances both tissue accumulation of lead and sensitivity to its effects; serum zinc is frequently low

Figure FP6A-2   Sources of Lead Exposure

Lead finds its way into the bodies of children when they ingest lead-containing foods, water, dust, or paint chips, or when they breathe lead-laden air.

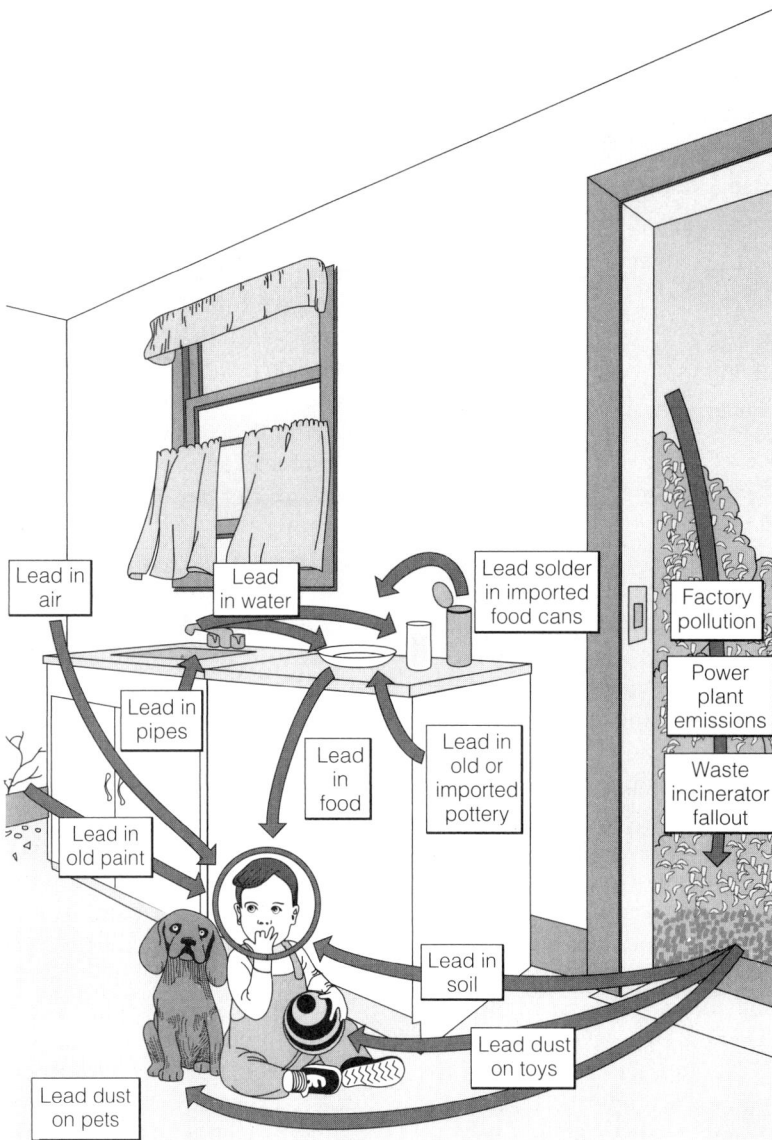

Lead in air

Lead in water

Lead solder in imported food cans

Factory pollution

Power plant emissions

Waste incinerator fallout

Lead in pipes

Lead in food

Lead in old or imported pottery

Lead in old paint

Lead in soil

Lead dust on toys

Lead dust on pets

in children with high blood lead. Prevention of lead toxicity rests primarily on reduced exposure, but parents can protect their children, at least to some degree, by making sure that they eat foods rich in iron, calcium, zinc, and other nutrients.

Lead in the Environment

All foods contain some lead. Most of it derives from industrial pollution. People are exposed to lead in some types of gasoline, paint, newspaper ink, batteries, shotgun ammunition, and pesticides as well as in the air and water that carry lead from industrial processes and landfills. Lead works its way through rainfall and soil into plants and animals that people use for food. Lead also enters food from containers such as tin cans sealed with lead solder. (Manufacturers in the United States no longer use lead solder in canning, but many in foreign countries still do.) Lead can leach into foods from old, handmade, or imported pottery decorated with lead glazes.

Pipelines soldered with lead or coupled with brass fixtures also release lead into drinking water, in which it is the nation's most significant contaminant. A recent sampling by the EPA found unsafe levels of lead in the public water systems that serve 30 million people. Exposures are highest in older communities along the nation's east coast; in urban and industrial areas; near highways; and in slums where old leaded paint peels from the buildings. People suffering the effects of lead exposure are most often black, male, and from low-income families, but the effects are also seen in upper-middle-class families as they move into inner-city areas and renovate older homes.

Federal law has mandated reductions in the use of leaded gasolines, lead-based solder, and other products in recent years. These efforts have helped reduce the amounts of lead in the environment—and in children's blood. The decline in blood lead in children during the late 1970s paralleled exactly the decline in the nation's use of leaded gasoline, leaded house paint, and lead-soldered food cans. Even so, lead still contaminates the blood of some 4 percent of our nation's children. Paint remains the primary source;[13] 70 percent of homes built before 1960 are covered with lead paint and are likely to cause lead poisoning, especially during times of renovation. A routine question asked when screening children for lead poisoning is whether they have ever lived in a house built before 1960.[14] If leaded surfaces in these homes are peeling and deteriorating, the children are either already poisoned or face immediate danger of lead poisoning.

Strategies for Protection

Three major discoveries about lead toxicity occurred simultaneously: Lead poisoning has *subtle* effects, the effects are *permanent*, and they occur at *low levels of exposure*. Consumers would be wise to take ultraconservative measures to protect themselves, and especially their infants and young children, from lead poisoning.

Defensive strategies include the following:

- The American Academy of Pediatrics recommends testing children for lead poisoning; effective screening with an appropriate questionnaire is essential to identifying high-risk children and thus treating the devastating effects.[15] About half the pediatricians surveyed report universal screening.[16]

- In contaminated environments, keep small children from putting dirty or old painted objects in their mouths, and make sure children wash their hands before eating. Similarly, keep small children from eating any nonfood items. Lead poisoning has been reported in young children who have eaten pool cue chalk.[17]

- Be aware that other countries do not have the same regulations protecting consumers against lead. Children have been poisoned by eating crayons made in China and drinking fruit juice canned in Mexico.

- Make infant formula from lead-free ingredients. Do not use lead-contaminated water.

- Once you have opened canned food, immediately move it to a lead-free storage container to prevent lead migration into the food.

- Do not store acidic foods or beverages (such as orange juice) in ceramic dishware.

- Many manufacturers are now making lead-safe products.* Old, handmade, or imported ceramic cups and bowls may contain lead and should not be used to heat coffee or tea or acidic foods such as tomato soup.

- Feed children nutritious meals regularly (see Chapter 6 for more details).

A Shopper's Guide to Low-Lead China is available from the Environmental Defense Fund, 257 Park Avenue South, New York, NY 10010; telephone (212) 505-2100.

■ Before using your newspaper to wrap food, mulch garden plants, or add to your compost, confirm with the publisher that the paper uses no lead in its ink.

The EPA also publishes a booklet, *Lead and Your Drinking Water*, in which the following cautions appear:

■ Have the water in your home tested by a competent laboratory.

■ Use only cold water for drinking, cooking, and making formula (cold water absorbs less lead).

■ When water has been standing in pipes for more than two hours, flush the cold-water pipes by running water through them for at least a minute before using it for drinking, cooking, or mixing formulas.

■ If lead contamination of your water supply seems probable, obtain additional informa-

tion and advice from the EPA and your local public health agency.[18]

By taking these steps, parents can protect themselves and their children from this preventable danger.*

This Focal Point may appear to have been just about lead, but lead's actions typify the ways all heavy metals behave in the body: They interfere with nutrients that are trying to do their jobs. The "good guy" nutrients are shoved aside by the "bad guy" contaminants. Then the contaminants—whether lead, mercury, cadmium, or some other—cannot perform the roles of the nutrients and health declines. To safeguard our health, we must defend ourselves against contamination by eating nutrient-rich foods and preserving a clean environment.

FOCAL POINT 6A NOTES

1. Committee on Drugs, American Academy of Pediatrics, Treatment guidelines for lead exposure in children, *Pediatrics* 96 (1995): 155–160.
2. P. A. Baghurst and coauthors, Environmental exposure to lead and children's intelligence at the age of seven years, *New England Journal of Medicine* 327 (1992): 1279–1284; D. C. Bellinger, K. M. Stiles, and H. L. Needleman, Low-level lead exposure, intelligence and academic achievement: A long-term follow-up study, *Pediatrics* 90 (1992): 855–961; H. L. Needleman and coauthors, The long-term effects of exposure to low doses of lead in childhood: An 11-year follow-up report, *New England Journal of Medicine* 322 (1990): 83–88.
3. K. N. Dietrich, O. G. Berger, and P. A. Succop, Lead exposure and the motor developmental status of urban six-year-old children in the Cincinnati Prospective Study, *Pediatrics* 97 (1993): 301–307.
4. J. Murphy, Federal agencies gearing up for

new efforts against lead, *Nation's Health*, May–June 1991, pp. 1, 23.
5. A. Greeley, Getting the lead out of just about everything, *FDA Consumer*, July–August 1991, pp. 27–36.
6. FDA seeks lower lead levels in food additives and GRAS ingredients, *Journal of the American Dietetic Association* 94 (1994): 495.
7. C. A. Huseman, M. M. Varma, and C. R. Angle, Neuroendocrine effects of toxic and low blood lead levels in children, *Pediatrics* 90 (1992): 186–189.
8. Update: Blood lead levels—United States, 1991–1994, *Morbidity and Mortality Weekly Report* 46 (1997): 141–146.
9. M. W. Shannon and J. W. Graef, Lead intoxication in infancy, *Pediatrics* 89 (1992): 87–90.
10. H. L. Needleman and coauthors, Bone lead levels and delinquent behavior, *Journal of the American Medical Association* 275 (1996): 363–369.
11. P. Mushak and A. F. Crocetti, Lead and nutrition: Biological interactions of lead with

*The National Lead Information Center provides two hotlines; call (800) LEAD-FYI (532-3394) for general information or (800) 424-LEAD (424-5323) with specific questions.

nutrients, *Nutrition Today*, January–February 1996, pp. 12–17.

12. R. A. Goyer, Nutrition and metal toxicity, *American Journal of Clinical Nutrition* 61 (1995): 646S–650S.

13. Childhood lead poisoning: A disease for the history texts (editorial), *American Journal of Public Health* 81 (1991): 685.

14. H. J. Binns, Is there lead in the suburbs? Risk assessment in Chicago suburban pediatric practices, *Pediatrics* 93 (1994): 164–171.

15. Committee on Environmental Health, American Academy of Pediatrics, Lead poisoning: From screening to primary prevention, *Pediatrics* 92 (1993): 176–183; S. J. Schaffer and coauthors, Lead poisoning risk determination in a rural setting, *Pediatrics* 97 (1996): 84–90; M. N. Haan, M. Gerson, and B. A. Zishka, Identification of children at risk for lead poisoning: An evaluation of routine pediatric blood lead screening in an HMO-insured population, *Pediatrics* 97 (1996): 79–83; D. C. Snyder and coauthors, Development of a population-specific risk assessment to predict elevated blood lead levels in Santa Clara County, California, *Pediatrics* 96 (1995): 643–648.

16. J. R. Campbell and coauthors, Blood lead screening practices among U.S. pediatricians, *Pediatrics* 98 (1996): 372–377.

17. Pool cue chalk: A source of environmental lead, *Pediatrics* 97 (1996): 916–917.

18. Office of Water, U.S. Environmental Protection Agency, *Lead and Your Drinking Water*, publication no. OPA 87–006 (Washington, D.C.: U.S. Government Printing Office, April 1987).

Nutrition Care of Sick Infants and Children

The emphasis of this text is on wellness and the prevention of disease, but even children who are generally in good health get sick at one time or another. This discussion examines some of the common symptoms that most children experience at one time or another—infections and fever, diarrhea, and constipation. It then continues with a look at the special needs of any child who requires hospitalization.

With an illness, a healthy, well-nourished child can easily slip into poor nutrition status. Illness can affect a child's nutrition status by interfering with:

- Appetite.
- Chewing and swallowing.
- Digestion and absorption.
- Metabolism.
- Excretion.

Changes in any or all of these can occur during illness and compromise nutrition status. Competent medical care includes attention to nutrition.

Infections and Fever

Each day, the body confronts an environment teeming with disease-causing organisms. The body's remarkable capacity to survive such an environment is a tribute to its immune system. The immune system has no central organ of control, but rather depends on various organs and white blood cells. Their interactions and secretions defend the body against infectious organisms, such as bacteria and viruses. (The accompanying glossary defines related terms.) The immune system recognizes these infectious organisms and destroys or otherwise neutralizes them. When organisms do manage to penetrate the immune defenses, an infection, or an even more damaging disease, develops. Still, for every successful penetration of foreign organisms, the immune system averts thousands of attempts.

Malnutrition alters immune system components in ways that compromise their function, thus impairing the defense against infecting organisms. It is little wonder that malnourished children develop more infections than well-nourished children. Infection is a major cause of mortality and morbidity in children with protein-energy malnutrition. A vicious cycle develops in which malnutrition reduces resistance to infection, and infection further aggravates malnutrition. This synergistic relationship between malnutrition and infection threatens a child's survival.

An infection generally progresses as follows. Disease-causing organisms invade the body, overcoming initial immune defenses. A few days after exposure to these infective organisms, symptoms begin to appear, with fever developing shortly thereafter.

Fever. People fear fever and rush to treat it because it accompanies many dangerous diseases, but the fever itself may actually assist the immune system. Clearly, fever stresses the body—it raises the heart rate and increases the tissues' demand for oxygen. Elevated temperatures (over 104 degrees Fahrenheit) can cause convulsions and demand medical attention not only to control the fever, but to determine and treat the underlying condition. Moderate temperatures (between 102 and 104 degrees Fahrenheit) may be cause for concern and require a consultation with a health care provider. Generally, though, a mild fever should be allowed to do its job of assisting the immune system. Fever causes the cells of the immune system to proliferate, thus enhancing the system's activity. When researchers suppress fever experimentally, sick animals are more likely to die from infection than when fever is allowed to run its course. Furthermore, animals kept continuously at fever temperatures resist viruses better than nonfeverish ones.

After the onset of fever, catabolic changes begin. As cells break down, the body loses nitrogen (from protein catabolism) and intracellular elec-

GLOSSARY

acetaminophen: an antipyretic, analgesic drug used to reduce fever or relieve pain. Pediatricians recommend acetaminophen instead of aspirin because of aspirin's association with Reye's syndrome, a rare disease that primarily affects children and adolescents, generally following flu or chicken pox. Symptoms begin with tiredness and vomiting, progress to permanent brain damage, and result in death in 20 to 30 percent of the cases.

acute: sharp or severe; disease or conditions that develop rapidly, have severe symptoms, and are of short duration (*acutus* = sharp).

anemia of infection: a condition in which iron moves from the blood to the liver to help fight infection, resulting in a decline in hemoglobin synthesis.

Celsius: the scientific name for the centigrade temperature scale. To convert from centigrade to Fahrenheit (and vice versa), use these equations:

$$t_F = 9/5t_C + 32$$
$$t_C = 5/9(t_F - 32)$$

chronic: long duration; diseases or conditions that develop slowly, show little change, and last a long time (*chronos* = time).

constipation: the condition of having painful or difficult bowel movements (elapsed time between movements is not relevant).

dehydration: loss of too much fluid from the body.

Fahrenheit scale: a temperature scale with the freezing point of water at 32°F and the boiling point at 212°F.

fever: an increase of body temperature of more than 1°F above normal (98.6°F).

immune system: the body's natural defense system against foreign materials.

immunity: the body's ability to recognize and eliminate foreign materials.

interleukin-1: a protein released by the immune system that mediates many responses to infection.

intractable diarrhea: severe, chronic diarrhea.

intravenous (IV) fluid therapy: the administration of nutrient solutions through a vein.

oral rehydration therapy (ORT): the administration of a simple solution of sugar, salt, and water taken by mouth, to treat dehydration caused by diarrhea. A simple ORT recipe:

> 1 c boiling water.
> 2 tsp sugar.
> A pinch of salt.

sodium salicylate: a compound used as an analgesic and antipyretic; aspirin is acetylsalicylic acid.

steatorrhea (stee-ah-toe-REE-ah): fatty diarrhea characteristic of fat malabsorption; stools are foamy, greasy, and malodorous.

water intoxication: the condition in which body water content is too high.

---

trolytes. These losses are great enough to result in negative balances. If diarrhea and vomiting accompany the infection, the body's losses are even greater.

Energy and Nutrient Needs. Fever is a major factor in determining the energy needs of an infected child. The basal metabolic rate increases (roughly 10 to 13 percent for each degree Celsius, 7 percent for each degree Fahrenheit) as the temperature rises above normal (98.6 degrees Fahrenheit; 37 degrees Celsius). Additional food energy may be needed to support physical activity—for example, if a child is restless, coughing, crying, or the like.

Ironically, during this time of heightened energy needs, appetite diminishes. Infection-related anorexia aggravates negative nutrient balances and contributes to weight loss. Researchers have questioned whether this anorexia is a result of the fever itself. When they blocked the fever of infected rats with sodium salicylate, appetite remained depressed. Additional investigation revealed that interleukin-1, a protein released by the immune system in response to infection, both induces fever and suppresses food intake.

The immune system functions best when protein status is optimal. The protein requirements of a child with an infection depend primarily on total energy consumption. At a minimum, protein intake should meet the RDA for age, weight, and sex. With fever, protein allowances increase above the RDA by 10 percent per each degree Celsius (5.5 percent for each degree Fahrenheit) of fever.

Little information is available regarding specific vitamin and mineral requirements during infection. The need for B vitamins increases with increasing energy and protein intakes, but remains consistent with the RDA. Foods can generally cover the vitamin and electrolyte losses incurred during catabolism. Children who have lost their appetites or are feeling nauseated may prefer liquids to solid foods. Liquids are an acceptable alternative to solid foods, provided that the caregiver selects those that offer energy, vitamins, and minerals. Drinking liquids also prevents dehydration, of course.

Children with infections may develop a type of anemia referred to as the anemia of infection. At the onset of the infection, the blood concentration of iron rapidly declines as iron moves into the liver for storage. This shift of iron from the blood to the liver helps fight the infection by making the body's iron unavailable for the infecting bacteria, which require iron to perform their metabolic functions. Iron in storage is also unavailable for hemoglobin synthesis, however, so anemia results. This type of anemia is the body's normal physiological response to infection and does not respond to iron, folate, or vitamin B_{12} supplements, nor to any other dietary or medical treatment. Instead, the situation corrects itself; iron returns to the blood from the liver as the infection resolves.

Fluid Balance. Maintaining fluid balance is of particular concern during an infection. In some infectious diseases, especially those that involve vomiting, diarrhea, or considerable sweating, fluid needs may be as high as 3 to 4 liters per day. If the child does not drink fluids at this rate, dehydration can quickly develop. On the other hand, the child may retain water due to the hormonal changes that accompany fever. In the rare case that a child's fluid intake is excessive, water intoxication can develop. Table FP6B-1 lists the symptoms associated with dehydration and water intoxication.

Medical Attention and TLC. Physicians' primary concerns for children with infections are to identify and eliminate the infecting organisms. Quite often, they prescribe antibiotics, but these can interfere with nutrient absorption, thus compromising nutrition status.

When the infection ends, a well-balanced diet best restores the body to its normal status. The time it takes to reach a positive balance depends on the extent of nutrient deficiencies, the quality of the diet, and the quantity of food intake. A well-balanced diet that restores the nutrient reserves also helps to defend against future infections. In some cases, physicians may prescribe a multivitamin-mineral supplement to augment nutrient intake.

Quite often, a child with a fever does not require medical attention, but may benefit from tender loving care. The child's caregiver can comfort the child by:

■ Helping the child's body to maintain its own body temperature by keeping the room temperature moderate and bed coverings to a minimum.

■ Sponging the child with lukewarm water to increase heat loss by evaporation.

■ Providing the child with plenty of fluids.

■ Providing the child with acetaminophen if a physician recommends drug intervention.

Fever is the body's signal that something is wrong and it is trying to defend itself. In many cases, the body is successful without medical attention, but infants with any degree of fever and older children with fevers of 103 degrees Fahrenheit or greater require medical attention. In addition, fevers that go away and recur, that persist for more than 72 hours, or that accompany a rash or marked irritability and confusion require consultation with a physician.

Table FP6B-1 Dehydration and Water Intoxication Symptoms

Dehydration Symptoms	Water Intoxication Symptoms
Thirst	Low plasma sodium concentrations
Muscle cramps	
Weakness, fatigue, exhaustion	Headache
	Muscular weakness and fatigue
Delirium	
Death	Lack of concentration, poor memory, delirium
	Loss of appetite
	Seizure
	Death

Diarrhea

Diarrhea is characterized by frequent, loose, watery stools. Such bowel movements indicate that the chyme has moved too quickly through the intestines for fluid absorption to take place, or that water has been drawn from the cells lining the intestinal tract and added to the food residue. In both cases, the result is the same—extensive fluid and electrolyte losses. If diarrhea continues without treatment, an infant or young child can quickly become dehydrated. The smaller and younger the child, the more life-threatening the condition. Some 300 to 500 infants and young children in the United States die from complications of diarrhea each year.[1]

Well-Nourished Children. Nearly every child suffers from diarrhea at one time or another. Many times an acute case of diarrhea develops and remits in 24 to 48 hours. Well-nourished children with acute diarrhea can usually endure the uncomfortable symptoms without medical treatment. Caregivers can support children during such episodes of diarrhea by eliminating food irritants from the diet, offering clear liquids, and encouraging rest. Fruit juices may aggravate diarrhea and therefore are inappropriate to offer; foods with soluble fiber, such as oatmeal and bananas, may help relieve diarrhea. Children

may enjoy such clear liquids as gelatin dessert, carbonated beverages, and Popsicles, but these treats fall short of correcting for dehydration. Their low electrolyte content and high osmolality make them unsuitable for rehydration therapy.

Nutrient reserves of well-nourished children protect them from the detrimental effects of diarrhea for a short while. The availability of medical treatment and high-quality food in industrialized countries offers children the opportunity to recover, both from fluid and electrolyte losses and from growth losses.

Malnourished Children. The story is quite different in developing countries where acute diarrhea in a malnourished child threatens life and requires immediate medical attention. In developing countries, more children suffer from malnutrition, and acute diarrhea seriously threatens their tenuous nutritional status. With repeated episodes of diarrhea falling close together, these children have little time to recover between bouts. Without full restoration of nutrients, children are progressively less able to defend against future infections.

Diarrhea is the most common cause of dehydration and malnutrition among young children in developing countries. Millions of children die from the complications of diarrhea each year. With reduced dietary intake, impaired intestinal absorption, and the increased nutrient requirements that accompany diarrhea, malnutrition is inevitable.

Children with diarrhea eat less, and therefore their energy intakes are low, averaging between 15 and 50 percent less than their usual intakes. Factors interfering with food intake include anorexia, nausea, and vomiting. In addition, parents or health care workers may withhold food in an effort to resolve the diarrhea. (The wisdom of such a practice is discussed in upcoming paragraphs.)

Maldigestion and malabsorption accompany diarrhea. Normally, hormones and nerves orchestrate the digestive and absorptive processes by signaling several organs to respond at the appropriate times with contractions that move the intestinal contents along, secretions that dismantle nutrients into absorbable molecules, and receptors that transport these molecules into the body. When the contents of the intestine pass too rapidly, digestion is incomplete. When intestinal cells that allow the transport of nutrients into the body are damaged, absorption is limited. Absorption of protein, carbohydrate, and fat in children with diarrhea can be 10 to 30 percent less than in healthy children. On the positive side, 70 to 90 percent of these nutrients *is* getting absorbed, a fact that emphasizes the value of feeding children with diarrhea. Without food intake, pancreatic function and intestinal cell production and maturation remain low, thus limiting the supply of enzymes for digestion and the surface area for absorption.

To Feed or Not to Feed. Clinicians focus on the most appropriate way to minimize and replace nutrient losses incurred by diarrhea. Controversy surrounds the question of whether to withhold or provide nutrients during episodes of diarrhea. The advantages of both delayed feeding and continued feeding are worthy of consideration.

The traditional practice of withholding food is based on the premise that the bowel needs to rest and that ingested food is malabsorbed. Malabsorption results from the rapid intestinal transit time and damage to the intestinal mucosa. Inflammation of the intestinal cells diminishes mucosal surface area, alters villus structure, and lowers enzyme concentrations. Lactose intolerance is a common, usually temporary, consequence of diarrhea. The intestinal cells that produce lactase are located on the delicate fringe of microvilli that form the brush border of the intestinal villi. Any condition (such as diarrhea) that damages the brush border can lead to lactose intolerance. For this reason, dairy products are reintroduced into the diet gradually following diarrhea.

The obvious consequence of malabsorption is the loss of potential nutrients, but unabsorbed nutrients in the intestine present other complications as well. For one, they have an osmotic effect, drawing water and electrolytes into the gut; this can intensify the diarrhea beyond that caused by the original infection. Unabsorbed nu-

trients in the intestine may also bind with bile acids, thus preventing their normal conservation via enterohepatic circulation. Such losses deplete the bile acid pool and can contribute to steatorrhea.

Even considering the problems of malabsorption, the argument in favor of feeding the child appears to carry more weight. For well-nourished children, short-term fasting may be appropriate therapy, but for malnourished children, fasting throughout the course of diarrhea can be devastating. Consider that the annual prevalence rate of diarrhea in children under the age of three in Bangladesh is 55 days a year. To fast for close to two months a year is to lose a significant percentage of a year's nutrient intakes. It is unreasonable to expect that the diet during times without diarrhea could replace losses incurred by such prolonged fasting. Health care professionals argue that suboptimal absorption of some food is better than nothing.

Oral Rehydration Therapy. The potential for accelerated deterioration of nutrition status demands rapid replacement of fluids and nutrients. Health care workers around the world treat diarrhea with oral rehydration therapy (ORT). The American Academy of Pediatrics recommends ORT as the "preferred treatment of fluid and electrolyte losses caused by diarrhea in children with mild to moderate diarrhea."[2] Traditionally, glucose-based solutions have been the standard treatment for acute diarrhea. These solutions provide an optimal glucose concentration, which favors rapid intestinal absorption of water and sodium. As a result, they effectively reverse dehydration, but may not correct the diarrhea. Rice-based solutions have recently proven more effective in correcting diarrhea and therefore may become the preferred therapy.[3] If diarrhea continues, the child receives water and the solution alternately. Infants with diarrhea usually tolerate breast milk well, and breastfeeding can alternate with supplements of the solution. Table FP6B-2 lists the World Health Organization (WHO) standards for ORT formulas; colas, juices, broths, and sports drinks are not appropriate for oral rehydration therapy.

Table FP6B-2 ORT Formula Recommended by the World Health Organization (WHO)

Nutrient	Concentration
Sodium	90 mmol/L
Chloride	80 mmol/L
Potassium	20 mmol/L
Glucose	111 mmol/L
Citrate tribasic or bicarbonate	30 mmol/L

The components of ORT formulas provide needed energy and electrolytes. Except in cases of severe dehydration, ORT can replace the traditional treatment of intravenous (IV) fluid therapy. Intravenous therapy is still most valuable in treating severely dehydrated children; ORT is useful in treating mild to moderate cases and following initial IV therapy.

Perhaps the most significant value of ORT is that it is oral—it does not require hospitalization, as with intravenous feedings. In addition, if the specific WHO-recommended solution is unavailable, caregivers can make other suitable ORT solutions. A mother who lives miles from the nearest pharmacy or clinic and who may not have the resources to purchase medicines can still prepare a solution to refeed her dehydrated infant. A properly prepared rice powder solution facilitates the absorption of electrolytes and water, as well as limiting the duration of diarrhea. The primary role of health care providers then becomes one of educating those who care for children. The people of a community must learn how to prepare a rehydration solution from ingredients available locally. They must learn to measure ingredients carefully and to use sanitary water. Parents will also want to learn how to recognize diarrhea and dehydration symptoms.

The success of ORT in developing countries is central to the WHO's effort to counter the dehydration and death commonly associated with diarrhea. Yet, in developed countries, physicians are reluctant to adopt ORT, recommending hospitalization and IV fluids instead. However, treating well-nourished children in developed countries with ORT is as effective as, and less expensive and invasive than, IV therapy.

Transition to Food. Children who are not dehydrated can continue to eat an appropriate diet. Those who are dehydrated can resume eating foods once they have been rehydrated. At first, they tolerate frequent, small meals best. If food consumption intensifies the diarrhea and threatens dehydration, then they should again go without food temporarily. Clinical observation and the child's willingness are the best determinants of whether to offer food.

The nutrient needs of an undernourished child with diarrhea are exceptionally high. Convalescence time must include food consumption of greater quantity and higher quality than normal to replenish nutrient losses and to allow for catch-up growth. The nutrient intake must compensate for impaired intestinal absorption, support a raised metabolic rate, repair damaged tissue, and make up for growth losses. A conservative goal during convalescence is to provide at least 25 percent more energy than the average requirement for healthy children. A protein intake twice as high as the recommendation for healthy children covers the catabolic and malabsorption losses incurred and the inefficient use of protein when energy intake or absorption is inadequate. These recommendations serve as guidelines and should be adjusted according to the child's growth response.

Constipation

The metabolic consequences of constipation are far less life-threatening than those of diarrhea. Each child's digestive tract responds to food in its own way, with its own rhythm. Food is digested and the waste is made ready for excretion in a predictable number of hours. Each GI (gastrointestinal) tract thus has its own cycle, which depends on its owner's physical makeup and such environmental considerations as the type of food eaten, when it was eaten, and when the person's schedule allows time to defecate. When a child receives the signal to defecate and ignores it (as active children having fun will do), the signal may not return for several hours. During this time, water continues to be withdrawn from the fecal matter, so that when the child does take time to defecate, the bowel movement is dry and hard. Bowel movements that are hard and passed with difficulty, discomfort, or pain define constipation. The time that has elapsed since the previous bowel movement is irrelevant. In the case of painful bowel movements, a parent will want to consult with a physician in order to rule out the presence of organic disease.

Dietary Fiber. One dietary measure that may be appropriate is to increase dietary fiber. Some fibers—those found in cereal products—help to prevent constipation by increasing fecal mass. In the GI tract, fiber attracts water, creating soft, bulky stools that stimulate bowel contractions to push the contents along. These contractions strengthen the intestinal muscles. The improved muscle tone, together with the water content of the stools, eases elimination, reducing pressure in the rectal veins and helping prevent hemorrhoids. Although the major impact of dietary fiber is on the colon, fiber acts as a bulking agent all along the intestine.

One group of pediatricians recommends one to two quarts of popped popcorn per day to relieve constipation. This "treatment" softens stools and increases stool volume, thus providing an enjoyable and inexpensive solution to the problem of constipation.

Fluid Intake. Drinking plenty of water in conjunction with eating high-fiber foods also helps relieve constipation by augmenting stool weight and softness. The increased bulk physically stimulates the upper GI tract, promoting peristalsis throughout. Children may prefer fruit juices to water, and these are acceptable alternatives.

Hospitalized Child

At times, children may require hospitalization. Health care professionals encounter unique problems when feeding children in the hospital.

Foods taste better when you're not alone and scared.

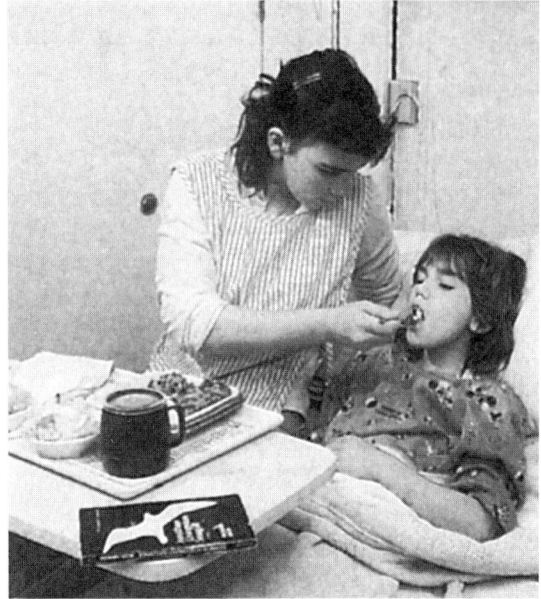

To effectively solve these problems, they need an understanding of the concepts underlying diet therapy and child development. This discussion does not examine the specific dietary treatments of diseases, but does recognize a child's developmental needs.

To work effectively with an infant or young child, a health care professional must work effectively with the parents or guardians as well. Family members must feel comfortable and be able to communicate openly with hospital staff. Their participation in the child's care helps family members to feel needed and the child to recover.

Much of the discussion in Chapter 6 on feeding children applies to hospitalized children as well. Hospitalized children require careful attention to ensure that their nutrient needs are met. Health care professionals and parents must also be sensitive to the child's emotional needs. Pointers from people experienced in working with hospitalized children include the following:

■ Notice the child's posture. Body language will tell you if a child feels fear, pain, or discomfort.

- Touch the child often and lovingly. Touch communicates more than words.
- Allow the child to choose what to eat as much as possible. If permissible, foods brought from outside the hospital can help stimulate appetite.
- Serve portions that are appropriate for the child's age.
- Notice whether the child eats the food. Putting a tray of food in front of a child is not enough.
- Stay with the child during the meal, or make sure a caring person is present. The child will eat and digest food better if a caring person soothes away anxiety and loneliness.
- Encourage the child to eat the most nutritious foods first before becoming too full to complete the meal.
- Allow children to eat with other children, if possible. They will enjoy mealtimes more, accept more food, and eat for longer periods.
- Avoid painful procedures near mealtimes. The stress of pain or fear shuts down digestion and turns off interest in food.
- Serve foods attractively, selecting a colorful variety of foods. Cut foods into different shapes. Arrange foods in patterns such as faces or vehicles.

Nutrition care contributes importantly to a child's recovery. Every effort should be made to provide optimal nourishment.

The care of children in times of sickness requires careful attention to their nutrition needs. To feed children is to provide them with much more than nutrients and fluids alone. Foods carry both a physical and an emotional comfort. Chicken noodle soup does offer fluids, some energy, a little protein, and a variety of vitamins and minerals, but its healing power also comes from the caregiver's concern about the child. Children given tender loving care recover more quickly than those deprived of it. As Dr. F. W. Peabody of Harvard University said, "The secret of the care of the patient is in the caring for the patient."

FOCAL POINT 6B NOTES

1. P. E. Kilgore and coauthors, Trends of diarrheal disease-associated mortality in US children, 1968 through 1991, *Journal of the American Medical Association* 274 (1995): 1143–1148; R. D. Williams, Preventing dehydration in children, *FDA Consumer*, July–August 1996, pp. 19–22.
2. Provisional Committee on Quality Improvement, Subcommittee on Acute Gastroenteritis, American Academy of Pediatrics, Practice parameter: The management of acute gastroenteritis in young children, *Pediatrics* 97 (1996): 424–433.
3. E. D. Goldberg and J. R. Saltzman, Rice inhibits intestinal secretions, *Nutrition Reviews* 54 (1996): 36–37.

Early Development of Chronic Diseases

The childhood years are a parent's last chance to influence food choices. As Chapter 6 described, parents who want to promote nutritious choices and healthful habits provide access to nutrient-dense foods and opportunities for active play at home. Sound food choices and regular physical activity not only promote healthy growth, but can also help prevent the degenerative diseases of later life. Many experts agree that early childhood is the time to put into effect practices that, until recently, were recommended only for adults.[1]

Disease of the heart and blood vessels, known as cardiovascular disease, or CVD, is the leading cause of death in the United States and Canada each year. Most CVD involves atherosclerosis (see the accompanying glossary for definitions). Compelling evidence shows that atherosclerosis has its beginnings in childhood

Take care of your body, and your body will take care of you.

and that its development is related to dietary fats and blood lipids.[2] The disease is not inevitable, however; people can reach advanced age with very little atherosclerosis. Strategies begun early in life offer the most promise of prevention.

Much of our knowledge about the early development of CVD comes from the Bogalusa Heart Study, a long-term epidemiological study of some 14,000 young people. For nearly three decades, researchers have been observing how changes in body weight, blood lipids, blood pressure, and individual behaviors correlate with the development of CVD over time—from infancy to childhood through adolescence and into young adulthood. Some major findings have emerged from this research:

- Changes inside the arteries—changes predictive of CVD—are evident in childhood.
- Obesity in children affects these changes.
- Behaviors that influence the development of obesity and of CVD are learned and begin in childhood. These behaviors include overeating, eating high-fat foods, being physically inactive, and smoking cigarettes.

This Focal Point focuses on efforts to prevent childhood obesity and CVD, but the benefits extend to prevention of cancer, diabetes, and other chronic diseases as well. The earlier in life health-promoting habits become established, the better they will stick.

Early Development of CVD

Invariably, questions arise as to what extent genetics is involved in CVD development. Genetics does not appear to play a *determining* role in CVD; that is, a person is not simply destined at birth to develop CVD.[3] Instead, genetics appears to play a *permissive* role—the potential is inherited and then will develop, if given a push by poor health choices such as excessive weight gain, poor diet, sedentary lifestyle, and cigarette smoking.

GLOSSARY

atherosclerosis (ath-er-oh-scler-OH-sis): a type of artery disease characterized by accumulations of lipid-containing material on the inner walls of the arteries.

cardiovascular disease (CVD): a general term for all diseases of the heart and blood vessels. Atherosclerosis is the main form of CVD.

HDL (high-density lipoprotein): the type of lipoprotein that transports cholesterol back to the liver from peripheral cells; composed primarily of protein.

LDL (low-density lipoprotein): the type of lipoprotein derived from very-low-density lipoproteins (VLDL) as cells remove triglycerides from them; composed primarily of cholesterol.

plaques (atheromatous): mounds of lipid material, mixed with smooth muscle cells and calcium, which develop in the artery walls in atherosclerosis.

Atherosclerosis. Infants are born with healthy, smooth, clear arteries, but within the first decade of life, fatty streaks may begin to appear. During adolescence, these fatty streaks may begin to turn to fibrous plaques. By early adulthood, the fibrous plaques may begin to calcify and become raised lesions, especially in boys and young men.[4] As the lesions grow more numerous and enlarge, the heart disease rate begins to rise, and the rise becomes dramatic at about age 45 in men and 55 in women.[5] From this point on, arterial damage and blockage progress rapidly, and heart attacks and strokes threaten life. In short, the consequences of atherosclerosis, which become apparent only in adulthood, have their beginnings in the first decades of life.[6]

As mentioned earlier, atherosclerosis is not inevitable; people can age with relatively clear arteries. Early lesions may either progress or regress, depending on several factors, many of which reflect lifestyle behaviors. Smoking, for example, is strongly associated with the prevalence of raised lesions, even in young adults.[7]

Blood Cholesterol. Atherosclerotic lesions reflect blood cholesterol: As blood cholesterol increases, lesion coverage increases.[8] Cholesterol values at birth are similar in all populations; differences emerge in early childhood. In countries where the adults have high blood cholesterol and high rates of CVD, the children also tend to have high blood cholesterol. Conversely, in countries where the adults have low blood cholesterol and low rates of CVD, the children tend to have low blood cholesterol, suggesting that adult heart disease tracks early trends and that early preventive efforts might reduce the incidence of later CVD.[9]

These findings reflect population trends, but individual cholesterol status reveals similar trends. At one year, cholesterol values predict the values that will be seen later in childhood, especially for those with high blood cholesterol.[10] Studies examining children for more than a decade have found that the best predictor of blood cholesterol is earlier baseline values: Childhood values correlate with values in young adulthood.[11] Quite simply, if you want to know a child's future cholesterol, measure it now.

Blood cholesterol also correlates with obesity, especially central obesity. LDL cholesterol correlates positively, and HDL negatively.[12] These relationships are apparent throughout childhood, and their magnitude increases with age.

Research has also confirmed an association between blood lipids and physical activity in children, similar to that seen in adults.[13] Inactive children have higher total cholesterol and LDL and lower HDL than physically active children.

Blood Pressure. An elevated blood pressure accelerates the development of CVD. On the average, children's blood pressure is lower than adults', but blood pressure increases as children grow, rising sharply at puberty and then leveling off. Like blood cholesterol, blood pressure correlates with obesity, especially central obesity.[14] Blood pressure tends to increase at a slower rate in children who participate in regular aerobic activity or who have either lost weight or maintained their weight as they grew taller.[15]

Obesity in Children

Many experts agree that preventing or treating obesity in childhood will reduce the rate of CVD in adulthood.[16] Without intervention, overweight children become overweight adolescents who become overweight adults, and being overweight exacerbates every chronic disease that adults face.[17]

Growing Fatter. Children are heavier today than they were 10 to 30 years ago. They weigh 4 to 8 pounds more than children did three decades ago. This pattern is a secular trend—that is, one that cannot be explained by genetics.[18] Diet and physical activity must be responsible.

Not Eating More. Reports from the Bogalusa Heart Study indicate that children's energy intakes have remained relatively stable over the past 15 years. There has even been a slight decline in fat intake, from 38 to 36 percent of kcalories from fat daily.[19] This slight decline in dietary fat is not enough, however, to have influenced body weight, nor is it enough to meet current dietary recommendations.

Children's dietary fat intakes vary, of course, and some children do eat high-fat diets. Children who prefer high-fat foods tend to consume a relatively large percentage of their energy intake from fat.[20] They also tend to be more overweight than their peers.[21] Particularly noteworthy is the finding that the children's fat preferences and consumption correlate with their parents' obesity as well.[22] Such findings confirm the significant roles parents play—teaching children about healthy food choices, providing children with low-fat selections, and serving as role models.

Growing Less Active. Most likely, children have grown more overweight because of their lack of physical activity.[23] An inactive child can become obese even while eating less food than an active child. Today's children are more sedentary and less physically fit than children were 20 to 30 years ago.

Watching television accounts for some 24 hours a week of sedentary behavior. Beyond these 24 hours, children spend more sedentary time working at computers and playing video games. Some studies have found that both obesity and blood cholesterol correlate with hours of television viewed.[24]

Just as blood cholesterol and obesity track over the years, so does a person's level of physical activity. A study of almost 1000 teenagers reported that over half of those who were initially described as inactive remained inactive six years later.[25] Similarly, almost half of those who were physically active remained so. Compared with inactive teens, those who were physically active weighed less, smoked less, ate a diet lower in saturated fats, and had a better blood lipid profile. The message is clear: Physical activity offers numerous health benefits, and children who are active today are most likely to be active for years to come.

Dealing with Childhood Obesity. Preventing childhood obesity is ideal, of course, but clearly not always possible. The child who is already obese needs careful management. Weight loss is not ordinarily recommended because restrictive diets can easily impair growth in children. Instead, aim to maintain a constant weight while the child grows taller. The object is to support normal lean body development, while letting children "grow out" of their obesity.

Cholesterol Screening for Children

Many children in the United States are not only overweight but also have high blood cholesterol.[26] The question of whether to screen children for high blood cholesterol is controversial.[27] Currently, selective screening for children and adolescents whose parents or grandparents have CVD is recommended.[28] Since blood cholesterol in children is a good predictor of adult values, however, some experts recommend universal screening to identify all children with high blood cholesterol.[29] They note that many children who have high blood cholesterol do not have family histories of CVD and would be missed under current screening criteria.[30] Opponents argue that some children with high blood cholesterol may reach adulthood with normal blood cholesterol and that treating adults who have high blood cholesterol should be sufficient. They believe screening will create unnecessary anxiety and lead to an overuse of drug therapy and overly restrictive dieting during childhood and adolescence.[31] Furthermore, studies have found that few children follow up with additional testing or dietary changes anyway.[32] Standard values for cholesterol screening in children and adolescents are listed in Table FP6C-1.

In some cases, parents are too young to have a CVD history. In many other cases, children, or their parents, may not know their family histories. For these reasons, it may be most effective for physicians of adult heart patients to refer the children and grandchildren of these patients for cholesterol screening.[33]

Some research shows that overweight children should also be considered for cholesterol screening, even if they do not satisfy the current criteria.[34] The incidence of high blood cholesterol in obese children with no other criteria is similar to that of nonobese children with family histories of CVD.

Considering the many lifestyle factors that accompany the development of CVD, questions regarding a child's health behaviors might also be informative. Health care professionals should determine whether children smoke and how physically active they are, especially when family history is unknown.[35]

Early—but not advanced—atherosclerotic lesions are reversible, making screening and education a high priority. Both those with family histories of CVD and those with multiple risk factors need intervention. Children with the highest risks of developing CVD are sedentary and obese, with high blood pressure and high blood cholesterol. In contrast, children with the lowest risks of heart disease are physically active and of normal weight, with low blood pressure and favorable lipid profiles. Routine pe-

Table FP6C-1 Cholesterol Values for Children and Adolescents

Disease Risk	Total Cholesterol (mg/dL)	LDL Cholesterol (mg/dL)
Acceptable	<170	<110
Borderline	170–199	110–129
High	≥200	≥130

Source: Reprinted with permission of National Cholesterol Education Program: Report of the Expert Panel on Blood Cholesterol Levels in Children and Adolescents, *Pediatrics* 89 (1992): 527. Copyright 1992 by the American Academy of Pediatrics.

diatric care should identify these known risk factors and provide education when needed (see Table FP6C-2).

Dietary Recommendations for Children

An expert panel on blood cholesterol in children and adolescents recommends that, regardless of family history, all children over age two should eat a variety of foods and maintain desirable weight.[36] Children should receive an average daily intake of 30 percent of total energy from fat, less than 10 percent from saturated fat, and less than 300 milligrams of cholesterol per day. The American Academy of Pediatrics agrees, but cautions against fat intakes of less than 30 percent of total kcalories for growing children.

In fact, some experts on children's nutrition believe that children need more fat in their diets than adults do. These experts contest the recommendation to limit children's fat intakes to 30 percent of total kcalories, arguing that the recommendation is "unreasonable and potentially dangerous to children."[37] They support the recommendation for adults because 30 percent of kcalories from fat will not compromise growth and because the guideline derived from studies of adults (mostly men). These experts believe the guideline should not apply to children. They endorse the Canadian recommendation for children, which states that from the age of two to the end of linear growth (adolescence) there should be a transition from the high-fat diet of

Table FP6C-2   Health Professional's Schedule of Cardiovascular Disease Assessment in Children

Birth	• Family history for early heart disease, high blood lipids (if positive, discuss risk factors and refer parents to health care).
	• Start growth chart.
	• Parental smoking history (if positive, refer to smoking cessation program).
0–2 years	• Update family history, growth chart.
	• With introduction of solids, begin teaching about healthy diet (nutritionally adequate, low in salt, low in saturated fats).
	• Recommend healthy snacks as finger foods.
	• Change to whole milk from formula or breastfeeding at approximately 1 year of age.
2–6 years	• Update family history, growth chart (review growth chart[a] with family and discuss concept of weight for height).
	• Introduce moderately low-fat diet.
	• Change to low-fat milk.
	• Start blood pressure chart at approximately 3 years of age[b]; review for concept of lower salt intake.
	• Encourage active parent–child play.
	• Lipid determination in children with positive family history or with parental cholesterol >240 mg/dL (if abnormal, initiate nutrition counseling).
6–10 years	• Update family history, blood pressure, and growth charts.
	• Complete cardiovascular health profile with child; determine family history, smoking history, blood pressure percentile, weight for height, fingerstick cholesterol, and level of activity and fitness.
	• Reinforce low-fat diet.
	• Begin active antismoking counseling.
	• Introduce fitness for health and encourage lifelong sport activities for child and family.
	• Discuss role of watching television in sedentary lifestyle and obesity.
>10 years	• Update family history, blood pressure, and growth charts annually.
	• Review low-fat diet, risks of smoking, fitness benefits whenever possible.
	• Consider lipid profile in all patients.
	• Final review of personal cardiovascular health status.

[a]If weight is >120% of normal for height, diagnosis of obesity should be considered and the subject addressed with the child and family.
[b]If three consecutive interval blood pressure measurements exceed the 90th percentile and blood pressure is not explained by height or weight, diagnosis of hypertension should be made and appropriate evaluation considered.
Source: Adapted with permission from W. B. Strong and coauthors, Integrated cardiovascular health promotion in childhood: A statement for health professionals from the Subcommittee on Atherosclerosis and Hypertension in Childhood of the Council on Cardiovascular Disease in the Young, American Heart Association, *Circulation* 85 (1992): 1638–1650. Copyright 1992 American Heart Association.

infancy to a diet that includes no more than 30 percent of kcalories as fat.[38] It should be noted that the Dietary Guidelines Advisory Committee of the United States specifically addresses the application of the 30 percent fat guideline for children. The committee recommends transition to a 30 percent fat diet between the ages of two and five, so that by the time children enter school, they consume diets that follow the *Dietary Guidelines.*[39]

Moderation, Not Deprivation. Healthy children over age two can begin the transition to eating according to recommendations. Even then, meals can include moderate amounts of a child's favorite foods, even if they are high-fat selections such as french fries and ice cream.[40] Without such additions, diets might be too low in fat, not to mention unappetizing and boring.

Balanced meals need to provide lean meat, poultry, fish, and legumes; fruits and vegetables; whole grains; and low-fat milk products. Such meals can provide enough food energy and nutrients to support growth and maintain blood cholesterol within a healthy range.[41] Pediatricians warn parents to avoid extremes; they caution that while intentions may be good, excessive food restriction may create nutrient deficiencies and impair growth. Furthermore, parental control over eating may instigate battles and foster attitudes about foods that can lead to inappropriate eating behaviors.

Diet First, Drugs Maybe. Experts agree that children with high blood cholesterol should first be treated with diet. If, in children 10 years old and older, blood cholesterol remains high after 6 to 12 months of dietary intervention, then drugs may be used to lower blood cholesterol.[42] Pharmacological doses of niacin effectively lower LDL cholesterol in children, but adverse effects are common; such treatment should be reserved only for severe cases.[43]

Smoking

Another risk factor for CVD that starts in childhood and carries over into adulthood is cigarette smoking. Each day 5000 children begin to use tobacco, 40 percent of them in grade school.

Children and teenagers are not likely to consider the long-term health consequences of tobacco use. They are more likely to be struck by the immediate health consequences, such as shortness of breath when playing sports, or social consequences, such as having bad breath. Whatever the context, the message to all children and teens should be clear: Don't start smoking. If you've already started, quit.

In conclusion, *adult* CVD is a major *pediatric* problem. Without intervention, some 60 million children are destined to suffer its consequences within the next 30 years. Optimal prevention efforts focus on children, especially on those who are overweight.

Just as young children receive vaccinations against infectious diseases, they need screening for, and education about, CVD. Many health education programs have been implemented in schools around the country.[44] These programs are most effective when they include education in the classroom, heart-healthy meals in the lunchroom, fitness activities on the playground, and parental involvement at home.

FOCAL POINT 6C NOTES

1. National Cholesterol Education Program, Report of the Expert Panel on Blood Cholesterol Levels in Children and Adolescents, The population approach: Nutrition recommendations for healthy children and adolescents, *Pediatrics* (supplement) 89 (1992): 537–544.
2. National Cholesterol Education Program, Report of the Expert Panel on Blood Cholesterol Levels in Children and Adolescents, Overview and summary, *Pediatrics* (supplement) 89 (1992): 525–527.
3. W. B. Kannel, R. B. D'Agostino, and A. Belanger, Concept of bridging the gap from youth to adulthood—The Framingham Study, an address presented at the Recognition and Prevention of Heart Disease: State of the Art conference, New Orleans, Louisiana, April 27 and 28, 1994.
4. G. S. Berenson and coauthors, Atherosclerosis of the aorta and coronary arteries and cardiovascular risk factors in persons aged 6 to 30 years and studied at necropsy (the Bogalusa Heart Study), *American Journal of Cardiology* 70 (1992): 851–858.
5. Kannel, D'Agostino, and Belanger, 1994.
6. Committee on Nutrition, American Academy of Pediatrics, Statement on cholesterol, *Pediatrics* 90 (1992): 469–473; National Cholesterol Education Program, 1992.
7. Pathobiological Determinants of Atheroscle-

rosis in Youth (PDAY) Research Group, Relationship of atherosclerosis in young men to serum lipoprotein cholesterol concentrations and smoking: A preliminary report from the Pathobiological Determinants of Atherosclerosis in Youth (PDAY) Research Group, *Journal of the American Medical Association* 264 (1990): 3018–3024; C. E. Bartecchi, T. D. MacKenzie, and R. W. Schrier, The human costs of tobacco use, *New England Journal of Medicine* 330 (1994): 907–912.

8. Pathobiological Determinants of Atherosclerosis in Youth (PDAY) Research Group, 1990.

9. L. Snetselaar and R. M. Lauer, Childhood, diet and the atherosclerotic process, *Nutrition Today*, January–February 1992, pp. 22–28.

10. M. J. T. Kallio and coauthors, Tracking of serum cholesterol and lipoprotein levels from the first year of life, *Pediatrics* 91 (1993): 949–954.

11. S. Guo and coauthors, Serial analysis of plasma lipids and lipoproteins from individuals 9–21 years of age, *American Journal of Clinical Nutrition* 58 (1993): 61–67.

12. W. A. Wattigney and coauthors, Increasing impact of obesity on serum lipids and lipoproteins in young adults: The Bogalusa Heart Study, *Archives of Internal Medicine* 151 (1991): 2017–2022.

13. E. Suter and M. R. Hawes, Relationship of physical activity, body fat, diet, and blood lipid profile in youths 10–15 yr, *Medicine and Science in Sports and Exercise* 25 (1993): 748–754.

14. C. L. Shear and coauthors, Body fat patterning and blood pressure in children and young adults: The Bogalusa Heart Study, *Hypertension* 9 (1987): 236–244.

15. S. Shea and coauthors, The rate of increase in blood pressure in children 5 years of age is related to changes in aerobic fitness and body mass index, *Pediatrics* 94 (1994): 465–470.

16. Update on the 1987 Task Force Report on High Blood Pressure in Children and Adolescents: A working group report from the National High Blood Pressure Education Program, *Pediatrics* 98 (1996): 649–658; M. Knip and O. Nuutinen, Long-term effects of weight reduction on serum lipids and plasma insulin in obese children, *American Journal of Clinical Nutrition* 57 (1993): 490–493.

17. S. S. Guo and coauthors, The predictive value of childhood body mass index values for overweight at age 35 y, *American Journal of Clinical Nutrition* 59 (1994): 810–819.

18. D. S. Freedman and coauthors, Secular increases in relative weight and adiposity among children over two decades: The Bogalusa Heart Study, *Pediatrics* 99 (1997): 420–426.

19. T. A. Nicklas and coauthors, Secular trends in dietary intakes and cardiovascular risk factors of 10-year-old children: The Bogalusa Heart Study (1973–1988), *American Journal of Clinical Nutrition* 58 (1993): 930–937.

20. J. O. Fisher and L. L. Birch, Fat preferences and fat consumption of 3- to 5-year-old children are related to parental obesity, *Journal of the American Dietetic Association* 95 (1995): 759–764.

21. J. M. Gazzaniga and T. L. Burns, Relationship between diet composition and body fatness, with adjustment for resting energy expenditure and physical activity, in preadolescent children, *American Journal of Clinical Nutrition* 58 (1993): 21–28.

22. V. T. Nsuyen and coauthors, Fat intake and adiposity in children of lean and obese parents, *American Journal of Clinical Nutrition* 63 (1996): 507–513.

23. S. A. Schlicker, S. T. Borra, and C. Regan, The weight and fitness of United States children, *Nutrition Reviews* 52 (1994): 11–17.

24. E. Obarzanek and coauthors, Energy intake and physical activity in relation to indexes of body fat: The National Heart, Lung, and Blood Institute Growth and Health Study, *American Journal of Clinical Nutrition* 60 (1994): 15–22.

25. O. T. Raitakari and coauthors, Effects of persistent physical activity on coronary risk factors in children and young adults: The Cardiovascular Risk in Young Finns Study, *American Journal of Epidemiology* 140 (1994): 195–205.

26. G. S. Berenson, S. R. Srinivasan, and L. S. Webber, Cardiovascular risk prevention in children: A challenge or a poor idea? *Nutrition, Metabolism and Cardiovascular Diseases* 4 (1994): 46–52.

27. S. S. Gidding, The rationale for lowering serum cholesterol levels in American children, *American Journal of Diseases of Children* 147 (1993): 386–392; P. T. Einhorn and B. M. Rifkind, Cholesterol measurement in children, *American Journal of Diseases of Children*

147 (1993): 373–375; G. S. Berenson, Cholesterol: Myth vs. reality in pediatric practice, *American Journal of Diseases of Children* 147 (1993): 371–373.

28. Report of the Expert Panel on Blood Cholesterol Levels in Children and Adolescents, *Pediatrics* 89 (1992): Entire supplement.

29. Berenson, Srinivasan, and Webber, 1994.

30. S. J. Wadowski and coauthors, Family history of coronary artery disease and cholesterol: Screening children in disadvantaged inner-city population, *Pediatrics* 93 (1994): 109–113; K. Resnicow and D. Cross, Are parents' self-reported total cholesterol levels useful in identifying children with hyperlipidemia? An examination of current guidelines, *Pediatrics* 92 (1993): 347–354.

31. National Cholesterol Education Program, Overview and summary, 1992.

32. C. M. Lannon and J. Earp, Parents' behavior and attitudes toward screening children for high serum cholesterol levels, *Pediatrics* 89 (1992): 1159–1163.

33. L. E. Muhonen and coauthors, Coronary risk factors in adolescents related to their knowledge of familial coronary heart disease and hypercholesterolemia: The Muscatine Study, *Pediatrics* 93 (1994): 444–451.

34. M. S. Glassman and S. M. Schwarz, Cholesterol screening in children: Should obesity be a risk factor? *Journal of the American College of Nutrition* 12 (1993): 270–273.

35. Committee on Nutrition, 1992.

36. NCEP Expert Panel on Blood Cholesterol Levels in Children and Adolescents, National Cholesterol Education Program (NCEP): Highlight of the report of the Expert Panel on Blood Cholesterol Levels in Children and Adolescents, *Pediatrics* 89 (1992): 495–501.

37. R. E. Olson, The folly of restricting fat in the diet of children, *Nutrition Today* 30 (1995): 234–244; J. McCormick and P. Skrabanek, Coronary heart disease is not preventable by population intervention, *Lancet* 2 (1988): 839–841.

38. Report of the Joint Working Group of the Canadian Pediatric Society and Health Canada, Nutrition Recommendations Update: Dietary fats and children, *Nutrition Reviews* 53 (1995): 367–375.

39. Report of the Dietary Guidelines Advisory Committee, *Dietary Guidelines for Americans*, 1995, *Nutrition Reviews* 53 (1995): 376–379.

40. M. Sigman-Grant, S. Zimmerman, and P. M. Kris-Etherton, Dietary approaches for reducing fat intake of preschool-age children, *Pediatrics* 91 (1993): 955–960.

41. Timely statement on NCEP report on children and adolescents, *Journal of the American Dietetic Association* 91 (1991): 983; Is there a relationship between dietary fat and stature or growth in children three to five years of age? *Pediatrics* 92 (1993): 579–586.

42. Committee on Nutrition, 1992.

43. R. B. Colletti and coauthors, Niacin treatment of hypercholesterolemia in children, *Pediatrics* 92 (1993): 78–82.

44. A. M. Downey, J. L. Cresanta, and G. S. Berenson, Cardiovascular health promotion in children: "Heart Smart" and the changing role of physicians, *American Journal of Preventive Medicine* 5 (1989): 279–295.

Nutrition during Adolescence

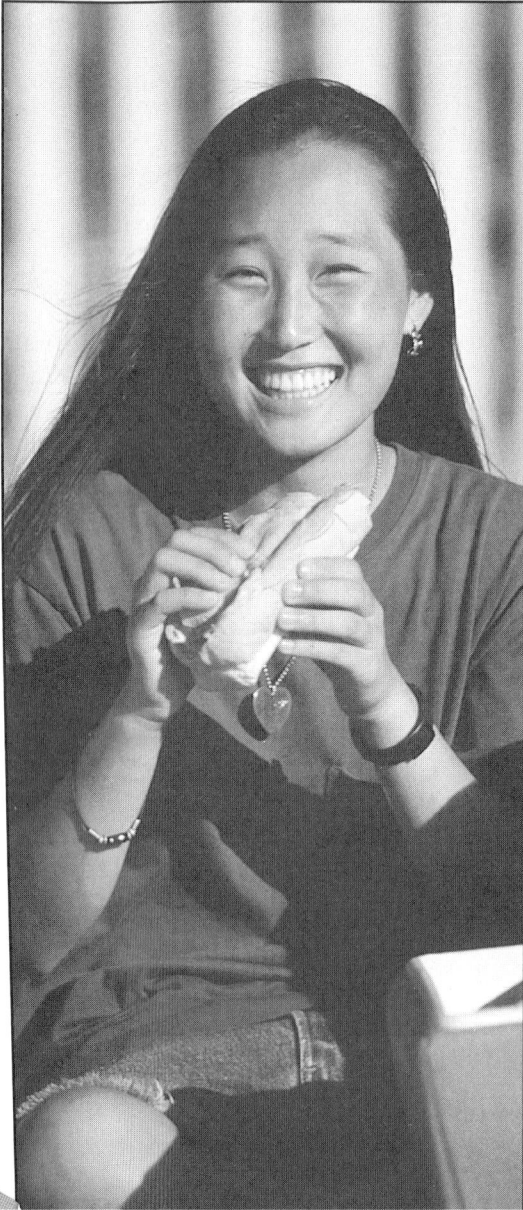

Growth and Development during Adolescence
Physical Growth and Development
Psychological Development

Energy and Nutrient Needs during Adolescence
Energy and Energy Nutrients
Vitamins
Minerals
Nutrient Supplements for Adolescents

Nutrition Assessment of Adolescents
Anthropometric Data
Historical Information
Biochemical Analyses

Nutrition-Related Concerns during Adolescence
Adolescent Obesity
Behaviors Incompatible with Health
Acne

Food Choices and Fitness during Adolescence
Diets Tailored to Individual Preferences
Fitness during Adolescence

FOCAL POINT 7A   *Eating Disorders*
FOCAL POINT 7B   *The Menstrual Cycle*

As children pass through adolescence on their way to becoming adults, they change in many ways. Their physical changes make their nutrient needs high, and their emotional, intellectual, and social changes make meeting those needs a challenge.

Teenagers make many more choices for themselves than they did as children. They are not fed, they eat; they are not sent out to play, they choose to go. At the same time, social pressures thrust choices at them: whether to drink alcoholic beverages and whether to develop their bodies to meet extreme ideals of slimness or athletic prowess. Their interest in nutrition derives from personal, immediate experiences. They are concerned with how diet can improve their lives now—they engage in crash dieting in order to buy a new bathing suit, avoid greasy foods in an effort to clear acne, or eat a pile of spaghetti to prepare for a big sporting event. In presenting information on the nutrition and health of adolescents, this chapter includes these many topics of interest to teens.

Growth and Development during Adolescence

With the onset of adolescence, the steady growth of childhood speeds up abruptly and dramatically, and the growth patterns of females and males become distinct (see Figure 7-1). Hormones direct the intensity and duration of the adolescent growth spurt, profoundly affecting every

adolescence: the period of growth from the beginning of puberty until full maturity.

Figure 7-1 Physical Changes of Adolescence

FEMALES

Growth
Rapid gains peak around age 12, then growth slows to a stop at maturity.

Hair
Hair grows on underarms and genital area; other body hair may grow coarser and longer.

Skin
Acne may develop.

Body shape and composition
Hips widen, fat deposits collect, and breasts develop.

Hormonal changes
Ovaries produce more estrogens and progesterone.

Reproductive organs
Uterus and ovaries enlarge; genitals enlarge; ovum ripening begins; normal vaginal secretions begin, including a mucuslike daily secretion and monthly menstruation.

MALES

Growth
Rapid gains peak around age 14, then growth slows to a stop at maturity.

Hair
Hair of forehead begins to move upward (recede). Hair grows on face, underarms, and around genitals; other body hair may grow coarser and longer.

Skin
Acne may develop.

Body shape and composition
Muscle tissue develops.

Hormonal changes
Testicles produce more testosterone.

Reproductive organs
Penis and testicles enlarge; sperm production begins; ejaculations begin.

organ of the body, including the brain. Hormones also bring mood changes and sexual feelings. After two to three years of intense growth and a few more at a slower pace, physically mature adults emerge.

Physical Growth and Development

In general, the adolescent growth spurt begins at age 10 or 11 for females, and at 12 or 13 for males. The duration lasts about 2 1/2 years, but of course, wide variations are seen for individuals because the schedule of maturation is determined by genetics.

In the early stage, adolescent growth is linear; the child "shoots up." On the average, males grow approximately 8 inches taller, with a peak of growth velocity at 14 years, and females, 6 inches taller, with a peak velocity at 12 years of age. Skeletal growth ceases with the closure of the epiphyses. Growth in the later stage of the spurt is lateral; the child "fills out." Males gain approximately 45 pounds, and females, about 35 pounds.

epiphyses: the end segments of long bones that contain a thin area of active bone growth; when this growth eventually stops, no further significant growth in bone length can occur.

Many changes in body shape and posture become evident during adolescence. The first areas of accelerated growth are the feet and hands, then the calves and forearms, followed by the hips and chest, and then the shoulders. The trunk is the last part of the body to go through a growth spurt. This sequence of development takes teenagers through an awkward phase of having large and ungainly limbs compared with the rest of their bodies.

puberty: the period in life in which a person becomes physically capable of reproduction.

Body Composition. Before puberty, the differences between male and female body composition are slight. Gender differences become apparent in the skeletal system, lean body mass, and fat stores during the adolescent spurt. In males, the lean body mass—principally muscle and bone—increases much more than in females. In females, fat assumes a larger percentage of the total body weight. Additional fat deposition dramatically alters a young girl's body shape. Why females deposit fat in specific locations is unknown, although the fat in the breasts does serve to protect the mammary glands.

Sexual Maturation. Hormones control the secondary sex changes experienced by both males and females. The earliest sign of puberty in males is the growth of the testicles and penis. Changes in the larynx, skin, and hair distribution follow. The earliest internal sign of puberty for females is growth of the ovaries. Externally, the breasts enlarge, and pubic hair appears. These secondary sexual characteristics are developing at younger ages than previously noted in girls, especially in African-American girls.[1] Menarche occurs after the peak of growth acceleration in height and coincides with growth deceleration. A female can expect to grow only about three more inches in height after menarche.

menarche (MEN-ark): the initial menstrual period, normally occurring between the ages of 10 and 16, with the average age at 12 years (*men* = month; *arche* = beginning).

With sexual development comes an awareness of sexual activity; slightly more than half of all high school students report having had sexual intercourse.[2] Teenage girls who become pregnant while their own needs for growth are still high may find meeting nutrient needs difficult (see Chapter 3).

Effects of Malnutrition. Nutrient and food energy deficiencies during adolescence can retard growth and delay sexual maturation. In fact, a classic symptom of anorexia nervosa is amenorrhea. Development during adolescence depends not only on the individual's present nutrition status but also on previous nutrient intake.[3] Undernutrition in early childhood may diminish a person's eventual body size. Malnourished children are typically shorter and lighter than their well-nourished peers at the end of the adolescent growth spurt. Given adequate food, however, previously undernourished children gain more height during the adolescent growth spurt than children who were previously well nourished. Weight gains are comparable in both groups. In other words, adolescence can serve as a catch-up period to regain at least some of the height losses due to undernutrition in the early years.

A possible explanation for the greater height gain seen in previously undernourished females could be the delay in menarche. Menarche may be delayed by as much as a couple of years in females who were undernourished as children. With delayed menarche, closure of the epiphyses is delayed, and this allows for a longer period of growth. Delayed sexual maturation is also evident in males who were undernourished as children. Their attainment of sexual maturity may be postponed by up to three years.

The delay of menarche is known as **primary amenorrhea.** *The cessation of menses in an adolescent who has previously menstruated is known as* **secondary amenorrhea.**

Psychological Development

The many physical changes that occur with puberty bring psychological and emotional changes with them. Adolescents may feel awkward and sensitive about their newly emerging bodies, especially when their own maturation process differs from that of their peers. All teens grow taller and heavier during this growing phase, and for some, becoming larger is uncomfortable. Such feelings may underlie the eating disorder anorexia nervosa, which has its highest incidence at the transition between childhood and adolescence. (Eating disorders are discussed fully in Focal Point 7A.)

Other psychological factors are also at work. Teenagers especially value their independence, and one way they express it is through food selection. Deciding what to eat, or even whether to eat, is an expression of autonomy. At the same time teens are developing their independence, they are learning how to fit into their social circle. Many of their behaviors reflect the opinions and actions of their peers, including their food and health choices. When others perceive milk as "babyish," a teen will choose soft drinks instead; when others skip lunch and hang out in the school parking lot, a teen may join in for the camaraderie, regardless of hunger. Adults need to remember that teenagers have the right to express their own opinions—even if those opinions are contrary to the adults' views. Caregivers can make sure that nutritious foods are available and can stand by with reliable nutrition information and advice, but the rest is up to the adolescents. Ultimately, they make the choices.

Energy and Nutrient Needs during Adolescence

The rapid growth that occurs during adolescence is reflected in the high nutrient needs of this stage in the life cycle. Energy and nutrient needs are greater during adolescence than at any other time of life, except for pregnancy and lactation. In general, nutrient needs rise throughout childhood, peak in adolescence, and then level off or even diminish as the teen becomes an adult.

Chronological age provides an inappropriate basis for determining the energy and nutrient requirements of adolescents; growth would be preferable. Yet age is most often used in establishing guidelines, because it is easy to determine and because it works in a general way. The nutrient needs of males and females diverge at puberty, reflecting the sex-related differences in their growth and development. The 1989 RDA for males and females begin to differ at age 11, and the 1997 DRI split into sex categories at age 9.

The RDA and the DRI split the teen years not only into male–female groups, but also into two age groups. These age groups assume that the physical changes that occur during those periods are similar for most adolescents. During adolescence, needs depend on the age of puberty onset, the velocity of growth, individual activity level, and the length of time required to complete the maturation process.

Energy and Energy Nutrients

Energy RDA during adolescence:
2500 kcal/day (11 to 14 yr, males).
2200 kcal/day (11 to 14 yr, females).
3000 kcal/day (15 to 18 yr, males).
2200 kcal/day (15 to 18 yr, females).

Canadian RNI during adolescence:
2500 kcal/day (10 to 12 yr, males).
2200 kcal/day (10 to 15 yr, females).
2800 kcal/day (13 to 15 yr, males).
3200 kcal/day (16 to 18 yr, males).
2100 kcal/day (16 to 18 yr, females).

The energy needs of adolescents vary greatly. Boys' energy needs may be especially high; they typically grow faster than girls and, as mentioned, develop a greater proportion of lean body mass. An active boy of 15 may need 4000 kcalories or more a day just to maintain his weight. Girls start growing earlier than boys and attain lower heights and weights, so their energy needs peak sooner and decline earlier than those of their male peers. An inactive girl of 15 whose growth is nearly at a standstill may need fewer than 2000 kcalories a day if she is to avoid excessive weight gain. Thus adolescent girls need to pay special attention to being physically active and selecting foods of high nutrient density in order to meet their nutrient needs without exceeding their energy needs.

Vitamins

The RDA for most vitamins increase during the teen years (see the RDA table on the inside front cover). Several of the vitamin recommendations for adolescents are similar to those for adults, including the new DRI for vitamin D. During puberty, both the activation of vitamin D and the absorption of calcium are enhanced, thus supporting the intense skeletal growth of the adolescent years without additional vitamin D.

On average, adolescent males have higher energy needs than females of the same age.

Minerals

Many of the recommended intakes for minerals increase during adolescence. The minerals of greatest concern during adolescence are calcium and iron.

Calcium. Adolescence is a crucial time for achieving maximum bone density, and the requirement for calcium reaches its peak during these years.[4] Unfortunately, many adolescents have calcium intakes below current recommendations.[5] Low calcium intakes during the adolescent growth spurt, especially if paired with physical inactivity, may compromise the development of peak bone density. In contrast, increasing milk products in the diet to meet calcium recommendations greatly increases bone density.[6] The attainment of maximal bone mass is considered the best protection against age-related bone loss and fractures.[7] Teenage girls are most vulnerable, for their milk—and therefore calcium—intakes begin to decline at the time when their calcium needs are greatest. Furthermore, women have much greater bone losses than men in later life. In addition to dietary calcium, sports activities during adolescence build strong bones.[8]

Calcium DRI during adolescence:
1300 mg/day (9 to 18 yr).

Iron. Iron status is most affected during four times of life: in infancy, during growth spurts, during the female reproductive years, and in pregnancy. All adolescents are in growth spurts, and half of them are females in their reproductive years. During periods of growth, blood volume and muscle mass increase, thus increasing the need for iron for hemoglobin and myoglobin synthesis. During the reproductive years, blood losses through menstruation increase iron needs.

Iron RDA during adolescence:
  12 mg/day (11 to 18 yr, males).
  15 mg/day (11 to 18 yr, females).

The need for iron during adolescence differs for males and females. Iron needs increase in females as they start to menstruate and in males as their lean body mass develops. Iron intakes often fail to keep pace with increasing needs, especially for females, who typically consume fewer iron-rich foods such as meat, and fewer total kcalories than males. For females, the RDA rises at adolescence and remains high into late adulthood. For males, the RDA returns to preadolescent values in early adulthood.

The sex difference in iron sufficiency is generally true of other nutrients as well. It primarily reflects the males' consumption of larger quantities of food. Females' lower food energy intakes make more evident the less-than-perfect nutrient density of their diets.

Nutrient Supplements for Adolescents

No across-the-board recommendations for supplements are made for teenagers. Like adults, teenagers can get the nutrients they need from foods. Approximately 10 percent of adolescents, however, report taking vitamin and mineral supplements. They are more likely to take single-nutrient supplements than multivitamin and mineral preparations, and their choices are not likely to match their specific nutrient inadequacies. If a teenager takes a supplement, the guidelines are the same as for others: The supplement should provide near RDA amounts of the whole spectrum of vitamins and minerals. Iron status should be monitored by a health care professional, and supplementation or dietary modification instituted as needed.

Nutrition Assessment of Adolescents

Rates and patterns of adolescent growth exhibit such wide variations that standards used for children must be abandoned when the signs of puberty begin to appear. Age in years indicates little about development. To record developmental changes during puberty, health care professionals use standard rating scales based on stages of adolescent development.

Anthropometric Data

One way to monitor teenage growth is to compare height and weight with previous measures. As growth spurts take place, wide variations in weight are seen and can be expected to smooth out as the steadier time of adulthood approaches. Obesity or underweight can set in, and adult norms are useful as rough indicators; deviations, however, are not unusual, and great significance need not be attached to them unless they persist. To evaluate underweight and overweight in adolescents, health care professionals may use measures of body mass index (BMI).[9] Figure 7-2 presents BMI values for adolescents.

Figure 7-2 Recommended BMI Cutoff Values for Adolescents

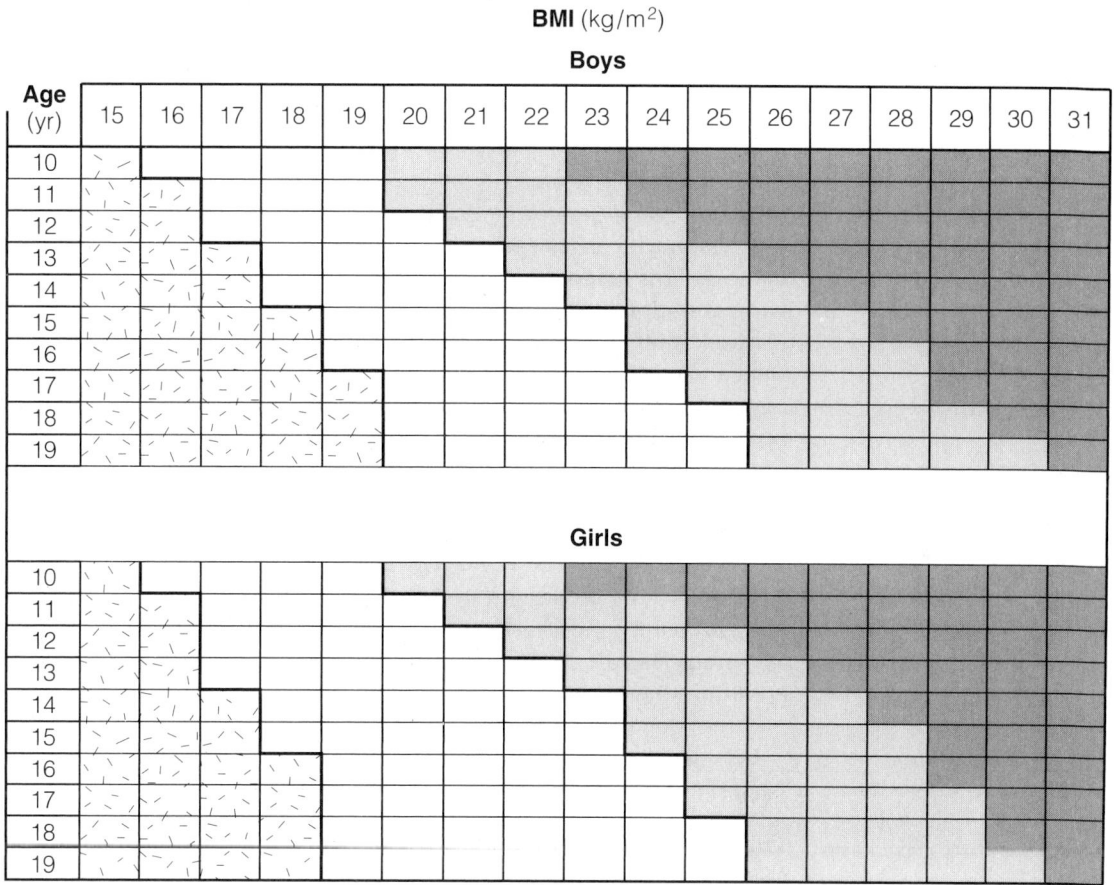

BMI (kg/m²)

Boys

Girls

Key:

- ▦ Underweight (≤15 percentile)
- ☐ Acceptable weight
- ☐ Overweight (≥85 percentile)
- ▨ Obese (≥95 percentile)

Note: Values for BMI are rounded to the nearest integer for simplicity; any loss of accuracy is unlikely to be clinically significant.

Source: Adapted from A. Must, G. E. Dallal, and W. H. Dietz, Reference data for obesity: 85th and 95th percentiles of body mass index (wt/ht²) and triceps skinfold thickness, *American Journal of Clinical Nutrition* 53 (1991): 839–846.

Historical Information

Overweight adolescents need further assessment to determine the presence of other health risks. A health history will reveal:

- ■ A family history of heart disease, diabetes, or obesity.
- ■ Elevated blood pressure.
- ■ Elevated blood cholesterol.

■ Increase in BMI by two points or more over the previous year.

■ Unusual concerns about weight.

If any of these factors are present, then the adolescent may need further in-depth medical assessment or intervention.

Biochemical Analyses

When assessing the nutrient status of adolescents, the health care professional must consider the many developmental changes and individual variations that occur during this time. For example, the stage of sexual maturation affects assessment of zinc status; adolescents who have reached adult sexual maturity have higher serum zinc concentrations than adolescents of the same age at a less mature stage. For another example, the increase in red blood cell production and hemoglobin concentration during adolescent growth affects assessment of iron status. The changes in serum ferritin and transferrin that are typical of dwindling iron stores also appear during adolescent growth, but do not necessarily indicate an iron deficiency.[10]

Few adolescents pay attention to the adequacy of their nutrient stores, but fortunately, most teens in the United States are nourished reasonably well. They do have other concerns, however, that may draw their attention to nutrition.

Nutrition-Related Concerns during Adolescence

Teens' efforts to define themselves are a major element in their lives. Many teens are overly concerned with their own body images and self-images. Some become concerned about their weight, especially if they are overweight or even if they only think they are overweight. Some become involved with substance abuse that can influence their lives for years to come. Some become concerned about their complexions, especially if they have acne (or even one pimple). The following sections deal with these concerns and their relationships with nutrition.

Adolescent Obesity

For boys, overweight was defined as BMI:
 ≥23.0 for 12 to 14 years.
 ≥24.3 for 15 to 17 years.
 ≥25.8 for 18 to 19 years.
For girls, overweight was defined as BMI:
 ≥23.4 for 12 to 14 years.
 ≥24.8 for 15 to 17 years.
 ≥25.7 for 18 to 19 years.

The insidious problem of obesity becomes ever more apparent in adolescence and often continues into adulthood. One in every five teens is overweight.[11] The problem is most evident in females, especially in African-American females.[12] Without intervention, overweight teens will face numerous physical and socioeconomic consequences for years to come. The consequences of obesity are so dramatic and our society's attitude toward obese people is so negative that even teens of normal weight perceive a need to lose weight. When taken to extremes, restrictive diets bring dramatic physical consequences of their own, as Focal Point 7A explains.

Perceptions of Body Size. About one-third of high school students perceive themselves as overweight.[13] Such weight concerns, more common among teenage girls than teenage boys, rarely reflect reality.[14] Quite often, adolescents who are concerned about their weight and are "dieting" to lose weight are not even overweight—they only think they are.[15] This perception of overweight and the need to diet is more apparent in girls than in boys, and more evident in white girls than in black girls. When normal-weight adolescent girls with similar BMI values are shown profiles of various body sizes, black girls select a larger ideal size than white girls do.[16] Such a cultural difference in accepting heavier weights may partially explain the higher prevalence of obesity among black women.

At any given time, two out of five high school students are trying to lose weight.[17] Interestingly, their approaches to weight loss differ depending on their sex. Girls typically restrict their dietary intakes to lose weight; consequently, they are prone to suffer from eating disorders and nutrient inadequacies. Boys, in contrast, typically increase their physical activities to lose weight, which helps explain why they have fewer nutrition problems. Adolescents need to learn to appreciate appropriate body weights and adopt eating and activity behaviors that will allow them to stop obsessing with weight control and "dieting."

Physical Maturation. The rate of physical maturation during the adolescent growth spurt influences the development of obesity. In general, children who mature rapidly in adolescence are more obese than those who mature slowly.[18] The reverse also seems to be true: Obesity influences the rate of physical maturation. Childhood obesity is associated with the early onset of menarche in young girls, for example.

Excessive weight gains during this time of rapid growth may have long-reaching health consequences. Fat tends to be deposited abdominally during adolescence, even with normal weight gains.[19] With excessive gains, central obesity becomes more problematic and the development of chronic diseases more likely. On a positive note, the health profile (blood cholesterol and blood pressure, for example) of adolescents with central obesity improves dramatically with weight loss.[20]

Physical and Health Consequences. The *immediate* health hazards of obesity in adolescence are few and rare.[21] Without intervention, however, most obese adolescents will become obese adults, and obesity is a key risk factor in the development of several chronic diseases, including cardiovascular disease, some cancers, and diabetes. Overweight in adolescence does have long-term health effects of its own, however, independent of adult weight.[22] Men who were overweight 50-plus years earlier as teens but of normal weight as adults were more than twice as likely to die from heart disease as men who were of normal weight as teens. The risks of gout and colon cancer were also increased for men; the risk of arthritis was increased for women.

Social and Economic Consequences. The immediate consequences of obesity are found in the social and economic obstacles overweight teenagers face. Our society places such enormous value on thinness that many overweight people face prejudice and discrimination; overweight people are judged on their appearance more than on their character. Socially, overweight people are stereotyped as lazy, stupid, and lacking in self-control. Young adults who are overweight are less likely to be married than those who are not overweight.[23] Overweight teens are less likely to be admitted to college or hired for employment, even when they are qualified. This is especially true for women. Psychologically, fat people may suffer embarrassment when others treat them with hostility and contempt, and some have even learned to view their own bodies as grotesque and loathsome.[24] Parents and friends may chide them about their weight and lack of discipline to resolve the problem. All this hurts self-esteem. An adolescent experiencing the many social stigmas of obesity may readily adopt a negative self-image that will remain throughout adulthood.

Inactivity. Chapter 6 already discussed the relationships between television watching and childhood obesity. The sedentary lifestyle of television watching may continue through the teen years, further contributing to obesity. The teen years also bring a new behavior that fosters inactivity—driving. Children who ride bicycles or walk for transportation may become less active as they begin to drive cars or ride with friends who are driving. Physical activity is an important component of adolescent obesity treatment.[25]

Lack of physical activity fosters obesity.

Weight Loss Plans. Prevention is always preferable to intervention. In both situations, though, the focus needs to be on establishing lifelong patterns of good nutrition and regular physical activity. Such habits, formed early in life, serve a person well through old age.

Other habits developed during the teen years also have long-term effects on health. Whether the habits benefit or harm health depends on the choices made.

Behaviors Incompatible with Health

Physical maturity and growing independence present adolescents with new choices to make. The consequences of those choices will influence their nutritional health both today and throughout life. Some teenagers begin using tobacco, alcohol, and drugs; others wisely refrain. Information about the use of these substances is presented here because most people are first exposed to them during adolescence, but it actually applies to people of all ages.

Smoking. A folder written by young people for young people describes how smokers try to explain their choice to smoke:

- ■ I'm young now. I can quit later.
- ■ I don't inhale. Smoking can't hurt me.
- ■ Smoking makes me look grown-up.
- ■ I smoke filter cigarettes. Filters protect me.
- ■ My parents (friends) smoke. Why shouldn't I?
- ■ If I don't spend the money on cigarettes, I'll spend it on something else.
- ■ It keeps me from biting my nails, putting on weight, or being mad or bored or hurt or unhappy.

All these reasons are unrealistic, but the new smoker believes them.

A person whose mind is open to using tobacco begins by trying it once. That one time may be followed by another. In a short while, the person becomes a regular user, because tobacco's active ingredient, nicotine, is a powerfully addictive drug. Tobacco companies know that once they've got people started, they've got them hooked. They must attract children and young teenagers in order to replace the more than 2 million adult smokers who die each year worldwide from lung cancer and other smoking-related diseases.

Statistics on smoking are disturbing. Each day, 5000 youngsters in a hurry to grow up light up for the first time—some of them are only seven or eight years old. Two out of three high school students have tried smoking, and one in seven smokes regularly.[26] Approximately 90 percent of all adult smokers began smoking before the age of 18.[27] If the current rate of tobacco use by young people continues, the surgeon

nicotine (NICK-oh-teen): an addictive drug present in tobacco.

general warns that 5 million of today's children will die of smoking-related illnesses in their later years.

Cigarette smoking is a pervasive health problem causing thousands of people to suffer from cancer and diseases of the cardiovascular, digestive, and respiratory systems. These effects are beyond the scope of nutrition, but smoking cigarettes does influence hunger, body weight, and nutrient status.

Smoking a cigarette eases feelings of hunger. When smokers receive a hunger signal, they can quiet it with cigarettes instead of food. Such behavior ignores natural body signals and postpones energy and nutrient intake. Studies on rats confirm that nicotine reduces food intake and increases the rate of energy expenditure, causing weight loss.[28]

Indeed, smokers tend to weigh less than nonsmokers and to gain weight when they stop smoking.[29] Weight gain is often a concern for people contemplating giving up cigarettes. They should know that the average person who quits smoking gains less than 10 pounds. Smokers wanting to quit need to prepare for this possibility and adjust their diet and activity habits so as to maintain weight during and after quitting. Smoking cessation programs need to include strategies for weight management.

Nutrient intakes of smokers and nonsmokers differ. Smokers tend to have lower intakes of dietary fiber, vitamin A, beta-carotene, folate, and vitamin C.[30] The association between smoking and low vitamin intakes may be noteworthy, considering the altered metabolism of vitamin C in smokers and the protective effect of vitamin A and beta-carotene against lung cancer.

The vitamin C RDA for people who regularly smoke cigarettes is 100 mg/day. The Canadian RNI suggests smokers should add 50% to the vitamin C recommendation.

Research shows that compared to nonsmokers, smokers require almost twice as much vitamin C to maintain steady body pools. Oxidants in cigarette smoke accelerate vitamin C metabolism and deplete body stores of this antioxidant; this depletion is even evident to some degree in nonsmokers who are exposed to passive smoke.[31]

Beta-carotene enhances the immune response and protects against some cancer activity.[32] Specifically, the risk of lung cancer is greatest for smokers who have the lowest intakes of carotene. Of course, such evidence should not be misinterpreted. It does not mean that as long as people eat their carrots, they can safely use tobacco. Nor does it mean that beta-carotene supplements would be beneficial. Smokers are 10 times more likely to get lung cancer than nonsmokers. Both smokers and nonsmokers can, however, reduce their cancer risks by eating fruits and vegetables rich in carotene.

Smokeless Tobacco. Nationwide, one in ten high school students reports having used smokeless tobacco products.[33] Like cigarettes, smokeless tobacco use is linked to many health problems, from minor mouth sores to tumors in the nasal cavities, cheeks, gums, and throat. The risk of mouth and throat cancers is even greater than for smoking tobacco. Other drawbacks to tobacco chewing and snuff dipping include bad breath, stained teeth, and blunted senses of smell and taste. Tobacco chewing also damages the gums, tooth surfaces, and jaw bones, making it likely that users will lose their teeth in later life.

Alcohol. Sooner or later all teenagers face the decision whether to drink alcohol. The law forbids the sale of alcohol to people under 21, but most adolescents who seek alcohol can obtain it. Four out of five high school students have had at least one alcoholic beverage; about half drink regularly; and one in three students drinks heavily (defined as five or more drinks on at least one occasion in the previous month).[34]

Drinkers give lots of reasons why they drink alcohol: to celebrate, to unwind, because they like the taste of alcoholic beverages, because it's the custom. Young people drink because peer pressure encourages it and parents permit it. Teenagers think drinking makes them look grown up or provides a way of rebelling against authority. They drink when they are upset, when they are alone, when they are bored, and when they want to feel high. The advertisers of alcohol also promote drinking, of course, when they appeal to many aspects of people's self-image.

For whatever reasons people drink, they derive drug effects by doing so. Like other addictive drugs, alcohol produces euphoria, changes mood, relieves pain, and releases tension.

Drinking may encourage people to relax and be social, but people will benefit more from learning how to relax in the face of life's pressures without alcohol. Dependency on any drug severely impairs development and deserves attention, but is beyond the scope of this text.

With regard to nutrition, alcohol affects nutrition status in numerous ways. To briefly summarize, alcohol provides energy but no nutrients and it can displace nutritious foods from the diet. Alcohol alters nutrient absorption and metabolism, so that imbalances develop. People who cannot keep their alcohol use moderate must abstain to maintain their health. Appendix F lists resources for people with alcohol-related problems.

Focal Point 9 features the relationships between alcohol and nutrition.

Marijuana. Many adolescents find that alcohol and marijuana serve similar purposes, and the pattern of substance abuse indicates parallel consumption, not a displacement of one by the other. One out of every three high school students reports having at least tried marijuana.[35] The chemicals in marijuana that produce euphoria are all related to the chemical delta-9-tetrahydrocannabinol, or THC for short, which the plant produces as it grows. When inhaled by smoking, the active chemicals are rapidly and almost completely (90 percent) absorbed from the lungs. They then travel in the blood to the various body tissues that process them—the brain, liver, and kidneys.

The THC from a single marijuana cigarette can linger in the body's fat for a month before being excreted in the urine. Drug tests can detect trace amounts of THC for as long as six months afterward.

Marijuana is unique among drugs in that it seems to enhance the enjoyment of eating, especially of sweets, a phenomenon commonly known as "the munchies." Why or how this effect occurs is not known;

Table 7-1 Consequences of Marijuana Use by Adolescents

Physical Consequences	*Behavioral Consequences*
• Accelerated heart rate. • Elevated blood pressure. • Bronchodilation, broncho-constriction, and eventually airway obstruction. • Cancerlike changes in the lungs. • Diminished sperm motility, decreased sperm count, and decreased testosterone levels.	• Euphoria, relaxation, disinhibition. • Poor coordination; poor ability to judge time, speed, and distance; poor ability to track a moving object; and slow reaction time. • Accidental injuries and deaths. • Impaired learning and memory.

Source: Adapted from Committee on Adolescents, Committee on Substance Abuse, American Academy of Pediatrics, Marijuana: A continuing concern for pediatricians, *Pediatrics* 88 (1991): 1070–1072.

it may be a social effect induced by suggestibility, or perhaps the drug stimulates appetite. Whatever the reason, adolescents who smoke marijuana report eating more snack foods and less fruit, vegetables, and milk than nonusers.

Adolescents may think that because they usually smoke fewer marijuana cigarettes in a day than they would tobacco cigarettes, their lungs will incur fewer harmful, long-term effects. This is not true (see Table 7-1). One marijuana cigarette is as bad for the body as four or five tobacco cigarettes, because people who smoke marijuana inhale more smoke and hold it in their lungs longer. People who regularly smoke several marijuana cigarettes a day face the same risk of lung cancer as people who smoke a pack of tobacco cigarettes a day.[36]

Cocaine. One in 20 high school seniors reports having used cocaine at least once.[37] Cocaine stimulates the nervous system and brings on the stress response—constricted blood vessels, raised blood pressure, widened pupils of the eyes, and increased body temperature. It also drives away feelings of fatigue. Cocaine occasionally causes immediate death—usually by heart attack, stroke, or seizure in an already damaged body system.

Weight loss is common, and cocaine abusers often develop eating disorders. Notably, the craving for cocaine replaces hunger; rats given unlimited cocaine will choose it over food until they starve to death.[38] Thus, unlike marijuana use, cocaine use has major nutritional consequences.

Cocaine abuse is a problem for over 1 million people in the United States. Between 60 and 80 percent of users surveyed believe themselves to be addicted, unable to turn the drug down if offered, and unable to limit their abuse of cocaine. A person who is addicted to cocaine loses the ability to work, to play, to keep a job, or to stop using the drug.

Drug Abuse, in General. The effects of other addictive drugs vary in degree but are similar in kind to those caused by cocaine. Drug abusers may become malnourished when:

- They spend money for drugs that could be spent on food.
- They lose interest in food during "highs."
- They use drugs that depress the appetite.
- Their lifestyle fails to promote good eating habits.
- They use intravenous (IV) drugs. They may contract AIDS, hepatitis, or other infectious diseases, which increase their nutrient needs. Hepatitis also causes taste changes and loss of appetite.
- Medicines used to treat drug abuse alter nutrition status.

During withdrawal from drugs, an important part of treatment is to identify and correct these nutrition problems.

Acne

Compared with obesity and drug abuse, acne may seem trivial, but it is extremely important to a teen with a pimply face. No one knows why some people get acne while others do not, but heredity is one factor. In addition, the hormones of adolescence increase the activity of the sebaceous glands in the skin (see Figure 7-3).

acne: a chronic inflammation of the skin's follicles and oil-producing glands that involves the accumulation of sebum inside the ducts that surround hairs; usually associated with the maturation of young adults.

Many a teenager would pay dearly for a sure way to prevent acne. Some teenagers hope that if they stop eating certain foods and drinking certain beverages, they can prevent acne. Among foods charged with aggravating acne are chocolate, cola beverages, fatty or greasy foods, milk, nuts, sugar, and foods or salt containing iodine. None of these foods has proven to worsen acne. Teenagers could omit these foods, with the exception of milk and foods or salt containing iodine, without harming their health, of course, and to do so might actually benefit their nutrition status.

Not all dietary practices attempting to treat acne are innocuous. Misinformed teenagers taking vitamin A supplements will obtain no relief from acne, but may induce vitamin A toxicity. Vitamin A is needed for the health of the skin, as are other nutrients, but excesses are toxic.

Dermatologists sometimes treat cystic acne with prescription drugs that are related to vitamin A but are chemically different from over-the-counter supplements. Excellent results have been obtained with the topical use of a salve containing the vitamin A relative tretinoin (retinoic acid); the trade name is Retin-A. Retin-A reddens the skin and causes tenderness and peeling, but the subsequent healing produces healthier skin.

Another vitamin A relative, isotretinoin (13-*cis*-retinoic acid), is available in soft gelatin capsules for oral administration under the trade name Accutane. Its mechanism of action is not completely understood, although it seems to inhibit sebaceous gland function. Accutane is responsible for a number of adverse effects, including birth defects in infants of women using it during pregnancy. As mentioned in Chapter 3,

Figure 7-3 Acne

The skin's natural oil, sebum, is made in deep sebaceous glands and is supposed to flow out through the tiny ducts around the hairs to the skin surface. In acne, the oily secretions exceed the skin's clearance capacity.

Inside each of the ducts is a skinlike lining that regularly sheds cells. These cells mix with the oil and then are pushed to the surface of the skin. When acne develops, they stick together, forming a plug that blocks the duct. The duct enlarges, allowing oil and the skin surface bacteria to leak into the surrounding skin. The oil and bacterial enzymes are irritating and cause redness, swelling, pus formation—and the beginning of a whitehead, or pimple. A cyst may form, or the skin may open above the plug, revealing an accumulation of dark skin pigments just below the surface—a blackhead.

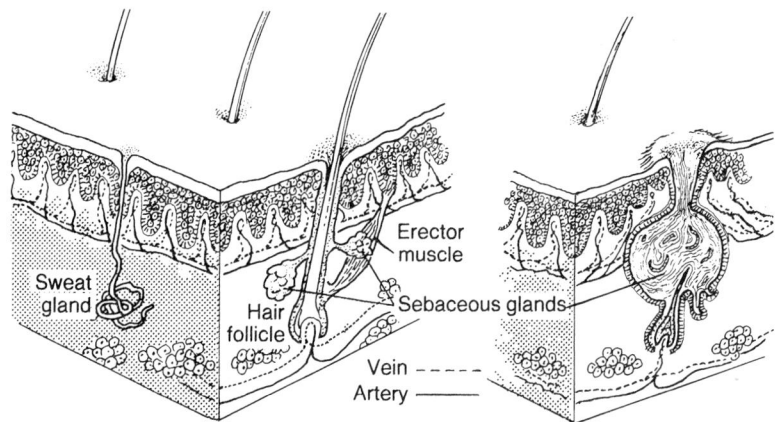

Source: Adapted from *Acne*, a pamphlet (May 1980) available from the National institute of Allergy and Infectious Diseases, Bethesda, MD 20205, NIH publication no. 80-188; *Stubborn and Vexing, That's Acne*, a pamphlet (May 1980) available from the Food and Drug Administration, 5600 Fishers Lane, Rockville, MD 20857, HHS publication no. (FDA) 80-3107.

an effective form of contraception is advised beginning one month before, and continuing until one month after, Accutane's use. Other side effects include raised blood lipids; nose bleeds; and dry skin, eyes, and mouth.

Because stress worsens acne, and because adolescents easily develop guilt feelings over what they eat, perhaps the best nutrition advice is to dispense with food-related guilt. The dietary guidelines for the acne-plagued teenager, as for anyone, are to eat nutritious foods in abundance and to enjoy sweet treats and high-fat snacks in moderation.

The many physical and emotional changes of adolescence combined with a variety of social and behavioral choices make attaining optimal nutrition during the teen years a challenge. The remainder of this chapter focuses on healthful food choices and physical activities.

Food Choices and Fitness during Adolescence

Teenagers like the freedom to come and go as they choose and eat what they want when they have time. With a multitude of after-school, social, and job activities, they almost inevitably fall into irregular eating habits. At any given time on any given day, a teenager may be skipping a meal, eating a snack, preparing a meal, or consuming food prepared by a parent or restaurant.

Diets Tailored to Individual Preferences

To receive all the nutrients necessary to support the rapid growth of adolescence, a teenager needs to select several servings from each of the five food groups daily. The number of daily servings recommended during adolescence represents the middle to upper end of the ranges suggested by the USDA Food Guide Pyramid (see Table 7-2). In keeping with the differences in energy requirements between males and females, the middle of the range of daily servings provides about the right amount of food energy for teenage girls, and the upper end meets the needs of most teenage boys. Teenagers who overindulge may exceed their energy needs and become overweight, whereas those who restrict their intakes may fall short in meeting their nutrient needs.

Eating the recommended quantities of grains and meats is rarely a problem for most teens. Meeting the goals for fruits, vegetables, and milk, however, seems to be more difficult. Researchers report that 1 in 3 adolescents did not eat any fruit and 1 in 15 did not eat any vegetable during the three days before the survey.[39] Milk consumption was also low, causing calcium intakes to fall below recommendations.

✶ Healthy People 2000

Increase calcium intake so that at least 50% of youth aged 12 through 24 years consume three or more servings of calcium-rich foods daily.

Vegetarian Diets during Adolescence. Few studies have focused on adolescent vegetarians, but at least one study reports that vegetarian children and adolescents who eat a balanced diet grow at least as tall as their meat-eating peers.[40] This finding confirms the position of the American Dietetic Association stating that well-planned vegetarian diets can support growth and maintain health.[41] Adolescents who are interested in following a vegetarian diet may need some guidance to ensure an adequate intake of all nutrients, most notably vitamin B_{12}, vitamin D, calcium, and iron.

Table 7-2 Daily Food Guide for Adolescents

	Number of Servings	
Food Group	Girls	Boys
Breads/cereals	9	11
Vegetables	4	5
Fruits	3	4
Meats/meat alternatives[a]	6	7
Milk/milk alternatives	3	3
kCalories	2200	2800

Note: Figure 1-2 in Chapter 1 provides a detailed summary of the Food Guide Pyramid. The 2200-kcalorie plan assumes a total of 73 grams fat and allows 12 teaspoons added sugar. The 2800-kcalorie plan assumes a total of 93 grams fat and allows 18 teaspoons added sugar.

[a]Meat group amounts are in total ounces.

Nutritious snacks play an important role in an active teen's diet.

For perspective, an 8-ounce cup of drip-brewed coffee contains about 184 mg of caffeine. A pharmacologically active dose of caffeine is defined as 200 mg. Table 6-4 on p. 236 lists caffeine contents of selected foods, beverages, and drugs.

Snacks. Snacks are a part of today's lifestyle. They fit in nicely when socializing, studying, working, playing, and relaxing. Because they are a part of the daily food intake, snacks need to provide nutrients. The teenage snacker who finds plenty of nutritious foods around the house is well provided for.

Snacks typically provide at least a fourth of an average teenager's daily food energy intake. Most often, favorite snacks are high in fat and low in iron, calcium, vitamin A, vitamin C, and folate.[42] Most adolescents need to eat a greater variety of foods to obtain these nutrients. To get extra iron, a teenager might snack on bran muffins, hard-boiled eggs, or crackers with peanut butter. Calcium is available from yogurt, cheese, and ice cream (if the kcalories can be afforded). Good vitamin-rich snack foods include carrot and broccoli pieces, fresh fruits, and fruit juices. Table 6-6 on p. 247 shows how to combine foods from different food groups to create healthy snacks. Vending machines rarely offer nutrient-dense options, and nutrition information alone does not convince people to make healthy choices.[43]

Beverage Choices. Teenagers frequently drink soft drinks with lunch, supper, and snacks. About the only time they select fruit juices is at breakfast. When they drink milk, they are more likely to consume it with a meal (especially breakfast) than as a snack. Because of their greater food intakes, boys are more likely to drink enough milk to meet their calcium needs, whereas girls typically fall short of calcium recommendations.

For teenagers who can afford the kcalories and are meeting their calcium needs, soft drinks are an acceptable part of the diet. Soft drinks may present a problem, however, when caffeine intake becomes excessive. Caffeine is a stimulant added during the manufacture of many soft drinks; on the average, caffeine-containing soft drinks deliver between

30 and 55 milligrams of caffeine per 12-ounce can.[44] Caffeine increases the respiration rate, heart rate, blood pressure, and secretion of stress and other hormones. Caffeine seems to be relatively harmless, however, when used in moderate doses (the equivalent of fewer than, say, four 12-ounce cola beverages a day). In greater amounts, it can cause the symptoms associated with anxiety—sweating, tenseness, and the inability to concentrate.

Eating Away from Home. Inevitably, adolescents do much of their eating away from home, and their nutritional welfare is enhanced or hindered by the choices they make. Fast foods need not mean total abandonment of nutrition. They can have an acceptable place in a teenager's diet, provided that they do not dominate the diet and that their short-comings are compensated for at other meals. An occasional fast-food meal has little impact on a teenager's overall nutrition status. Teenagers who consume fast-food meals daily, however, need to vary their menu selections and pay close attention to the nutrient contributions of their other meals.

A lunch of a hamburger, a chocolate shake, and french fries supplies substantial quantities of many nutrients at a kcalorie cost of 820, an energy intake many adolescents can afford (see Table 7-3). When they eat this sort of lunch, teens can adjust their breakfast and dinner choices to include fruits and vegetables for vitamin A, vitamin C, folate, and fiber and lean meats for iron and zinc.

Fitness during Adolescence

During the teen years, outdoor play shifts to sports for many, but is replaced by sedentary indoor life for many more. Ideally, physical activity should become a habit as regular as eating, practiced in a way that will bring optimal benefits. Regular, frequent activity does much to promote both physical and mental health, prevent obesity, and retard or reverse the degenerative conditions of later life, such as osteoporosis and heart disease.

Table 7-3 Selected Nutrients in a Hamburger, Chocolate Shake, and Small Serving of French Fries

	Percent RDA	
Nutrient	*Male[a]*	*Female[a]*
Energy	30	35
Protein	47	64
Fat[b]	26	31
Calcium[c]	35	35
Iron	30	24
Zinc	16	20
Vitamin A	9	11
Thiamin	41	48
Riboflavin	42	49
Niacin	35	41
Folate	18	20
Vitamin C	7	7
Sodium[b]	34	34

[a]RDA for a 15- to 18-year-old, moderately active person of average height and weight.
[b]Daily Values used for fat and sodium.
[c]DRI used for calcium.

★ Healthy People 2000

Increase to at least 50% the proportion of overweight people aged 12 years and older who have adopted sound dietary practices combined with regular physical activity to attain an appropriate body weight.

Diets for Physically Active People. No one diet best supports physical performance. Active people who choose foods within the framework of the diet-planning principles presented in Chapter 1 can design many excellent diets.

A variety of foods is the best source of nutrients for athletes.

In planning a diet for physically active people, remember that the body is depleted more rapidly of water than any other nutrient. A diet to support fitness must provide water, energy, and all the essential nutrients.

Water recommendation: 1.0 to 1.5 mL/kcal expended. Note: 1 mL = 0.03 fluid oz. Easy estimation: ≈1/2 c/ 100 kcal.

Water. Even casual exercisers must attend conscientiously to their fluid needs. During physical activity the body loses water as sweat and as vapor in exhaled breath. When losses are significant, dehydration becomes a threat. Dehydration's first symptom is fatigue; a water loss of even 1 to 2 percent of body weight can reduce a person's capacity to do muscular work. When water loss reaches about 7 percent of body weight, a person is likely to collapse.[45] To prevent dehydration and the fatigue that accompanies it, physically active people need to drink plenty of liquids before, during, and after exercise.

Energy. Active people need to eat both for nutrient adequacy and for energy. Their diet should be high in carbohydrate, low in fat, and adequate in protein.

Carbohydrate. Glucose, stored in the liver and muscles as glycogen, is vital to physical activity. During exertion, the liver releases its glucose into the bloodstream. The muscles use both this glucose and glucose from their own private glycogen stores to fuel their work. Glycogen supplies can easily support everyday activities, but are limited. The more glycogen the muscles store, the longer the stores will last during physical activity. To ensure adequate glucose to fuel prolonged physical activity, the diet needs to be high in carbohydrate (60 percent of total kcalories or more).

Carbohydrate recommendation for athletes in heavy training: 8 g/kg body weight.

On two occasions, the active person's regular high-carbohydrate, fiber-rich diet may require temporary adjustment. Both these exceptions involve training for competition rather than fitness. During intensive training, energy needs are high—so high that they may exceed the person's capacity to eat enough food. At that point, added sugar and fat may be needed. The person can add concentrated carbohydrate foods such as dried fruits, sweet potatoes, nectars, and even high-fat foods such as avocados, nuts, cookies, and ice cream. Still, a nutrient-rich diet remains central for adequacy's sake. Although vital, energy alone is not enough to support performance. The other special occasion is the pregame meal, when fiber-rich, bulky foods are best avoided. (The pregame meal is discussed in a later section.)

Protein recommendation for athletes in heavy training: 1.0 to 1.5 g/kg body weight.

Protein. In addition to needing carbohydrate and some fat (for the energy they provide), physically active people need protein. How much of what kinds of foods supply enough protein to meet their needs? Meats and milk products are rich protein sources, but to recommend that active people emphasize these foods would be narrow advice. Legumes, grains, and vegetables also provide protein, with the added benefits of abundant carbohydrate and little fat.

A Performance Diet Example. A person who engages in vigorous physical activity on a daily basis can easily require more than 3000 kcalories per day. To meet this need, the person can choose a variety of nutrient-dense foods. Figure 7-4 shows one example of a day's meals that provide about 3300 kcalories. These meals supply over 130 grams protein, more than enough protein for most athletes. The meals also provide almost 550 grams carbohydrate, or over 60 percent of total kcalories. Athletes who train exhaustively for endurance events may want to aim for somewhat higher carbohydrate intakes. Beyond these specific concerns of total energy, protein, and carbohydrate, the diet most beneficial to ath-

Figure 7-4 An Athlete's Meal Selections

Total kcal: 3300
    63% kcal from carbohydrate
    22% kcal from fat
    15% kcal from protein

All vitamin and mineral intakes exceed the RDA for both men and women.

Breakfast:
1 c shredded wheat.
1 c 1% low-fat milk.
1 small banana.
2 slices whole-wheat toast.
4 tsp. jelly.
1 1/2 c orange juice.

Lunch:
2 turkey sandwiches.
1 1/2 c 1% low-fat milk.
Large bunch of grapes.

Snack:
3 c plain popcorn.
A smoothie made from:
    1 1/2 c apple juice.
    1 1/2 frozen banana.

Dinner:
Salad: 1 c spinach, carrots and mushrooms.
    1/2 c garbanzo beans.
    1 tbs sunflower seeds.
    1 tbs ranch salad dressing.
1 c spaghetti with meat sauce.
1 c green beans.
1 corn on the cob.
2 slices Italian bread.
4 tsp butter.
1 piece angel food cake.
1 1/4 c fresh strawberries.
1 tbs whipping cream.
1 c 1% low-fat milk.

letic performance is remarkably similar to the diet recommended for most people.

Meals before and after Competition. No single food improves speed, strength, or skill in competitive events, although some *kinds* of foods do support performance better than others. Still, a competitor may eat a particular food before or after an event for psychological reasons. One eats a steak the night before wrestling, another takes some honey five minutes after diving. Such practices may not be beneficial, but as long as they remain harmless, they should be respected. The following paragraphs describe pregame and postgame meals that may be beneficial to an athlete's performance.

Pregame Meals. Science indicates that the pregame meal or snack should include plenty of fluids and be easy to digest. It should provide between 300 and 800 kcalories, primarily from carbohydrate-rich foods that are familiar and well tolerated by the athlete. The meal should end three to five hours before competition to allow plenty of time for the stomach to empty before exertion.

High-carbohydrate, liquid pregame meal ideas:
- *Apple juice, frozen banana, and cinnamon.*
- *Papaya juice, frozen strawberries, and mint.*
- *Nonfat milk, frozen banana, and vanilla.*

Breads, potatoes, pasta, and fruit juices—that is, carbohydrate-rich foods low in fat, protein, and fiber—form the basis of the best pregame meal. Bulky, fiber-rich foods such as raw vegetables or high-bran cereals, although usually desirable, are best avoided just before competition. Fiber in the digestive tract attracts water out of the blood and can cause stomach discomfort during performance. Liquid meals are easy to digest, and many such meals are commercially available. Alternatively, athletes can mix nonfat milk or juice, frozen fruits, and flavorings in a blender.

Postgame Meals. Eating high-carbohydrate foods *after* physical activity enhances glycogen storage. Since people are usually not hungry immediately following physical activity, carbohydrate-containing beverages such as sports drinks or fruit juices may be preferred. If an active person does feel hungry after an event, then foods high in carbohydrate and low in protein, fat, and fiber are the ones to choose—the same ones recommended prior to competition. Foods high in protein and fat should be avoided during the first few hours after activity, as these foods may suppress hunger and thus limit carbohydrate intake.[46]

Supplements for Athletes. Research confirms that nutrient supplements do not enhance the performance of well-nourished people.[47] Studies do consistently find, though, that deficiencies of vitamins and minerals impede performance. In general, active people who eat enough nutrient-dense foods to meet energy needs also meet their vitamin and mineral needs. After all, active people eat more food; it stands to reason that with the right choices, they'll get more nutrients.

Some athletes mistakenly believe that taking vitamin or mineral supplements directly before competition will enhance performance. These beliefs are contrary to scientific findings. Most vitamins and minerals function as small parts of larger working units. After entering the blood, they have to wait for the tissues to combine them with their appropriate other parts so they can do their work. This takes time—hours or days. Vitamins or minerals taken right before an event are useless for improving performance, even when the person actually is suffering deficiencies of them.

In general, then, active people need no vitamins or minerals in supplement form. Iron may be an exception to this rule, however, as the following paragraphs explain.

For perfect functioning, every nutrient is needed.

Iron. Physical activity can affect iron status in several ways. For one thing, iron losses in sweat can contribute to iron deficiency. For another, red blood cell destruction can lead to iron loss; blood cells are squashed when body tissues (such as the soles of the feet) make high-impact contact with an unyielding surface (such as the ground). Perhaps more significant than these losses are deficits caused by poor iron absorption in some athletes and the high demands of muscles for the iron-containing molecules of the mitochondria and the muscle protein myoglobin. In addition, physical activity may cause small blood losses through the digestive tract, at least in some athletes.

Iron Deficiency. Iron deficiency affects more young women than men, and physically active people are no exception. Habitually low intakes of iron-rich foods and high iron losses through menstruation, as well as through the other routes mentioned, can cause iron deficiency in physically active young women. Even short-term moderate aerobic activity has been shown to compromise women's iron status.[48]

aerobic activity: exercise that requires the use of oxygen in the production of energy.

Iron Deficiency Anemia. Evidence is equivocal as to whether marginal iron deficiency impairs physical performance. Iron deficiency anemia, however, clearly does dramatically impair physical performance. The hemoglobin in the red blood cells is indispensable for delivering oxygen for the processes that use it in releasing energy from nutrients. Without adequate oxygen, an active person cannot perform aerobic activities and tires easily.

Sports Anemia. Early in training, athletes may develop low blood hemoglobin for a while. This condition, sometimes called "sports anemia," is not a true iron deficiency condition. Sports anemia does not respond to iron supplementation, as iron deficiency anemia does. Strenuous aerobic activity promotes destruction of the more fragile, older red blood cells, reducing the blood's iron concentration temporarily. Strenuous activity also expands the blood plasma volume, thereby reducing the red blood cell count per unit of blood, but the red blood cells do not diminish in size or number as in anemia, so oxygen-carrying capacity is not hindered. Most researchers view sports anemia as an *adaptive*, temporary response to endurance training. The increase in plasma volume may even help in training; it dilutes the blood and augments the amount of blood leaving the heart per minute.

sports anemia: a transient condition of low hemoglobin in the blood, associated with the early stages of sports training or other strenuous activity.

Iron Recommendations for Athletes. The best strategy concerning iron depends on the individual. Many menstruating women probably border on iron deficiency even without the iron losses incurred by physical activity. Active teens of both sexes also have high iron needs because they are growing. Especially for women and teens, then, prescribed supplements may be needed to correct deficiencies of iron, but nutrition assessment should guide decisions on supplementation.

Anabolic Steroids. Athletes gravitate to promises that they can enhance their performance by taking pills, powders, and potions. Unfortunately, they often hear such promises from their coaches and peers, who advise them to use nutrient supplements, take drugs, or follow procedures that claim to deliver results without effort. When such aids are harmless, they are only a waste of money; when they impair performance or harm health, they waste athletic potential and cost lives. Among the most dangerous and illegal practices is the taking of anabolic steroids. These drugs are derived from the male sex hormone testosterone, which promotes the development of male characteristics and lean body mass. Athletes take steroids to stimulate muscle bulking.

To athletes struggling to excel, the promise of bigger, stronger muscles than training alone can produce has been tempting. Athletes who normally would not be able to break into the elite ranks can, with the help of steroids, suddenly compete with true champions. Especially in professional circles, where monetary rewards for excellence are sky-high, steroid use is common despite its illegality and side effects.

The American Academy of Pediatrics and the American College of Sports Medicine condemn athletes' use of anabolic steroids, and the International Olympic Committee bans their use. These authorities cite the known toxic side effects and maintain that taking these drugs is a form of cheating. Other athletes are put in the difficult position of either conceding an unfair advantage to competitors who use steroids, or taking them and accepting the risk of harmful side effects. Young athletes should not be forced to make such a choice; unfortunately, 1 in 50 high school students reports using steroids without a physician's prescription.[49]

The list of adverse reactions to steroids is long and continues to grow amid only a slight decline in use of the drugs. Table 7-4 lists the side effects of steroids.

The price for the potential competitive edge that steroids confer is high—sometimes it is life itself. The commissioner of the Food and Drug Administration (FDA) warns that steroids are not simple pills that build bigger muscles, but complex chemicals to which the body reacts in many ways, particularly when bodybuilders and other athletes take large amounts. The safest, most effective way to build muscle has always been through hard training and a sound diet, and—despite popular misconceptions—it still is.

Some manufacturers push specific herbs as legal substitutes for steroid drugs. They falsely claim that these herbs contain hormones, enhance the body's hormonal activity, or both. In some cases, an herb may

"I lied to a lot of people for a lot of years when I said I didn't use steroids. . . . If you're on steroids or human growth hormone, stop. I should have." Lyle Alzado, former NFL football player who died of cancer in May 1992; as quoted by S. Smith, I'm sick and I'm scared, *Sports Illustrated*, July 8, 1991, pp. 21–25.

Table 7-4 Anabolic Steroids: Side Effects and Adverse Reactions

---

Mind

- Extreme aggression with hostility ("steroid rage"); mood swings; anxiety; dizziness; drowsiness; unpredictability; insomnia; psychotic depression; personality changes; suicidal thoughts

---

Face and Hair

- Swollen appearance; greasy skin; severe, scarring acne; mouth and tongue soreness; yellowing of whites of eyes (jaundice)
- In females, male pattern hair loss and increased growth of face and body hair

---

Voice

- In females, irreversible deepening of voice

Chest

- In males, breathing difficulty; breathing stoppage; breast development
- In females, breast atrophy

Heart

- Heart disease; elevated or reduced heart rate; heart attack; stroke; hypertension; increased LDL; drastic reduction in HDL

Abdominal Organs

- Nausea; vomiting; bloody diarrhea; pain; edema; liver tumors (possibly cancerous); liver damage, disease, or rupture leading to fatal liver failure (peliosis hepatitis)[a]; kidney stones and damage; gallstones; frequent urination; possible rupture of aneurysm or hemorrhage

---

Blood

- Blood clots; high risk of blood poisoning; those who share needles risk contracting HIV (the AIDS virus) or other disease-causing organisms; septic shock (from injections)

Reproductive System

- In males, permanent shrinkage of testes; prostate enlargement with increased risk of cancer; sexual dysfunction; loss of fertility; excessive and painful erections
- In females, loss of menstruation and fertility; permanent enlargement of external genitalia; fetal damage, if pregnant

(continued)

Table 7-4 Anabolic Steroids: Side Effects and Adverse Reactions (*cont.*)

Muscles, Bones, and Connective Tissues

- Increased susceptibility to injury with delayed recovery times; cramps; tremors; seizurelike movements; injury at injection site
- In adolescents, failure to grow to normal height

Other

- Fatigue; increased risk of cancer

[a]In peliosis hepatitis, excess of bile causes destruction of liver cells. Blood pools form, and liver failure causes death.

Source: K. L. Ropp, No-win situation for athletes, *FDA Consumer*, December 1992, pp. 8–12; National Academy of Sports Medicine policy statement and position paper: Anabolic androgenic steroids, growth hormones, stimulants, ergogenics, and drug use in sports, an appendix to B. Goldman and R. Klatz, *Death in the Locker Room II: Drugs and Sports* (Chicago: Elite Sports Medicine Publications, 1992), pp. 328–373.

contain plant sterols, such as oryzanol, but these compounds are poorly absorbed. Even if absorption occurs, the body cannot convert herbal compounds to anabolic steroids. Ironically, injections of oryzanol in rats alter metabolic pathways to favor *catabolism.*[50] None of these products has any proven anabolic steroid activity, and none enhances muscle strength. Some contain natural toxins, but toxins are harmful, even when they are "natural."

Many young teens who become interested in nutrition make choices that will benefit their fitness. Others may become unhealthily obsessed with weight control and develop eating disorders (see Focal Point 7A). Eating disorders illustrate how a desired end (in this case, weight loss) can alter a person's eating habits and food choices, and therefore nutrition status. Teenagers who are willing to go to considerable trouble and expense to achieve their perception of physical fitness or beauty risk jeopardizing their health.

Yet the teen years can be an exciting time of preparation for adulthood. The lifestyle choices people make as teenagers can facilitate the transition into adulthood. Habits established during the teen years lay a foundation for health throughout the rest of life.

CHAPTER SEVEN NOTES

1. M. E. Herman-Giddens and coauthors, Secondary sexual characteristics and menses in young girls seen in office practice: A study from the Pediatric Research in Office Settings Network, *Pediatrics* 99 (1997): 505–512.

2. L. Kann and coauthors, Youth risk behavior surveillance—United States, 1993, *Journal of School Health* 65 (1995): 163–171.

3. J. L. Cameron, Nutritional determinants of puberty, *Nutrition Reviews* 54 (1996): S17–S22.

4. S. M. Ott, Bone density in adolescents, *New England Journal of Medicine* 325 (1991): 1646–1647.

5. S. I. Barr, Associations of social and demographic variables with calcium intakes of high school students, *Journal of the American*

Dietetic Association 94 (1994): 260–266, 269.

6. G. M. Chan, K. Hoffman, and M. McMurry, Effects of dairy products on bone and body composition in pubertal girls, *Journal of Pediatrics* 126 (1995): 551–556.

7. V. Matkovic, Diet, genetics, and peak bone mass of adolescent girls, *Nutrition Today*, March–April 1991, pp. 21–24; R. B. Sandler and coauthors, Postmenopausal bone density and milk consumption in childhood and adolescence, *American Journal of Clinical Nutrition* 42 (1985): 270–274.

8. A. M. Fehily and coauthors, Factors affecting bone density in young adults, *American Journal of Clinical Nutrition* 56 (1992): 579–586.

9. J. H. Himes and W. H. Dietz, Guidelines for overweight in adolescent preventive services: Recommendations from an expert committee, *American Journal of Clinical Nutrition* 59 (1994): 307–316.

10. R. Anttila and M. A. Siimes, Serum transferrin and ferritin in pubertal boys: Relations to body growth, pubertal stage, erythropoiesis, and iron deficiency, *American Journal of Clinical Nutrition* 63 (1996): 179–183.

11. Prevalence of overweight among adolescents, United States, 1988–1991, *Morbidity and Mortality Weekly Report* 43 (1994): 819–821.

12. M. G. Melnyk and E. Weinstein, Preventing obesity in black women by targeting adolescents: A literature review, *Journal of the American Dietetic Association* 94 (1994): 536–540; T. A. Wadden and coauthors, Obesity in black adolescent girls: A controlled clinical trial of treatment by diet, behavior modification, and parental support, *Pediatrics* 85 (1990): 345–352.

13. Kann and coauthors, 1995.

14. R. C. Casper, Fear of fatness and anorexia nervosa in children, in *Child Health, Nutrition, and Physical Activity*, eds. L. W. Y. Cheung and J. B. Richmond (Champaign, Ill.: Human Kinetics, 1995), pp. 211–234.

15. L. Emmons, Predisposing factors differentiating adolescent dieters and nondieters, *Journal of the American Dietetic Association* 94 (1994): 725–728, 731.

16. K. Parnell and coauthors, Black and white adolescent females' perceptions of ideal body size, *Journal of School Health* 66 (1996): 112–118.

17. Kann and coauthors, 1995.

18. F. J. van Lenthe, H. C. G. Kemper, and W. van Mechelen, Rapid maturation in adolescence results in greater obesity in adulthood: The Amsterdam Growth and Health Study, *American Journal of Clinical Nutrition* 64 (1996): 18–24.

19. M. L. Hediger and coauthors, One-year changes in weight and fatness in girls during late adolescence, *Pediatrics* 96 (1995): 253–258.

20. M. Wabitsch and coauthors, Body-fat distribution and changes in the atherogenic risk-factor profile in obese adolescent girls during weight reduction, *American Journal of Clinical Nutrition* 60 (1994): 54–60.

21. A. Must, Morbidity and mortality associated with elevated body weight in children and adolescents, *American Journal of Clinical Nutrition* 63 (1996): 445S–447S.

22. A. Must and coauthors, Long-term morbidity and mortality of overweight adolescents: A follow-up of the Harvard Growth Study of 1922 to 1935, *New England Journal of Medicine* 327 (1992): 1350–1355.

23. S. L. Gortmaker and coauthors, Social and economic consequences of overweight in adolescence and young adulthood, *New England Journal of Medicine* 329 (1993): 1008–1012.

24. A. J. Stunkard and T. A. Wadden, Psychological aspects of human obesity, in *Obesity*, eds. P. Björntorp and B. N. Brodoff (Philadelphia: Lippincott, 1992), pp. 352–360.

25. L. H. Epstein, K. J. Coleman, and M. D. Myers, Exercise in treating obesity in children and adolescents, *Medicine and Science in Sports and Exercise* 28 (1996): 428–435.

26. Kann and coauthors, 1995.

27. Tobacco use and usual source of cigarettes among high school students—United States, 1995, *Journal of the American Medical Association* 276 (1996): 184–185.

28. S. R. Schwid, M. D. Hirvonen, and R. E. Keesey, Nicotine effects on body weight: A regulatory perspective, *American Journal of Clinical Nutrition* 55 (1992): 878–884.

29. D. F. Williamson and coauthors, Smoking cessation and severity of weight gain in a national cohort, *New England Journal of Medicine* 324 (1991): 739–745.

30. T. A. B. Sanders and coauthors, Essential fatty acids, plasma cholesterol, and fat-soluble vitamins in subjects with age-related maculopathy and matched control subjects, *American Journal of Clinical Nutrition* 57 (1993): 428–433; A. F. Subar, L. C. Harlan, and M. E. Matt-

son, Food and nutrient intake differences between smokers and non-smokers in the U.S., *American Journal of Public Health* 80 (1990): 1323–1329.

31. D. L. Tribble, L. J. Giuliano, and S. P. Fortmann, Reduced plasma ascorbic acid concentrations in nonsmokers regularly exposed to environmental tobacco smoke, *American Journal of Clinical Nutrition* 58 (1993): 886–890.

32. G. van Poppel, S. Spanhaak, and T. Ockhuizen, Effect of β-carotene on immunological indexes in healthy male smokers, *American Journal of Clinical Nutrition* 57 (1993): 402–407; T. V. Ringer and coauthors, Beta-carotene's effects on serum lipoproteins and immunologic indices in humans, *American Journal of Clinical Nutrition* 53 (1991): 688–694.

33. Kann and coauthors, 1995.

34. Kann and coauthors, 1995.

35. Kann and coauthors, 1995.

36. T. C. Wu and coauthors, Pulmonary hazards of smoking marijuana as compared with tobacco, *New England Journal of Medicine* 318 (1988): 347–351.

37. American Council on Science and Health, *Cocaine: Facts and Dangers* (New York: American Council on Science and Health, 1990); Kann and coauthors, 1995.

38. M. A. Bozarth and R. A. Wise, Toxicity associated with long-term intravenous heroin and cocaine self-administration in the rat, *Journal of the American Medical Association* 254 (1985): 81–85.

39. Life Sciences Research Office, Federation of American Societies for Experimental Biology, Executive summary from the third report on nutrition monitoring in the United States, *Journal of Nutrition* 126 (1996): 1907S–1936S; Committee on the Nutritional Status of Pregnancy and Lactation, *Nutrition during Pregnancy* (Washington, D.C.: National Academy Press, 1990), pp. 299–317.

40. J. Sabaté and coauthors, Attained height of lacto-ovo vegetarian children and adolescents, *European Journal of Clinical Nutrition* 45 (1991): 51–58.

41. Position of The American Dietetic Association: Vegetarian diets, *Journal of the American Dietetic Association* 93 (1993): 1317–1319.

42. J. G. Dausch and coauthors, Correlates of high-fat/low-nutrient-dense snack consumption among adolescents: Results from two national health surveys, *American Journal of Health Promotion* 10 (1995): 85–88.

43. S. M. Hoerr and V. A. Louden, Can nutrition information increase sales of healthful vended snacks? *Journal of School Health* 63 (1993): 386–390.

44. A. N. Grand and L. N. Bell, Caffeine content of fountain and private-label store brand carbonated beverages, *Journal of the American Dietetic Association* 97 (1997): 179–182; International Food Information Council, Caffeine and health: Clarifying the controversies, *IFIC Review*, May 1993.

45. J. E. Greenleaf, Problem: Thirst, drinking behavior, and involuntary dehydration, *Medicine and Science in Sports and Exercise* 24 (1992): 645–656.

46. E. F. Coyle, Timing and method of increased carbohydrate intake to cope with heavy training, competition, and recovery, in *Food, Nutrition, and Sports Performance: An International Scientific Consensus*, eds. C. Williams and J. T. Devlin (London: E and F Spon, 1992), pp. 37–61.

47. A. Singh, F. M. Moses, and P. A. Deuster, Chronic multivitamin-mineral supplementation does not enhance physical performance, *Medicine and Science in Sports and Exercise* 24 (1992): 726–732.

48. R. M. Lyle and coauthors, Iron status in exercising women: The effect of oral iron therapy vs. increased consumption of muscle foods, *American Journal of Clinical Nutrition* 56 (1992): 1049–1055.

49. Kann and coauthors, 1995.

50. K. B. Wheeler and K. A. Garleb, Gamma oryzanol—Plant sterol supplementation: Metabolic, endocrine, and physiologic effects, *International Journal of Sport Nutrition* 1 (1991): 170–177.

Eating Disorders

An estimated 2 million people in the United States, primarily girls and young women, suffer from the eating disorders anorexia nervosa and bulimia nervosa. Many more suffer from un-specified eating disorders—conditions that do not meet the strict criteria for anorexia nervosa or bulimia nervosa, but that still imperil a per-son's well-being. Two-thirds of adolescent girls and one-third of adolescent boys are dissatisfied with their body weight.[1] Other characteristics of disordered eating such as restrained eating, binge eating, purging, fear of fatness, and distortion of body image are also common, especially among young middle-class girls.[2] In most other soci-eties, these behaviors and attitudes are much less prevalent.

Why do so many young people in our society suffer from eating disorders? Excessive pressure to be thin is at least partly to blame. By making thinness the ideal, society pushes people to view the healthy body of normal weight as too fat. Healthy people then take unhealthy actions to lose weight. Severe restriction of food intake may create intense hunger that leads to binges. Re-search confirms this theory, showing that un-healthy or dangerous diets often precede binge eating in adolescent girls.[3] Energy restriction fol-lowed by bingeing can set in motion a pattern of weight cycling, which may make weight loss and maintenance more difficult over time.[4]

People who attempt extreme weight loss are dissatisfied with their bodies to begin with; they may also be depressed or suffer social anxiety. As weight loss becomes more and more difficult, psychological problems worsen and the likeli-hood of developing full-blown eating disorders intensifies. Athletes are particularly likely to de-velop eating disorders.

The Female Athlete Triad

At age 14, Suzanne was a top contender for a spot on the state gymnastics team. Each day her coach reminded team members that they must weigh no more than a few ounces above their assigned weights in order to qualify for competition. The coach chastised gymnasts who gained weight, and Suzanne was terrified of being singled out. She was convinced that the less she weighed, the better she would per-form. She weighed herself several times a day to confirm that she had not exceeded her 80-

---

GLOSSARY

anorexia nervosa: an eating disorder character-ized by a refusal to maintain a minimally normal body weight, self-starvation to the extreme, and a disturbed perception of body weight and shape, seen (usually) in teenage girls and young women (*anorexia* = without appetite; *nervosa* = of nervous origin).

bulimia (byoo-LEEM-ee-uh) **nervosa:** recurring episodes of binge eating combined with a morbid fear of becoming fat, usually followed by self-induced vomiting or purging.

cathartic: a strong laxative.

eating disorder: a disturbance in eating behavior that jeopardizes a person's physical or psychological health.

emetic (em-ETT-ic): an agent that causes vom-iting.

female athlete triad: a potentially fatal triad of medical problems: disordered eating, amenorrhea, and osteoporosis.

unspecified eating disorders: eating disorders that do not meet the criteria for specific eating dis-orders previously defined.

pound limit. Driven to excel in her sport, Suzanne kept her weight down by eating very little and training very hard. Unlike many of her friends at school, Suzanne never began menstruating. A few months before her fifteenth birthday, Suzanne's coach dropped her back to the second-level team. Suzanne blamed her poor performance on a slow-healing stress fracture. Mentally stressed and physically exhausted, she quit gymnastics and began overeating between her periods of self-starvation. Suzanne had developed the dangerous combination of problems—disordered eating, amenorrhea, and osteoporosis—collectively known as the female athlete triad. (see Figure FP7A-1).

Disordered Eating. At least part of the reason many athletic women engage in self-destructive eating behaviors is that they and their coaches have adopted unsuitable weight standards. An athlete's body must be heavier for height than a nonathlete's body because the athlete's bones and muscles are denser. When athletes consult standard weight-for-height tables and see that they are on the heavy side, they may mistakenly believe that they are too fat. Weight standards that may be appropriate for others are inappropriate for athletes. Measures such as fatfold measures yield more useful information about body composition.

Many young athletes severely restrict energy intakes to improve performance, enhance the aesthetic appeal of their performance, or meet the weight guidelines of their specific sports.[5] They fail to realize that the loss of lean tissue that accompanies energy restriction actually impairs their physical performance. The increasing incidence of abnormal eating habits among athletes, especially young women, is causing concern. Male athletes, especially wrestlers and gymnasts, are affected by these disorders as well, but research shows that females are most vulnerable.[6] Risk factors for this triad include the following:

■ Young age (adolescence).

■ Pressure to excel at a chosen physical activity.

■ Focus on achieving or maintaining an "ideal" body weight or body fat percentage.

■ Participation in endurance sports or competitions that judge performance on aesthetic appeal such as gymnastics, figure skating, or dance.

■ Dieting at an early age.

■ Unsupervised dieting.

Amenorrhea. The prevalence of amenorrhea among premenopausal women in the United States is about 2 to 5 percent overall, but among female athletes it may be as high as 66 percent.[7] Contrary to previous notions, amenorrhea is *not* a normal adaptation to strenuous physical train-

Figure FPA-1 The Female Athlete Triad

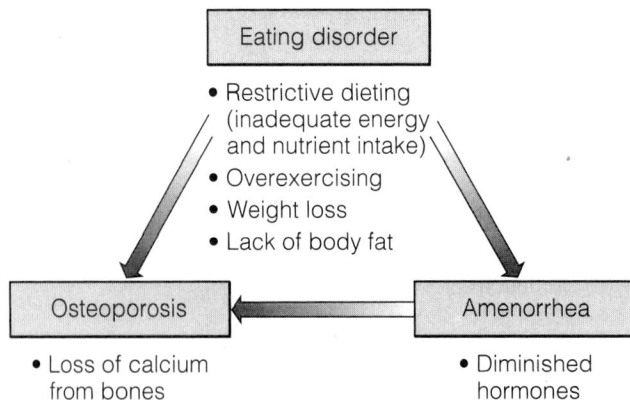

ing: It is a symptom of something going wrong.[8] Amenorrhea is characterized by low blood estrogen, infertility, and often bone mineral losses. Some research seems to indicate that low body fat contributes to amenorrhea;[9] other studies indicate that the percentage of body fat is not associated with normal menstruation in athletes.[10] However amenorrhea develops, amenorrheic athletes are more likely to suffer bone loss than other adolescents.

Osteoporosis. In general, weight-bearing physical activity, dietary calcium, and the hormone estrogen protect against the bone loss of osteoporosis, but in women with disordered eating and amenorrhea, strenuous activity may impair bone health. Vigorous training combined with low food energy intakes and other life stresses seems to trigger amenorrhea and promote bone loss. Low estrogen leads to diminished bone mass and increased bone fragility. Many amenorrheic athletes have decreased bone density, similar to that of 50- to 60-year-old women when they should have dense, strong bones. Amenorrheic athletes should be encouraged to consume at least 1500 milligrams of calcium each day, to eat nutrient-dense foods, and to obtain enough energy to cover the energy expended in physical activity. Future research will focus on the question of hormone replacement therapy for these women.

Preventing Eating Disorders in Athletes. To prevent eating disorders in athletes and dancers, both the performers and their coaches must be educated about links between inappropriate body weight ideals, improper weight loss techniques, eating disorder development, adequate nutrition, and safe weight control methods. Coaches and dance instructors should never encourage unhealthy weight loss to qualify for competition or to conform with distorted artistic ideals. Frequent weighings can push young people who are striving to lose weight into a cycle of starving to confront the scale, then bingeing uncontrollably afterward. The erosion of self-esteem that accompanies these events can interfere with the normal psychological develop-

Table FP7A-1 Tips for Combatting Eating Disorders

General Guidelines

- Never restrict food servings to below the numbers suggested for adequacy by the Food Guide Pyramid.

- Eat frequently. People often do not eat frequent meals because of time constraints, but eating can be incorporated into other activities, such as snacking while studying or commuting. The person who eats frequently never gets so hungry as to allow hunger to dictate food choices.

- If not at a healthy weight, establish a reasonable weight goal based on a healthy body composition.

- Allow a reasonable time to achieve the goal. A reasonable loss of excess fat can be achieved at the rate of about 1 percent of body weight per week.

- Establish a weight maintenance support group with people who share interests.

Specific Guidelines for Athletes and Dancers

- Remember that eating disorders impair physical performance. Seek confidential help in obtaining treatment if needed.

- Restrict weight loss activities to the off season.

- Focus on proper nutrition as an important facet of your training, as important as proper technique.

ment of the teen years and set the stage for serious problems later on.

Table FP7A-1 provides some suggestions to help athletes and dancers protect themselves against developing eating disorders. The next sections describe eating disorders that anyone, athlete or nonathlete, may experience.

Anorexia Nervosa

Julie is 18 years old and a superachiever in school. She watches her diet with great care, and she exercises daily, maintaining a heroic schedule of self-discipline. She is thin, but she is determined to lose more weight. She is 5 feet, 6 inches tall and weighs 85 pounds. She has anorexia nervosa.

Julie is unaware that she is undernourished, and she sees no need to obtain treatment. She stopped menstruating several months ago and is moody and chronically depressed. She insists that she is too fat, although her eyes are sunk in deep hollows in her face. Julie denies that she is ever tired, although she is close to physical exhaustion, and she no longer sleeps easily. Her family is concerned, and though reluctant to push her, they have finally insisted that she see a psychiatrist. Julie's psychiatrist has diagnosed anorexia nervosa (see Table FP7A-2) and has prescribed group therapy as a start, but warns that if Julie does not begin to gain weight soon, she will need to be hospitalized.

Characteristics of Anorexia Nervosa. As mentioned in the introduction, most anorexia nervosa victims are females; males account for only about 1 in 20 cases.[11] Most come from middle- or upper-class families. Family patterns often include parents who oppose one another's authority, and who vacillate between defending and condemning the anorexic child's behavior, confusing the child and disrupting normal parental control.[12] In the extreme, parents may even be abusive. Julie is a perfectionist, and her parents expect perfection. She cannot get in touch with her own identity and rejects food as a means of gaining control.

How can a person as thin as Julie continue to starve herself? Julie uses tremendous discipline against her hunger to strictly limit her portions of low-kcalorie foods. She will deny her hunger, saying she is full after having eaten only a half-dozen carrot sticks. She can recite the kcalorie contents of dozens of foods and the kcalorie costs of as many exercises. If she feels that she has gained an ounce of weight, she runs or jumps rope until she is sure she has exercised it off. If she fears that the food she has eaten outweighs the exercise, she takes laxatives to hasten the passage of food from her system. She is desperately hungry. In fact she is starving, but she

Table FP7A-2 Criteria for Diagnosis of Anorexia Nervosa

A person with anorexia nervosa demonstrates the following:

A. Refusal to maintain body weight at or above a minimal normal weight for age and height, e.g., weight loss leading to maintenance of body weight less than 85% of that expected; or failure to make expected weight gain during period of growth, leading to body weight less than 85% of that expected.

B. Intense fear of gaining weight or becoming fat, even though underweight.

C. Disturbance in the way in which one's body weight or shape is experienced; undue influence of body weight or shape on self-evaluation, or denial of the seriousness of the current low body weight.

D. In females past puberty, amenorrhea, i.e., the absence of at least three consecutive menstrual cycles. (A woman is considered to have amenorrhea if her periods occur only following hormone, e.g., estrogen, administration.)

Two types:

- **Restricting type:** during the episode of anorexia nervosa, the person does not regularly engage in binge eating or purging behavior (i.e., self-induced vomiting or the misuse of laxatives, diuretics, or enemas).

- **Binge eating/purging type**: during the episode of anorexia nervosa, the person regularly engages in binge eating or purging behavior (i.e., self-induced vomiting or the misuse of laxatives, diuretics, or enemas).

Source: Reprinted with permission from the *Diagnostic and Statistical Manual of Mental Disorders*, 4th ed. (Washington, D.C.: American Psychiatric Association, 1994). Copyright 1994 American Psychiatric Association.

doesn't eat because her need for self-control dominates.

Many people, on learning of this disorder, say they wish they had "a touch" of it to get thin. They mistakenly think that people with anorexia nervosa feel no hunger. They also fail to recognize the pain of the associated psychological and physical trauma.

Women with anorexia nervosa see themselves as fat, even when they are dangerously underweight.

Central to the diagnosis of anorexia nervosa is a distorted body image that overestimates body fatness. When Julie looks at herself in the mirror, she sees a "fat" 85-pound body. The more Julie overestimates her body size, the more resistant she is to treatment, and the more unwilling to examine her values and misconceptions. Vitamin deficiencies are known to affect brain functioning and judgment in this way, causing lethargy, confusion, and delirium.

Physical Effects of Anorexia Nervosa. Anorexia nervosa damages the body much as starvation does. Victims are dying to be thin—quite literally. In young people, growth ceases and normal development falters. They lose so much lean tissue that basal metabolic rate slows, an effect that may remain even after treatment and regain of weight.[13] In addition, the heart pumps ineffi-

ciently and irregularly, the heart muscle becomes weak and thin, the chambers diminish in size, and the blood pressure falls. Electrolytes that help to regulate heartbeat become unbalanced. Many deaths in people with anorexia nervosa are due to heart failure.

Starvation brings other physical consequences as well: impaired immune response, anemia, and a loss of digestive functions that worsens malnutrition. Digestive functioning becomes sluggish, the stomach empties slowly, and the lining of the intestinal tract shrinks. The ailing digestive tract fails to provide sufficient digestion of any food the victim may eat. The pancreas slows its production of digestive enzymes. The person may suffer from diarrhea, further worsening malnutrition.

Treatment of Anorexia Nervosa. Treatment of anorexia nervosa requires a multidisciplinary approach that addresses two sets of issues and behaviors: those relating to food and weight, and those involving relationships with oneself and others.[14] Teams of physicians, nurses, psychiatrists, family therapists, and dietitians work together to treat people with anorexia nervosa. Appropriate diet is crucial and must be tailored individually to each client's needs. Seldom are clients willing to eat for themselves, but if they are, chances are they can recover without other interventions.

High-risk clients may require hospitalization and may need to be force-fed by tube at first to prevent death. This step causes psychological trauma. Drugs are commonly prescribed, but to date, they play a limited role in treatment.[15]

Denial runs high among those with anorexia nervosa. Few seek treatment on their own. Almost half of the women who are treated can maintain their body weight within 15 percent of a healthy weight; at that weight, many of them begin menstruating again. The other half have poor or fair outcomes of treatment, and two-thirds of those treated continue a mental battle with recurring morbid thoughts about food and body weight.[16] Many relapse into abnormal eating behaviors to some extent. About 5 percent die during treatment, 1 percent by suicide.

Before drawing conclusions about someone

who is extremely thin or who eats very little, remember that diagnosis of anorexia nervosa requires professional assessment. People who are seeking help with anorexia nervosa, either for themselves or for others, can speak with school guidance counselors or call the National Association of Anorexia Nervosa and Associated Disorders for information.* Appendix F provides addresses for other resources as well.

Bulimia Nervosa

Beth is a charming, intelligent, 16-year-old cheerleader of normal weight who thinks constantly about food. She alternately starves herself and then secretly binges; when she has eaten too much, she vomits. Most readers recognize these symptoms as those of bulimia nervosa.

Bulimia nervosa is distinct from anorexia nervosa and is more prevalent, although the true incidence is difficult to establish because bulimia nervosa is not as physically apparent.[17] More men suffer from bulimia nervosa than from anorexia nervosa, but bulimia nervosa is still more common in women than in men. The secretive nature of bulimic behaviors makes recognition of the problem difficult, but once it is recognized, diagnosis is based on the criteria listed in Table FP7A-3.

Families of bulimic adolescents have often established unusually close emotional ties between members, and they may be overcontrolling and intermeshed in ways that stifle individual growth and development.[18] Any changes in the family structure often meet with resistance, even when the changes would greatly benefit the person with bulimia nervosa. In addition, the family may have "secrets" that are hidden from outsiders; many bulimic women report having been sexually or physically abused by family members or family friends.

Characteristics of Bulimia Nervosa. Like the typical person with bulimia nervosa, Beth is single, female, and white. She is well educated and close to her ideal body weight, although her weight fluctuates over a range of 10 pounds or so every few weeks. As a cheerleader, she prefers to

Table FP7A-3   Criteria for Diagnosis of Bulimia Nervosa

A person with bulimia nervosa demonstrates the following:

A. Recurrent episodes of binge eating. An episode of binge eating is characterized by both of the following:

   1. eating, in a discrete period of time (e.g., within any two-hour period), an amount of food that is definitely larger than most people would eat during a similar period of time and under similar circumstances, and,

   2. a sense of lack of control over eating during the episode (e.g., a feeling that one cannot stop eating or control what or how much one is eating).

B. Recurrent inappropriate compensatory behavior in order to prevent weight gain, such as self-induced vomiting; misuse of laxatives, diuretics, enemas, or other medications; fasting; or excessive exercise.

C. Binge eating and inappropriate compensatory behaviors that both occur, on average, at least twice a week for three months.

D. Self-evaluation unduly influenced by body shape and weight.

E. The disturbance does not occur exclusively during episodes of anorexia nervosa.

Two types:

- **Purging type:** the person regularly engages in self-induced vomiting or the misuse of laxatives, diuretics, or enemas.

- **Nonpurging type:** the person uses other inappropriate compensatory behaviors, such as fasting or excessive exercise, but does not regularly engage in self-induced vomiting or the misuse of laxatives, diuretics, or enemas.

Source: Reprinted with permission from the *Diagnostic and Statistical Manual of Mental Disorders*, 4th ed. (Washington, D.C.: American Psychiatric Association, 1994). Copyright 1994 American Psychiatric Association.

weigh less than the weight that her body maintains naturally.

Beth seldom lets her bulimia nervosa interfere with her school or other activities, although

*Telephone (708) 831-3438.

a third of all victims do so. From early childhood she has been a high achiever, emotionally dependent on her parents. Beth cycles on and off crash diets. She feels anxious at social events and cannot easily establish close relationships. She is usually depressed, is often impulsive, and has low self-esteem.

A bulimic binge is unlike normal eating, and the food is not consumed for its nutritional value. During a binge, Beth's eating is accelerated by her hunger from previous food restriction. She may take in anywhere from 1000 to 10,000 kcalories of easy-to-eat, high-fat, and, especially, high-carbohydrate foods. Typically, she chooses cookies, cakes, and ice cream; and she eats the entire bag of cookies, the whole cake, and every spoonful in a carton of ice cream.

The binge is a compulsion and usually occurs in several stages: "anticipation and planning, anxiety, urgency to begin, rapid and uncontrollable consumption of food, relief and relaxation, disappointment, and finally shame or disgust." Then, to purge the food from her body, Beth may use a cathartic—a strong laxative that can injure the lower intestinal tract. Or she may induce vomiting, using an emetic—a drug intended as first aid for poisoning.

On first glance, purging seems to offer a quick and easy solution to the problems of unwanted kcalories and body weight. Many people perceive such behavior as neutral or even positive, when, in fact, bingeing and purging have serious physical consequences.[19] Fluid and electrolyte imbalances caused by vomiting or diarrhea can lead to metabolic alkalosis, a condition characterized by apathy, confusion, and muscle spasms. Vomiting causes irritation and infection of the pharynx, esophagus, and salivary glands; erosion of the teeth; and dental caries. The esophagus may rupture or tear, as may the stomach. Overuse of emetics depletes potassium concentrations and can lead to death by heart failure.[20]

Unlike Julie, Beth is aware that her behavior is abnormal, and she is deeply ashamed of it. She wants to recover, and this makes recovery more likely for her than for Julie, who clings to denial. Feeling inadequate ("I can't even control my eating"), Beth tends to be passive and to look to others, primarily men, for confirmation of her

Table FP7A-4 Diet Strategies for Combatting Bulimia Nervosa

- Avoid finger foods; eat foods that require the use of utensils.
- Enhance satiety by eating warm foods.
- Include vegetables, salad, and/or fruit at meals to prolong eating time.
- Choose whole-grain and high-fiber breads and cereals to maximize bulk.
- Eat a well-balanced diet and meals consisting of a variety of foods.
- Use foods that are naturally divided into portions, such as potatoes (rather than rice or pasta); 4- and 8-oz containers of yogurt, ice cream, or cottage cheese; precut steak or chicken parts; and frozen entrees.
- Include foods containing ample complex carbohydrates (for satiety) and some fat (to slow gastric emptying).
- Eat meals and snacks sitting down.
- Plan meals and snacks, and record plans in a food diary prior to eating.

sense of worth. When she experiences rejection, either in reality or in her imagination, her bulimia nervosa becomes worse. If Beth's depression deepens, she may seek solace in drug or alcohol abuse or other addictive behaviors. As many as 50 percent of women with bulimia nervosa are alcohol dependent.[21]

Treatment of Bulimia Nervosa. To help clients gain control over food and establish regular eating patterns requires adherence to a structured eating plan. Restrictive weight loss dieting almost always precedes and may even trigger bingeing. Weight maintenance, rather than cyclic gains and losses, is the treatment goal. Many a victim has taken a major step toward recovery by learning to eat enough food to satisfy hunger needs (at least 1600 kcalories a day). Table FP7A-4 offers some ways to begin correcting the eating problems of bulimia nervosa.

Anorexia nervosa and bulimia nervosa are distinct eating disorders, yet they sometimes overlap in important ways. Anorexia victims may purge, and victims of both conditions share an

A person may consume up to 10,000 kcalories during an eating binge.

overconcern with body weight and the tendency to drastically undereat. The two disorders can also appear in the same person, or one can lead to the other.

Societal pressure to be thin is no doubt a factor in the development of eating disorders. Most experts agree, however, that the disorders are multifactorial: sociocultural, psychological, and perhaps neurochemical.

Perhaps a young person's best defense against these disorders is to learn to appreciate his or her own uniqueness. When people discover and honor the body's real needs, they become unwilling to sacrifice health for conformity.

FOCAL POINT 7A NOTES

1. D. C. Moore, Body image and eating behavior in adolescents, *Journal of the American College of Nutrition* 12 (1993): 975–980.
2. L. M. Mellin, C. E. Irwin, and S. Scully, Prevalence of disordered eating in girls: A survey of middle-class children, *Journal of the American Dietetic Association* 92 (1992): 851–853.
3. D. Neumark-Sztainer, R. Butler, and H. Palti, Dieting and binge eating: Which dieters are at risk? *Journal of the American Dietetic Association* 95 (1995): 586–588.
4. D. Neumark-Sztainer, Excessive weight preoccupation, *Nutrition Today*, March–April 1995, pp. 68–74.
5. J. H. Wilson, Nutrition, physical activity and bone health in women, *Nutrition Research Reviews* 7 (1994): 67–91.
6. K. K. Yeager and coauthors, The female athlete triad: Disordered eating, amenorrhea, osteoporosis, *Medicine and Science in Sports and Exercise* 25 (1993): 775–777.
7. Yeager and coauthors, 1993.
8. C. L. Otis, American College of Sports Medicine's Ad Hoc Task Force on Women's Issues in Sports Medicine, as quoted in A. A. Skolnick, "Female athlete triad" risk for women, *Journal of the American Medical Association* 270 (1993): 921–923.
9. J. E. Benson and coauthors, Relationship between nutrient intake, body mass index, menstrual function, and ballet injury, *Journal of the American Dietetic Association* 89 (1989): 58–63.

10. J. T. Baer and L. J. Taper, Amenorrheic and eumenorrheic adolescent runners: Dietary intake and exercise training status, *Journal of the American Dietetic Association* 92 (1992): 89–91; P. M. Howat and coauthors, The influence of diet, body fat, menstrual cycling, and activity upon the bone density of females, *Journal of the American Dietetic Association* 89 (1989): 1305–1307.
11. "Anorexia athletica," Special report on nutrition and the athlete, *Sports Medicine Digest* (1989): p. 10; S. N. Steen and K. D. Brownell, Patterns of weight loss and regain in wrestlers: Has the tradition changed? *Medicine and Science in Sports and Exercise* 22 (1990): 762–768.
12. G. Szmukler and C. Dare, Family therapy of early-onset, short-history anorexia nervosa, in *Family Approaches in Treatment of Eating Disorders*, eds. D. B. Woodside and L. Shekter-Wolfson (Washington, D.C.: American Psychiatric Press, 1991), pp. 25–47.
13. R. C. Casper and coauthors, Total daily energy expenditure and activity level in anorexia nervosa, *American Journal of Clinical Nutrition* 53 (1991): 1143–1150; L. Scalfi and coauthors, Bioimpedance analysis and resting energy expenditure in undernourished and refed anorectic patients, *European Journal of Clinical Nutrition* 47 (1993): 61–67.
14. Position of The American Dietetic Association: Nutrition intervention in the treatment of anorexia nervosa, bulimia nervosa, and

binge eating, *Journal of the American Dietetic Association* 94 (1994): 902–907.

15. L. G. Tolstoi, The role of pharmacotherapy in anorexia nervosa and bulimia, *Journal of the American Dietetic Association* 89 (1989): 1640–1646.

16. American Psychiatric Association Workgroup on Eating Disorders, Practice guidelines for eating disorders, I. Disease definition, epidemiology, and natural history, *American Journal of Psychiatry* 150 (1993): 212–228.

17. D. M. Stein, The prevalence of bulimia: A review of empirical research, *Journal of Nutrition Education* 23 (1991): 205–213.

18. L. G. Roberto, Impasses in the family treatment of bulimia, in *Family Approaches in Treatment of Eating Disorders*, eds. D. B. Woodside and L. Shekter-Wolfson, (Washington, D.C.: American Psychiatric Press. 1991, pp. 69–85.

19. P. W. Meilman, F. A. von Hippel, and M. S. Gaylor, Self-induced vomiting in college women: Its relation to eating, alcohol use, and Greek life, *College Health* 40 (1991): 39–41.

20. J. E. Mitchell, Medical complications of bulimia nervosa, in *Eating Disorders and Obesity: A Comprehensive Handbook*, eds. K. D. Brownell and C. G. Fairburn (New York: Guilford Press, 1995), pp. 271–275.

21. C. M. Bulik and coauthors, Drug use in women with anorexia and bulimia nervosa, *International Journal of Eating Disorders* 11 (1992): 213–225.

The Menstrual Cycle

Chapter 7 described how the physiological changes and hormonal shifts that adolescents experience create special nutrient needs. This Focal Point focuses on a time that begins in early adolescence and lasts some 40 years. It is the nonpregnant, reproductive time in a woman's life—the time of menstrual cycles.

The average women experiences 500 menstrual cycles in her lifetime, losing more than 17 liters of blood and 6500 milligrams of iron. No doubt, the physical losses incurred by menstrual cycles impose on women tremendous needs, not only of iron, but of all nutrients. Furthermore, the constantly changing hormonal and metabolic activities of the cycle change

energy and nutrient needs accordingly. Understanding the interrelationships among the hormones and organs in a typical menstrual cycle can lay the foundation for understanding the nutrition needs of women during this phase of life.

Changes in Hormones and Physiology

During puberty, a woman's monthly cycles begin—an event known as *menarche*. Hormones synchronize and coordinate the events of the monthly menstrual cycle. These hormones affect one another's activities and elicit responses from the sex organs and from other organs and tissues. Figure FP7B-1 shows how fluctuating

---

GLOSSARY

corpus luteum (CORE-pus LOO-te-um): a mass of glandular tissue that develops from a ruptured follicle and secretes hormones (*corpus* = body; *luteum* = yellow).

endometrium (en-doe-MEE-tree-um): the membrane lining the inner surface of the uterus.

estrogen (ESS-tro-jen): one of the female sex hormones produced in the ovaries, responsible for sexual development and for regulation of the menstrual cycle.

follicle: a small, saclike structure in the ovary, consisting of an ovum surrounded by epithelial cells that secrete the hormone estrogen.

FSH or **follicle-stimulating hormone:** a hormone from the pituitary gland that stimulates the growth of follicles.

LH or **luteinizing** (LOO-tin-eye-zing) **hormone:** a hormone from the pituitary gland that stimulates development of the ruptured follicle into the corpus luteum and signals ovulation.

menarche (men-ARK): the onset of menstruation; the first menstrual period, usually occurring between the ages of 10 and 16 (*men* = month; *arche* = beginning).

menopause: the time in a woman's life when menstrual activity ceases, usually by age 55 (*pause* = cessation).

menses: monthly flow of bloody fluid from the uterine membrane.

ovaries: the two glands that produce ova and female hormones.

ovulation: the ripening and rupturing of the mature follicle and subsequent release of the ovum.

placebo: an inert drug or treatment used for its psychological effect.

premenstrual syndrome (PMS): a cluster of physical, emotional, and psychological symptoms that occur before menstruation and diminish during or after menstruation.

progesterone (pro-JESS-teh-rone): a female hormone produced in the ovaries, responsible for changes in the uterine endometrium in the luteal phase of the menstrual cycle.

uterus: the muscular organ within which the embryo and fetus develop from the time of implantation to birth.

Figure FP7B-1   The Menstrual Cycle

Changes in hormone concentrations elicit development of the ovarian follicle and changes in the uterine lining during the menstrual cycle.

Several names apply to the various phases of the menstrual cycle. One set of names reflects events in the ovaries; another, those in the uterus. Note that the ovary's follicular, or preovulatory, phase coincides with menstrual flow and with the uterus's proliferative phase. Similarly, the ovary's luteal, or postovulatory, phase coincides with the uterus's secretory phase.

Events within the ovaries:

follicular (preovulatory) phase: ovarian events of follicle development before ovulation; coincides with uterine menstrual flow and the uterine proliferative phase.

luteal (postovulatory) phase: ovarian events of corpus luteum development from ovulation to menstrual flow; coincides with uterine secretory phase.

Events within the uterus:

menstrual flow: the periodic discharge of blood, disintegrated endometrial cells, and gland secretions from the uterus.

proliferative phase: uterine events of endometrium development before ovulation; coincides with ovarian follicular phase.

secretory phase: uterine events of endometrium disintegration that lead to menstrual flow; coincides with ovarian luteal phase.

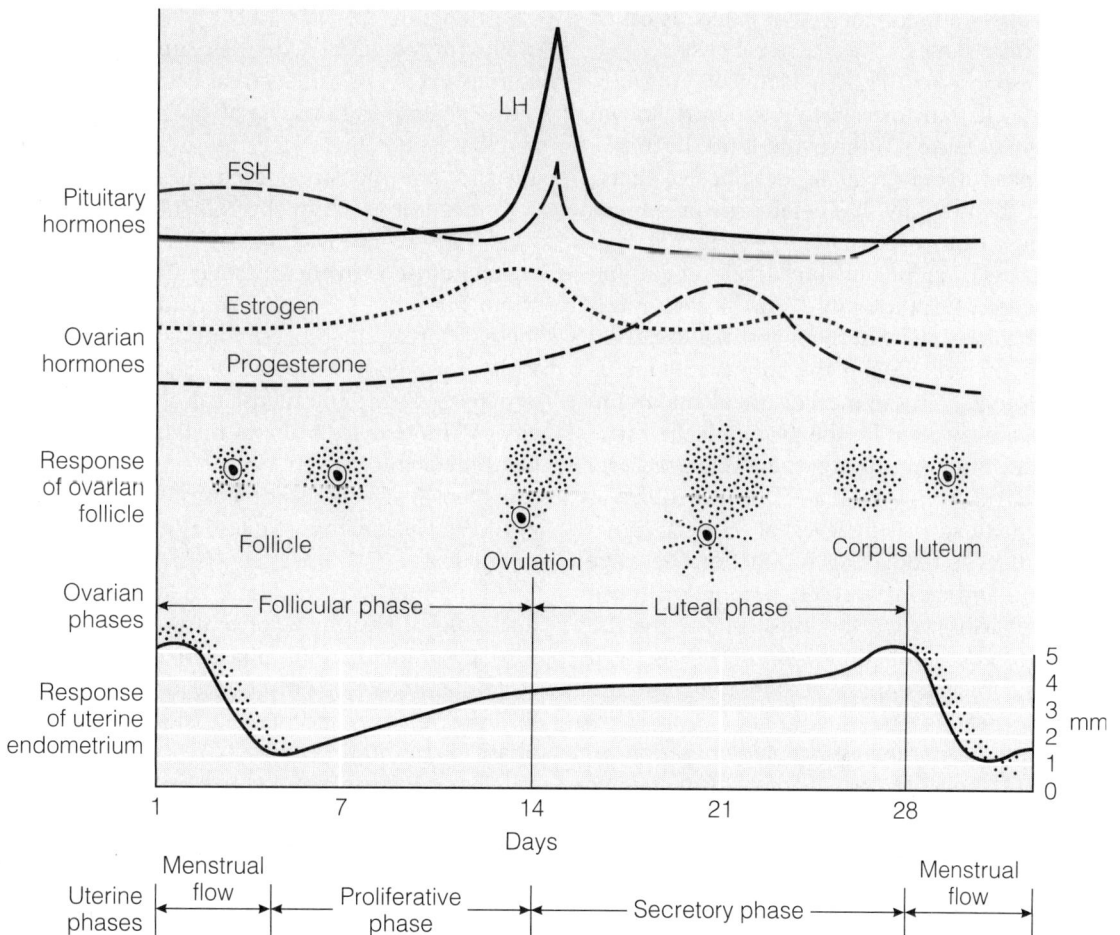

hormone concentrations account for the events of the month. The glossary defines related terms.

The most dramatic physical changes occur in the ovaries and the uterus. Female infants are born with about 1 million ova in each ovary. Of these ova, relatively few reach maturity and participate in a menstrual cycle. Fewer still undergo fertilization and develop into newborns. Most ova degenerate over the years, leaving none at the time of menopause.

In the ovary, an ovum, surrounded by a layer of follicular cells, enlarges and develops into a mature follicle. The hormone FSH (follicle-stimulating hormone), which dominates the start of the cycle, encourages the growth of the follicle. When the follicle reaches maturity, a sharp rise in LH (luteinizing hormone) triggers its rupture, thus releasing the ovum. This event, called *ovulation*, occurs about 14 days before the next menstrual flow. Follicle development, which takes approximately two weeks, is called the *follicular phase* of the menstrual cycle.

In response to the high LH levels, the ruptured follicle transforms into a structure known as the *corpus luteum*. If the ovum is not fertilized, the corpus luteum gradually develops to maturity and then rapidly degenerates. (If pregnancy does occur, the corpus luteum remains active.) The process of corpus luteum development and degeneration takes roughly 10 to 14 days and is called the *luteal phase* of the menstrual cycle.

Both the follicle and the corpus luteum secrete estrogen. The times of their maximum activity are evident in the two rises in estrogen concentration during the cycle (see Figure FP7B-1). During the last half of the follicular phase, the rapidly rising levels of estrogen signal the pituitary to reduce FSH production (negative feedback) and to increase LH production (positive feedback). During the luteal phase, the rise in estrogen does not stimulate an LH surge, as it did in the follicular phase, because the corpus luteum is secreting progesterone as well as estrogen. The progesterone-estrogen combination inhibits FSH and LH release, preventing follicle development. As the corpus luteum degenerates, progesterone and estrogen levels diminish, FSH and LH levels begin to rise, and the cycle repeats.

In the uterus, progesterone and estrogen prompt profound changes. Rising estrogen levels during the first half of the menstrual cycle stimulate the growth of the endometrium (the uterine lining). The endometrium thickens and develops an extensive vascular system to prepare for pregnancy should fertilization occur. The proliferative phase of endometrium development lasts about ten days, ceasing with ovulation. Without fertilization, the endometrium begins to disintegrate—the secretory phase. Finally, the endometrium lining begins to be shed. Menstrual "blood" is the endometrium lining, which leaves the body within four or five days. At the end of menstruation, the endometrium has returned to its minimal state.

The average menstrual cycle lasts 28 days, give or take a few days. The days are counted from the first day of bleeding.

Changes in Nutrition Needs

Many physiological activities support the numerous events of a menstrual cycle, some with nutrition implications. Basal body temperature and basal metabolic rate, for example, fluctuate with the menstrual cycle. Basal body temperature drops during the follicular phase and rises during the luteal phase. The temperature begins to rise with the LH surge. It continues to climb after the LH peak, coinciding with the increase of serum progesterone concentrations, which raise body temperature. Secretion of progesterone during the menstrual cycle also coincides with changes in basal metabolic rate. The basal metabolic rate rises about a week or so before menses and drops at menses. It remains low until the week before ovulation, when it begins to rise again.

Hormones are not alone in their monthly fluctuations. The blood concentrations of several substances fluctuate by phase of the menstrual cycle. Carotene concentrations, for example, are lowest at menses and peak in the late follicular phase (about mid-cycle).[1]

These changes in blood concentrations, basal body temperature, and basal metabolic rate may affect food intake and body weight. Women tend to eat more food per day for the half-month after ovulation (during the luteal phase prior to menstruation) than during the half-month before ovulation (during the follicular phase after menstruation). Studies have reported mean energy intakes prior to menstruation ranging from 100 to 300 kcalories per day greater than after menstruation.[2] These studies seem to suggest that eating more food before menstruation may be a woman's natural response to the hormonal and physiological changes that occur then. A woman can best respond to her body's signals by carefully selecting a well-balanced diet that provides the extra food energy needed before menstruation and returning to a lower food energy intake afterward.

Premenstrual Syndrome

Premenstrual syndrome (PMS) is a cluster of physical, psychological, and behavioral symptoms that as many as one in three women experiences before her menstrual period. A woman suffering from PMS may complain of any or all of the following symptoms: cramps or aches in the abdomen; back pain; headaches; acne; bloating; weight gain; food cravings, especially for sweets; constipation; breast swelling and tenderness; diarrhea; fatigue; and mood changes, including nervousness, anxiety, irritability, and depression. In about 5 percent of women, at least one physical or psychological symptom can be so severe as to be temporarily disabling.[3]

Specific PMS symptoms and combinations of symptoms vary from woman to woman. The distinguishing feature of PMS is the timing of the symptoms with a woman's menstrual cycle. Most women begin to experience symptoms seven to ten days before menstruation, during the luteal phase of their cycles. Symptoms disappear, or at least diminish significantly, with menstruation.

The cause or causes of PMS remain undefined, although researchers generally agree that the hormonal changes of the menstrual cycle must be responsible. The hormones that regulate the menstrual cycle also link nutrition and PMS. The most common dietary changes recommended to help relieve PMS discomfort involve reducing consumption of sugar, fat, salt, alcohol, and caffeine. Diets that emphasize complex carbohydrates are generally lower in simple

sugar, fat, and salt than diets that do not. For this and many other reasons, everyone—including those with PMS—benefits from eating ample complex carbohydrates every day.

Recommendations for women with PMS to reduce their alcohol intakes are based partly on the improvement in general health and partly on logic: Alcohol is a depressant and can worsen depression in some women. Research also shows that PMS is more prevalent and severe among women who consume caffeine than among those who do not, and is worst among the highest consumers.[4]

Despite a lack of evidence showing that nutrient deficiencies cause PMS, unproven nutrient treatments for PMS are still promoted by supplement manufacturers. Among the most popular nutrients advocated to treat PMS have been vitamin B_6, vitamin E, and zinc.

Trials of vitamin B_6 in PMS have resulted in contradictory findings. In one study, vitamin B_6 did improve premenstrual symptoms related to autonomic reactions (such as dizziness and vomiting) and behavioral changes (such as poor performance and a tendency to withdraw from social activities), but a significant number of physical symptoms remained.[5] In another study, neither diet modification nor vitamin B_6 supplements (250 milligrams per day) improved symptom severity.[6] Doses of vitamin B_6 as low as 50 milligrams (a dose sometimes suggested for treatment of PMS) are potentially toxic. Possible benefits of such therapy must be weighed against possible harm from megadoses. Megadoses of vitamin B_6 can cause defective muscle coordination and severe nerve damage.

Vitamin E is another popular candidate for treatment of PMS, and many extravagant claims have been made for it over the years. Vitamin E has little effect on the many symptoms of PMS, however, and until further research is done, the evidence for vitamin E as a treatment for PMS remains inconclusive.[7]

Because tiny amounts of zinc help regulate the secretion of hormones in the menstrual cycle, researchers have compared blood zinc concentrations of women with and without PMS.[8] Women with PMS had significantly lower blood zinc concentrations than women without PMS.

Perhaps low blood zinc impairs the secretion of menstrual hormones such as progesterone, as well as natural opiates, or endorphins, which are the body's own painkillers. Zinc is toxic in excess, so women should not self-diagnose zinc deficiency or self-prescribe zinc as a treatment for PMS. Further research is needed to confirm a relationship between zinc deficiency and PMS.

Many nutrient-based remedies have been advocated to treat PMS, and new claims continue to surface. So far, though, preventing or alleviating PMS remains a challenge. One obstacle researchers encounter in their efforts to find a treatment is that many women with PMS respond favorably to placebo treatments. Until a treatment can be found, the same advice holds for women with PMS as for women or men with any other health problem or need: Be sure to get adequate sleep and adequate physical activity. Eat well, and be sensible about intakes of sugar, fat, caffeine, salt, and alcohol. If your nutrient intake is inadequate and cannot be rectified by food alone, use a daily supplement for a while. Finally, be skeptical. Don't trust anyone who wants money for a product claimed to relieve symptoms.

FOCAL POINT 7B NOTES

1. M. R. Forman and coauthors, The fluctuation of plasma carotenoid concentrations by phase of the menstrual cycle: A controlled diet study, *American Journal of Clinical Nutrition* 64 (1996): 559–565.
2. S. I. Barr, K. C. Janelle, and J. C. Prior, Energy intakes are higher during the luteal phase of ovulatory menstrual cycles, *American Journal of Clinical Nutrition* 61 (1995): 39–43; M. C. Martini and coauthors, Effect of the menstrual cycle on energy and nutrient intake, *American Journal of Clinical Nutrition* 60 (1994): 895–899; A. K. H. Fong and M. J. Kretsch, Changes in dietary intakes, urinary nitrogen, and urinary volume across the menstrual cycle, *American Journal of Clinical Nutrition* 57 (1993): 43–46; V. Tarasuk and G. H. Beaton, Menstrual-cycle patterns in energy and macronutrient intake, *American Journal of Clinical Nutrition* 53 (1991): 442–447.
3. R. L. Reid, Premenstrual syndrome, *New England Journal of Medicine* 324 (1991): 1208–1210.
4. A. M. Rossignol and H. Bonnlander, Caffeine-containing beverages, total fluid consumption, and premenstrual syndrome, *American Journal of Public Health* 80 (1990): 1106–1110; A. M. Rossignol and coauthors, Tea and premenstrual syndrome in the People's Republic of China, *American Journal of Public Health* 79 (1989): 67–69.
5. K. E. Kendall and P. P. Schnurr, The effects of vitamin B_6 supplementation on premenstrual symptoms, *Obstetrics and Gynecology* 2 (1987): 145–149.
6. M. K. Berman, M. L. Taylor, and E. Freeman, Vitamin B_6 and premenstrual syndrome, *Journal of the American Dietetic Association* 90 (1990): 859–861.
7. R. S. London and coauthors, Efficacy of alpha-tocopherol in the treatment of the premenstrual syndrome, *Journal of Reproductive Medicine* 32 (1987): 400–404.
8. PMS: Hints of a link to lunchtime and zinc, *Science News* 27 (1990): 263.

Nutrition during Early Adulthood

As people enter their twenties, they make more choices than ever before and often reassess some of those they made as teens. In early adulthood people begin to make life-shaping decisions such as which career to pursue or whether to marry. Young adults are often completely on their own for the first time—not living in a dorm or relying on frequent meals at home. As a result, they may, for the first time, have to decide whether to learn to cook, depend on somebody else to cook for them, or rely on convenience and fast foods for the next few years. The decisions people make in their twenties—even the seemingly less important ones—reverberate far into the future, enhancing or undermining health for many years to come. This chapter focuses on nutrition and other lifestyle choices during early adulthood.

Growth and Development

The rapid and sometimes sporadic changes in growth and development that mark the teen years slow down and smooth out as early adulthood approaches. Ideally, the person who emerges during the mid to late twenties has a keen sense of identity and is ready to face the challenges of adulthood.

Physical Growth

Growth as most people think of it—growth in stature—is achieved by age 20, or even earlier. But growth in other ways is by no means complete. For example, although peak bone mass is largely achieved by age 20, bone mass continues to accumulate until age 25 or beyond.[1] This ongoing mineralization of bone during the third decade of life has important implications for calcium intake during early adulthood. Furthermore, with notable exceptions such as the central nervous system, in which cell division has ceased, the cells of all body systems continue to undergo cycles of growth, division, degeneration, death, and replacement. Cells become damaged at various rates depending on the body system and time of life, and their replacement occurs at different rates and to various degrees. Studies of different organ systems indicate that many attain maximum functional efficiency between 20 and 30 years of age. An adequate intake of all the nutrients during early adulthood therefore helps the entire body realize its fullest possible health potential.

Psychological Development

The years between 18 and 22 are a time when most people are making a major transition—from the small family circle to the huge adult world. According to one expert on adult development, a person's major tasks during this time are to gain a secure sense of identity and to establish a peer group, a sex role, an occupation, and a world view.[2] Obviously, the tasks are enormous, but also stimulating.

Typically, the twenties bring welcome relief from the inner turmoil of the teen years, and the focus shifts to the more external question of how to step into the adult world. The early adult years are not entirely tranquil, however: They are rightly called the "trying twenties." Perhaps one of the greatest challenges in establishing one's own identity and view of the world is to do so without relying too heavily on the comfort and safety of family or close relationships. People who fail to break away from home or who marry too early in this stage of development often impede their own self-sufficiency.

With self-sufficiency comes the responsibility of nourishing and caring for one's body. Those who accept this responsibility establish sound eating patterns and engage in frequent, vigorous physical activity to maintain a healthy body weight and prevent the gain of excess fat with its accompanying health risks. The next section focuses on energy and nutrient needs during early adulthood.

Energy and Nutrient Needs during Early Adulthood

As is true at any stage of life, energy and nutrient needs during early adulthood vary depending on genetics, physical activity, sex, and other factors. As the RDA table shows, the need for most nutrients changes little, if any, from adolescence to the early years of adulthood.* For calcium, recommended intakes decline slightly for both men and women but remain high enough to ensure maximum bone density during these years. Men's iron needs decrease slightly as the rapid growth of adolescence comes to an end, but women's stay the same. The next sections offer details about differences in nutrient needs for young men and women.

Energy

Differences in energy needs between men and women reflect differences in body composition. A normal-weight man may have from 10 to 25 percent body fat; a woman, because of her greater quantity of indispensable fat, 18 to 32 percent. The man's greater lean body mass uses more energy to support his body's metabolic activities. Energy needs also reflect physical activity. Energy recommendations during early adulthood assume light to moderate activity; sedentary people's energy needs may be lower, while those of physically active people may be higher.

*As Table 1-2 on p. 10 shows, the 1989 RDA groups young adults ages 19 to 24, whereas the 1997 DRI age grouping for young adults is 19 to 30.

Energy RDA during early adulthood:
  2900 kcal/day (19 to 24 yr, males).
  2200 kcal/day (19 to 24 yr, females).
Canadian RNI during early adulthood:
  3000 kcal/day (19 to 24 yr, males).
  2100 kcal/day (19 to 24 yr, females)

Vitamins and Minerals

The diets young men and women typically eat provide most nutrients in ample amounts. Other nutrients—notably calcium and iron—seem harder to get, especially for women. The importance of calcium to bone health for both men and women and of adequate iron for women during the reproductive years merits attention to these nutrients here.

Calcium DRI during early adulthood:
  1000 mg/day.

Nonmilk sources of calcium include:
■ *Corn tortillas.*
■ *Almonds and sesame seeds.*
■ *Oysters and sardines.*
■ *Mustard and turnip greens, bok choy, kale, parsley, watercress, and broccoli.*

Milk and milk products are rightly famous for their calcium contents.

Calcium. Figure 8-1 shows the three phases of bone development throughout life. From birth to approximately age 20, the bones are actively growing by modifying their length, width, and shape. This rapid growing phase overlaps with the next period of peak bone mass development, which occurs between the ages of 12 and 30. During this period, skeletal mass increases. Bones grow both thicker and denser by remodeling, a maintenance and repair process involving the loss of existing bone and the deposition of new bone. The final phase, which begins between 30 and 40 years of age and continues throughout life, finds bone loss exceeding new bone formation.

Bone loss is a natural process that people cannot completely prevent, but they may be able to forestall its impact by achieving maximal bone mass before and during early adulthood. Many factors determine a person's bone mass; important among them are heredity, calcium intake, and the extent to which a person puts demands on the bones through physical activity. Even modest increases in calcium intake and physical activity during young adulthood help maximize bone mass and minimize bone loss later in life.[3]

🏃 Healthy People 2000

Increase calcium intake so that at least 50% of people aged 25 years and older consume two or more servings of calcium-rich foods daily.

Figure 8-1 Phases of Bone Development throughout Life

The active growth phase occurs from birth to approximately age 20. The next phase of peak bone mass development occurs between the ages of 12 and 30. The final phase, when bone resorption exceeds formation, begins between age 30 and 40 and continues throughout the remainder of life.

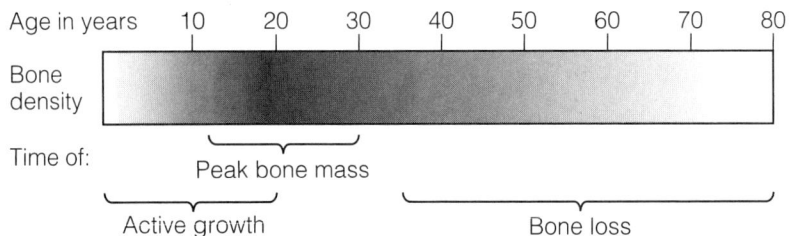

Figure 8-2 Daily Calcium Intakes of Females Compared with Their DRI*ᵃ*

Women's average calcium intakes tend to fall lower throughout life. In hopes of preventing this decline, and in the hope that increased calcium intakes *may* help forestall the adult bone loss that leads to osteoporosis, the DRI Committee set the adequate intake values for adults higher than the 1989 RDA.

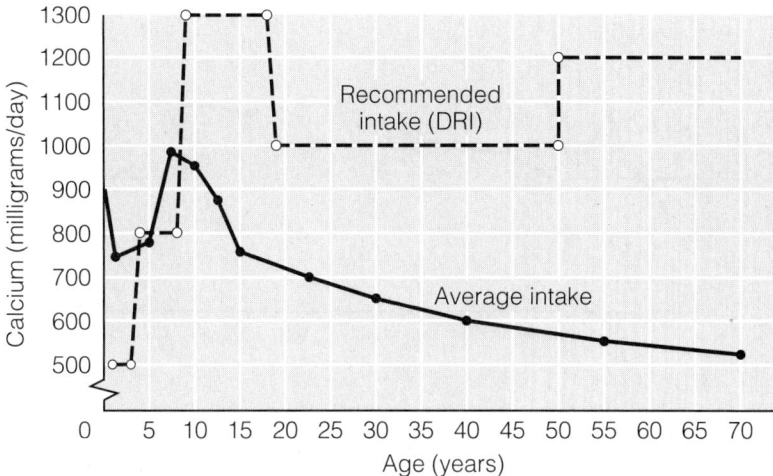

*ᵃ*Mean calcium intakes from food in the U.S. population have not changed substantially in the past two decades. Life Sciences Research Office, Federation of American Societies for Experimental Biology, *Third Report on Nutrition Monitoring in the United States: Executive Summary* (Washington, D.C.: U.S. Government Printing Office, 1995).

Source: Adapted from National Center for Health Statistics, *Dietary Intake Source Data: United States, 1976–1980,* DHHS publication no. (PHS) 83 1681 (Washington, D.C.: U.S. Government Printing Office, March 1983).

Men's intakes of calcium meet the DRI until age 60, but the average calcium intakes of young women fall far short after age 11, as Figure 8-2 illustrates.[4] The calcium intakes of many *individual* young women are even lower than the average. Evidence shows that achievement of peak bone mass during youth can significantly reduce the risk of extensive adult bone loss or osteoporosis. This finding led the National Institutes of Health to suggest calcium intakes as high as 1500 milligrams per day for young women.[5]

The best way to ensure a generous calcium intake is to eat three to four servings of calcium-rich foods daily. A later section on supplements describes calcium supplement usage, absorption, and risks.

Iron. During early and middle adulthood, women are in a precarious state with respect to iron sufficiency. Their normal iron losses exceed those of men because of repeated menstrual blood losses. In addition, the average iron intakes of women are consistently lower than the recommended intake, as Figure 8-3 illustrates. The combination of inade-

Iron RDA during early adulthood:
 10 mg/day (19 to 24 yr, males).
 15 mg/day (19 to 24 yr, females).
Canadian RNI during early adulthood:
 9 mg/day (19 to 24 yr, males).
 13 mg/day (19 to 24 yr, females).

Figure 8-3 Daily Iron Intakes of Females Compared with Their RDA[a]

Women's average iron intakes throughout early and middle life fall below recommendations.

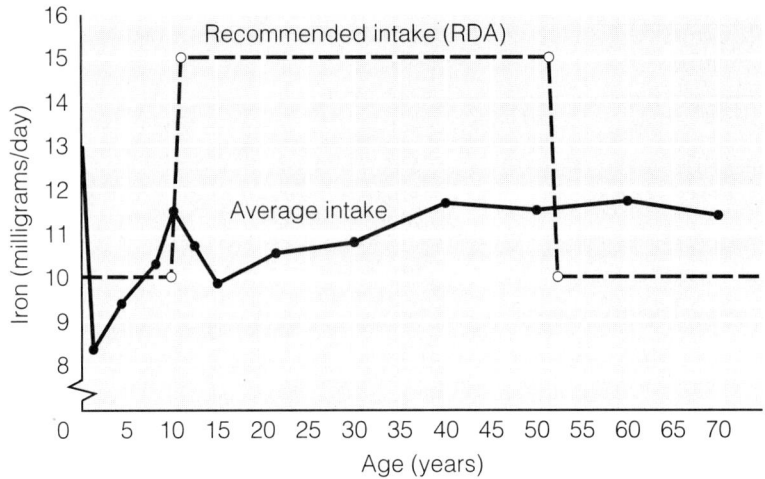

[a]Mean iron intakes from food in the U.S. population have not changed substantially in the past two decades. Life Sciences Research Office, Federation of American Societies for Experimental Biology, *Third Report on Nutrition Monitoring in the United States: Executive Summary* (Washington, D.C.: U.S. Government Printing Office, 1995).

Source: Adapted from National Center for Health Statistics, *Dietary Intake Source Data: United States, 1976–1980,* DHHS publication no. (PHS) 83–1681 (Washington, D.C.: U.S. Government Printing Office, March 1983).

This chili dinner provides heme and nonheme iron and MFP from meat, nonheme iron from legumes, and vitamin C from tomatoes. The combination of heme iron, nonheme iron, MFP, and vitamin C helps a person achieve maximum iron absorption.

quate iron intake and the natural iron loss that occurs in the procreative years makes it a challenge to obtain sufficient iron.

As is true of calcium, the best way to meet iron needs is to rely on foods, because the iron in supplements is less well absorbed. Because women eat less food than men, however, women must select iron-rich, low-kcalorie foods from each food group.

Nutrient Supplements for Adults

Nutrition experts and recommendations urge people to rely on foods for necessary nutrients, using vitamin and mineral supplements only under specific circumstances. Determining whether an individual has a nutrient deficiency requires a complete nutrition assessment. When such an assessment so indicates, a health care provider can recommend an appropriate supplement. The following list acknowledges that in these specific conditions, adults may need to take supplements:

- People with low food energy intakes (less than 1200 kcalories per day).
- People who eat bizarre or monotonous diets.

■ People with illnesses that interfere with the intake, absorption, metabolism, or excretion of nutrients.

■ People taking medications that interfere with the body's use of specific nutrients (see Table FP10-1).

■ People who have diseases, infections, or injuries, or who have undergone surgery resulting in increased metabolic needs.

■ People who eat all-plant diets (vegans).

Except for people in these circumstances, most adults can normally get all the nutrients they need by eating a varied diet of nutrient-dense foods. Unfortunately, many people do not eat this way, for one reason or another. People who do not receive the nutrients they need from food may benefit from multivitamin-mineral supplements. A rule of thumb when selecting a supplement is to find one that provides all the nutrients in amounts smaller than, equal to, or very close to the RDA.

Vitamins and minerals in abundance are best obtained from foods, not supplements.

Calcium Supplements. As mentioned, inadequate calcium intake may contribute to age-related bone loss. People who do not consume milk products or other calcium-rich foods in amounts that provide even half the recommendation may benefit from calcium supplements. Supplementation may be a way—in fact, the only way—that some people, especially women, can meet their recommended intakes of calcium. In the absence of final answers on this question, it may, at least for the present, be prudent to add to the previous list of justifications for supplement use:

■ People whose calcium intakes are too low to forestall extensive bone loss.

Most healthy people absorb calcium as well from calcium carbonate, calcium acetate, calcium lactate, calcium gluconate, and calcium citrate as they do from milk.[6] A consumer can refine the choice among these by comparing other variables such as cost and number of tablets to be ingested. Calcium carbonate contains the highest percentage of calcium (40 percent) and therefore can meet daily needs with the smallest number of tablets. Calcium from supplements is more efficiently absorbed when the supplements are taken with meals and in divided doses of less than 500 milligrams each. Consumers should avoid calcium supplements prepared from bone meal or dolomite (limestone), because they may contain unsafe levels of contaminants such as lead, arsenic, cadmium, or mercury. To avoid toxic doses, consumers using calcium supplements combined with vitamin D must limit the quantity of vitamin D they are taking to no more than the recommended intake.

Iron Supplements. Iron is like many other nutrients in that foods meet the body's requirements better than supplements do. Still, many women cannot meet their iron requirements with foods alone, especially if they are limiting their food energy intakes. Such women may benefit from supplements providing the recommended 15 milligrams. The body absorbs the ferrous form of iron efficiently, and taking such a supplement with meals enhances absorption. We can add to the previous list of circumstances that may justify the taking of a supplement:

■ Women who bleed excessively during menstruation.

Supplement Cautions. The taking of individual mineral supplements requires caution. Minerals compete for binding sites and thus interfere with one another's bioavailability. For example, a calcium phosphate dibasic supplement inhibits magnesium absorption. If the supplement is fortified with magnesium, then it interferes with both calcium and iron absorption.

In concluding this section on supplements, for the best possible health people need to pay attention to their nutrient intakes. A thorough nutrition assessment can give important information about nutrition status. Few people, however, receive a complete nutrition assessment until problems begin to surface. More commonly, nutrition assessment measures for people in this age group focus on body weight.

Weight Control during Early Adulthood

As people in their early twenties set out to establish their own place in the world, their choices and actions begin to lay the foundation for their remaining years. Poor habits and lifestyle choices, such as eating one fast-food meal after another or being sedentary day after day, may seem to matter little in terms of nutrition and health at present, but can prove costly in the long run. Positive choices, such as eating nutrient-dense foods and being physically active, can enhance nutrition status and help prevent the major nutrition-related problem of early adulthood: excess weight gain.

■ *BMI <20 = underweight.*
■ *BMI 20 to 25 = normal.*
■ *BMI 25 to 30 = over-weight.*
■ *BMI >30 = obese.*

As Chapter 9 explains, the health risks associated with excess body fat underscore the importance of preventing excess weight gain during early adulthood. Young men and women will want to achieve and maintain a weight that supports good health, and for women, an optimal pregnancy should it occur. Table 8-1 presents suggested weight standards for adults. An expert panel on healthy weight stresses prevention of excess weight gain after early childhood, so that by age 21, a person's weight should be stable.[7] Because many people enter early adulthood at an appropriate weight, it is the ideal time to establish habits that will maintain this weight and prevent the health risks that accompany weight gain.

Unfortunately, many U.S. adults gain weight throughout early and middle adulthood.[8] More than one-third of people over age 20 are 20 percent or more overweight.[9] Americans spend $33 billion a year trying to control weight.[10] Adult weight gain is so prevalent in the United States that many think it is normal and inevitable. In fact, not all U.S.

adults gain weight—some actually lose weight by middle adulthood. Individual variances become obscured, however, when viewed among population weight curves.

In pursuing good health, keep in mind that it is a lifelong journey. Most adults are keenly aware of their body weights and shapes and realize that what they eat and do can make a difference. Those who are most successful at preventing weight gain or maintaining weight loss seem to have fully incorporated healthful eating and physical activity into their lives. Their approach is not "on again, off again," but a routine part of their daily lives. And it is a multifaceted approach, involving diet, physical activity, and behavior.

✶ Healthy People 2000

Increase to at least 50% the proportion of worksites with 50 or more employees that offer nutrition education and/or weight management programs for employees.

Eating Plans

No one food plan is magical, and no specific food must be included or avoided. In designing a plan, people need only consider foods that they like or can learn to like, that are available, and that are within their means.

Energy intake should provide nutritional adequacy without excess—that is, somewhere between kcalorie-restricted dieting and complete freedom to eat everything in sight. A rule of thumb for those trying to lose weight is that a person needs at least 10 kcalories per pound of current weight each day to lose fat efficiently while retaining lean tissue. For example, a 140-pound woman would start by setting her energy intake at 1400 kcalories a day. Then, as she lost weight, she would adjust her energy intake downward.

Nutritional Adequacy. Nutritional adequacy is difficult to achieve on fewer than 1200 kcalories a day, and most healthy adults should not consume fewer than that. Consider that 1200 kcalories a day allows a person who weighs 120 pounds to lose weight, yet rarely would an adult who weighs 120 pounds be overweight. People weighing more can lose weight on higher energy intakes. A plan that provides an adequate intake supports a healthier and more successful weight loss than a restrictive plan that creates feelings of starvation and deprivation, which can lead to an irresistible urge to binge.

Take a look at Table 8-2, and notice that the 1200-kcalorie food plan represents the minimum servings suggested in the Food Guide Pyramid (introduced in Chapter 1) and allows a teaspoon of fat at each of three meals. Such an intake allows most people to lose weight gradually and

Table 8-1 Suggested Weights for Adults

| Height[a] | Weight (lb)[a] | |
	Mid-point	Range
4'10"	105	91–119
4'11"	109	94–124
5'0"	112	97–128
5'1"	116	101–132
5'2"	120	104–137
5'3"	124	107–141
5'4"	128	111–146
5'5"	132	114–150
5'6"	136	118–155
5'7"	140	121–160
5'8"	144	125–164
5'9"	149	129–169
5'10"	153	132–174
5'11"	157	136–179
6'0"	162	140–184
6'1"	166	144–189
6'2"	171	148–195
6'3"	176	152–200
6'4"	180	156–205
6'5"	185	160–211
6'6"	190	164–216

Note: The higher weights in the ranges generally apply to men, who tend to have more muscle and bone; the lower weights more often apply to women, who have less muscle and bone.

[a]Without shoes or clothes.

Source: *Report of the Dietary Guidelines Advisory Committee on the Dietary Guidelines for Americans,* 1995.

Table 8-2 Recommended Number of Servings for Different Energy Intakes

Food Group	Energy Level (kcal)						
	1200	1500	1800	2000	2200	2600	3000
Breads and cereals	6	7	8	9	11	13	15
Meat (lean)[a]	4	5	6	6	6	7	8
Vegetable	3	4	5	5	5	6	6
Fruit	2	3	4	4	4	5	6
Milk (nonfat)	2	2	2	3	3	3	3
Fat (tsp)	3	5	6	7	8	10	12

[a]Meat servings are given in ounces.

Note: These patterns follow the Food Guide Pyramid and supply less than 30 percent of kcalories as fat.

still meet their nutrient needs with careful, nutrient-dense food selections. (As mentioned earlier, women may need an iron supplement.) The other patterns provide for higher energy intakes.

Small Portions. Overweight people usually need to learn to eat less food at each meal—one piece of chicken for dinner instead of two, one teaspoon of butter on the vegetables instead of a tablespoon, and one cookie for dessert instead of six. The goal is to eat enough food for energy, nutrients, and pleasure, but not so much as to have an excess. This amount should leave a person feeling satisfied—not necessarily full. Keep in mind that even low-fat foods can deliver a lot of kcalories when a person eats large quantities. A low-fat cookie or two can be a sweet treat even on a weight loss diet, but a whole box is clearly excessive.

Carbohydrates, Not Fats. Healthful meals and snacks center on complex carbohydrate foods. Fresh fruits, vegetables, legumes, and whole grains offer abundant vitamins, minerals, and fiber but little fat. They also require effort to eat—an added bonus. People who eat these foods in abundance spontaneously eat for longer times and take in fewer kcalories than when eating foods of high energy density. The satiety signal indicating fullness is sent after a 20-minute lag, so a person who slows down and savors each bite eats less before the signal reaches the brain.

Satiety plays a key role in weight loss diets. Researchers compared a diet in which fat was restricted, but complex carbohydrates were eaten freely, with a more conventional energy-restricted diet.[11] Moderately obese women in both diet groups lost weight, but those who ate the low-fat, complex carbohydrate-rich diet rated it higher in terms of satiety and taste. They also gave higher ratings to "quality of life," perhaps because they felt less deprived when allowed to eat unrestricted quantities of low-fat, complex carbohydrate foods.

Findings from several other studies agree: Restricting fat is effective in restricting energy for weight loss.[12] Furthermore, body composition may improve on a low-fat diet. Rats who had been on a high-fat diet were switched to a diet that provided fewer kcalories, with each group receiving a low-, or medium-, or a high-fat diet.[13] Compared with rats on the low-fat diet, the rats on the high-fat diet lost less weight and kept much more of their body fat. This study suggests that even when weight is lost, body *fat* is not lost by following a low-kcalorie diet alone—dietary fat must also be reduced.

Similarly, women who followed a low-fat diet (20 percent kcalories from fat) lost both weight and body fat.[14] In fact, the only way the women could *maintain* weight was to raise their total energy intakes. Clearly, then, to lose weight and improve body composition, measure fat with extra caution. A slip of the butter knife adds more kcalories than a slip of the sugar spoon. Less fat in the diet means less fat in the body. Be careful not to take this advice to the extreme, however; too little fat in the diet or in the body carries health risks as well.

Speaking of empty kcalories, a person trying to achieve or maintain a healthy weight needs to pay attention not only to fat, but also to sugar and alcohol. Using them for pleasure on occasion is compatible with health as long as most daily choices are of nutrient-dense foods.

Adequate Water. Water fills the stomach between meals and dilutes the metabolic wastes generated from the breakdown of fat. It meets the water need that was formerly met by eating extra food (remember that foods provide water).

In short, young adults need to adopt a lifelong "eating plan for good health" rather than a "diet for weight loss." In this way, they will be able to maintain a healthy weight.

Drinking water is a healthy habit.

Physical Activity

Because one's focus turns to career and family in early adulthood, an active adolescent may all too easily become a sedentary young adult. Weight gain occurs when a decrease in energy expenditure is not accompanied by a decrease in energy intake.

Physical activity can shift body composition toward more lean and less fat tissue. It can ensure that weight loss is mostly fat loss and that weight gain is mostly muscle gain. People find it difficult to lose body fat or maintain a loss of fat without physical activity.[15] People who include physical activity in their weight control program seem to follow their eating plan more closely than those who do not exercise.[16] Consequently they benefit from both a little less energy input and the added energy output of physical activity. In addition, physical activity contributes to a healthy body composition and weight control because:

- Energy expenditure increases.
- Resting metabolic rate increases (slightly) over the long term.

- Body fat diminishes.
- Appetite remains under better control.
- Stress and stress-induced overeating or undereating become less likely.
- Self-esteem improves.

If physical activity is to help with fat loss, it must involve the voluntary moving of muscles, not passive motion such as being massaged. Passive activity neither increases energy expenditure nor builds muscles, but the more actively muscles move, and the more time spent moving, the more kcalories spent and the more muscle tissue built. Table 8-3 shows the approximate numbers of kcalories people of various weights spend on activities.

Some people have the impression that, to aid in fat loss, physical activity must be of long duration; others think it must be intense. Actually, both types of exercise are effective. Health care professionals often recommend activities of low to moderate intensity for long duration, such as fast-paced, hour-long walks, for those who want to lose body fat. They emphasize that low- to moderate-intensity activity burns more fat than high-intensity activity. The greater the intensity, the less contribution fat makes to the fuel mixture. Within a given time period, however, even though the *proportion* of fat used for energy is greater at lower intensities, more *total energy* is expended at higher intensities, and so more *total energy from fat* is used. Thus, in a given time frame, total energy spent, as well as total energy from fat, is greater with higher-intensity activity.[17]

The bottom line on how much fat loss physical activity will promote seems to be total energy expenditure, regardless of intensity. To lose fat, expend as much energy as time and stamina allow.

Some people enjoy walking, others want to dance or ride bikes. For those who want to be stronger and firmer as well, lift weights or do push-ups, pull-ups, and stomach crunches. Remember this benefit: Muscle is more metabolically active than fat, so the more muscle built, the more energy expended in a given time.

Behavior

Behavior is an important supporting factor in achieving and maintaining appropriate body weight and composition. Behavior modification changes the hundreds of small behaviors of overeating and underexercising that lead to, and perpetuate, obesity. Healthy eating and activity choices are not intolerable tasks that demand herculean willpower to achieve. They are an essential part of healthy living and should simply be incorporated into the day—much like brushing one's teeth or wearing a safety belt. Table 8-4 applies behavior modification principles to help people plan and adjust habits concerning eating plans and physical activity.

Table 8-3 Energy Demands of Activities

Activity	Energy per Pound of Body Weight per Minute kcal/lb/min[a]	Body Weight (lb)				
		110	125	150	175	200
		kcal/min[b]				
Aerobic dance (vigorous)	.062	6.8	7.8	9.3	10.9	12.4
Basketball (vigorous, full court)	.097	10.7	12.1	14.6	17.0	19.4
Bicycling						
13 miles per hour	.045	5.0	5.6	6.8	7.9	9.0
15 miles per hour	.049	5.4	6.1	7.4	8.6	9.8
17 miles per hour	.057	6.3	7.1	8.6	10.0	11.4
19 miles per hour	.076	8.4	9.5	11.4	13.3	15.2
21 miles per hour	.090	9.9	11.3	13.5	15.8	18.0
23 miles per hour	.109	12.0	13.6	16.4	19.0	21.8
25 miles per hour	.139	15.3	17.4	20.9	24.3	27.8
Canoeing (flat water, moderate pace)	.045	5.0	5.6	6.8	7.9	9.0
Cross-country skiing (8 miles per hour)	.104	11.4	13.0	15.6	18.2	20.8
Golf (carrying clubs)	.045	5.0	5.6	6.8	7.9	9.0
Handball	.078	8.6	9.8	11.7	13.7	15.6
Horseback riding (trot)	.052	5.7	6.5	7.8	9.1	10.4
Rowing (vigorous)	.097	10.7	12.1	14.6	17.0	19.4
Running						
5 miles per hour	.061	6.7	7.6	9.2	10.7	12.2
6 miles per hour	.074	8.1	9.2	11.1	13.0	14.8
7.5 miles per hour	.094	10.3	11.8	14.1	16.4	18.8
9 miles per hour	.103	11.3	12.9	15.5	18.0	20.6
10 miles per hour	.114	12.5	14.3	17.1	20.0	22.9
11 miles per hour	.131	14.4	16.4	19.7	22.9	26.2
Soccer (vigorous)	.097	10.7	12.1	14.6	17.0	19.4
Studying	.011	1.2	1.4	1.7	1.9	2.2
Swimming						
20 yards per minute	.032	3.5	4.0	4.8	5.6	6.4
45 yards per minute	.058	6.4	7.3	8.7	10.2	11.6
50 yards per minute	.070	7.7	8.8	10.5	12.3	14.0
Table tennis (skilled)	.045	5.0	5.6	6.8	7.9	9.0
Tennis (beginner)	.032	3.5	4.0	4.8	5.6	6.4
Walking (brisk pace)						
3.5 miles per hour	.035	3.9	4.4	5.2	6.1	7.0
4.5 miles per hour	.048	5.3	6.0	7.2	8.4	9.6

[a]Use this column if you want to calculate kcalories spent for your own exact body weight. Multiply kcal/lb/min by your exact weight and then multiply that number by the number of minutes spent in the activity. For example, if you weigh 142 pounds, and you want to know how many kcalories you spent doing 30 minutes of vigorous aerobic dance: .062 × 142 = 8.8 kcalories per minute, 8.8 × 30 (minutes) = 264 total kcalories spent.

[b]Use this column if you weigh 110, 125, 150, 175, or 200 pounds. This eliminates the need to calculate from column 1.

Source: Values for swimming, bicycling, and running have been adapted with permission of Ross Laboratories, Columbus, Ohio 43216, from G. P. Town and K. B. Wheeler, Nutrition concerns for the endurance athlete, *Dietetic Currents* 13 (1986): 7–12. Copyright 1986 Ross Laboratories. Values for all other activities have been adapted from *Physical Fitness for Practically Everybody: The Consumers Union Report on Exercise.* Copyright 1983 by Consumers Union of U.S., Inc., Yonkers, NY 10703–1057. Reprinted by permission from CONSUMER REPORTS BOOKS, 1983.

Table 8-4 Behaviors to Help Control Body Fatness[a]

1. Eliminate inappropriate eating cues:
   - Don't buy problem foods.
   - Eat only in one room at the designated time.
   - Shop when not hungry.
   - Turn off television food commercials.
   - Avoid vending machines, fast-food restaurants, and convenience stores.

2. Suppress the cues you cannot eliminate:
   - Serve individual plates; don't serve "family style."
   - Make small portions look large by spreading them over the plate.
   - Create obstacles to consuming problem foods—wrap them and freeze them, making them less quickly accessible.
   - Control deprivation; plan and eat regular meals.

3. Strengthen cues to appropriate behaviors:
   - Share appropriate foods with others.
   - Store appropriate foods in convenient spots in the refrigerator.
   - Learn appropriate portion sizes.
   - Plan appropriate snacks.
   - Keep sports and play equipment by the door.

4. Repeat desired behaviors:
   - Slow down eating—put down utensils between bites.
   - Always use utensils.
   - Leave some food on your plate.
   - Move more—shake a leg, pace, stretch often.
   - Join groups of active people and participate.

5. Arrange negative consequences for negative behavior:
   - Ask that others respond neutrally to your deviations (make no comments—even negative attention is a reward).
   - If you slip, don't punish yourself.

6. Reward yourself personally and immediately:
   - Buy tickets to sports events, movies, concerts, or other nonfood amusement.
   - Indulge in a new small purchase.
   - Get a massage; buy some flowers.
   - Take a bubble bath; read a good book.
   - Join a card game; listen to music.
   - Praise yourself; visit friends.
   - Nap; relax.

[a]These behaviors are based on the behavior modification technique of controlling actions by controlling the environment.

Food Choices and Fitness during Early Adulthood

Many young adults are on their own for the first time. With this new independence comes responsibility, and the responsibility of nourishing and caring for one's body should not be taken lightly. Taking responsibility lays the foundation for lifelong health. It is central to well-being.

Diets Tailored to Individual Preferences

Young adults may choose a diet low in fat to protect themselves against cardiovascular disease in their later years, but for now, such guidelines help mostly to maintain a healthy body weight. In limiting fat intake, women need to be particularly careful to include enough lean meats and low-fat dairy products to protect themselves against iron and calcium deficiencies. Because men generally eat more food than women, nutrient adequacy is easier to achieve for young men. In short, men and women face different nutrition- and diet-related problems.

Planning Healthy Diets. Chapter 1 described the Food Guide Pyramid in detail. The number of daily servings recommended for young men and women depends on their level of physical activity. The middle of the range of servings provides about 2200 kcalories and is appropriate for most moderately active women. The upper end of the range meets the needs of most moderately active young men or exceptionally active young women. Table 7-2 on p. 294, showing the Daily Food Guide for Adolescents, is appropriate for young adults as well.

Young adults today have access to more nutrition information and misinformation than ever before. And although many young adults choose diets low in fat and high in fiber and nutrients, many still eat more high-fat foods and fewer grains, fruits, and vegetables than recommended. In addition, newly independent young adults may decide to cast tradition aside in favor of eating patterns that are new and different to them. For example, some may decide to stop eating meat or other animal-derived foods. And although many different eating patterns can meet nutrient needs, attention to diet planning is central to sound nutrition. The next few sections offer ideas and strategies for planning and preparing meals that are both enjoyable and nutritious. Some sections are written directly to you, the consumer.

Planning Vegetarian Diets. Young adults, exploring their newfound independence in cooking, may choose to exclude meat and other animal-derived foods for many different reasons—health, economics, ecology, ethics, or religion. Whatever their reasons, people must be aware of the nutrition and health implications of vegetarian diets. Vegetarians are usually categorized by the foods they choose to exclude. Some exclude red meat only; some also exclude chicken or fish; others also exclude eggs; and still others exclude milk and milk products as well. The foods a person *excludes*, however, are not nearly as important as the foods a person *includes* in the diet. Vegetarian diets that include a variety of grains, vegetables, legumes, and fruits offer abundant complex carbohydrates and fibers, an assortment of vitamins and minerals, and little fat—characteristics that reflect current dietary recommendations aimed at promoting health and preventing obesity. Well-planned vegetarian diets can offer sound nutrition and health benefits to adults.[18] Table 8-5 offers the Daily Food Guide for Vegetarians.

Table 8-5 Daily Food Guide for Vegetarians

Food Group	Suggested Daily Servings	Serving Sizes
Breads, cereals, and other grain products	6 or more	1 slice bread
		1/2 bun, bagel, or English muffin
		1/2 c cooked cereal, rice, or pasta
		1 oz dry cereal
Vegetables	4 or more[a]	1/2 c cooked or 1 c raw
Fruits	3 or more	1 piece fresh fruit
		3/4 c fruit juice
		1/2 c canned or cooked fruit
Legumes and other meat alternates	2 to 3	1/2 c cooked beans
		4 oz tofu or tempeh
		8 oz soy milk
		2 tbs nuts or seeds (these tend to be high in fat, so use sparingly)
		1 egg or 2 egg whites
Milk and milk products	2 to 3[b]	1 c low-fat or nonfat milk
		1 c low-fat or nonfat yogurt
		1 1/2 oz low-fat cheese

[a]Include 1 cup of dark green vegetables daily to help meet iron requirements.

[b]People who do not use milk or milk products: use soy milk fortified with calcium, vitamin D, and vitamin B_{12}.

Source: Adapted with permission from Position of The American Dietetic Association: Vegetarian diets, *Journal of the American Dietetic Association* 93 (1993): 1318.

From Guidelines to Groceries

Only you can design a healthy diet for yourself, but how do you begin? Start with the foods you enjoy eating. Then try to make improvements, little by little. When shopping, think of the food groups, and choose nutrient-dense foods within each group.

Breads, Cereals, and Other Grain Products. When shopping for grain products, you will find them described as *refined, enriched,* or *whole grain.* These terms refer to the milling process and the making of products, and they have different nutrition implications. Refined foods may have lost many nutrients during processing; enriched products may have had some nutrients added back; and whole-grain products may be rich in all nutrients found in the original grain.

Whole-grain products, such as brown rice or oatmeal, not only provide more nutrients and fiber than refined or enriched products, but do not contain the added salt and sugar of flavored, processed rice or sweet-

With its many grains (including wheat, rye, oats, corn, and rice) and types of foods (such as pastas, breads, and cereals), this group does more than its share for variety.

ened cereals. So, when grocery shopping, choose whole-grain breads and cereals often.

Vegetables. Choose fresh vegetables, especially green and yellow-orange vegetables like spinach, broccoli, and sweet potatoes. Cooked or raw, vegetables are good sources of vitamins, minerals, and fiber. Frozen and canned vegetables without added salt are acceptable alternatives to fresh. To control fat, energy, and sodium intakes, limit butter, salad dressing, and salt on vegetables.

Fruit. Choose fresh fruits often, especially citrus fruits and yellow-orange fruits such as cantaloupes and apricots. Fruits supply valuable vitamins, minerals, and fibers. They add flavors, colors, and textures to meals, and their natural sweetness makes them enjoyable as snacks or desserts.

Fruit juices are healthy beverages, but contain little dietary fiber compared with whole fruits. Whole fruits satisfy the appetite better than juices and are a better selection for people who need to limit food energy intakes. Juices, in contrast, are a good choice for people who need extra food energy. Frozen, dried, and canned fruits without added sugar are acceptable alternatives to fresh. Be aware that sweetened fruit "drinks" or "-ades" contain mostly water, sugar, and a little juice for flavor. Some may have been fortified with vitamin C, but lack any other significant nutritional value.

Legumes. Choose often from the legumes (beans and peas such as pinto beans, split peas, lima beans, and black beans). They are available fresh, frozen, dried, or canned. Whether you eat legumes by themselves or incorporate them into a recipe, use them often as they are an economical, low-fat, nutrient- and fiber-rich food choice.

Meat, Fish, and Poultry. Meat, fish, and poultry provide essential minerals, such as iron and zinc, and abundant B vitamins as well as protein. To buy and prepare these foods without excess energy, fat, and sodium takes skill. Choose fish, poultry, and lean meats when shopping in the meat department. Lean cuts of beef and pork are named "round" or "loin" (as in top round or pork tenderloin). As a guide, "prime" and "choice" cuts generally have more fat than "select" cuts. Restaurants usually serve prime cuts. Ground beef, even "lean" ground beef, derives most of its food energy from fat. Have the butcher trim and grind a lean round steak instead.

Weigh meat after it is cooked and the bones and fat are removed. In general, 4 ounces of raw meat is equal to about 3 ounces of cooked meat. Some examples of 3-ounce portions of meat include 1 medium pork chop, 1/2 chicken breast, or 1 steak or hamburger about the size of a deck of cards. To keep fat intake down, bake, roast, broil, grill, or braise meats (but do not fry them in fat); remove the skin from poultry; trim visible fat before cooking; and drain fat after cooking.

Legumes include:
- *Black beans.*
- *Black-eyed peas.*
- *Garbanzo beans.*
- *Great northern beans.*
- *Kidney beans.*
- *Lentils.*
- *Navy beans.*
- *Peanuts.*
- *Pinto beans.*
- *Soybeans.*
- *Split peas.*

Milk. Shoppers will find fortified foods in the dairy case. Examples are milk, to which vitamins A and D have been added, and soy milk, to which calcium, vitamin D, and vitamin B_{12} have been added. In addition, shoppers may find imitation foods (such as cheeses) and food substitutes. As food technology advances, many such foods offer low-fat alternatives. For example, egg substitutes help people to reduce their fat and cholesterol intakes.

When shopping, choose low-fat or nonfat milk, yogurt, and cheeses. They are important sources of calcium, but can provide too much sodium and fat if selections aren't made with care.

✶ Healthy People 2000

Increase to at least 5000 brand items the number of processed food products that are reduced in fat and saturated fat.

Read Food Labels

Recall from Chapter 1 that food labels offer valuable nutrition information. Consumers can use packaged foods in diet planning if they can interpret the foods' labels. For example, by knowing that the first ingredient named is the one that predominates by weight, consumers who read ingredient lists can glean much information. Compare these products: A cereal that contains "puffed milled corn, sugar, corn syrup, molasses, salt . . ." versus one that contains "100 percent rolled oats." In this comparison, consumers can tell that the second product is the more nutrient dense.

Serving Sizes. Because labels present nutrient information per serving, they must identify the size of a serving. The FDA has established specific serving sizes that reflect amounts that people customarily consume and requires that all labels for a given product use the same serving size. For example, the serving size for all ice creams is a half cup and for all beverages, 8 fluid ounces. This facilitates comparison shopping. Consumers can see at a glance whether one brand or another has more or fewer kcalories or grams of fat. Standard serving sizes are expressed in both common household measures, such as cups, and metric measures, such as milliliters, to accommodate users of both types of measures.

Nutrition Facts. In addition to the serving size and the servings per container, the "Nutrition Facts" panel on a label shows quantities of energy (in kcalories), of fat (in both kcalories and grams), and of certain other nutrients (in grams or milligrams) in a serving:

■ Total food energy (kcalories).

■ Food energy from fat (kcalories).

■ Total fat (grams).

- Saturated fat (grams).
- Cholesterol (milligrams).
- Sodium (milligrams).
- Total carbohydrate, including starch, sugar, and fiber (grams).
- Dietary fiber (grams).
- Sugars (grams).
- Proteins (grams).

In addition, labels must present nutrient content information as compared with a standard for the following vitamins and minerals:

- Vitamin A.
- Vitamin C.
- Iron.
- Calcium.

Comparing nutrient amounts against ,a standard helps make them meaningful to label readers. A label reader might wonder, for example, whether 1 milligram of iron or calcium is a little or a lot. Well, the standard value for iron is 18 milligrams, so 1 milligram of iron is enough to take notice of: It is over 5 percent. But the standard value for calcium is 1000 milligrams, so 1 milligram of calcium is essentially nothing.

It would be nice for consumers if food labels could express each food's nutrient contents as a percentage of each individual's recommended intakes. Unfortunately, though, recommended intakes, such as the RDA, are not the same for everybody; they depend on age and sex. Manufacturers can't know who will be reading the label—an 8-year-old boy, a 70-year-old woman, or a pregnant teenage girl. Label makers do the best they can with this variability, though: They use one set of standard values to represent the needs of a "typical consumer." These standard values, developed by the FDA for use on food labels, are called the Daily Values (see inside front cover, right).

Daily Values (DV): reference values developed by the FDA specifically for use on food labels.

From Groceries to Meals

The person who knows how to plan nutritious meals, which foods to buy, and how to read food labels also needs to know where to shop to get the most for money spent and how to prepare the foods for health and enjoyment. The next few sections offer some suggestions.

Learning to Prepare Foods. Young adults living on their own would be wise to learn how to cook for themselves if they do not already know how. People who do not know how to prepare the foods they eat must rely on others to do so. Many wonderful cookbooks are available today, and with a little time and effort, anyone can learn the basics for preparing nutritious, low-fat foods. Table 8-6 offers practical tips for lowering fat intake and the following paragraphs present strategies for preparing meals for singles.

Table 8-6 How to Lower Fat Intake—by Food Group

Meat, Fish, and Poultry

- Fat adds up quickly, even with lean meat; limit intake to about 6 ounces (cooked weight) daily.
- Choose fish, poultry, or lean cuts of pork or beef; look for un-marbled cuts named *round* or *loin* (eye of round, top round, round tip, tenderloin, sirloin, and top loin).
- Trim the fat from pork and beef; remove the skin from poultry.
- Grill, roast, broil, bake, stir-fry, stew, or braise meats; don't fry. When possible, place meat on a rack so that fat can drain.
- Use lean ground turkey or lean ground beef in recipes; brown ground meats without added fat, then drain off fat.
- Refrigerate meat pan drippings and broth; when the fat solidifies, remove it and use the defatted broth in recipes.
- Select tuna and other canned meats packed in water; rinse oil-packed items with hot water to remove much of the fat.
- Fill kabob skewers with lots of vegetables and slivers of meat; create main dishes and casseroles by combining a little meat, fish, or poultry with a lot of pasta, rice, or vegetables.
- Make meatless spaghetti sauces and casseroles; use legumes often.
- Eat a meatless meal or two daily.

Milk and Cheeses

- Drink nonfat and low-fat milk instead of whole milk.
- Use nonfat and low-fat cheeses (such as part-skim ricotta and low-fat mozzarella) instead of regular cheeses.
- Use nonfat or low-fat yogurt or fat-free sour cream instead of regular sour cream.
- Use evaporated nonfat milk instead of cream.
- Enjoy nonfat frozen yogurt, sherbet, or ice milk instead of ice cream.

Fruits and Vegetables

- Enjoy the natural flavor of steamed vegetables for dinner and fruits for dessert.
- Use butter-flavored granules on vegetables instead of butter or margarine.
- Use nonfat yogurt or nonfat salad dressing instead of sour cream, cheese, mayonnaise, or other sauces on vegetables and in casseroles.
- Select nonfat or low-fat salad dressings, or use herbs, lemon juice, and spices instead of regular salad dressing.
- Add a little water to thick, bottled salad dressing to dilute the amount of fat each serving provides.
- Eat at least two vegetables (in addition to a salad) with dinner.
- Snack on raw vegetables or fruits instead of high-fat items like potato chips.

Breads and Cereals

- Use fruit butters or jellies on bread instead of butter or margarine.
- Select breads, cereals, and crackers that are low in fat (for example, bagels instead of croissants).

(continued)

Table 8-6   How to Lower Fat Intake—by Food Group (*continued*)

Other Foods and Cooking Tips

- Use a nonstick pan or coat the pan lightly with vegetable oil.
- Use egg substitutes in recipes instead of whole eggs or use 2 egg whites in place of each whole egg.
- Use half the margarine, butter, or oil called for in a recipe. (The minimum amount of fat for muffins, quick breads, and biscuits is 1 to 2 tablespoons per cup of flour; for cakes and cookies, 2 tablespoons per cup).
- Use less butter or margarine. Select the whipped types of butter, margarine, or cream cheese for use at the table; they contain half the kcalories of the regular types.
- Use butter replacers instead of butter.
- For sandwiches and salads, use spicy mustard, nonfat salad dressing, lemon juice, flavored vinegar, salsa, or the nonfat versions instead of regular mayonnaise, salad dressing, or sour cream.
- Use wine; lemon, orange, or tomato juice; herbs; spices; fruits; or broth instead of butter, margarine, or oil when cooking.
- Stir-fry in a small amount of oil; add moisture and flavor with broth, tomato juice, or wine.
- Use variety to enhance enjoyment of the meal: vary colors, textures, and temperatures—hot cooked versus cool raw foods— and use garnishes to complement food.

Spend Wisely. People who have the means to shop and cook for themselves can cut their food bills just by being wise shoppers. The first decision a person with a tight grocery budget must make is where to shop. Large supermarkets are usually less expensive than convenience stores. A grocery list helps reduce impulse buying, and specials and coupons can save money when the items featured are those that the shopper needs and uses.

Buying the right amount so as not to waste any food is a challenge for people eating alone. They can buy fresh milk in the size best suited for personal needs. Pint-size and even cup-size boxes of milk are available and can be stored unopened on a shelf for up to three months without refrigeration.

Be Creative. Creative chefs think of various ways to use foods when only large amounts are available. For example, a head of cauliflower can be divided into thirds. Then one-third is cooked and eaten hot. Another third is put into a vinegar and oil marinade for use in a salad. And the last third can be used in a casserole or stew.

Chefs also experiment with stir-fried foods. A large frying pan often works as well or better than a wok on modern ranges. A variety of vegetables and meats can be enjoyed this way; inexpensive vegetables such as cabbage, celery, and onion are delicious when crisp cooked in a little oil with soy sauce or lemon added. Interesting frozen vegetable mix-

Buy only what you will use.

Invite guests to share a meal.

tures are available in larger grocery stores. Cooked, leftover vegetables can be dropped in at the last minute. A bonus of a stir-fried meal is that there is only one pan to wash. Similarly, a microwave oven allows a chef to use fewer pots and pans. Meals can also be frozen or refrigerated in microwavable containers to reheat as needed.

Also, single people shouldn't hesitate to invite someone to share meals with them whenever there is a lot of food. It's likely that that person will return the invitation, and both parties will get to enjoy companionship and a meal prepared by others.

Many frozen dinners that are now available are low in fat and nutritious. Adding a fresh salad, a whole-wheat roll, and a glass of milk can make a nice meal.

Fitness during Early Adulthood

Young adults who make regular physical activity part of their lifestyle not only help protect themselves from excess fat gain and the accompanying health risks, but also establish a habit that will benefit them in many other ways for years to come. Physical activity fosters a positive self-image, a sense of well-being, and a positive attitude in general—all important to a person who is making decisions about the future.

Physical activity also improves nutrition status. Bones do not passively accumulate calcium from foods; physical activity helps bones to store calcium and become dense, strong, and able to carry more weight. Physical activity also develops the lean body tissue that serves as a storage site for iron and other nutrients. Table 8-7 summarizes the benefits of fitness.

Table 8-7 Benefits of Fitness (Summary)

- Sound, beneficial rest and sleep
- Improved nutritional health
- Reduced fatness and increased lean body tissue
- Strengthened immunity
- Reduced risk of disease (see Table 9-6, p. 372)
- Reduced probability of accidents; fewer and less severe injuries
- Reduced incidence and severity of anxiety and depression
- Freedom from drug (including alcohol) abuse
- Improved self-image and self-confidence
- Better learning ability
- Greater interpersonal, social, and spiritual strengths
- Improved quality of life in the later years
- Longer life

In conclusion, young adults can either enhance or impair the quality of their lives by the choices they make each day. For each poor choice, a positive alternative can maximize the probability of enjoying life well into a healthy old age.

CHAPTER EIGHT NOTES

1. Optimal calcium intake, NIH Consensus Development Panel on Optimal Calcium Intake, *Journal of the American Medical Association* 272 (1994): 1942–1948.

2. G. Sheehy, *Passages: Predictable Crises of Adult Life* (New York: Bantam Books, 1976), pp. 48–61.

3. D. Teegarden and coauthors, Previous physical relates to bone mineral measures in young women, *Medicine and Science in Sports and Exercise* 28 (1996): 105–113; R. R. Recker and coauthors, Bone gain in young adult women, *Journal of the American Medical Association* 268 (1992): 2403–2406; ACSM Position stand on osteoporosis and exercise, *Medicine and Science in Sports and Exercise* 27 (1995): i–vii.

4. C. M. Weaver, Age-related calcium requirements due to changes in absorption and utilization, *Journal of Nutrition* (supplement) 124 (1994): 1418–1425.

5. D. U. Porter, Washington Update: NIH Consensus Development Conference Statement Optimal Calcium Intake, *Nutrition Today* 29 (1994): 37–40.

6. D. I. Levenson and R. S. Bockman, A review of calcium preparations, *Nutrition Reviews* 52 (1994): 221–232; L. Mortensen and P. Charles, Bioavailability of calcium supplements and the effect of vitamin D: Comparisons between milk, calcium carbonate, and calcium carbonate plus vitamin D, *American Journal of Clinical Nutrition* 63 (1996): 354–357.

7. J. G. Meisler and S. St. Jeor, Summary and Foundation's Expert Panel on Healthy Weight, *American Journal of Clinical Nutrition* 63 (1996): 474S–477S.

8. S. M. Garn, Fractioning healthy weight, *American Journal of Clinical Nutrition* 63 (1996): 412S–414S.

9. J. M. Rippe, Overweight and health: Communications, challenges, and opportunities, *American Journal of Clinical Nutrition* 62 (1996): 470S–473S.

10. J. Foreyt and K. Goodrick, The ultimate tri-

umph of obesity, *Lancet* 346 (1995): 134–135.

11. M. Shah and coauthors, Comparison of a low-fat ad libitum complex-carbohydrate diet with a low-energy diet in moderately obese women, *American Journal of Clinical Nutrition* 59 (1994): 980–984.

12. L. Lissner, Dietary fat and the regulation of energy intake in human subjects, *American Journal of Clinical Nutrition* 46 (1987): 886–892; A. Kendall and coauthors, Weight loss on a low-fat diet: Consequence of the imprecision of the control of food intake in humans, *American Journal of Clinical Nutrition* 53 (1991): 1124–1129; L. Sheppard, A. R. Kristal, and L. H. Kushi, Weight loss in women participating in a randomized trial of low-fat diets, *American Journal of Clinical Nutrition* 54 (1991): 821–828.

13. C. N. Boozer, A. Brasseur, and R. L. Atkinson, Dietary fat affects weight loss and adiposity during energy restriction in rats, *American Journal of Clinical Nutrition* 58 (1993): 846–852.

14. Sheppard, Kristal, and Kushi, 1991; T. E. Prewitt and coauthors, Changes in body weight, body composition, and energy intake in women fed high- and low-fat diets, *American Journal of Clinical Nutrition* 54 (1991): 304–310.

15. K. P. G. Kempen, W. H. M. Saris, and K. R. Westerterp, Energy balance during an 8-wk energy-restricted diet with and without exercise in obese women, *American Journal of Clinical Nutrition* 62 (1995): 722–729.

16. S. B. Racette and coauthors, Exercise enhances dietary compliance during moderate energy restriction in obese women, *American Journal of Clinical Nutrition* 62 (1995): 345–349.

17. C. J. Zelasko, Exercise for weight loss: What are the facts? *Journal of the American Dietetic Association* 95 (1995): 1414–1417.

18. Position of The American Dietetic Association: Vegetarian diets, *Journal of the American Dietetic Association* 93 (1993): 1317–1319.

Nutrition and Contraception

Each menstrual cycle presents a woman and her partner with the responsibility for making a choice of whether to start a pregnancy. Looking at the nutrition implications of some contraceptive methods can help a couple in their decision. The two methods discussed here are those with known nutrition relationships—oral contraceptives and intrauterine devices. The accompanying glossary defines related terms.

Oral Contraceptives

Millions of women use oral contraceptives, popularly known as the pill, to prevent pregnancy. In its 30-year history, the pill has become the most studied drug in the United States. Early identification of a number of risk factors prompted changes in dosages and formulations to produce an effective contraceptive with a wide margin of safety.

Oral contraceptives today contain one-fifth the estrogen and one-tenth the progesterone originally in the pills of the 1960s, making them as risk free as possible, yet still effective. The hormone concentrations are low and may vary throughout the month to roughly simulate the changes that normally occur during a menstrual cycle (see Focal Point 7B). Such pills are called *multiphasic* and may avert some minor side effects, such as breakthrough bleeding or, in some women, blood lipid changes. Manufacturers may claim that multiphasic pills are more natural or physiologically superior to fixed-dose pills, but

Honest talk about contraception leads to both wise planning and closeness.

this claim is fallacious. After all, the purpose of oral contraceptives is to alter the natural events of the menstrual cycle.

Combination Pill. There are two types of oral contraceptives—the combination pill and the minipill. Of the women using oral contraceptives in the United States, 99 percent take the combination pill, which contains synthetic versions of the hormones estrogen and progesterone. Recall from the discussion on the menstrual cycle (in Focal Point 7B) that the ovaries naturally produce estrogen after ovulation to

GLOSSARY

oral contraceptives: pills that prevent pregnancy by stopping ovulation or by changing conditions in the uterus; often called birth control pills or "the pill."

combination pill: an oral contraceptive that contains progestin and synthetic estrogen.

intrauterine device (IUD): a device inserted into the uterus to prevent conception or implantation.

progestin: a synthetic version of progesterone used in contraceptives.

minipill: a progestin-only oral contraceptive.

suppress ovulation until after the next menstruation. The estrogen in the combination pills prevents pregnancy the same way, by suppressing ovulation.

Minipill. The other type of oral contraceptive, the minipill, contains only synthetic progesterone, known as *progestin*. The progestin in the minipill makes the mucus surrounding the uterine opening (the cervix) less penetrable than normal to spermatozoa and may deactivate them. It also interrupts the normal preparation of the uterine endometrium and so prevents zygote implantation. The minipill is rarely used now that long-acting progestin injections are available for women who cannot use estrogen.

Side Effects. Oral contraceptives not only prevent pregnancy but may also benefit a woman's reproductive system physiologically. (Table FP8-1 lists the positive and negative side effects of oral contraceptives.) Apparently, the suppression of ovulation reduces the likelihood of ovarian cancer, ovarian cysts, painful menstrual periods, and premenstrual syndrome. The reduced risk of ovarian cancer is seen in women who use the pill for as little as three months and continues for 15 years after use ends. The progesterone in the pill offers protection against endometrial cancer and benign breast disease. The protective effect against endometrial cancer is seen in women who use the pill for at least 12 months and persists for 15 years after use stops.

Research results on the relationship, if any, between oral contraceptives and breast cancer are inconclusive. For now, it appears that even long-term oral contraceptive use in a young, prepregnant woman does not present a risk of breast cancer.

The association between cardiovascular disease and oral contraceptive use has been extensively studied. Research shows some risk for women over age 35 who smoke. These women should find alternative contraceptive methods. For young, healthy women who do not smoke, the association between oral contraceptives and coronary artery disease disappears.[1]

The pill may be responsible for high blood pressure in some users, especially women who

Table FP8-1 Side Effects of Oral Contraceptives

Reduced Risk of	Increased Risk of
Benign breast disease[a]	Coronary artery disease
Endometrial cancer	Gallbladder disease
Iron deficiency anemia	Glucose intolerance
Ovarian cancer	High blood pressure
Ovarian cysts	Raised cholesterol and triglycerides
Painful menstrual periods	Reduced high-density lipoproteins (HDL)
Premenstrual syndrome	
Pregnancy	

[a]The effect of oral contraceptives on breast cancer is inconclusive; it appears that oral contraceptive use does not increase the risk for breast cancer.

are older, multiparous, and obese. This high blood pressure reverts to normal when users stop taking the pill.

The progesterone in oral contraceptives alters carbohydrate metabolism: Glucose tolerance diminishes, and blood glucose and insulin rise, especially in women predisposed to diabetes. For most women, the changes in carbohydrate metabolism are minimal and return to normal with cessation of pill use. Because these metabolic effects are transient and reversible, they present a small risk of diabetes compared with other nonreversible factors, such as genetics.

Nutrition Concerns

Oral contraceptives reduce menstrual blood flow and thereby conserve for the body all nutrients normally lost in menstrual blood—most notably iron. They may also alter the status of many vitamins and minerals by increasing demand in metabolic pathways, impairing absorption, increasing excretion, or altering tissue distribution. Whatever the mechanism, the effects on vitamins and minerals are reflected in higher or lower blood concentrations. In most cases, however, the clinical significance of these changes

Before choosing a method of contraception, seek accurate information.

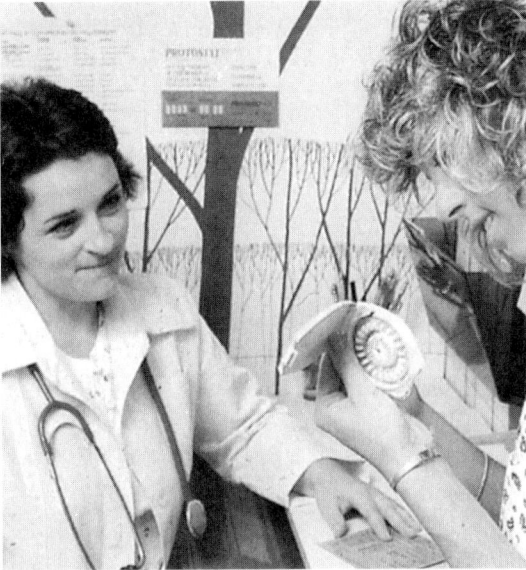

Oral contraceptives seem to produce the same fluid retention effect commonly seen around the time of menstrual periods, temporarily adding a few pounds to body weight. A health care provider can determine whether this is a problem and either recommend a diuretic if warranted or prescribe a different brand of pill.

The Intrauterine Device (IUD)

The intrauterine device (IUD, or coil) is a small piece of molded plastic or plastic and metal that is inserted into the uterus. One type of IUD contains progestin. The IUD's effectiveness is not completely understood. Apparently, it induces a change in the uterine lining that interferes with the implantation of a zygote.

IUDs can cause heavy menstrual bleeding. This can be inconvenient and distressing, but rarely signifies a serious disorder. It does entail a nutrition risk, however: the development of iron deficiency anemia due to loss of blood. Therefore, the iron status of a woman using an IUD requires regular assessment.

Before choosing a method of contraception, a young couple would be wise to seek advice from a health care professional. Knowing how, and choosing, to prevent pregnancy until the time is right can prevent lifelong regrets.

in nutrient concentrations is minor, given the many factors affecting women's nutrition status.

Using supplements to correct the nutrient fluctuations incurred by oral contraceptive use is inappropriate. If nutrient deficiencies and excesses are evident in women taking oral contraceptives, the remedy is to improve the diet.

FOCAL POINT 8 NOTES

1. R. T. Burkman, Oral contraceptive use and coronary and cardiovascular risk, *Medicine and Science in Sports and Exercise* 28 (1996): 11–12.

Nutrition during Middle Adulthood

Growth and Development

Physical Development

Psychological Growth

Energy and Nutrient Needs during Middle Adulthood

Energy

Vitamins

Minerals

Nutrient Supplements

Diet and Health

Risk Factors

Dietary Recommendations and Risk Factors

Food Choices and Fitness during Middle Adulthood

Food Choices and Eating Habits

Fitness during Middle Adulthood

FOCAL POINT 9   *Alcohol and Nutrition*

B eyond the childbearing years, a person's primary nutrition-related responsibility is to self. The time has passed when one's nutrition choices will physically affect the next generation. The quality of an adult's own life is now paramount, and he or she can do much to enhance it. The right choices, made throughout adulthood, can support a person's ability to meet physical, emotional, and mental challenges. Two goals motivate adults to pay attention to their diets: promoting health and slowing aging. This chapter on middle adulthood features health promotion; the next chapter on late adulthood focuses on aging.

The chronological ages identifying the transition points between early, middle, and late adulthood are ambiguous. It seems reasonable that early adulthood ends when growth ceases, usually by the middle to late twenties. And, although unsure of the exact ages of middle adulthood, society seems to agree that 65 marks the beginning of late adulthood. Consequently, middle adulthood could cover all the years between 25 and 65, but four decades is a long age span. The physical, social, and psychological differences between people at each end of the range can be dramatic. To better address the nutrient needs of adults from age 19 on, the current revision of the RDA proposes four age groupings instead of three.[1] In this revised plan, middle adulthood falls between the ages of 30 and 50, a reasonable range to feature in this chapter as well.

Growth and Development

By age 25, physical growth is over, but physical changes continue. Some physical changes, such as weight gain, profoundly influence health. Others, such as graying hair, are simply cosmetic, but may still influence a person's social and emotional well-being.

Physical Development

After growth ceases, lean body mass begins to diminish at a rate of 2 to 3 percent per decade. Basal metabolic rate declines proportionately. Without an increase in physical activity or a decrease in food energy intake, weight gain follows. Although it is not inevitable, most people experience a notable weight gain between the ages of 25 and 55 or so (see Figure 9-1).[2] As a later section of this chapter explains, excess body fat is implicated in many of today's chronic diseases, including cardiovascular disease, cancer, and diabetes.

Between the ages of 45 and 55, a woman experiences menopause. Estrogen secretion from the ovaries diminishes, ovulation stops, menstrual cycles cease, and she becomes infertile. The decline in estrogen triggers the many physical symptoms associated with menopause, the most notable being hot flashes. Some women combat the physical discomforts of menopause with estrogen replacement therapy. In addition to alleviating menopausal symptoms, estrogen replacement therapy reduces the risks of heart disease and osteoporosis, but increases the risks

chronic diseases: degenerative diseases characterized by deterioration of the body organs; also called chronic, **noncommunicable diseases (NCD).** Examples include heart disease, cancer, and diabetes.

menopause (MEN-oh-pawz): the period in a woman's life (on average, between the ages of 50 and 55) marked by the permanent cessation of menstrual activity.

Figure 9-1   Weight Gain Patterns during Middle Adulthood

Weight increases progressively with each decade until about age 60 and then decreases at older ages.

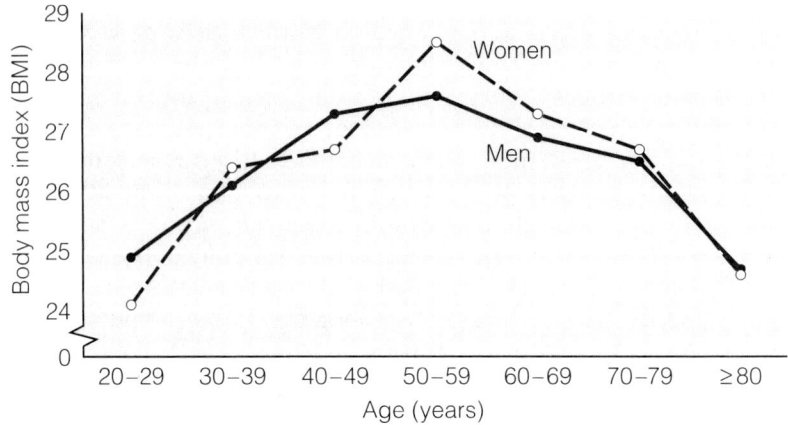

Source: Adapted from R. J. Kuczmarski and coauthors, Increasing prevalence of overweight among U.S. adults: The National Health and Nutrition Examination Surveys, 1960–1991, *Journal of the American Medical Association* 272 (1994): 205–211.

of some cancers. For this reason, a woman must carefully consider her health history when balancing the apparent benefits against the potential risks of such treatment.[3]

Psychological Growth

Middle age has the advantages of experience and, for most people, an improved economic status. Housing and career choices are more secure for this age group than for any other. It is a time of settling down and changing dreams into realities. Yet even as a middle-aged adult looks back and appreciates having reached the prime of life, a look forward brings a sense of urgency. Youth and physical prowess begin to fade and can no longer be taken for granted. People in this age group who established healthy habits in their younger years reap the rewards. Others begin to have chronic health problems, and some die prematurely. Such events heighten the need for breaking bad health habits and establishing good ones. Diet is one of the easiest health habits to improve.

Physical growth may cease, but physical changes continue through the middle adulthood years.

Energy and Nutrient Needs during Middle Adulthood

The RDA for most nutrients remain the same from early to middle adulthood. Even so, a few nutrients deserve brief mention in this section—

the antioxidant vitamins C and E; the bone minerals calcium, phosphorus, and magnesium; and the mineral iron.

Energy

To avoid "middle-aged spread," adults need to balance the food they eat with physical activity. This can be quite a challenge, especially when money may be available for purchasing and eating food in abundance and time is tight for squeezing in exercise regularly.

Recommended energy allowances assume light to moderate physical activity and do not change as a young adult passes into middle adulthood. Unfortunately, many moderately active young adults become sedentary middle-age adults. Given the propensity toward weight gain, the energy RDA for adults may be set too high.

Vitamins

All vitamins support health, but two vitamins deserve special mention for their roles in fighting disease—vitamins C and E, the antioxidant vitamins. Much research has focused on the theory that antioxidant nutrients act as scavengers of oxygen-derived free radicals, thereby helping to prevent cell and tissue damage that otherwise would give rise to degenerative diseases. The beneficial effects of fruits, vegetables, and grains in fighting degenerative diseases have been attributed, in part, to the antioxidants they provide. The antioxidant roles of vitamins C and E are under especially extensive study. Other vitamins may not serve as antioxidants, but play supporting roles. For example, riboflavin forms part of the coenzymes that are required by several enzymes active in oxidative metabolism.[4]

Vitamin C. Vitamin C is the most abundant water-soluble antioxidant in the body and is active primarily in extracellular fluid. Its actions are most notable in combatting the free radicals of polluted air and cigarette smoke. Not only does vitamin C scavenge many free radicals, but it helps return vitamin E to its active form.

A variety of physical stresses deplete the body's vitamin C supply and may make higher intakes desirable. Among the stresses known to increase vitamin C needs are the chronic use of certain medications, including aspirin, barbiturates, and oral contraceptives, and cigarette smoking. Cigarette smoke contains oxidants, which greedily consume this potent antioxidant. Exposure to cigarette smoke, especially when accompanied by low intakes of vitamin C, depletes the body's pool in both active and passive smokers.[5] Whereas the RDA for nonsmokers is 60 milligrams a day, the RDA for people who smoke cigarettes regularly is 100 milligrams; the Canadian RNI provides a similar increase for those who smoke.

Vitamin E. Vitamin E is the most abundant fat-soluble antioxidant and one of the body's primary defenders against oxidation. It protects the

---

Energy RDA during middle adulthood:
  2900 kcal/day (men).
  2200 kcal/day (women).
Canadian RNI during middle adulthood:
  2700 kcal/day (men).
  1900 kcal/day (women).

Quick and Easy Estimates Energy intake for weight loss:
  10 kcal/lb body weight.
Energy intake for weight maintenance:
  12 kcal/lb body weight.

antioxidant: a compound that protects other compounds from oxidation by being oxidized itself. An antioxidant donates electrons to another substance; that substance becomes reduced as the antioxidant simultaneously becomes oxidized.

free radical: an atom or molecule that has one or more unpaired electron(s) in the outer orbital. This electron imbalance makes free radicals unstable and highly reactive. Radicals typically arise during oxidation reactions and readily attack other molecules with which they come in contact.

Vitamin C RDA during middle adulthood:
  60 mg/day (adults).
  100 mg/day (smokers).
Canadian RNI during middle adulthood:
  30 mg/day (women).
  40 mg/day (men).
  +50% (smokers).

peroxidation: the production of unstable molecules containing more than the usual amount of oxygen. Hydrogen peroxide, H_2O_2, for example, may be produced from water, H_2O.

polyunsaturated fatty acids, all other lipids, and related fat-soluble compounds such as vitamin A.

Vitamin E serves as the body's first line of defense against lipid peroxidation by effectively breaking the chain reaction. In fact, vitamin E is one of the most efficient chain-breaking antioxidants available, reacting 200 times faster than the antioxidant BHT (butylated hydroxytoluene) commonly used in commercial bakery products. For this reason, a small amount of vitamin E can protect a large amount of lipid. Of course, in protecting other substances, vitamin E is used up and so needs to be replenished from dietary sources such as vegetable oils, seeds, and nuts.

Minerals

Calcium DRI during middle adulthood:
1000 mg/day.

Phosphorus DRI during middle adulthood:
700 mg/day.

Magnesium DRI during middle adulthood:
420 mg/day (men).
320 mg/day (women).

Iron RDA during middle adulthood:
10 mg/day (men).
15 mg/day (women).
Canadian RNI during middle adulthood:
9 mg/day (men).
13 mg/day (women).

Recommended intakes for the key minerals in bone development—calcium, phosphorus, and magnesium—were recently revised and reflect new criteria in determining adequacy. In the past, the RDA addressed average daily losses and absorption rates, but not the prevention of osteoporosis. One of the main goals of the revised recommendations (the DRI) is to reduce the risk of chronic diseases. Consequently, adequate intakes of the bone minerals were estimated based on the amounts that provide full retention of calcium in the bone, which is expected to reduce the risk of osteoporosis. The estimated adequate intake for calcium is 1000 milligrams per day for adults up to 50 years of age. The DRI Committee used similar criteria in setting recommended intakes for phosphorus and magnesium.

Iron continues to be a problem nutrient for most women, with average intakes falling short of recommendations (review Figure 8-3 on p. 324). The RDA for iron remains elevated for women from early adulthood until age 50, when many women enter menopause. With the cessation of iron losses from menstruation, the daily iron needs of women are similar to those of men.

Nutrient Supplements

Many adults take supplements as dietary insurance—as kind of an all-purpose extra food in case they are not meeting their nutrient needs from foods alone. Others take supplements as health insurance—as kind of an all-purpose extra drug to protect against certain diseases. In either case, supplement taking is risky, and the higher the dose, the greater the risk of harm. People's tolerances for high doses of nutrients vary, just as their risks of deficiencies do. Amounts that some can tolerate may be harmful for others, and no one knows who falls where along the spectrum. It is impossible to say just how much of a nutrient is enough—or too much. Assuming, however, that it is best to err on the conservative side, supplement doses should be limited to the recommended intake. Chapter 8 (pp. 324–326) provides a full discussion of nutrient supplements for adults.

In the United States and Canada, adults rarely suffer nutrient deficiencies. They are far more likely to be *over*nourished than *under*nourished. Overconsumption of foods—especially foods high in fat—is today's major diet and health concern.

Diet and Health

A century ago our ancestors feared infectious and communicable diseases such as smallpox—diseases that claimed many children's lives and limited the average life expectancy of adults. Today far fewer infectious diseases threaten us, thanks to medical science's ability to identify disease-causing microorganisms and to develop vaccines. In developed countries, water purification systems prevent the spread of infection, and immunizations protect individuals. Most people live well into their later years, and today's average life expectancy far exceeds that of our ancestors.

Still, some infectious diseases threaten many lives despite public health measures and medical treatments. Perhaps the most infamous is AIDS (acquired immune deficiency syndrome). AIDS develops from infection by the human immunodeficiency virus (HIV), which attacks the immune system and disables the body's defenses against other diseases. Then these diseases—which would produce only mild, if any, illness in people with healthy immune systems—destroy health and life. The only saving grace is that in most cases, HIV infection can be prevented. Transmission of the virus requires sexual activity, direct blood contact, or passage of the infection from a mother to her infant during pregnancy, birth, or breastfeeding.

Although some infectious diseases remain serious threats in developed countries, most of the diseases that threaten life develop as a result of metabolic abnormalities induced by such factors as genetics, age, sex, lifestyle, and environment. Diet is among the many lifestyle factors that influence the risks of developing chronic diseases. Before discussing the specific dietary factors that influence the progression of chronic diseases, it is important to understand risk factors in general.

acquired immune deficiency syndrome (AIDS): the end stage of HIV infection, in which severe complications are manifested. In the early, symptomless stages, the person is said to have an HIV infection.

human immuno-deficiency virus (HIV): the virus that causes AIDS. The infection progresses to become an immune system disorder that leaves its victims defenseless against numerous infections.

Risk Factors

Figure 9-2 shows the ten leading causes of illness and death in the United States.[6] Four of these causes, including the top three, have some relationship with diet. Taken together, these four conditions account for two-thirds of the nation's 2 million deaths each year.

risk factors: certain conditions or behaviors associated with an elevated frequency of a disease but not proven to be causal.

✗ Healthy People 2000

Reduce coronary heart disease deaths to no more than 100 per 100,000 people.

Figure 9-2 The Ten Leading Causes of Illness and Death in the United States

Diet influences the development of several chronic diseases—notably, heart disease, some types of cancer, stroke, and diabetes. Taken together, these four diseases account for about two-thirds of the nation's 2 million deaths each year.

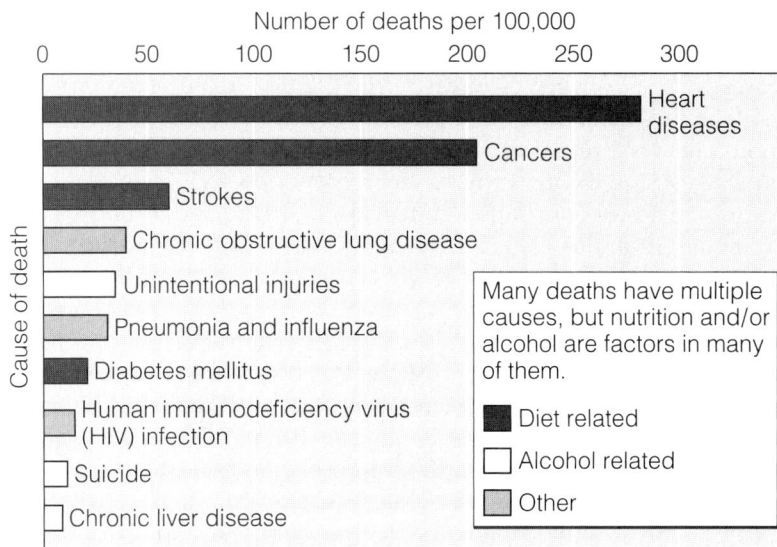

Source: National Center for Health Statistics, *Monthly Vital Statistics Report*, October 1996, p. 31.

⚐ Healthy People 2000

Reverse the rise in cancer deaths to achieve a rate of no more than 130 per 100,000 people.

These "causes" are stated as if a single condition such as heart disease caused death, but most chronic diseases arise from multiple factors over many years. A person who died of heart disease may have had pre-existing conditions such as overweight and high blood pressure, may have been a cigarette smoker, and may have spent years eating a high-fat diet and getting too little exercise.

Of course, not all people who die of heart disease fit this description, nor do all people with these characteristics die of heart disease. People who are overweight may die from the complications of diabetes, instead; those who smoke may die of cancer. They may even die from something totally unrelated to any of these factors, such as an automobile accident. Still, statistical studies show that certain conditions and behaviors increase or reduce the risk of developing certain chronic diseases. On the average, the more risk factors in a person's life, the greater that person's chances of developing the disease. Conversely, the fewer risk factors in a person's life, the better the chances for good health.

Risk Factors Persist. Risk factors tend to persist over time. Without intervention, a young adult with high blood pressure will most likely continue to have high blood pressure as an older adult, for example. To minimize the damage, then, early intervention is most effective.

Risk Factors Cluster. Risk factors also tend to cluster. For example, a person who is overweight is likely to be physically inactive, to have high blood pressure, and to have high blood cholesterol—all risk factors associated with heart disease. Intervention that focuses on one risk factor often benefits the others as well. For example, physical activity can help reduce weight. Both the physical activity and the weight loss help lower blood pressure and blood cholesterol.

Figure 9-3 illustrates some relationships between risk factors and degenerative diseases. Notice how many of the diseases have a genetic component. A family history of a certain disease is a powerful indicator of a person's tendency to contract that disease. Still, lifestyle factors are often pivotal in determining whether that tendency will be expressed. Genetics and lifestyle factors often work synergistically; for instance, cigarette smoking is especially likely to bring on heart disease in people who are genetically predisposed to develop it.

Risk Factors in Perspective. An estimated *half* of all deaths each year can be attributed to specific risk factors, many of which reflect personal behaviors.[7] The most prominent factor contributing to death in the United States is tobacco use, followed by diet and activity patterns and alcohol use (see Table 9-1). For the two out of three Americans who do not smoke or drink alcohol excessively, the one choice that can influence long-term health prospects more than any other is diet.[8]

Some risk factors, such as smoking, dietary habits, physical activity, and alcohol consumption, are personal behaviors that can be changed. Decisions to not smoke, to eat a well-balanced diet, to engage in regular physical activity, and to drink alcohol in moderation (if at all) improve the likelihood that a person will enjoy good health. Other risk factors, such as genetics, sex, and age, also play important roles in the development of chronic diseases, but they cannot be altered. Health recommendations acknowledge the influence of such factors on the development of disease, but must focus on those that are modifiable. The following section highlights health recommendations and some important relationships between specific dietary components and disease development.

Dietary Recommendations and Risk Factors

People vary in their hereditary susceptibility to diseases and their responsiveness to dietary measures. Unlike nutrient deficiency diseases, which develop when nutrients are lacking and disappear when the

Other risk factors for heart disease include smoking, high blood pressure, diabetes, family history, sex, race, and obesity.

When multiple factors act together in such a way that their combined effects are greater than the sum of their individual effects, they are working synergistically.

Table 9-1 Actual Causes of Death in the United States (1990)

Cause	Percentage of Total Deaths
Tobacco	19
Diet /activity	14
Alcohol	5
Microbial agents	4
Toxic agents	3
Firearms	2
Sexual behavior	1
Motor vehicles	1
Illicit drugs	<1
Total	50

Note: Half of all deaths each year can be attributed to specific risk factors.

Source: J. M. McGinnis and W. H. Foege, Actual causes of death in the United States, *Journal of the American Medical Association* 270 (1993): 2207–2212.

Figure 9-3 Risk Factors and Degenerative Diseases

The chart at the top shows that the same risk factor can affect many chronic diseases. Notice, for example, how many diseases have been linked to a high-fat diet. The chart also shows that a particular disease, such as atherosclerosis, may have several risk factors.

The flow chart at the bottom shows that many of these conditions are themselves risk factors for other chronic diseases. For example, a person with diabetes is likely to develop atherosclerosis and hypertension. These two conditions, in turn, worsen each other. Notice how all these diseases are linked to obesity.

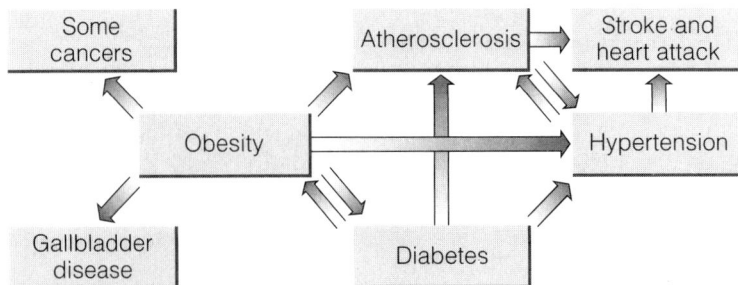

Degenerative diseases	Diet risk factors						Other risk factors					
	High-fat diet	Excessive alcohol intake	Low complex carbohydrate/fiber intake	Low vitamin and/or mineral intake	High sugar intake	High intake of salty or pickled foods	Genetics	Age	Sedentary lifestyle	Smoking and tobacco use	Stress	Environmental contaminants
Cancers	✔	✔	✔	✔		✔	✔	✔	✔	✔		✔
Hypertension	✔	✔		✔		✔	✔	✔	✔	✔	✔	
Diabetes (NIDDM)	✔		✔				✔	✔	✔			
Osteoporosis		✔		✔			✔	✔	✔	✔		
Atherosclerosis	✔	✔	✔	✔			✔	✔	✔	✔	✔	✔
Obesity	✔	✔	✔		✔		✔		✔			
Stroke	✔	✔	✔				✔	✔		✔	✔	
Diverticulosis	✔		✔	✔					✔	✔		
Dental and oral disease	✔				✔	✔	✔			✔		

Flow chart: Obesity → Some cancers; Obesity → Gallbladder disease; Obesity → Atherosclerosis; Obesity ↔ Diabetes → Hypertension; Atherosclerosis ↔ Hypertension; Atherosclerosis → Stroke and heart attack; Hypertension → Stroke and heart attack; Diabetes → Atherosclerosis.

nutrients are provided, chronic diseases are neither caused nor prevented by diet alone. Many people have followed dietary advice and developed heart disease or cancer anyway; others have ignored all advice and lived long and healthy lives. For many people, though, diet does influence the time of onset and course of some chronic diseases, and many health care professionals urge dietary measures as part of a disease prevention strategy.

After considering the results of over 7000 studies examining these relationships between various dietary components and health, the Committee on Diet and Health concluded that most people can gain some disease prevention benefits by making dietary changes.[9] To that end, the *Diet and Health* recommendations were developed (see Table 9-2).[10]

The *Diet and Health* recommendations were established for "the potential public health benefit, and the likelihood of minimal risk."[11] They are aimed at reducing disease risks and describe the kinds of foods people should include, limit, or avoid. They also address weight maintenance and physical activity and pinpoint trouble areas surrounding specific nutrients. Like the *Diet and Health* report, the *Nutrition Recommendations for Canadians* report makes recommendations that will supply enough nutrients, while reducing the risk of chronic disease (see Table 9-3).

Fat Of all the nutrients, fat is most often linked with chronic diseases. A high-fat diet raises the risks of heart disease, some types of cancer, and obesity. Fortunately, the same recommendation can help with all these health problems: Choose a diet that provides no more than 30 percent of total kcalories from fat.

Most people realize that elevated blood cholesterol is a major risk factor for cardiovascular disease. Cholesterol accumulates in plaques, restricting blood flow in the arteries and raising blood pressure. The consequences are deadly; in fact, heart disease is the nation's number one killer of adults. Blood cholesterol is often used to predict the likelihood of a person's suffering a heart attack or stroke; the higher the cholesterol, the earlier and more likely the tragedy. What most people don't realize, though, is that dietary fat, and specifically *saturated fat*, raises blood cholesterol even more dramatically than does cholesterol ingested in food. Recommendations to reduce total fat add a note to reduce saturated fat intake to less than 10 percent of total kcalories and cholesterol intake to less than 300 milligrams. The criteria for defining blood lipids, blood pressure, and obesity in relation to cardiovascular disease risk are shown in Table 9-4.

cardiovascular disease (CVD): a general term for all diseases of the heart and blood vessels. Atherosclerosis is the main cause of CVD. When the arteries that carry blood to the heart muscle become occluded, the heart suffers damage known as **coronary heart disease (CHD)** (*cardio* = heart; *vascular* = blood vessels).

plaques (PLACKS): mounds of lipid material, mixed with smooth muscle cells and calcium, that develop in the artery walls in atherosclerosis. This type of plaque is known as **atheromatous** plaque (*placken* = patch or plate).

✷ **Healthy People 2000**

Reduce dietary fat intake to an average of 30% of energy or less and average saturated fat intake to less than 10% of energy among people aged two years and older.

Table 9-2 *Diet and Health* Recommendations

- Reduce total *fat* intake to 30 percent or less of kcalories. Reduce saturated fatty acid intake to less than 10 percent of kcalories and the intake of cholesterol to less than 300 milligrams daily.
- Increase intake of starches and other *complex carbohydrates*.
- Maintain *protein* intake at moderate levels.
- Balance food intake and physical activity to maintain appropriate *body weight*.
- For those who drink *alcoholic beverages*, limit consumption to the equivalent of less than 1 ounce of pure alcohol in a single day. Pregnant women should avoid alcoholic beverages.
- Limit total daily intake of *salt* (sodium chloride) to 6 grams or less.
- Maintain adequate *calcium* intake.
- Avoid taking dietary *supplements* in excess of the RDA in any one day.
- Maintain an optimal intake of *fluoride*, particularly during the years of primary and secondary tooth formation and growth.

Note: Italics added to highlight the areas of concern.

Source: Adapted from Committee on Diet and Health, *Diet and Health: Implications for Reducing Chronic Disease Risk* (Washington, D.C.: National Academy Press, 1989).

Table 9-3 *Nutrition Recommendations for Canadians*

- The Canadian diet should provide energy consistent with the maintenance of *body weight* within the recommended range.
- The Canadian diet should include *essential nutrients* in amounts recommended.
- The Canadian diet should include no more than 30 percent of energy as *fat* (33 grams/1000 kcalories or 39 grams/5000 kilojoules) and no more than 10 percent as saturated fat (11 grams/1000 kcalories or 13 grams/5000 kilojoules).
- The Canadian diet should provide 55 percent of energy as *carbohydrate* (138 grams/1000 kcalories or 165 grams/5000 kilojoules) from a variety of sources.
- The *sodium* content of the Canadian diet should be reduced.
- The Canadian diet should include no more than 5 percent of total energy as *alcohol*, or two drinks daily, whichever is less.
- The Canadian diet should contain no more *caffeine* than the equivalent of four regular cups of coffee per day.
- Community water supplies containing less than 1 milligram per liter should be *fluoridated* to that level.

Note: Italics added to highlight areas of concern.

Source: Health and Welfare Canada, *Nutrition Recommendations: The Report of the Scientific Review Committee* (Ottawa: Canadian Government Publishing Centre, 1990).

Table 9-4 Standards for Heart Disease Risk Factors

LDL Cholesterol	Total Cholesterol[b]
<130 mg/dL = desirable.[a] 130–159 mg/dL = borderline high. ≥160 mg/dL = high.	<200 mg/dL = desirable. 200–239 mg/dL = borderline high. ≥240 mg/dL = high.

HDL Cholesterol	Triglycerides (Fasting)[d]

HDL: ≤35 mg/dL indicates risk.[c]
LDL-to-HDL ratio:[e]
  Men: >5.0 indicates risk.
  Women: >4.5 indicates risk.
	<200 mg/dL = desirable.
	200–400 mg/dL = borderline high.
	400–1000 mg/dL = high.
	>1000 mg/dL = very high.

Hypertension	Obesity

Diastolic pressure:[f]
  <85 = normal.
  85–89 = high-normal.
  90–99 = mild.
  100–109 = moderate.
  110–119 = severe.
  >120 = very severe.
	Body mass index:
	  Men: >27.8.
	  Women: >27.3.

[a]For people with existing heart disease, desirable LDL cholesterol values are lower (≤100 mg/dL).

[b]To convert cholesterol (mg/dL) to standard international units (mmol/L), multiply by 0.02586.

[c]This HDL value may be too low for women; no alternative value has yet been proposed. NIH Consensus Development Panel on Triglyceride, High-Density Lipoprotein, and Coronary Heart Disease, Triglyceride, high-density lipoprotein, and coronary heart disease, *Journal of the American Medical Association* 269 (1993): 505–510.

[d]High triglycerides alone normally do not indicate *direct* risk, but may reflect lipoprotein abnormalities associated with CHD. The risk of CHD increases as triglyceride levels increase in people with other risk factors. High triglycerides also occur in conditions such as kidney disease and diabetes, which suggest a high CHD risk.

[e]The LDL-to-HDL ratio compares the concentration of LDL to that of HDL. A ratio of 5.0 really means "5 to 1," or that the first value (in this case LDL) is 5 times greater than the second (HDL). Similarly, a ratio of 4.5 means that LDL is 4.5 times greater than HDL.

[f]Diastolic pressure is the lower of the numbers in the blood pressure reading—for example, the 70 in 105/70.

Source: Blood lipid standards adapted from The Expert Panel, Summary of the second report of the National Cholesterol Education Program (NCEP), Expert Panel on Detection, Evaluation, and Treatment of High Blood Cholesterol in Adults (Adult Treatment Panel), *Journal of the American Medical Association* 269 (1993): 3015–3023; Hypertension standards adapted from the Fifth Report of the Joint National Committee on Detection, Evaluation, and Treatment of High Blood Pressure, National High Blood Pressure Education Program, National Heart, Lung, and Blood Institute, National Institutes of Health, October 30, 1992, p. 5.

Other risk factors for cancer include smoking, alcohol, and environmental contaminants.

Enjoy low-fat foods for good heart health.

The evidence linking dietary fats with cancer is less conclusive than for heart disease, but it does suggest an association between total fat and some types of cancers. Dietary fat seems not to *initiate* cancer development but to *promote* cancer once it has arisen. Some epidemiological studies suggest a relationship between specific cancers and saturated fats or dietary fat from animal sources (which is mostly saturated). Thus health advice to reduce cancer risks parallels that given to reduce heart disease risks: Reduce total fat intake, especially saturated fat.

Animal studies suggest that *polyunsaturated fats* from vegetable oils (mostly omega-6 fatty acids) are more likely to promote cancer than are saturated fats. On the other hand, polyunsaturated fats from fish oils (mostly omega-3 fatty acids) are likely to delay cancer development and reduce the rate of growth and the size and number of tumors.[12]

Fat accounts for a lot of the energy in some foods, and removing fat from foods cuts energy intake dramatically. Fat contributes twice as many kcalories per gram as either carbohydrate or protein. Consequently, people who eat high-fat diets tend to exceed their energy needs and gain weight.

Complex Carbohydrates. Foods rich in complex carbohydrates tend to be low in fat and can therefore promote weight loss by providing less food energy per bite. They also provide satiety and delay hunger. In addition, fiber-rich foods slow the rate at which food leaves the stomach and draw water into the GI tract, prolonging the satiety enjoyed from carbohydrate-rich meals.

🏃 **Healthy People 2000**

Increase complex carbohydrate and fiber-containing foods in the diets of adults to five or more daily servings for vegetables (including legumes) and fruits and to six or more daily servings for grain products.

High-carbohydrate diets are associated with low blood cholesterol and a low risk of heart disease.[13] Sorting out the exact reasons why can be difficult. Such diets are low in animal fat and cholesterol and high in soluble fibers and vegetable proteins—all factors associated with a lower risk of heart disease.

Foods rich in soluble fibers (such as oat bran, barley, and legumes) lower blood cholesterol by binding with bile, the emulsifier that otherwise would assist with fat and cholesterol absorption.[14] With less bile available, less fat and cholesterol are absorbed, and blood cholesterol declines. Then the liver makes more bile from cholesterol to compensate for the bile bound to fiber and excreted in the GI tract, which reduces blood cholesterol further.[15]

Several researchers have speculated that fiber also exerts an indirect cholesterol-lowering effect by displacing fats in the diet.[16] Even when dietary fat is low, however, research shows that high intakes of soluble fibers exert a separate and significant cholesterol-lowering effect.[17]

A high-carbohydrate diet, especially one that includes plenty of green and yellow vegetables and citrus fruits, protects against some types of cancer. Again, it is unclear whether the beneficial effects of these fruits and vegetables derive from their fiber, their vitamins, or even other, nonnutrient compounds called *phytochemicals*. Like vitamins C and E, many phytochemicals (such as beta-carotene and flavonoids) act as antioxidants in the body. Table 9-5 summarizes the common food sources and actions of selected phytochemicals.

phytochemicals: nonnutrient compounds in plant-derived foods that have biological activity in the body (*phyto* = plant).

Fiber may help prevent colon cancer specifically by diluting, binding, and rapidly removing potentially cancer-causing agents from the colon. Alternatively, the protective effect may be due to the fermentation of resistant starch and fiber in the colon, which lowers the pH. A decreased pH in the colon is associated with decreased colon cancer risks.[18]

Populations eating high-carbohydrate diets often have low rates of diabetes, most likely because such diets are low in fat. High-carbohydrate, low-fat diets help control weight, and this is the most effective way to prevent the most common type of diabetes (NIDDM). Furthermore, when soluble fibers trap nutrients and delay their exit from the stomach, glucose absorption is slowed, and this helps to prevent the glucose surge and rebound that seem to be associated with diabetes onset.

Dietary fibers also enhance the health of the large intestine. The healthier the intestinal walls, the better they can block absorption of unwanted constituents, such as bacteria. Fibers enlarge the stools, easing passage, and they speed up transit time. Insoluble fibers such as cellulose (as in cereal brans, fruits, and vegetables) are most important in this regard. Their undigested residue, together with the microbial growth they stimulate, enlarges the stools, helping to alleviate or prevent constipation. Fibers also stimulate microbial digestion of absorbable products.

Protein. The relationships between protein and chronic diseases are not clear. Population studies have difficulty determining whether diseases correlate with animal proteins or with their accompanying saturated fats. Studies that rely on data from vegetarians must sort out the many lifestyle factors, other than a "no-meat diet," that might explain relationships between protein and health. Because overconsumption of protein offers no benefits and may pose health risks, recommendations aim at moderate intakes.

Body Weight. The health risks of excessive body fat are so many that it has been declared a disease: obesity.[19] Among the health risks of obesity are diabetes, hypertension, cardiovascular disease, sleep apnea

Table 9-5 Phytochemicals—Their Food Sources and Actions

Food Source	Name	Action in the Body
Deeply pigmented fruits and vegetables (carrots, sweet potatoes, tomatoes, spinach, broccoli, cantaloupe, pumpkin, apricots)	Carotenoids[a] (including beta-carotene)	Act as antioxidants, reducing the risk of cancer.
Citrus fruits	Limonene	Triggers enzyme production to facilitate carcinogen excretion.
	Phenols	Inhibit lipid oxidation; block formation of carcinogenic nitrosamines in the body.
Garlic/onions	Allyl sulfides	Trigger enzyme production to facilitate carcinogen excretion.
Broccoli and other cruciferous vegetables (cauliflower and brussels sprouts)	Sulforaphane	Protects against cancer.
	Dithiolthiones	Trigger enzyme production to block carcinogen damage to cells' DNA.
	Indoles	Trigger enzymes to inhibit estrogen action, reducing the risk of breast cancer.
	Isothiocyanates	Trigger enzyme production to block carcinogen damage of cells' DNA.
Grapes	Ellagic acid	Scavenges carcinogens.
Soy/legumes	Protease inhibitors	Suppress enzyme production in cancer cells, slowing tumor growth.
	Phytosterols	Inhibit cell reproduction in GI tract, preventing colon cancer.
	Isoflavones[b]	Block estrogen activity in cells, reducing the risk of breast and ovarian cancer.
	Saponins	Interfere with DNA reproduction, preventing cancer cell multiplication.
Flaxseed	Lignans[b]	Block estrogen activity in cells, reducing the risk of breast and ovarian cancer.
Fruits (blueberries, prunes, grapes), oats, soybeans	Caffeic acid	Triggers enzyme production to make carcinogens water-soluble, facilitating excretion.
	Ferulic acid	Binds to nitrates in stomach, preventing the conversion to nitrosamines.
Grains	Phytic acid	Binds to minerals, preventing cancer-causing free-radical formation.
Fruits, vegetables, tea, wine, oregano	Flavonoids	Act as antioxidants, reducing the risk of cancer.

[a]In addition to beta-carotene, other carotenoids include alpha-carotene, beta-cryptoxanthin, lectein, zeaxanthin, and lycopene.

[b]Isoflavones and lignans are types of phytoestrogens—compounds that bind to estrogen receptors and reduce estrogen activity.

(abnormal ceasing of breathing during sleep), osteoarthritis, abdominal hernias, some cancers, varicose veins, gout, gallbladder disease, respiratory problems (including Pickwickian syndrome, a breathing blockage linked with sudden death), liver malfunction, complications in pregnancy and surgery, flat feet, and even a high accident rate. The costs of these obesity-related illnesses were estimated at $45.8 billion in 1990.[20] The costs in terms of lives are also great. Mortality increases as excess weight increases (see Figure 9-4).[21] People with a body mass index greater than 35 are twice as likely to die prematurely as those with a BMI of 21 to 25.[22]

The distribution of fat on the body may be more critical than fatness alone. Visceral fat stored around the organs of the abdomen presents a greater risk to health than fat elsewhere on the body and increases the risk of premature death.[23] This distribution of fat is referred to as central obesity or upper-body fat and, independently of total body fat, is associated with increased risks of heart disease, stroke, diabetes, hypertension, and some types of cancer.[24]

Abdominal fat is common in women past menopause and even more common in men. Even when total body fat is similar, men have more abdominal fat than either premenopausal or postmenopausal women.[25] Interestingly, people with central obesity smoke more and drink alcohol more than the average. A smoker may weigh less than the average nonsmoker, but the smoker's central obesity may be greater, leading researchers to think that smoking may directly affect body fat distribution.[26] Physical activity, in contrast, correlates negatively with central obesity.

Figure 9-4 Body Mass Index and Mortality

This J-shaped curve describes the relationship between body mass index (BMI) and mortality and shows that optimal BMI is between 21 and 25 (some researchers extend this range from 19 to 27).

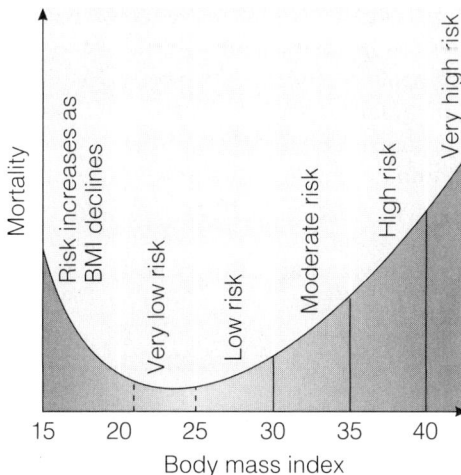

Fat around the hips and thighs, sometimes referred to as lower-body fat, is most common in women in their reproductive years and seems relatively harmless. In fact, people who are overweight, but who do not have excessive fat around the abdomen "seem robust" and less susceptible to health problems than overweight people with central obesity; theirs is a benign obesity.[27]

The relationship between obesity and cardiovascular disease risk is strong, with links to both blood cholesterol and blood pressure. Central obesity may raise the risk of heart disease as much as the leading three risk factors (high blood cholesterol, hypertension, and smoking) do.[28] Weight loss is effective in preventing and treating hypertension in overweight people.[29]

Diabetes (NIDDM) is three times more likely to develop in an obese person than in a nonobese person. Furthermore, the diabetic person often has central obesity. Central-body fat cells appear to be larger and more insulin-resistant than lower-body fat cells, and insulin resistance is a major risk factor for the development of NIDDM.[30] For those who are overweight, weight loss effectively improves glucose tolerance and lowers insulin resistance.

The risk of cancer increases with body weight, but researchers do not fully understand the relationship. One possible explanation may be that obese people have elevated levels of hormones that could influence cancer development. For example, adipose tissue is a significant site of estrogen synthesis in women (especially after menopause), obese women have elevated levels of estrogen, and estrogen has been implicated in the development of cancers of the female reproductive system. These cancers account for half of all cancers in women.[31]

Antioxidant Nutrients. Recommendations to increase complex carbohydrate intake parallel those to choose a diet with plenty of vegetables, fruits, and grain products. Among the many benefits of these foods is their abundance of antioxidant nutrients. Epidemiological reports indicate a correlation between low intakes of foods rich in antioxidant nutrients and high cancer rates. Antioxidant nutrients may reduce cancer risks by protecting cellular DNA from free-radical damage.[32]

Many studies using animals as subjects have shown that beta-carotene protects against cancer.[33] The protective effects vary, depending on the experimental conditions, the types of animals tested, and the cancer types and sites.[34]

Research on human beings produces diverse results. Researchers have compared groups of people with high cancer rates with groups that have low cancer rates, but are similar in other characteristics—for example, smoking history and age. They report a consistent relationship between low intakes of vegetables and fruits (specifically of those containing beta-carotene and its relatives) and high rates of lung cancer.[35] Such findings suggest that beta-carotene is protective, but researchers are quick to add that other constituents of fruits and vegetables may be responsible for the effect. When researchers collect blood samples, though, they find that low concentrations of beta-carotene consistently correlate with the development of both lung and breast cancers.[36]

Research has not always made it clear whether preformed vitamin A, beta-carotene, or both are protective against cancer.[37] Some research indicates that dietary vitamin A—from whatever source—seems to play a role in inhibiting the development of breast cancer.[38] Evidence of protection against cancers of the colon or prostate is less convincing.

Large-scale studies of populations suggest that vitamin C also helps protect against certain types of cancers, especially those of the mouth, larynx, and esophagus.[39] A dozen or so different studies have correlated high vitamin C intakes with low rates of cancer. Such a correlation may reflect the benefits of a diet rich in fruits and vegetables and low in fat and does not necessarily support the taking of vitamin C supplements to treat or prevent cancer.

Some research suggests that vitamin C protects against stomach cancer specifically, by preventing the formation of carcinogenic nitrite compounds in the stomach.[40] More research is needed, but the results so far are promising.

Evidence that vitamin E helps guard against cancer is less consistent than for beta-carotene and vitamin C. One large population study showed the highest risks of certain cancers in the people with lowest blood vitamin E.[41] The association was strongest for some gastrointestinal cancers and for cancers not related to smoking.

Much of the research on antioxidant nutrients has focused on cancer prevention, but antioxidants, especially vitamin E, may protect against cardiovascular disease as well.[42] Research confirms that high blood cholesterol carried in low-density lipoproteins (LDL) correlates directly with cardiovascular disease. Researchers are now asking how LDL exert their damage. Some of the most promising research suggests that LDL first undergo oxidation by free radicals inside the artery wall and then promote the formation of artery-clogging plaques.[43] Evidence thus far is persuasive but not conclusive.

If oxidized LDL are a factor in heart disease, might antioxidant nutrients offer some protection? Some research suggests that they do. Findings from an epidemiological study suggest a negative correlation between vitamin E status and death rates from heart disease.[44] Researchers selected groups of men in 16 European regions where rates of death from heart disease varied sixfold. The researchers measured plasma vitamin E, cholesterol, and blood pressure in men from each region. When the groups were compared, high death rates from heart disease correlated more strongly with low vitamin E concentrations than with either cholesterol or blood pressure. The authors cautioned that the evidence for the "antioxidant hypothesis" of heart disease was suggestive, but indirect.

Two other large epidemiological studies found that large doses of vitamin E supplements were associated with a significantly reduced risk of heart disease.[45] This correlation remained strong after the researchers analyzed for coronary risk factors and other dietary antioxidants.

Vitamin C may also affect the susceptibility of LDL to oxidation. Some epidemiological studies have found an association between vitamin C and cardiovascular disease; others have not.[46] Research suggests a synergism between vitamin C and vitamin E in defending LDL against oxidation: Vitamin C defends against free radicals in the watery compartments of cells, and vitamin E acts in lipid environments. Together, they effectively protect LDL against oxidation. In addition, vitamin C may regenerate vitamin E from its oxidized form, making it available to act as an antioxidant once again.[47] Some studies also suggest that vitamin C may raise HDL, lower total cholesterol, and improve blood pressure.[48]

Alcoholic Beverages. For many adults, two to three alcoholic drinks set in motion a series of destructive processes in the body, but the next day's abstinence reverses them. Recovery is probably complete as long as the doses are moderate, time between them is ample, and nutrition is adequate.

If the doses of alcohol are heavy and the time between them is short, complete recovery cannot take place. Repeated onslaughts of alcohol gradually take a toll on all parts of the body and increase the risk of several diseases, as Table FP9-3 in Focal Point 9 shows. Compared with nondrinkers, heavy drinkers have significantly greater risks of dying from all causes.[49]

Many studies over the last few decades suggest that *moderate* (one or two drinks a day) alcohol consumption reduces the risk of heart disease by raising HDL cholesterol and preventing blood clot formation.[50] A recent, large study of more than 22,000 men confirmed these findings and revealed that the benefit was most apparent when intake was even more moderate. Men who drank two to six drinks *per week* reduced their risk of death from all causes compared to those who either drank less than one drink per week or one or more drinks per day.[51] The absence of an overall benefit to cardiovascular mortality among those who drink one or more drinks per day should serve to calm enthusiasm about the promotion of alcohol's protection against heart disease.

Salt. For years, a high *sodium* intake was considered *the* primary factor responsible for high blood pressure. Then research pointed to *salt* (sodium chloride) as the dietary culprit. Salt has a greater effect on

---

✗ **Healthy People 2000**

Decrease salt and sodium intake so at least 65% of home meal preparers prepare foods without adding salt, at least 80% of people avoid using salt at the table, and at least 40% of adults regularly purchase foods modified or lower in sodium.

blood pressure than either sodium or chloride alone or in combination with other ions.[52]

Some individuals are genetically sensitive and experience high blood pressure from excesses in salt intake. People with chronic renal disease, those who have one or two parents with hypertension, blacks, and people over 50 years of age are most likely to be salt sensitive.* Salt avoidance promises to help prevent hypertension in salt-sensitive individuals, but for the majority of people with hypertension, salt restriction does not lower blood pressure. The most effective dietary treatment for hypertension is weight loss.

Calcium. Calcium may also be useful in both preventing and treating hypertension. Epidemiological studies show that low dietary calcium correlates with a high prevalence of hypertension.[53] Reports that an increase in calcium intake can lower blood pressure in people with and without hypertension further support this hypothesis. Some evidence also suggests relationships among dietary calcium and blood cholesterol, diabetes, and cancer.[54]

★ **Healthy People 2000**

Increase calcium intake so that at least 50% of people aged 25 and older consume two or more servings of calcium-rich foods daily.

Perhaps calcium's most famous role in disease prevention is in building strong bones to protect against osteoporosis. As important as calcium may be to bone health, osteoporosis is not a calcium deficiency disease comparable to iron deficiency anemia. In iron deficiency anemia, high iron intakes reliably reverse the condition; in osteoporosis, however, high calcium intakes alone during adulthood may do little or nothing to reverse bone loss. As mentioned in earlier chapters of this book, an adequate calcium intake early in life helps most to grow a healthy skeleton that can defend itself against bone loss in later life.

Supplements. Dietary recommendations have long advised against taking supplements in excess of the RDA in any one day. All nutrients can cause harm when taken in sufficiently large amounts. The latest research showing that the antioxidant nutrients might help prevent such life-threatening diseases as cancer and cardiovascular disease has sparked new controversies.[55] Some research suggests a protective effect from as little as a daily glass of orange juice and carrot juice (rich sources

---

*Salt-sensitive individuals have elevated concentrations of the enzyme renin in their blood, compared with others.

of vitamin C and beta-carotene, respectively).[56] Other intervention studies, however, are using levels of nutrients that far exceed the RDA and can only be achieved by taking supplements. What if research finds a true benefit in taking vitamin pills as opposed to eating a healthy diet alone? Members of the Food and Nutrition Board of the National Research Council are reconsidering their long-held bias against vitamin supplements. They realize that it may be necessary to broaden the concept of the RDA to include both a recommended daily intake to prevent classic deficiency diseases and another substantially higher intake to help protect against chronic diseases.[57]

While awaiting final answers, should people anticipate the go-ahead and start taking vitamin E or other antioxidant supplements now?[58] Most scientists agree that it is too early to make such a recommendation. Although fruits and vegetables that contain many antioxidant nutrients have been associated with a diminished risk of many cancers, supplements of beta-carotene and vitamins C and E have not always proven beneficial.[59] Clinical studies will take several years to complete, and until they prove a clear benefit from taking antioxidant supplements, it would be irresponsible for health care professionals to make such recommendations. We do not know the consequences of taking large doses of antioxidants, even naturally occurring ones, over the long term, much less over a lifetime. Without data to confirm the benefits, we cannot accept the potential risks. And the risks are real.

Consider the findings from a study to determine whether daily supplements of vitamin E, beta-carotene, or both would reduce the incidence of lung cancer among smokers.[60] After five to eight years of supplementation, there was no reduction in the incidence of lung cancer; in fact, the incidence of lung cancer was higher among those receiving the beta-carotene. Such findings were surprising, to say the least, especially given the association between high beta-carotene intakes and low rates of lung cancer reported in earlier epidemiological studies. These discrepancies highlight the importance of considering research findings from a variety of studies and the need for replication. The findings also suggest that remedies to life-threatening diseases such as lung cancer may not be as simple as taking daily pills. Smokers are much wiser to stop smoking than to rely on vitamin supplements to protect them from lung cancer.

Much more research is needed to define optimal and dangerous levels of intake. The Food and Drug Administration (FDA) is now in the process of establishing guidelines for the safe use of nutrient supplements in quantities greater than those needed to meet basic nutrient requirements. This much we know: Antioxidant nutrients behave differently at various levels of intake. At physiological levels typical of a healthy diet, they act as antioxidants, but at the pharmacological doses typical of supplements, they may act as *pro-oxidants, stimulating* the production of free radicals, especially when metal ions such as iron are present.[61] As long as the risks of supplement use remain unclear, the best way to supplement antioxidant nutrients is to eat generous servings of fruits and vegetables, especially citrus fruits and green and yellow vegetables.

Fluoride. Dental caries ranks as the nation's most widespread health problem: An estimated 95 percent of the population have decayed, missing, or filled teeth. Dental problems interfere with a person's ability to chew and eat a wide variety of foods, and this can lead to a multitude of nutrition problems. Where fluoride is lacking, dental decay is common. By fluoridating the drinking water, a community offers its residents a safe, economical, practical, and effective way to defend against dental caries.[62]

All normal diets contain some fluoride, but drinking water is usually the most significant source. The National Research Council of the National Academy of Sciences recommends fluoridation of drinking water to raise the concentration to about 1 part fluoride per 1 million parts water. Water with 1 part per million (1 ppm) fluoride offers the greatest protection against dental caries at virtually no risk of toxicity.

1 ppm = 1 mg per liter.

Recommendations for the Population. The *Diet and Health* recommendations focus on the general population in the hope that all people at all levels of risk may benefit. Such a strategy is similar to national efforts to vaccinate to prevent polio, fluoridate water to prevent dental caries, and fortify grains to prevent iron deficiency.

Recommendations for Individuals. To determine whether dietary recommendations are important to you personally, look at your family history to see which diseases are common to your relatives. In addition, examine your personal history, taking note of your blood pressure, blood test results, and lifestyle habits such as smoking. Figure 9-5 presents a hypothetical "medical family tree."

Recommendations that urge all people to make dietary changes believed to forestall or prevent diseases are taking a preventive or population approach. Alternatively, recommendations that urge dietary changes only for people who are known to need them are taking a medical or individual approach.

In learning about chronic diseases and their major links with nutrition, some people may be left with the mistaken impression of "one disease–one nutrient" relationships.[63] Indeed, valid links do exist between fiber and diabetes, fat and heart disease, calcium and osteoporosis, and antioxidant vitamins and cancer, but to focus only on these links oversimplifies the story. In reality, each nutrient may have connections with several diseases, because its role in the body is not specific to a disease, but to a body function. Fiber—because it binds substances in the GI tract—helps prevent both diabetes and cancer. Vitamin C—because it acts as an antioxidant—helps prevent both cancer and heart disease. Furthermore, each of the chronic diseases develops in response to multiple factors—including many nondietary factors such as genetics, physical activity, and smoking. An integrated and balanced approach to disease prevention therefore includes attention to all the many factors involved.

Figure 9-5 Hypothetical Medical Family Tree

A "medical family tree" notes the types of diseases family members have had, their ages at the times of major medical events or death, and their personal medical histories.

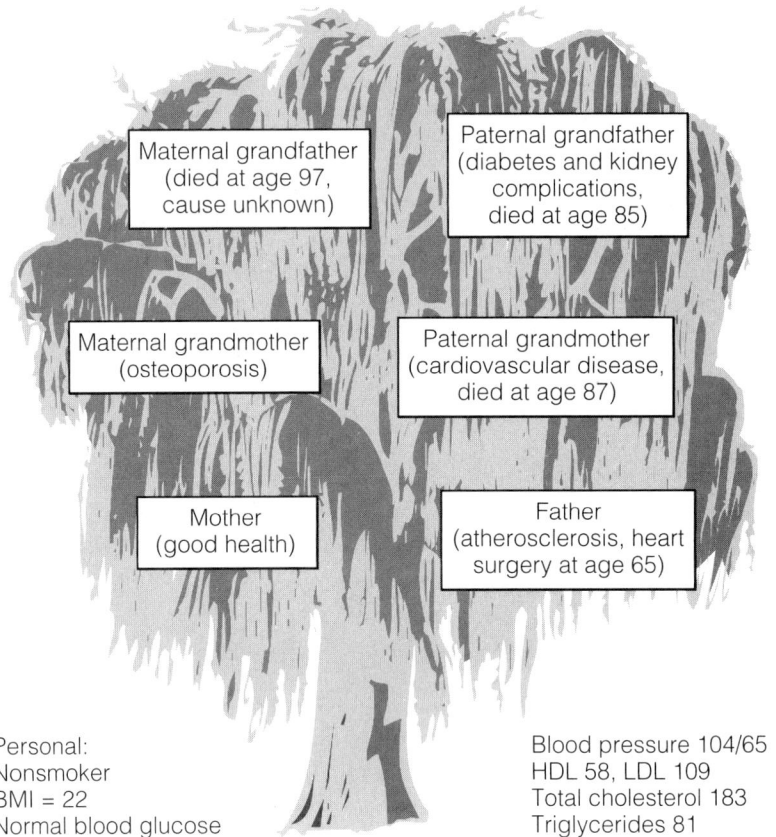

Maternal grandfather
(died at age 97,
cause unknown)

Paternal grandfather
(diabetes and kidney
complications,
died at age 85)

Maternal grandmother
(osteoporosis)

Paternal grandmother
(cardiovascular disease,
died at age 87)

Mother
(good health)

Father
(atherosclerosis, heart
surgery at age 65)

Personal:
Nonsmoker
BMI = 22
Normal blood glucose

Blood pressure 104/65
HDL 58, LDL 109
Total cholesterol 183
Triglycerides 81

Food Choices and Fitness during Middle Adulthood

The life-threatening diseases of heart disease, cancer, and diabetes have distinct sets of causes, yet all are responsive to diet, and in some ways the responses are similar. Dietary excesses, particularly excess food energy and fat intakes, increase the likelihood of these diseases. Not all diet recommendations apply equally to all the diseases, but fortunately for the consumer, dietary recommendations do not contradict one another. In fact, they support each other. A person designing a low-fat diet would need to select plenty of vegetables, fruits, and grain products—foods that supply fiber, vitamins, and minerals in abundance. Especially when combined with physical activity, such a diet would help a person to maintain a healthy weight—and weight control is central to disease prevention.

Clearly, a nutritious diet plays a key role in keeping people healthy.

Food Choices and Eating Habits

What should an adult eat to stay healthy? The *Dietary Guidelines for Americans* (p. 14) provide reasonable answers to that question, based on the *Diet and Health* recommendations. They focus on weight control and urge people to limit fat, increase complex carbohydrates, and balance food intake with activity.

Limit Fat. To reduce dietary fat, eliminate fat as a seasoning and in cooking; remove the fat from high-fat foods; replace high-fat foods with low-fat alternatives; and emphasize grains, fruits, and vegetables. Table 8-6 (pp. 338–339) provides additional tips for reducing fat in the diet, food group by food group. Selecting lean meats and nonfat milk products helps lower saturated fat intake. Eating less fat from meat, eggs, and milk products also helps lower dietary cholesterol intake (as well as total and saturated fat intakes). Over the past three decades, many adults have followed such advice, and both their fat intakes and their cholesterol levels have declined steadily.[64]

Increase Complex Carbohydrate. Choosing vegetables, fruits, cereals, and legumes increases complex carbohydrate intake and lowers fat intake. Vegetables and fruits contain no fat, and most grains contain only trace amounts. Some grain *products* such as fried taco shells, croissants, and granola cereal are high in fat, though, so consumers need to read product labels. A diet rich in vegetables, fruits, cereals, and legumes also offers abundant vitamin C, folate, vitamin A, beta-carotene, and dietary fiber—all important in supporting health. Consequently such a diet protects against disease by lowering energy intake, reducing fat, increasing nutrients, and providing valuable phytochemicals.[65]

health claim: any statement that characterizes the relationship between any nutrient or other substance in a food and a disease or health-related condition.

Read Food Labels. Chapter 1 introduced food labels, and Chapter 8 provided a quick lesson in reading food labels for nutrition information. Some food labels also feature *health* information. Health claims describe an association between a specific nutrient or food substance and a specific health problem. They are permitted only in cases where scientifically valid links between diet and health have been clearly established. The FDA has approved several health claims based on specified criteria, including:

■ The nutrient or food substance must be related to a disease or health condition for which most people or a specific group of people, such as the elderly, are at risk.

■ The claim is supported by scientific evidence from well-designed studies conducted with recognized scientific procedures and principles.

Health claims on products must emphasize the importance of the total diet and must not exaggerate the role of a particular food or diet in disease prevention. No one food possesses magical healing powers, and manufacturers must take care not to distort the roles of their products in promoting health.

Claims must be honest and balanced. For example, health claims can say that foods high in calcium "may" or "might" reduce the risk of osteoporosis. Claims must also explain that diseases develop in response to many factors. They may even mention beneficial factors, such as exercise. For example, a health claim may state that "Development of cancer depends on many factors. A diet low in total fat may reduce the risk of some cancers." This health claim is true, it acknowledges that diet is among factors influencing disease development, and it is phrased in terms of total diet, not in terms of the particular product. The following relationships for health claims on labels have been authorized:

■ *Calcium and osteoporosis.* Foods or supplements must be high in calcium (at least 20 percent of the Daily Value) and contain no more phosphorus than calcium.

■ *Sodium and hypertension (high blood pressure).* Foods must qualify as "low sodium" (140 milligrams or less per serving).

■ *Dietary saturated fat and cholesterol and risk of coronary heart disease.* Foods must qualify as "low saturated fat" (1 gram or less per serving), "low cholesterol" (20 milligrams or less per serving and 2 grams or less saturated fat per serving), and "low fat" (3 grams or less per serving) or, in the case of meat and poultry, as "extra lean" (less than 5 grams of fat, 2 grams of saturated fat, and 95 milligrams of cholesterol per serving and per 100 grams of meat, poultry, and seafood).

■ *Dietary fat and cancer.* Foods must qualify as "low fat" or, in the case of meat and poultry, as "extra lean." Claims may not specify types of fat and must speak in terms of "some types of cancers" or "some cancers."

■ *Fiber-containing grain products, fruits, and vegetables and cancer.* Grain products, fruits, or vegetables must qualify both as "low fat" and as "good sources" (provides between 10 and 19 percent of the Daily Value without fortification) of dietary fiber. Claims may not specify types of fiber and must speak in terms of "some types of cancer" or "some cancers."

■ *Fruits, vegetables, and grain products that contain fiber, particularly soluble fiber, and risk of coronary heart disease.* Fruits, vegetables, or grain products must qualify as "low saturated fat," "low cholesterol," and "low fat" and provide (without fortification) at least 0.6 grams of soluble fiber per serving.

■ *Fruits and vegetables and cancer.* Fruits or vegetables must qualify as "low fat" and as "good sources" (without fortification) of vitamin A, vitamin C, or dietary fiber.

■ *Folate and neural tube defects.* Foods must qualify as "good sources" (without fortification) of folate.

■ *Sugar alcohols and tooth decay.* Foods must contain sugar alcohols and not lower dental plaque pH below 5.7 by bacterial fermentation.

■ *Soluble fiber from whole oats and heart disease.* Products must provide at least 0.75 grams of soluble fiber per serving.

With the exception of sugar alcohols and dental caries, all other health claims must also meet two additional criteria. First, a food for which a producer is making a health claim must be a *naturally* good source (containing at least 10 percent of the Daily Value) of at least one of the following nutrients: vitamin A, vitamin C, iron, calcium, protein, or fiber. Second, producers of a food are disqualified from making health claims if a standard serving contains more than 20 percent of the Daily Value for total fat, saturated fat, cholesterol, or sodium. Thus producers may make a calcium and osteoporosis claim for milk, which is high in calcium, if it is nonfat or low-fat milk, but not if it is whole milk.

Balance Food Intake with Activity. The most effective way to maintain health and achieve a desirable body weight is to balance food intake with physical activity. People who perform heavy manual labor at their jobs daily may not need additional exercise, but those with sedentary occupations need to schedule physical activity into their leisure time. The remainder of this chapter features fitness during middle adulthood.

Fitness during Middle Adulthood

Extensive evidence confirms that regular physical activity promotes health and prevents disease.[66] Still, despite an increasing awareness of the health benefits that physical activity confers, more than 75 percent of adults in the United States are either irregularly active or completely inactive.[67] Physical inactivity is linked to the major degenerative diseases—heart disease, cancer, stroke, diabetes, and hypertension—that are the primary killers of adults in developed countries.[68]

Benefits of Physical Activity. Table 9-6 shows that physical activity helps defend against many physical conditions. People spend billions of

Table 9-6 Health Benefits of Regular Physical Activity

- Lowers mortality rates.
- Reduces the risk of cardiovascular disease.
- Reduces blood pressure.
- Reduces the risk of some cancers (notably colon cancer, and perhaps others).
- Reduces the risk of diabetes (non-insulin-dependent).
- Maintains muscle strength, joint structure, and joint function, protecting against arthritis.
- Increases peak bone mass, protecting against osteoporosis.
- Preserves independent living.
- Reduces the risk of injuries from falls.
- Improves body composition.
- Reduces the risk of obesity.
- Relieves symptoms of depression and anxiety.
- Enhances psychological well-being.

Source: Adapted from *The Surgeon General's Report on Physical Activity and Health* (Washington, D.C.: U.S. Government Printing Office, 1996).

dollars each year in hopes of obtaining these benefits, yet they won't spend the effort to exercise. They needn't run marathons to reap the health rewards of physical activity. Most experts agree that any physical activity, even moderate activity, provides health benefits. In fact, people who are extremely inactive stand to gain the greatest health benefits by engaging in regular moderate-intensity, endurance-type activity.[69] The authors of an extensive study on fitness and mortality concluded that "moderate levels of physical fitness that are attainable by most adults appear to be protective against early mortality."[70] It makes sense, then, to promote activities that can readily be performed by the least active people, since they may benefit most.

In 1990, the American College of Sports Medicine (ACSM) updated its position statement on the quantity and quality of exercise recommended for developing and maintaining fitness in healthy adults (see Table 9-7).[71] The main objective of these guidelines was to outline the types and amounts of physical activity needed to improve *physical fitness*. These familiar fitness guidelines have helped adults develop flexibility, muscle strength and endurance, and cardiorespiratory endurance. The types and amounts of physical activity needed to promote *fitness*, however, may differ from those needed to obtain *health* benefits. For health's sake, people should spend an accumulated minimum of 30 minutes in some sort of physical activity on most days of each week.[72] The activity need not be sports. Eight minutes spent climbing up stairs, another 10 spent pulling weeds, and 12 more spent walking the dog all contribute to the day's total. Such activities may help to pre-

Table 9-7 Physical Activity Guidelines

Guidelines for developing and maintaining *physical fitness*:

- **Frequency of activity:** three to five days per week.
- **Intensity of activity:** 60 to 90% of maximum heart rate.
- **Duration of activity:** 20 to 60 minutes of continuous activity.
- **Mode of activity:** any activity that uses large muscle groups.
- **Resistance activity:** strength training of moderate intensity at least two times per week.

Guidelines for obtaining *health* benefits:

- **Frequency of activity:** every day.
- **Intensity of activity:** any level (can be minimal).
- **Duration of activity:** at least 30 minutes total of activity (can be intermittent).
- **Mode of activity:** any activity.

Note: Duration and intensity are inversely related. To obtain similar fitness benefits, a person may exercise either at a low intensity for a long duration or at a high intensity for a short duration. For example, a person may choose to walk briskly for 40 to 50 minutes (lower intensity and longer duration) or to jog for 20 to 30 minutes (higher intensity and shorter duration).

Source: Adapted from R. R. Pate and coauthors, Physical activity and public health: A recommendation from the Centers for Disease Control and Prevention and the American College of Sports Medicine, *Journal of the American Medical Association* 273 (1995): 402–407.

vent disease, but do little toward developing fitness. The guidelines for fitness are optimal because health improves as a person becomes physically fit.

Risks of Physical Activity. Rarely does life hand you the good without the bad. With the many benefits of physical activity come a few risks. Most common among the problems that accompany physical activity are musculoskeletal injuries, which typically occur when people exceed their usual activity patterns. More serious problems, such as heart attack, sometimes occur when unconditioned people engage in strenuous activity.

Before beginning a fitness program, make sure it is safe to do so. The American College of Sports Medicine classifies individuals into three groups based on major coronary risk factors: "apparently healthy" individuals have no more than one of the risk factors listed in the margin, "individuals at higher risk" have two or more of the risk factors and/or symptoms suggestive of disease, and "individuals with disease" are known to have cardiac, pulmonary, or metabolic disease.[73] Most apparently healthy people can begin moderate exercise programs such as walking or increasing daily activities without medical examination; they need to start with short durations of moderate-intensity activity and gradually increase duration and/or intensity until they reach their goals. People in either of the other two classifications need medical advice before beginning an exercise program.

Major coronary risk factors:
1. *Hypertension.*
2. *Serum cholesterol ≥240 mg/dL.*
3. *Cigarette smoking.*
4. *Diabetes mellitus.*
5. *Family history of heart disease.*

moderate exercise: activity that can be sustained comfortably for 60 minutes or so.

The strategies adults use to meet the two goals mentioned at the start of this chapter—promoting health and slowing aging—are actually very much the same. What to eat, when to sleep, how physically active to be, and other lifestyle choices greatly influence both physical health and the aging process. Over the years, the effects of these choices accumulate.

A person's *physiological* age reflects his or her health status and may or may not reflect the person's *chronological* age. Quite simply, some people seem younger, and others older, than their years. Six lifestyle practices seem to have the greatest influence on people's health and, therefore, their physiological age:

- Sleeping regularly and adequately.
- Eating regular meals, including breakfast.
- Engaging in regular physical activity.
- Not smoking.
- Not using alcohol, or using it in moderation.
- Keeping weight under control.

The next chapter explores these factors and their roles in the aging process, as well as the special nutrition needs of older adults.

CHAPTER NINE NOTES

1. Committee on Dietary Reference Intakes, *Dietary Reference Intakes for Calcium, Phosphorus, Magnesium, Vitamin D, and Fluoride* (Washington, D.C.: National Academy Press, 1977).

2. R. J. Kuczmarski and coauthors, Increasing prevalence of overweight among U.S. adults: The National Health and Nutrition Examination Surveys, 1960–1991, *Journal of the American Medical Association* 272 (1994): 205–211.

3. N. E. Davidson, Hormone-replacement therapy—Breast versus heart versus bone, *New England Journal of Medicine* 332 (1995): 1638–1639.

4. R. S. Rivlin and P. Dutta, Vitamin B_2 (riboflavin)—Relevance to malaria and antioxidant activity, *Nutrition Today* 30 (1995): 62–67.

5. D. L. Tribble, L. J. Giuliano, and S. P. Fortmann, Reduced plasma ascorbic acid concentrations in nonsmokers regularly exposed to environmental tobacco smoke, *American Journal of Clinical Nutrition* 58 (1993): 886–890.

6. National Center for Health Statistics, *Monthly Vital Statistics Report*, October 1996, p. 31.

7. J. M. McGinnis and W. H. Foege, Actual causes of death in the United States, *Journal of the American Medical Association* 270 (1993): 2207–2212.

8. *The Surgeon General's Report on Nutrition and Health: Summary and Recommendations*, DHHS (PHS) publication no. 88-50211 (Washington, D.C.: U.S. Government Printing Office, 1988).

9. Committee on Diet and Health, *Diet and Health: Implications for Reducing Chronic Disease Risk* (Washington, D.C.: National Academy Press, 1989).

10. Committee on Diet and Health, 1989.

11. Committee on Diet and Health, 1989, pp. 665–710.

12. M. Anti and coauthors, Effect of w-3 fatty acids on rectal mucosal cell proliferation in subjects at risk for colon cancer, *Gastroenterology* 103 (1992): 883–891; L. A. Sauer, R. T. Dauchy, and A. S. Hurtubise, Effects of omega-6 and omega-3 fatty acids on rate of ^{3}H-thymidine incorporation in hepatoma, *FASEB Journal* 4 (1990): A508; D. Magrane and M. Philley, Effects of dietary corn oil and menhaden oil on rat mammary tumorigenesis and PGE_2 levels, *FASEB Journal* 4 (1990): A1176.

13. A. S. Truswell, Food carbohydrates and plasma lipids—An update, *American Journal of Clinical Nutrition* 59 (1994): 710S–718S.

14. G. H. McIntosh and coauthors, Barley and wheat foods: Influence on plasma cholesterol concentrations in hypercholesterolemic men, *American Journal of Clinical Nutrition* 53 (1991): 1205–1209; M. Kestin and coauthors, Comparative effects of three cereal brans on plasma lipids, blood pressure, and glucose metabolism in mildly hypercholesterolemic men, *American Journal of Clinical Nutrition* 52 (1990): 661–666; J. W. Anderson and coauthors, Serum lipid response of hypercholesterolemic men to single and divided doses of canned beans, *American Journal of Clinical Nutrition* 51 (1990): 1013–1019; L. P. Bell and coauthors, Cholesterol-lowering effects of soluble-fiber cereals as part of a prudent diet for patients with mild to moderate hypercholesterolemia, *American Journal of Clinical Nutrition* 52 (1990): 1020–1026; J. W. Anderson and N. J. Gustafson, Hypocholesterolemic effects of oat and bean products, *American Journal of Clinical Nutrition* 48 (1988): 749–753.

15. Y. A. Kesaniemi, S. Tarpila, and T. A. Miettinen, Low vs high dietary fiber and serum, biliary, and fecal lipids in middle-aged men, *American Journal of Clinical Nutrition* 51 (1990): 1007–1112.

16. J. F. Swain and coauthors, Comparison of the effects of oat bran and low-fiber wheat on serum lipoprotein levels and blood pressure, *New England Journal of Medicine* 322 (1990): 147–152; W. Denmark-Wahnefried, J. Bowering, and P. S. Cohen, Reduced serum cholesterol with dietary change using fat-modified and oat bran supplemented diets, *Journal of the American Dietetic Association* 90 (1990): 223–229.

17. D. J. A. Jenkins and coauthors, Effect on blood lipids of very high intakes of fiber in diets low in saturated fat and cholesterol, *New England Journal of Medicine* 329 (1993): 21–26.

18. I. P. Munster and coauthors, Effect of resistant starch on breath-hydrogen and methane excretion in healthy volunteers, *American Journal of Clinical Nutrition* 59 (1994): 626–630.

19. F. X. Pi-Sunyer, Health implications of obesity, *American Journal of Clinical Nutrition* (supplement) 53 (1991): 159–160.

20. A. M. Wolf and G. A. Colditz, Social and economic effects of body weight in the United States, *American Journal of Clinical Nutrition* 63 (1996): 466S–469S.

21. T. B. Vanitallie, Body weight, morbidity, and longevity, in *Obesity*, eds. P. Björntorp and B. N. Brodoff (Philadelphia: Lippincott, 1992), pp. 361–369.

22. L. V. Sjöström, Mortality of severely obese subjects, *American Journal of Clinical Nutrition* (supplement) 55 (1992): 516–523.

23. P. Björntorp, Regional adiposity, in *Obesity*, eds. P. Björntorp and B. N. Brodoff (Philadelphia: Lippincott, 1992), pp. 579–586.

24. M. Zamboni and coauthors, Obesity and regional body-fat distribution in men: Separate and joint relationships to glucose tolerance and plasma lipoproteins, *American Journal of Clinical Nutrition* 60 (1994): 682–687; E. M. Emery and coauthors, A review of the association between abdominal fat distribution, health outcome measures, and modifiable risk factors, *American Journal of Health Promotion* 7 (1993): 342–353; Pi-Sunyer, 1991.

25. S. Lemieux and coauthors, Sex differences in the relation of visceral adipose tissue accumulation to total body fatness, *American Journal of Clinical Nutrition* 58 (1993): 463–467; C. J. Ley, B. Lees, and J. C. Stevenson, Sex- and menopause-associated changes in body-fat distribution, *American Journal of Clinical Nutrition* 55 (1992): 950–954.

26. R. J. Troisi, Cigarette smoking, dietary intake, and physical activity: Effects on body fat distribution—The Normative Aging study, *American Journal of Clinical Nutrition* 53 (1991): 1104–1111.

27. Björntorp, 1992.

28. C. Bouchard, G. A. Bray, and V. S. Hubbard, Basic and clinical aspects of regional fat distribution, *American Journal of Clinical Nutrition* 52 (1990): 946–950.

29. J. Wylie-Rosett and coauthors, Trial of Antihypertensive Intervention and Management: Greater efficacy with weight reduction than with a sodium-potassium intervention, *Journal of the American Dietetic Association* 93 (1993): 408–415; S. A. Corrigan and coauthors, Weight reduction in the prevention and treatment of hypertension: A review of representative clinical trials, *American Journal of Health Promotion* 5 (1991): 208–214.

30. S. Lillioja and coauthors, Insulin resistance

and insulin secretory dysfunction as precursors of non-insulin-dependent diabetes mellitus: Prospective Studies of Pima Indians, *New England Journal of Medicine* 329 (1993): 1988–1992.

31. A. P. Simopoulos, Characteristics of obesity, in *Obesity*, eds. P. Björntorp and B. N. Brodoff (Philadelphia: Lippincott, 1992), pp. 309–319.

32. I. T. Johnson, G. Williamson, and S. R. R. Musk, Anticarcinogenic factors in plant foods: A new class of nutrients? *Nutrition Research Reviews* 7 (1994): 175–204; B. N. Ames, M. K. Shigenaga, and T. M. Hagen, Oxidants, antioxidants, and the degenerative diseases of aging, *Proceedings of the National Academy of Sciences* 90 (1993): 7915–7922.

33. N. I. Krinsky, Effects of carotenoids in cellular and animal systems, *American Journal of Clinical Nutrition* 52 (1991): 238S–246S.

34. T. Byers and G. Perry, Dietary carotenes, vitamin C, and vitamin E as protective antioxidants in human cancers, *Annual Review of Nutrition* 12 (1992): 139–159.

35. R. G. Ziegler, Vegetables, fruits, and carotenoids and the risk of cancer, *American Journal of Clinical Nutrition* 53 (1991): 251S–259S.

36. H. B. Stahalein and coauthors, Beta-carotene and cancer prevention: The Basel Study, *American Journal of Clinical Nutrition* 53 (1991): 265S–269S; N. Potischman and coauthors, Breast cancer and dietary and plasma concentrations of carotenoids and vitamin A, *American Journal of Clinical Nutrition* 52 (1990): 909–915.

37. W. C. Willett and D. J. Hunter, Vitamin A and cancers of the breast, large bowel, and prostate: Epidemiologic evidence, *Nutrition Reviews* 52 (1994): S53–S59.

38. D. J. Hunter and coauthors, A prospective study of the intake of vitamins C, E, and A and the risk of breast cancer, *New England Journal of Medicine* 329 (1993): 234–240.

39. G. Block, Vitamin C and cancer prevention: The epidemiologic evidence, *American Journal of Clinical Nutrition* 53 (1991): 270S–282S.

40. S. R. Tannenbaum, J. S. Wishnok, and C. D. Leaf, Inhibition of nitrosamine formation by ascorbic acid, *American Journal of Clinical Nutrition* 53 (1991): 247S–250S.

41. P. Knekt and coauthors, Vitamin E and cancer prevention, *American Journal of Clinical Nutrition* 53 (1991): 283S–286S.

42. T. Byers, Vitamin E supplements and coronary heart disease, *Nutrition Reviews* 51 (1993): 333–336; K. F. Gey and coauthors, Increased risk of cardiovascular disease at suboptimal plasma concentrations of essential antioxidants: An epidemiological update with special attention to carotene and vitamin C, *American Journal of Clinical Nutrition* 57 (1993): 787S–797S.

43. B. Halliwell, Oxidation of low-density lipoproteins: Questions of initiation, propagation, and the effect of antioxidants, *American Journal of Clinical Nutrition* 61 (1995): 670S–677S.

44. K. F. Gey and coauthors, Inverse correlation between plasma vitamin E and mortality from ischemic heart disease in cross-cultural epidemiology, *American Journal of Clinical Nutrition* 53 (1991): 326S–334S; Gey and coauthors, 1993.

45. M. J. Stampfer and coauthors, Vitamin E consumption and the risk of coronary disease in women, *New England Journal of Medicine* 328 (1993): 1444–1449; E. B. Rimm and coauthors, Vitamin E consumption and the risk of coronary disease in men, *New England Journal of Medicine* 328 (1993): 1450–1456.

46. Gey and coauthors, 1993; D. L. Trout, Vitamin C and cardiovascular risk factors, *American Journal of Clinical Nutrition* 53 (1991): 322S–325S; Stampfer and coauthors, 1993; Rimm and coauthors, 1993.

47. D. Kritchevsky, Antioxidant vitamins in the prevention of cardiovascular disease, *Nutrition Today*, January–February 1992, pp. 30–33.

48. Trout, 1991; J. P. Moran and coauthors, Plasma ascorbic acid concentrations relate inversely to blood pressure in human subjects, *American Journal of Clinical Nutrition* 57 (1993): 213–217.

49. A. L. Klatsky, M. A. Armstrong, and G. D. Friedman, Alcohol and mortality, *Annals of Internal Medicine* 117 (1992): 646–654.

50. J. M. Gaziano and coauthors, Moderate alcohol intake, increased levels of high-density lipoprotein and its subfractions, and decreased risk of myocardial infarction, *New England Journal of Medicine* 329 (1993): 1829–1834; P. R. Ridker and coauthors, Association of moderate alcohol consumption and plasma concentration of endogenous tissue-type plasminogen activator, *Journal of the*

American Medical Association 272 (1994): 929–933.

51. C. A. Camargo and coauthors, Prospective study of moderate alcohol consumption and mortality in U.S. male physicians, *Archives of Internal Medicine* 157 (1997): 79–85.

52. T. A. Kotchen and J. M. Kotchen, Nutrition, diet, and hypertension, in *Modern Nutrition in Health and Disease*, 8th ed., eds. M. E. Shils, J. A. Olson, and M. Shike (Philadelphia: Lea & Febiger, 1994), pp. 1287–1297.

53. D. A. McCarron and coauthors, Dietary calcium and blood pressure: Modifying factors in specific populations, *American Journal of Clinical Nutrition* 54 (1991): 215S–219S.

54. J. Sharlin and coauthors, Nutrition and behavioral characteristics and determinants of plasma cholesterol levels in men and women, *Journal of the American Dietetic Association* 92 (1992): 434–440; G. A. Golditz and coauthors, Diet and risk of clinical diabetes in women, *American Journal of Clinical Nutrition* 55 (1992): 1018–1023; C. F. Garland, F. C. Garland, and E. D. Gorham, Can colon cancer incidence and death rates be reduced with calcium and vitamin D? *American Journal of Clinical Nutrition* 54 (1991): 193S–201S; K. K. Carroll and coauthors, Calcium and carcinogenesis of the mammary gland, *American Journal of Clinical Nutrition* 54 (1991): 206S–208S.

55. J. Blumberg, Are antioxidants at an awkward age? *Journal of the American College of Nutrition* 13 (1994): 218–219.

56. M. Abbey, M. Noakes, and P. J. Nestel, Dietary supplementation with orange and carrot juice in cigarette smokers lowers oxidation products in copper-oxidized low-density lipoproteins, *Journal of the American Dietetic Association* 95 (1995): 671–675.

57. W. A. Pryor, The antioxidant nutrients and disease prevention—What do we know and what do we need to find out? *American Journal of Clinical Nutrition* 53 (1991): 391S–393S.

58. D. Steinberg, Antioxidant vitamins and coronary heart disease, *New England Journal of Medicine* 328 (1993): 1487–1489.

59. E. R. Greenberg and coauthors, A clinical trial of antioxidant vitamins to prevent colorectal adenoma, *New England Journal of Medicine* 331 (1994): 141–147.

60. O. P. Heinonen, J. K. Huttunen, and D. Albanes (and other participants in the alpha-tocopherol, beta carotene cancer prevention study group), The effect of vitamin E and beta carotene on the incidence of lung cancer and other cancers in male smokers, *New England Journal of Medicine* 330 (1994): 1029–1035.

61. T. Repka and R. P. Hebbel, Hydroxyl radical formation by sickle erythrocyte membranes: Role of pathological iron deposits and cytoplasmic reducing agents, *Blood* 78 (1991): 2753–2758; V. Herbert, The antioxidant supplement myth, *American Journal of Clinical Nutrition* 60 (1994): 157–158; B. Halliwell, Antioxidants: Sense or speculation? *Nutrition Today*, November–December 1994, pp. 15–19.

62. Position of The American Dietetic Association: The impact of fluoride on dental health, *Journal of the American Dietetic Association* 94 (1994): 1428–1431.

63. W. Mertz, A balanced approach to nutrition for health: The need for biologically essential minerals and vitamins, *Journal of the American Dietetic Association* 94 (1994): 1259–1262.

64. B. E. Miller and coauthors, Population nutrient intake approaches dietary recommendations: 1991 to 1995 Framingham Nutrition Studies, *Journal of the American Dietetic Association* 97 (1997): 742–749; Federation of American Societies for Experimental Biology, Executive summary from the third report on nutrition monitoring in the United States, *Journal of Nutrition* 126 (1996): 1907S–1936S.

65. A. F. Subar and coauthors, US dietary patterns associated with fat intake: The 1987 National Health Interview Survey, *American Journal of Public Health* 84 (1994): 359–366.

66. R. S. Paffenbarger and coauthors, The association of changes in physical-activity level and other lifestyle characteristics with mortality among men, *New England Journal of Medicine* 328 (1993): 538–545; L. Sandvik and coauthors, Physical fitness as a predictor of mortality among healthy, middle-aged Norwegian men, *New England Journal of Medicine* 328 (1993): 533–537.

67. U.S. Centers for Disease Control and Prevention and American College of Sports Medicine, Summary statement: Workshop on physical activity and public health, *Sports Medicine Bulletin* 28 (1993): 7.

68. American Heart Association Position State-

ment on Exercise: Benefits and recommendations for physical activity programs for all Americans, *Circulation* 86 (1992): 340–344; A. M. Bovens and coauthors, Physical activity, fitness, and selected risk factors for CHD in active men and women, *Medicine and Science in Sports and Exercise* 25 (1993): 572–576; R. R. Pate and coauthors, Physical activity and public health: A recommendation from the Centers for Disease Control and Prevention and the American College of Sports Medicine, *Journal of the American Medical Association* 273 (1995): 402–407.

69. W. L. Haskell, Health consequences of physical activity: Understanding and challenges regarding dose-response, *Medicine and Science in Sports and Exercise* 26 (1994): 649–660.

70. S. N. Blair and coauthors, Changes in physical fitness and all-cause mortality, *Journal of the American Medical Association* 273 (1995): 1093–1098.

71. American College of Sports Medicine, The recommended quality and quantity of exercise for developing and maintaining fitness in healthy adults, *Medicine and Science in Sports and Exercise* 22 (1990): 265–274.

72. Pate and coauthors, 1995.

73. American College of Sports Medicine, *Guidelines for Exercise Testing and Prescription*, 4th ed. (Philadelphia: Lea & Febiger, 1991).

Alcohol and Nutrition

Social gatherings offer opportunities for people to share conversation, food, and drink. Among the beverages available are those that contain alcohol, and people must choose whether to drink them. Most people who drink manage their relationships with alcohol relatively safely.[1] Unfortunately, some 18 million people in the United States abuse alcohol to the point that their personal relationships, work, and health become impaired. This discussion focuses on how the body handles alcohol, how alcohol affects the body, and how alcohol impairs people's health and nutrition.

Alcohol in Beverages

To the chemist, *alcohol* refers to a class of organic compounds containing hydroxyl (OH) groups. But to most people, *alcohol* refers to the intoxicating ingredient in beer, wine, and distilled liquor. The chemist's name for this particular alcohol is *ethyl alcohol*, or *ethanol*. The remainder of this Focal Point talks about the particular alcohol ethanol, but refers to it simply as *alcohol*. (The glossary on p. 380 defines related terms.)

Alcohols affect living things profoundly, partly because they act as lipid solvents. Their ability to dissolve lipids out of cell membranes allows them to penetrate rapidly into cells, destroying cell structures and thereby killing the cells. For this reason, most alcohols are toxic, or poisonous, in relatively small amounts; by the same token, because they kill microbial cells, they are useful as disinfectants.

Ethanol is less toxic than the other alcohols. Sufficiently diluted and taken in small enough doses, its action in the brain produces euphoria—a pleasing effect that people seek—not with zero risk, but with a low enough risk (if the doses are small enough) to be tolerable. Used to achieve this effect, alcohol is a drug—that is, a substance that modifies body functions. Like all drugs, alcohol offers both benefits and hazards. It must be used with caution, if used at all.

Shared conversations and meals sometimes include alcoholic beverages.

Taken in moderation, alcohol can be compatible with good health. The term *moderation* is important in describing alcohol use. How many drinks constitute moderate use, and how much is "a drink"? First, a drink is any alcoholic beverage that delivers 1/2 ounce of *pure ethanol*:

- 4 to 5 ounces of wine.
- 10 ounces of wine cooler.
- 12 ounces of beer.
- 1 1/4 ounce of distilled liquor (80-proof whiskey, scotch, rum, or vodka).

Second, because people have different tolerances to alcohol, it is impossible to name an exact amount of alcohol per day that is appropriate for everyone. Authorities have attempted to set limits that are acceptable for most healthy people. An accepted definition of moderation is not more than two drinks a day for the average-sized man and not more than one drink a day for the average-sized woman. Notice that this advice is stated as a maximum, not as an average; seven

GLOSSARY

alcohol: a class of organic compounds containing hydroxyl (OH) groups.

alcohol dehydrogenase: an enzyme that converts ethanol to acetaldehyde.

beer: an alcoholic beverage brewed by fermenting malt and hops.

distilled liquor: an alcoholic beverage made by fermenting and distilling grains; sometimes called *distilled spirits* or *hard liquor*. Examples include gin, vodka, rum, bourbon, and scotch whiskey.

drink: a dose of any alcoholic beverage that delivers 1/2 oz of pure ethanol:

- 4 to 5 oz of wine.
- 10 oz of wine cooler.
- 12 oz of beer.
- 1 1/4 oz of hard liquor (80-proof whiskey, scotch, rum, or vodka).

drug: a substance that can modify one or more of the body's functions.

ethanol: a particular type of alcohol found in beer, wine, and distilled spirits; also called *ethyl alcohol*. Ethanol is the most widely used—and abused—drug in our society. It is also the only legal, nonprescription drug that produces euphoria.

euphoria (you-FORE-eh-uh): a feeling of great well-being, which people often seek through the use of drugs such as alcohol (*eu* = good; *phoria* = bearing).

moderation: in relation to alcohol consumption, not more than two drinks a day for the average-sized man and not more than one drink a day for the average-sized woman.

proof: a way of stating the percentage of alcohol in distilled liquor. Liquor that is 100 proof is 50 percent alcohol; 90 proof is 45 percent, and so forth.

wine: an alcoholic beverage made by fermenting grape juice.

---

drinks one night a week would not be considered moderate, even though one a day would be. Doubtless some people could consume slightly more; others could not handle nearly so much without risk. The amount a person can drink safely is highly individual, depending on genetics, health condition, sex, body composition, age, and family history.

Alcohol in the Body

From the moment an alcoholic beverage enters the body, the body treats it as if it had special privileges. Unlike foods, which require time for digestion, alcohol needs no digestion and is quickly absorbed. About 20 percent is absorbed directly across the walls of an empty stomach and can reach the brain within a minute. Consequently, a person can immediately feel euphoric when drinking, especially on an empty stomach.

When the stomach is full of food, alcohol has less chance of touching the walls and diffusing through, so its influence on the brain is slightly delayed. This information leads to a practical tip: Eat snacks when drinking alcoholic beverages. Carbohydrate snacks slow alcohol absorption and high-fat snacks slow peristalsis, keeping the alcohol in the stomach longer. Salty snacks make a person thirsty; to quench thirst, drink water instead of more alcohol.

The stomach begins to break down alcohol with its alcohol dehydrogenase enzyme. This action can reduce the amount of alcohol entering the blood by about 20 percent. Research shows that women produce less of this stomach enzyme than men, which partially explains why women become more intoxicated on less alcohol than men.[2] Women absorb about one-third more alcohol than men of the same size who drink the same amount of alcohol.

Alcohol is rapidly absorbed in the small intestine. From this point on, alcohol receives VIP (Very Important Person) treatment: It gets absorbed and metabolized before most nutrients.

Table FP9-1 Consequences of Alcohol in the Body

In General

- Slowed immune system.

In the Liver

- Accumulated fatty acids.
- Damaged liver cells.
- Impaired vitamin D metabolism.
- Diminished bile production.
- Altered protein metabolism.

In the Brain

- Impaired judgment, relaxed inhibitions, altered mood, increased heart rate.
- Impaired coordination, delayed reaction time, exaggerated emotions, impaired peripheral vision, impaired ability to operate a vehicle.
- Slurred speech, blurred vision, staggered walk, seriously impaired coordination and judgment.
- Double vision, inability to walk.
- Uninhibited behavior, stupor, confusion, inability to comprehend.
- Unconsciousness, shock, coma, death (cardiac or respiratory failure).

To overcome these problems, a person needs to stop drinking alcohol.

If more alcohol reaches the liver than it can handle, consequences follow (see Table FP9-1).

Alcohol and Malnutrition

For many moderate drinkers, alcohol does not suppress food intake, and in some cases, it may actually stimulate appetite. When alcohol is consumed as *added* energy, it can contribute to body fat.[3] Metabolically, alcohol behaves like fat in promoting obesity; each ounce of alcohol represents about half an ounce of fat.[4] Chronic alcohol ingestion seems to have the opposite effect,

If you decide to drink alcohol, remember to drink slowly enough to allow the liver to process it: no more than one drink per hour.

however. Alcohol produces euphoria, which depresses appetite, so heavy drinkers tend to eat poorly and suffer malnutrition. Alcohol is rich in energy (7 kcalories per gram), but like pure sugar or fat, the kcalories are empty of nutrients. The more alcohol people drink, the less likely that they will eat enough food to obtain adequate nutrients. The more kcalories spent on alcohol, the fewer kcalories available to spend on nutritious foods. Table FP9-2 shows the kcalorie amounts of typical alcoholic beverages.

Chronic alcohol abuse not only displaces nutrients from the diet but also interferes with the body's metabolism of nutrients. Most dramatic is alcohol's effect on the B vitamin folate. When alcohol is present, the body behaves as if it were actively trying to expel folate. The liver, which normally contains enough folate to meet all

Table FP9-2 kCalories in Alcoholic Beverages and Mixers

Beverage	Amount (oz)	Energy (kcal)
Beer		
Regular	12	150
Light	12	100
Nonalcoholic	12	32–82
Distilled liquor (gin, rum, vodka, whiskey)		
80 proof	1 1/2	100
86 proof	1 1/2	105
90 proof	1 1/2	110
Liqueurs		
Coffee liqueur	1 1/2	175
Coffee and cream liqueur	1 1/2	155
Crème de menthe	1 1/2	185
Mixers		
Club soda	12	0
Cola	12	150
Cranberry juice cocktail	8	145
Diet drinks	12	2
Ginger ale	12	125
Grapefruit juice	8	95
Orange juice	8	110
Tomato or vegetable juice	8	45
Tonic	12	124
Wine		
Dessert	3 1/2	160
Nonalcoholic	8	14
Red	3 1/2	75
Rosé	3 1/2	75
White	3 1/2	70
Wine cooler	12	50

needs, now leaks folate into the blood. As blood folate rises, the kidneys are deceived into excreting it. Alcohol abuse causes a folate deficiency that devastates digestive system function. The intestine normally releases and retrieves folate continuously, but it becomes damaged by folate deficiency and alcohol toxicity, so it fails to retrieve its own folate and misses any that may enter from food as well. Alcohol also interferes with the action of what little folate is left, which inhibits the production of new cells, especially the rapidly dividing cells of the intestine and the blood. The combination of poor folate status and alcohol consumption has been implicated as promoting colorectal cancer.[5]

Malnutrition occurs not only because of lack of intake and altered metabolism, but because of direct toxic effects as well.[6] Alcohol causes stomach cells to oversecrete both gastric acid and histamine, an agent of the immune system that produces inflammation. Beer in particular stimulates gastric acid secretion, irritating the stomach and esophagus linings and making them vulnerable to ulcer formation.[7]

Over a lifetime, excessive drinking, regardless of dietary intake, creates deficits of all the nutrients. No diet can compensate for the damage caused by heavy alcohol consumption.[8]

Alcohol's Short-Term Effects

Heavy or binge drinking (defined as at least four to five drinks in a row) poses serious health and social consequences to both drinkers and nondrinkers alike.[9]* Alcohol is responsible for most accidental deaths, including automobile fatalities. Compared with nondrinkers or moderate drinkers, people who frequently binge drink (at least three times within two weeks) are more likely to engage in unprotected sex, damage property, and assault others.

Alcohol's Long-Term Effects

By far the longest-term effect of alcohol is the damage done to a child whose mother abused al-

---

*This definition of binge drinking, without specification of time elapsed, is consistent with standard practice in alcohol research.

Table FP9-3 Health Effects of Alcohol Consumption

Health Problem	Effects of Alcohol
Arthritis	Increases the risk of gouty arthritis.
Cancer	Increases the risk of cancer of the liver, pancreas, rectum, and breast; increases the risk of cancer of the mouth, pharynx, larynx, and esophagus, where alcohol interacts synergistically with tobacco.
Fetal alcohol syndrome	Causes physical and behavioral abnormalities in the fetus.
Heart disease	Raises blood pressure, blood lipids, and the risk of stroke and heart disease in heavy drinkers; when compared with those who abstain, heart disease risk is generally lower in light-to-moderate drinkers (see Chapter 9).
Hyperglycemia	Raises blood glucose.
Hypoglycemia	Lowers blood glucose, especially in people with diabetes.
Kidney disease	Enlarges the kidneys, alters hormone functions, and increases the risk of kidney failure.
Liver disease	Causes fatty liver, alcoholic hepatitis, and cirrhosis.
Malnutrition	Increases the risk of protein-energy malnutrition; low intakes of protein, calcium, iron, vitamin A, vitamin C, thiamin, vitamin B_6, and riboflavin; and impaired absorption of calcium, phosphorus, vitamin D, and zinc.
Nervous disorders	Causes neuropathy and dementia; impairs balance and memory.
Obesity	Increases energy intake, but is not a primary cause of obesity.
Psychological disturbances	Causes depression, anxiety, and insomnia.

Note: This list is by no means all inclusive. Alcohol has direct toxic effects on all body systems.

cohol during pregnancy. The devastating effects of alcohol on the unborn and the message that pregnant women should not drink alcohol were presented in Focal Point 3. Table FP9-3 presents a sampling of the long-term health effects of alcohol consumption.

Personal Strategies

One obvious option available to people attending social gatherings is to enjoy the conversation, eat the food, and drink nonalcoholic beverages. Several nonalcoholic beverages are available that mimic the look and taste of their alcoholic counterparts. For those who enjoy champagne or beer, sparkling ciders and beers are available without alcohol. Instead of drinking a cocktail, a person can sip tomato juice with a slice of lime and a stalk of celery or just a plain cola beverage. Any of these drinks can ease conversation.

The person who chooses to drink alcohol should sip each drink slowly with food. The alcohol should arrive at the liver cells slowly enough that the enzymes can handle the load. It is best to space drinks, too, allowing about an hour or so to metabolize each drink.

Judgment might tell a person to limit alcohol consumption to two drinks at a party, but if the first drink takes judgment away, many more drinks may follow. The failure to stop drinking as planned, on repeated occasions, is a danger sign warning that the person should not drink at all.*

The road back from alcohol addiction may take years. It's hard, and not everyone succeeds.

*Appendix F provides addresses for organizations that offer information about alcohol and alcohol abuse.

It helps if the person is in touch with other people who have recovered. No one is as powerful in helping a person who is recovering from substance abuse as another person who has already traveled that road. The worldwide self-help recovery group Alcoholics Anonymous (AA) works this way. The AA program takes a positive approach—12 steps to recovery and spiritual growth that end in the person's helping others. In AA, thousands of people have helped one another recover from alcoholism.

With heavy alcohol consumption, the potential for harm is great. The best way to escape the harmful effects of alcohol is, of course, to refuse alcohol altogether. If you do drink, do so with care and in moderation.

FOCAL POINT 9 NOTES

1. Secretary of Health and Human Services, *Eighth Special Report to the U.S. Congress on Alcohol and Health* (Rockville, Md.: U.S. Department of Health and Human Services, 1993).
2. M. Frezza and coauthors, High blood alcohol levels in women: The role of decreased gastric alcohol dehydrogenase activity and first-pass metabolism, *New England Journal of Medicine* 322 (1990): 95–99.
3. B. J. Sonko and coauthors, Effect of alcohol on postmeal fat storage, *American Journal of Clinical Nutrition* 59 (1994): 619–625; P. M. Suter, Y. Schutz, and E. Jequier, The effect of ethanol on fat storage in healthy subjects, *New England Journal of Medicine* 326 (1992): 983–987.
4. J. P. Flatt, Body weight, fat storage, and alcohol metabolism, *Nutrition Reviews* 50 (1992): 267–270.
5. Folate, alcohol, methionine, and colon cancer risk: Is there a unifying theme? *Nutrition Reviews* 52 (1994): 18–20; A. E. Rogers, Methyl donors in the diet and responses to chemical carcinogens, *American Journal of Clinical Nutrition* 61 (1995): 659S–665S.
6. C. S. Lieber, 1993, Herman Award Lecture, 1993: A personal perspective on alcohol, nutrition, and the liver, *American Journal of Clinical Nutrition* 58 (1993): 430–442.
7. M. V. Singer, S. Teyssen, and V. E. Eysselein, Action of beer and its ingredients on gastric acid secretion and release of gastrin in humans, *Gastroenterology* 101 (1991): 935–942.
8. Lieber, 1993.
9. H. Wechsler and coauthors, Health and behavioral consequences of binge drinking in college: A national survey of students at 140 campuses, *Journal of the American Medical Association* 272 (1994): 1672–1677.

Nutrition during Late Adulthood

T he U.S. population is graying. The majority is now middle aged, and the ratio of old people to young is becoming greater, as Figure 10-1 shows. Our society uses the arbitrary age of 65 years to define the transition point between middle age and old age, but growing "old" happens day by day, with changes occurring gradually over time. Since 1950 the population of people over 65 has more than doubled. Remarkably, the fastest-growing age group is people over 85 years (see Figure 10-2).[1] The U.S. Bureau of the Census projects that by the year 2040 there will be more than a million Americans 100 years old or older. This chapter presents information on aging and the nutrition needs of older adults.

Life expectancy for women in the United States is 79 years; for men, it is 73 years—both up from about 47 years in 1900.[2] Advances in medical science—antibiotics and other treatments—are largely responsible for almost doubling life expectancy in this century. Improved nutrition and an abundant supply of food have also lengthened life expectancy.[3] Still, human longevity appears to have an upper limit that even medicine and nutrition cannot extend. The human life span is about 115 years and has not changed much over the years.

Figure 10-1 The Aging of the U.S. Population

In 1940, 6.8 percent of the population was 65 or older. In 1990, 12.7 percent of us had reached age 65; by 2040, 21.7 percent will have reached age 65; and a century from now, nearly one of four Americans will be 65 and older. An estimated 25,000 Americans now living are 100 years old or older.

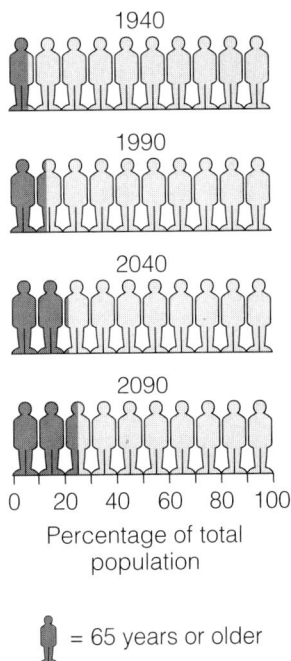

1940

1990

2040

2090

0   20   40   60   80   100

Percentage of total population

= 65 years or older

Figure 10-2 U.S. Population Growth, 1960 to 1990

The "oldest old"—those 85 years and older—are the fastest-growing age group in the United States. Between 1960 and 1990, the U.S. population grew 39 percent, but the population of those over 85 more than doubled.

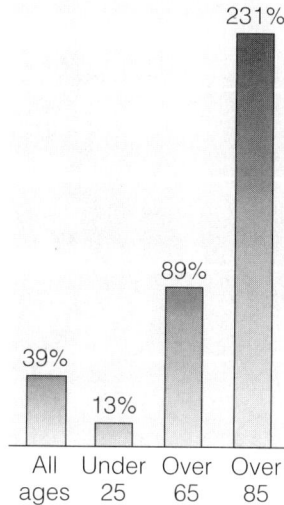

The Aging Process

Only in this century have human beings achieved a life expectancy that permits science to study aging. Research in the field is active—and difficult. Researchers are challenged by the diversity of older adults. When older adults experience health problems, it is hard to know whether to attribute these problems to normal, age-related processes or to other factors. The idea that nutrition can influence the aging process is particularly appealing, because people can control and change their eating habits. The questions researchers are asking include:

■ To what extent is aging inevitable, and can it be slowed through changes in lifestyle and environment?

■ What role does nutrition play in the aging process, and what role can it play in retarding aging?

With respect to the first question, aging is an inevitable, natural process, programmed into the genes at conception. People can, however, slow the process within the natural limits set by heredity. They need to adopt healthy lifestyle habits such as eating nutritious food and engaging in physical activity.

With respect to the second question, good nutrition helps to maintain a healthy body and can therefore ease the aging process in many significant ways. Clearly, nutrition can improve the quality of the later years.

Growing old is enjoyable for people who take care of their health and live each day fully.

Physiological Changes

As people get older, each person becomes less and less like anyone else. The older people are, the more time has elapsed for such factors as nutrition, genetics, physical activity, and everyday stress to influence physical and psychological aging. Each of these factors can interact with one or more of the others, and each can affect the aging process positively or negatively. Researchers studying healthy older adults find remarkable variation in the physiological changes of aging. The extent of age-related changes depends largely on lifestyle factors such as nutrition and physical activity. For example, research shows that age-related deterioration of lean body mass and muscle strength not only can be prevented with sound nutrition and strength training, but can even be reversed.[4]

stressor: an environmental element, physical or psychological, that causes stress.

stress: any threat to a person's well-being; a demand placed on the body to adapt.

stress response: the body's response to stress, mediated by both nerves and hormones.

Both physical stressors (such as alcohol abuse, other drug abuse, smoking, pain, heat, and illness) and psychological stressors (such as exams, divorce, moving, and death of a loved one) elicit the body's stress response. The body responds to such stressors with an elaborate series of physiological steps, using the nervous and hormonal systems to bring about defensive readiness in every body part. The effects all favor physical action—the classic fight-or-flight response. Stress that is prolonged or severe can drain the body of its reserves and leave it weakened, aged, and vulnerable to illness, especially if physical action is not taken. As people age, they lose their ability to adapt to both external and internal disturbances. When disease strikes, the reduced ability to adapt makes the aging individual more vulnerable to death than a younger person.

As aging progresses, inevitable changes in each of the body's organs contribute to the body's declining function. These physiological changes influence nutrition status, just as growth and development do in the earlier stages of the life cycle.

Body Composition. As mentioned earlier, optimal nutrition and physical activity can minimize the body composition changes associated with aging. In general, though, older people tend to lose bone and lean body mass and gain body fat.[5] Many of these changes in body composition occur because the activity of some hormones that regulate metabolism decreases with age, while the activity of others increases. The action of insulin, for example, diminishes with age as the pancreas begins to secrete less of the hormone and the cells lose their ability to respond efficiently.*

The Immune System. Changes in the immune system also bring declining function with age.[6] The immune system is also compromised by nutrient deficiencies, and so the combination of age and malnutrition makes older people vulnerable to infectious diseases.[7] Adding insult to injury, antibiotics often are not effective against infections in people with compromised immune systems.[8] Consequently, infectious diseases are a major cause of death in older adults.

The GI Tract. In the GI tract, the intestinal wall loses strength and elasticity with age, and this slows motility. Constipation is four to eight times more common in the elderly than in the young.[9] Atrophic gastritis, a condition that affects almost one-third of those over 60, is characterized by an inflamed stomach, bacterial overgrowth of the small intestine, and a lack of hydrochloric acid—all of which can impair the digestion and absorption of nutrients, most notably, vitamin B_{12}, biotin, calcium, and iron.

atrophic gastritis: a condition characterized by chronic inflammation of the stomach accompanied by a diminished size and functioning of the mucosa and glands.

Tooth Loss. Improvements in dental care over a lifetime may reduce the incidence of tooth loss and gum disease, which are common in old age. These conditions make chewing difficult or painful. Dentures, even when they fit properly, are less effective than natural teeth, and inefficient chewing can cause choking. People with tooth loss, gum disease, and ill-fitting dentures tend to limit their food selections to those that are soft. If foods such as corn on the cob, apples, and hard rolls are replaced by creamed corn, applesauce, and rice, then nutrition status may not be greatly affected, but when food groups are eliminated and variety is limited, nutrient deficiencies follow.

Sensory Losses and Other Physical Problems. A multitude of other physical problems can also interfere with an older person's ability to

*Other examples of hormones that change with age include growth hormone and androgens, which decline with advancing age, thus contributing to the decrease in lean body mass, and the hormone prolactin, which increases with age, helping to maintain body fat.

obtain adequate nourishment. Failing eyesight, for example, can make driving to the grocery store impossible and shopping for food a frustrating experience. It may become so difficult to read food labels and count money that the person doesn't buy the needed foods. Carrying bags of groceries may be an unmanageable task. Similarly, a person with limited mobility may find cooking and cleaning up too hard to do.

Sensory losses can interfere with a person's ability or willingness to eat. Taste and smell sensitivities tend to diminish with age and may make eating less enjoyable.[10] Loss of vision and hearing may contribute to social isolation.[11]

Other Changes

In addition to the physiological changes that accompany aging, adults are changing in many other ways that influence their nutrition status. Psychological, economic, and social factors play big roles in a person's ability and willingness to eat.

Shared meals can brighten the day and enhance the appetite.

Psychological Changes. Although not an inevitable component of aging, depression is common among older adults. Loss of appetite and of motivation to cook or even to eat frequently accompanies depression. An overwhelming sense of grief and sadness at the death of a spouse, friend, or family member may leave a person, especially an elderly person, feeling powerless to overcome the depression. When a person is suffering the heartache and loneliness of bereavement, cooking meals may not seem worthwhile. The support and companionship of family and friends, especially at mealtimes, can help overcome depression and enhance appetite.

Economic Changes. Overall, the older population today has a higher income than its cohorts of previous generations. Still, poverty is a major problem for about 20 percent of people over age 65. Factors such as living arrangements and income make significant differences in the food choices, eating habits, and nutrition status of older adults, especially those over age 80.[12] People of low socioeconomic status are likely to have inadequate food and nutrient intakes. For example, studies report a consistent relationship between low income and low intakes of vitamin B_6.[13]

Social Changes. Malnutrition among older adults is most likely to occur among those with the least education, those living alone in federally funded housing (an indicator of low income), and those who have recently experienced a change in lifestyle. The risk of nutrient deficiencies is high among people living alone, especially men.[14] One study on home-delivered meals confirmed that men living alone eat less than men living with others; interestingly, women living alone eat more than women living with others.[15] Adults who live alone do not necessarily make poor food choices, but they often consume too little food: Loneliness is directly related to inadequacies, especially of energy intake.[16]

Energy and Nutrient Needs during Late Adulthood

Knowledge about the nutrient needs and nutrition status of older adults has grown considerably in the last decade or so. The 1989 RDA, however, still combine all people over 50 into one group. As mentioned in Chapter 1, however, a revision of the RDA is presently under way and includes changes in age groupings (see Table 1-2 on p. 10). Older adults are grouped into two age categories—51 to 70 years and 71 and older.*[17] After all, the needs of people 50 to 60 years old may be very different from those of people over 80. The need for more age-specific recommendations is becoming more and more urgent as the population ages.

Setting standards for older people is difficult, though, because individual differences become more pronounced as people grow older. People start out with different genetic predispositions and ways of handling nutrients, and the effects of these become magnified with the years. Then, too, one person may tend to omit vegetables from his diet, and by the time he is old, he will have an associated set of nutrition problems. Another may have omitted milk and milk products—her nutrition problems will be different. Also, as people age, they may suffer different chronic diseases and take different drugs—both of which will have impacts on nutrient needs. For all these reasons, researchers have difficulty even defining "healthy aging," a prerequisite to developing recommendations to meet the "needs of practically all healthy persons."[18] Still, some generalizations are valid, and although new RDA may be needed, the present RDA for adults are of some use. The next sections give special attention to a few nutrients of concern.

Energy and Energy Nutrients

Energy needs decline with advancing age. As a rule of thumb, adult energy needs decline an estimated 5 percent per decade. One reason is that people usually reduce their physical activity as they age, although they need not do so. Another reason is that lean body mass diminishes, slowing the basal metabolic rate. At least one researcher contends that if older adults (up to age 75) could avoid changes in body composition (by way of physical activity), they would show little, if any, decrease in basal metabolic rate.[19] Physical activity not only increases energy expenditure but, along with sound nutrition, enhances bone density and supports many body functions.[20]

The lower energy expenditure of many older adults requires that they obtain less food energy to maintain their weights. Accordingly, the energy RDA for adults decreases slightly after age 50. Energy intakes typically decline in parallel with needs. Still, many older adults are overweight, indicating that their food intakes do not decline enough to compensate for reduced energy expenditure.[21]

---

*The current Canadian RNI divide older people into two age groups—50 to 74, and 75 and older.

On limited energy allowances, people must select mostly nutrient-dense foods. There is little leeway for sugars, fats, oils, or alcohol. The Food Guide Pyramid (Figure 1-2 on pp. 12–13) offers a dietary framework for adults of all ages.

Protein. The protein needs of older adults seem to be about the same as, or even greater than, those of younger people.[22] Because energy needs decrease, however, the protein must be obtained from low-kcalorie sources of high-quality protein, such as lean meats, poultry, fish, and eggs; nonfat and low-fat milk products; and legumes.

Carbohydrate and Fiber. As always, abundant carbohydrate is needed to protect protein from being used as an energy source. Sources of complex carbohydrates such as vegetables, whole grains, and fruits are also rich in fiber and essential vitamins and minerals.

The combination of ample water and high-fiber foods can alleviate constipation—a condition common among older adults, especially nursing home residents. Physical inactivity and medications probably contribute to the high incidence of constipation, but lack of water and fiber does, too. In fact, average fiber intakes among older adults are lower than current recommendations (25 grams per day).

It has been estimated that as many as 50 percent of nursing home residents may be malnourished and underweight.[23] For these people, a diet that emphasizes fiber-rich foods such as whole grains, fruits, and vegetables may be too low in protein and energy. Protein- and energy-dense snacks such as hard-boiled eggs, tuna fish and crackers, peanut butter on graham crackers, and hearty homemade soups are valuable additions to the diets of underweight or malnourished older adults.

Fat. As is true for people of all ages, fat needs to be limited in the diet of most older adults. Cutting fat may help retard the development of cancer, atherosclerosis, and other degenerative diseases. This should not be overdone, however; for some older adults, limiting fat intake too severely may lead to nutrient deficiencies and weight loss—two problems that carry greater health risks in the elderly than overweight.[24]

Water

Dehydration is a risk for older adults, who may not notice or pay attention to their thirst, or who find it difficult and bothersome to get a drink or to get to a bathroom. Older adults who have lost bladder control may be afraid to drink too much water. Despite real fluid needs, many older people do not seem to feel thirsty or notice mouth dryness.[25] Many nursing home employees say it is hard to persuade their elderly clients to drink enough water and fruit juices.

Total body water decreases as people age, so that even mild stresses such as fever or hot weather can precipitate rapid dehydration in older adults.[26] Dehydrated older adults seem to be more susceptible to urinary tract infections, pneumonia, pressure ulcers, and confusion and disorientation.[27] An intake of 6 to 8 glasses of water a day is recommended. Milk and juices may replace some of this water, but beverages containing alcohol or caffeine cannot, because of their diuretic effect.

Water recommendation for adults: 1 to 1 1/2 oz/kg actual body weight.

Vitamins and Minerals

As research reveals more about how specific vitamins and minerals influence disease prevention and how age-related physiological changes affect nutrient metabolism, optimal intakes of vitamins and minerals for different groups of older adults may eventually be defined. Until such time, all adults over 50 use the same recommendations.

Vitamin A. Only vitamin A is absorbed and stored *more* efficiently by the aging GI tract and liver, although its processing within the body slows slightly.[28] Several studies have reported that healthy older adults have normal concentrations of plasma vitamin A even when their dietary intakes fall below the RDA, suggesting that the current RDA may be too high.[29] The Committee on Dietary Allowances has hesitated in lowering the RDA, recognizing both the need to prevent vitamin A deficiency and the possibility that the vitamin A precursor beta-carotene might delay the onset of some age-related diseases.

Vitamin A RDA during late adulthood:
  800 μg RE/day.
Canadian RNI during late adulthood:
  1000 μg RE/day (men).
  800 μg RE/day (women).

Vitamin D. Older adults face a greater risk of vitamin D deficiency than younger people do. Only vitamin D-fortified milk provides significant vitamin D, and many older adults drink little or no milk. Further compromising the vitamin D status of many older people, especially those in nursing homes, is their limited exposure to sunlight. Finally, aging reduces the skin's capacity to make vitamin D and the kidneys' ability to convert it to its active form. Not only are older adults not getting enough vitamin D, but they may actually need more than the RDA to prevent bone loss and to maintain vitamin D status, especially in those who engage in minimal outdoor activity.[30] For this reason, the revised recommendations for vitamin D doubled the 1989 RDA of 5 to 10 micrograms per day.[31]

Vitamin D DRI during late adulthood:
  10 μg/day.

Vitamin B_6. Studies on vitamin B_6 reveal that its metabolism is altered with age, resulting in a higher requirement. Many older adults consume far less than the RDA for vitamin B_6.[32] Research suggests that vitamin B_6 deficiency impairs immune response. Such findings may have important implications for older adults, because the aging process itself seems to entail a decline in immune function.[33] The current vitamin B_6 RDA for people over 50 may need to be raised.[34]

Another approach to determining the vitamin B_6 requirements of older adults, as well as the requirements for vitamin B_{12} and folate, focuses on the amino acid homocysteine. Homocysteine is recognized as an independent risk factor for heart disease and stroke in the United States.[35] Homocysteine concentrations rise with vitamin B_6, vitamin

Vitamin B_6 RDA during late adulthood:
  2.0 mg/day (men).
  1.6 mg/day (women).

B_{12}, or folate deficiencies. Without vitamin B_6, the body cannot convert homocysteine to cystathionine, and without vitamin B_{12} and folate, homocysteine is not converted to methionine.

Vitamin B_{12}. As mentioned earlier, the prevalence of atrophic gastritis among those 60 years of age and over is high. People with atrophic gastritis are particularly vulnerable to vitamin B_{12} deficiency for two reasons. First, digestion in the inflamed stomach is inefficient. Second, the bacterial overgrowth that accompanies this condition uses up the vitamin. Many older adults take in less vitamin B_{12} than the RDA.[36] Given the devastating neurological effects of a vitamin B_{12} deficiency, an RDA higher than the current one may be appropriate.[37]

Folate. As is true of vitamin B_6 and vitamin B_{12}, folate intakes of older adults fall short of recommendations.[38] Folate absorption, too, may be compromised by atrophic gastritis.

Many older adults take medications that influence folate absorption, use, and excretion. Antacids, diuretics, and anti-inflammatory drugs may alter folate metabolism.

Iron. Among the minerals, iron deserves first mention. Iron deficiency anemia is less common in older adults than in younger people, but still occurs in some, especially in people with low food energy intakes. Aside from diet, other factors in many older people's lives make iron deficiency likely: chronic blood loss from disease conditions and medicines, and poor iron absorption due to reduced stomach acid secretion and antacid use. Anyone concerned with older people's nutrition should keep these possibilities in mind.

Zinc. Zinc intake is commonly low in older people, with many receiving less than half of the recommended amount.[39] In addition, older adults may absorb zinc less efficiently than younger people do. A number of different factors, including medications that older adults commonly use (diuretics, antacids, and laxatives), can impair zinc absorption or enhance its excretion and thus lead to deficiency.[40] Older adults who do not make special efforts to eat zinc-rich foods such as meats, fish, and poultry will no doubt fail to meet the zinc RDA. Some symptoms of zinc deficiency resemble symptoms associated with aging—for example, decline in taste acuity and dermatitis. Whether these symptoms are attributable to zinc deficiency remains unclear.

Calcium. The appropriate calcium intake for older adults remains controversial. A National Institutes of Health panel has concluded that women over 50 who are not on estrogen replacement therapy and all adults over 65 should receive 1500 milligrams of calcium daily.[41] The DRI Committee recently raised the 1989 RDA of 800 milligrams of calcium a day for older adults to 1200.[42]

As researchers attempt to reach agreement about the calcium requirements of older adults, especially those of women, one thing is

---

Vitamin B_{12} RDA during late adulthood:
  2 mg/day.
Canadian RNI during late adulthood:
  1 mg/day.

Folate RDA during late adulthood:
  200 μg/day (men).
  180 μg/day (women).
Canadian RNI during late adulthood:
  230 μg/day (men, 50 to 74).
  215 μg/day (men, 75+).
  195 μg/day (women, 50 to 74).
  200 μg/day (women, 75+).

Iron RDA during late adulthood:
  10 mg/day.
Canadian RNI during late adulthood:
  9 mg/day (men).
  8 mg/day (women).

Zinc RDA during late adulthood:
  15 mg/day (men).
  12 mg/day (women).
Canadian RNI during late adulthood:
  12 mg/day (men).
  9 mg/day (women).

Calcium DRI during late adulthood:
  1200 mg/day.

clear: The calcium intakes of many people, especially women, in the United States are well below recommended intakes.

Some older adults avoid milk and milk products either because they dislike them or because they associate these foods with stomach discomfort. These people need to eat other calcium-rich foods. For those who don't like to drink milk, milk and milk products can be concealed in foods. Powdered nonfat milk, which is an excellent and inexpensive source of protein, calcium, and other nutrients, can be added when preparing casseroles, meatloaf, and other mixed dishes; 5 heaping tablespoons offer the equivalent of a cup of milk.

Those who are allergic to milk or who are lactose intolerant need to find nonmilk sources of calcium to help meet their calcium needs. Among the vegetables, mustard and turnip greens, bok choy, kale, parsley, watercress, and broccoli are good sources of available calcium. Oysters and small canned fish prepared with their bones, such as sardines, are also rich sources of calcium. Some foods offer large amounts of calcium because they are fortified. Calcium-fortified orange juice, for example, provides as much calcium as regular milk.

Nutrient Supplements for Older Adults

People judge for themselves how to manage their nutrition, and some turn to supplements. Advertisers target older people with appeals to take supplements and eat "health" foods, claiming that these products prevent disease and promote longevity. About half of all women over 65 years of age take some type of nutrient supplement, and about one-fifth of older men do. Quite often those who take supplements are not deficient in the nutrients being supplemented.[43] Certain diseases or health problems may necessitate the taking of supplements, but often supplements have not been prescribed by health care professionals and are inappropriate.

When recommended by a physician, vitamin D and calcium supplements for osteoporosis or iron for iron deficiency anemia may be beneficial. In most cases, though, the money spent on supplements would be better spent on nutritious foods. Older adults with food energy intakes less than about 1500 kcalories per day, however, should probably take the once-daily type of vitamin-mineral supplements.

People with small energy allowances would do well to become more active so they can afford to eat more food. Food is the best source of nutrients for everybody. Supplements are just that—supplements to foods, not substitutes for them. For anyone who is motivated to obtain the best possible health, it is never too late to learn to eat well, exercise regularly, and adopt other lifestyle changes such as quitting smoking and moderating alcohol use.

Table 10-1 summarizes the nutrient concerns of aging. Although some nutrients need special attention in the diet, supplements are not routinely recommended.

Table 10-1 Summary of Nutrient Concerns in Aging

Nutrient	Effect of Aging	Comments
Water	Lack of thirst and decreased total body water make dehydration likely.	Mild dehydration is a common cause of confusion. Difficulty obtaining water or getting to the bathroom may compound the problem.
Energy	Need decreases.	Physical activity moderates the decline.
Fiber	Likelihood of constipation increases with low intakes and changes in the GI tract.	Inadequate water intakes and lack of physical activity, along with some medications, compound the problem.
Protein	Needs stay the same.	Low-fat, high-fiber legumes and grains meet both protein and other nutrient needs.
Vitamin A	Absorption increases.	RDA may be high.
Vitamin D	Increased likelihood of inadequate intake; skin synthesis declines.	Daily limited sunlight exposure may be of benefit.
Iron	In women, status improves after menopause; deficiencies are linked to chronic blood losses and low stomach acid output.	Adequate stomach acid is required for absorption; antacid or other medicine use may aggravate iron deficiency; vitamin C and meat increase absorption.
Zinc	Intakes may be low and absorption reduced; but needs may also decrease.	Medications interfere with absorption; deficiency may depress appetite and sense of taste.
Calcium	Intakes may be low; osteoporosis common.	Stomach discomfort commonly limits milk intake; calcium substitutes are needed.

Nutrition Assessment of Older Adults

Nutrition assessment of older adults must focus on the past as well as the present. Each person's long history of eating and health habits has contributed to the present nutrition status:

- The person's dietary history should reveal past as well as present intake habits. Of particular interest are any current habits that may jeopardize adequate nutrient intakes: Does the person skip meals or routinely exclude any food groups? Can the person get and prepare meals?

- The drug history is usually relevant to an older person's nutrition status. What drugs does the person take, and what are each drug's effects? Does the person use alcohol? How much?

- Socioeconomic factors may also affect nutrition status and may indicate the need for food assistance programs.

- The physical examination should be made with an eye to nutrition status. Hair, skin, and nails may offer clues to imbalances, excesses, or deficiencies. Inadequate fluid intake may be apparent from a high body temperature, a swollen tongue, reduced blood pressure, sunken eyeballs, or a reduced urine output. Are any chronic diseases, illnesses, or physical disabilities present?

■ Oral health should be assessed, as it may affect food and nutrient intake. Can the person chew and swallow foods satisfactorily? Does the person have dentures or need them? Are there any lesions in the mouth, on the lips, or on the tongue?

■ Anthropometric measures such as height, weight, and fatfold measures can reveal altered body composition that may indicate malnutrition. Loss of lean tissue is an early indicator of PEM. Obesity suggests a raised risk of chronic diseases.

■ Biochemical measures may be used to follow up on other assessment steps as appropriate. For example, serum albumin concentration has been used extensively as a marker of the degree of PEM. Albumin is the major protein produced by the liver, and its synthesis depends on an adequate protein intake.

The Nutrition Screening Initiative is part of a national effort to identify and treat nutrition problems in older persons; it uses a screening checklist (see Table 10-2). To *determine* the risk of malnutrition in older clients, health care providers can keep in mind the characteristics listed in the margin.[44]

Risk factors for malnutrition in older adults:
■ *Disease.*
■ *Eating poorly.*
■ *Tooth loss or oral pain.*
■ *Economic hardship.*
■ *Reduced social contact.*
■ *Multiple medications.*
■ *Involuntary weight loss or gain.*
■ *Needs assistance with self-care.*
■ *Elderly person older than 80 years.*

Nutrition-Related Concerns during Late Adulthood

Nutrition through the prime years may play a greater role than has been realized in preventing many changes once thought to be inevitable consequences of growing older. The following discussions of cataracts, arthritis, osteoporosis, and the aging brain show that nutrition may provide at least some protection against some of the conditions associated with aging.

Cataracts

Cataracts are age-related thickenings in the lenses of the eye that impair vision. If not surgically removed, they ultimately lead to blindness. Cataracts occur even in well-nourished individuals as a result of ultraviolet light exposure, oxidative damage, injury, viral infections, toxic substances, and genetic disorders. Many cataracts, however, are vaguely called senile cataracts—meaning "caused by aging." In the United States, some 46 percent of people between the ages of 75 and 85 have cataracts, compared to only 5 percent of those between the ages of 52 and 64.[45]

Oxidative stress appears to play a significant role in the development of cataracts, and antioxidant nutrients may help minimize the damage.[46] Studies have reported an inverse relationship between cataracts and dietary intakes of vitamin C, vitamin E, and carotenoids.[47] One

cataracts: thickenings of the eye lenses that impair vision and can lead to blindness.

Table 10-2 Nutrition Screening Initiative Checklist

Circle the number to the right if the statement applies to you.

Statement	Yes
I have an illness or condition that made me change the kind and/or amount of food I eat.	2
I eat fewer than 2 meals per day.	3
I eat few fruits or vegetables or milk products.	2
I have 3 or more drinks of beer, liquor, or wine almost every day.	2
I have tooth or mouth problems that make it hard for me to eat.	2
I don't always have enough money to buy the food I need.	4
I eat alone most of the time.	1
I take 3 or more different prescribed or over-the-counter drugs a day.	1
Without wanting to, I have lost or gained 10 pounds in the last 6 months.	2
I am not always physically able to shop, cook, and/or feed myself.	2

Total

SCORE:

0–2: Good. Recheck your score in 6 months.

3–5: Moderate nutritional risk. Visit your local office on aging, senior nutrition program, senior citizens center, or health department for tips on improving eating habits.

6 or more: High nutritional risk. See your doctor, dietitian, or other health care professional for help in improving your nutrition status.

study found that people who had no cataracts took significantly more supplements of vitamins C and E than those who had cataracts.[48]

Arthritis

arthritis: inflammation of a joint, usually accompanied by pain, swelling, and structural changes.

The most common type of arthritis that disables older people is osteoarthritis, a painful swelling of the joints. During movement, the ends of bones are normally protected from wear by cartilage and by small sacs of fluid that act as a lubricant. With age, bones sometimes disintegrate, and the joints become malformed and painful to move. Osteoarthritis afflicts millions of people around the world, especially the elderly. Nutrition quackery to treat osteoarthritis is abundant, but no known diet, food, or supplement prevents, relieves, or cures it. Table 10-3 presents some of the many *non*effective dietary treatments for osteoarthritis.

Table 10-3 *Noneffective Dietary Strategies for Arthritis*

- Alfalfa tea.
- Amino acid supplements.
- Blackstrap molasses.
- Burdock root.
- Calcium.
- Celery juice.
- Cod liver oil.
- Copper supplements.
- Dimethyl sulfoxide (DMSO).
- Fasting.
- Fresh fruit.
- Garlic.
- Honey.
- Inositol.
- Kelp.
- Lecithin.
- Para-amino benzoic acid (PABA).
- Raw liver.
- Aloe vera liquid.
- Superoxide dismutase (SOD).
- Vitamin D.
- Vitamin megadoses.
- Watercress.
- Yeast.

A known connection between osteoarthritis and nutrition is overweight. Weight loss is important for overweight people with osteoarthritis, partly because the joints affected are often weight-bearing joints that are stressed and irritated by having to carry excess poundage. Interestingly, though, weight loss often relieves the worst pain of osteoarthritis in the hands as well, even though they are not weight-bearing joints. Jogging and other weight-bearing activities do not worsen osteoarthritis, even in marathon runners.

Another type of arthritis, known as rheumatoid arthritis, has a possible link to diet through the immune system. In rheumatoid arthritis, the immune system mistakenly attacks the bone coverings as if they were made of foreign tissue.[49] The integrity of the immune system depends on adequate nutrition, and a poor diet may worsen this type of arthritis. It is also possible that in some individuals, certain foods may stimulate the immune system to attack. For example, milk and milk products seem to aggravate rheumatoid arthritis in some people.[50]

Another nutrient linked to rheumatoid arthritis is the omega-3 fatty acid found in fish oil, eicosapentaenoic acid (EPA). Research shows that the same diet recommended for heart health—one low in saturated fat from meats and milk products and high in polyunsaturated oils from fish—helps prevent or reduce the inflammation in the joints that makes arthritis so painful.[51] Researchers theorize that EPA probably interferes with the action of prostaglandins, compounds involved in inflammation.

Another possible link between nutrition and rheumatoid arthritis involves lipid peroxidation. Lipid peroxidation of the membranes within joints causes inflammation and swelling.[52] Vitamin E helps prevent peroxidation but it has not improved active cases of rheumatoid arthritis. This is not surprising, though, because the vitamin's role in lipid peroxidation is preventive, not restorative.

osteoarthritis: a painful, chronic disease of the joints caused when the cushioning cartilage in a joint breaks down; joint structure is usually altered, with loss of function; also called *degenerative arthritis*.

rheumatoid arthritis: a disease of the immune system involving painful inflammation of the joints and related structures.

Drugs used to relieve arthritis can impose nutrition risks.[53] Many drugs affect appetite and alter the body's use of nutrients, as Focal Point 10 explains.

Osteoporosis

Osteoporosis is one of the most prevalent diseases of aging, affecting more than 25 million people in the United States—most of them women. Women are most vulnerable because bone dwindles rapidly after menopause, when secretion of the hormone estrogen diminishes and menstruation ceases. Accelerated losses occur for six to eight years following menopause, then taper off, so that women again lose bone at the same rate as men their age. Losses of bone minerals continue throughout the remainder of a woman's lifetime, but not at the free-fall pace of the menopause years.

Researchers have identified two types of osteoporosis, which cause two types of bone breaks. Type I osteoporosis involves losses of trabecular bone (see Figure 10-3). These losses sometimes exceed three times the expected rate, and bone breaks may occur suddenly. Trabecular bones become so fragile that even the body's own weight can overburden the spine—vertebrae may suddenly disintegrate and crush down, painfully pinching major nerves. Wrists may break as bone ends weaken, and teeth may loosen or fall out as the trabecular bone of the jaw recedes. Women are most often the victims of this type of osteoporosis, six to one over men. Taking estrogen for at least seven years

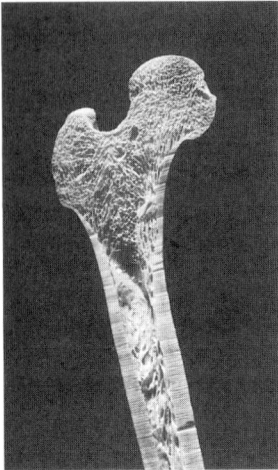

Trabecular bone is the lacy network of calcium-containing crystals that fills the interior. Cortical bone is the dense, ivorylike bone that forms the exterior shell.

type I osteoporosis: osteoporosis characterized by rapid bone losses, primarily of trabecular bone.

Figure 10-3   Healthy and Osteoporotic Trabecular Bones

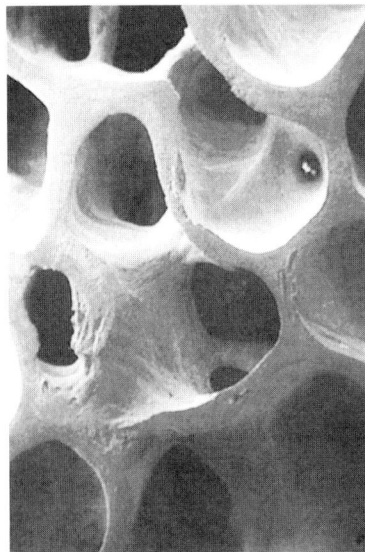

Electron micrograph of healthy trabecular bone.

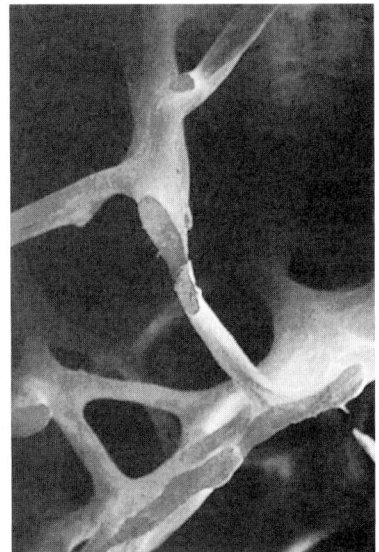

Electron micrograph of trabecular bone affected by osteoporosis

after menopause is the most effective preventive measure against this type of osteoporosis.[54]

In type II osteoporosis, the calcium of both cortical and trabecular bone is drawn out of storage, but slowly, over the years. As old age approaches, the vertebrae may compress into wedge shapes, forming what is often called "dowager's hump," the posture many older people assume as they "grow shorter." Figure 10-4 shows the effect of compressed spinal bone on a woman's height and posture. Because both the cortical shell and the trabecular interior weaken, breaks most often occur in the hip. A woman is twice as likely as a man to suffer type II osteoporosis. Table 10-4 summarizes the differences between the two types of osteoporosis.

In addition to genetics and hormones, many outside factors also influence the progression of osteoporosis. Nutrition and physical activity, for example, can maximize peak bone density during growth,

trabecular (tra-BECK-you-lar) **bone:** the lacy inner structure of calcium crystals that supports the bone's structure and provides a calcium storage bank.

type II osteoporosis: osteoporosis characterized by gradual losses of both trabecular and cortical bone.

cortical bone: the ivorylike outer bone layer that forms a shell surrounding trabecular bone and comprises the shaft of a long bone.

Figure 10-4 Loss of Height Caused by Osteoporosis in a Woman

The woman on the left is about 50 years old. On the right, she is 80 years old. Her legs have not grown shorter: Only her back has lost length, due to collapse of her spinal bones (vertebrae). Collapsed vertebrae cannot protect the spinal nerves from pressure that causes excruciating pain.

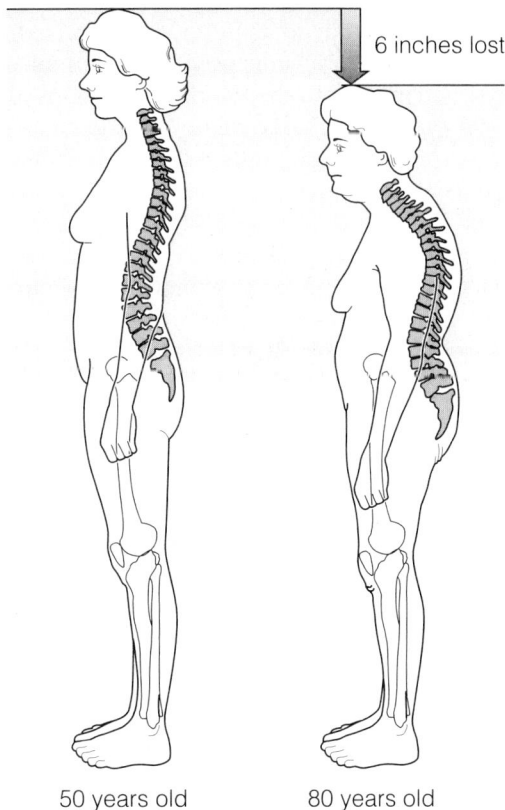

6 inches lost

50 years old   80 years old

Table 10-4 Types of Osteoporosis Compared

	Type I	Type II
Other Name	Postmenopausal osteoporosis	Senile osteoporosis
Age of Onset	50 to 70 years old	70 years and older
Bone Loss	Trabecular bone	Both trabecular and cortical bone
Fracture Sites	Wrist and spine	Hip
Gender Incidence	6 women to 1 man	2 women to 1 man
Primary Causes	Rapid loss of estrogen in women following menopause; loss of testosterone in men with advancing age	Reduced calcium absorption, increased bone mineral loss, increased propensity to fall

Source: Adapted from C. Niewoehner, Calcium and osteoporosis, *Cereal Foods World* 33 (1988): 784–787.

whereas alcohol and tobacco abuse can accelerate bone losses later in life. For example, smokers experience more fractures from slight injury than do nonsmokers. A study of twins reports that women who smoke a pack of cigarettes a day throughout adulthood lose an extra 5 to 10 percent of their bone density by menopause.[55] Although the mechanism of action remains undefined, both the lower body weights of smokers and the earlier menopause of female smokers may be factors.[56]

People who do not consume milk products or other calcium-rich foods in amounts that provide even half the recommended calcium may benefit from calcium supplements.[57] During the menopausal years, calcium supplements of 1 gram daily may slow, but cannot fully prevent, the inevitable bone loss.[58] Supplements are commonly used as a part of therapy for osteoporosis, along with gentle exercise, and, for women, estrogen replacement; supplements should not be used as a substitute for estrogen.[59] As a rule, women taking estrogen need no more calcium than the recommended intake.

The Aging Brain

The brain, like all the body's organs, responds to both inherited and environmental factors that can enhance or diminish its amazing capacities. One of the challenges researchers face when studying aging of the brain in human beings is to distinguish among normal age-related physiological changes, changes caused by diseases, and changes that result from cumulative, extrinsic factors such as diet.

The brain normally changes in some characteristic ways as it ages. For one thing, its blood supply decreases. For another, the number of

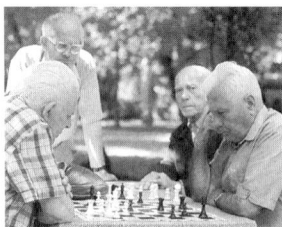

The brain is nourished by both foods and mental challenges.

neurons, the brain cells that specialize in transmitting information, diminishes as people age. When the number of nerve cells in one part of the cerebral cortex diminishes, hearing and speech are affected. Losses of neurons in other parts of the cortex can impair memory and cognitive function. When the number of neurons in the hindbrain diminishes, balance and posture are affected. Losses of neurons in other parts of the brain affect still other functions.

Clinicians now recognize that much of the cognitive loss and forgetfulness generally attributed to aging is due in part to extrinsic, and therefore controllable, factors such as nutrient deficiencies. In some instances, the degree of cognitive loss is extensive and attributable to a specific disorder such as a brain tumor. In cases such as Alzheimer's disease, deterioration may be genetically determined and will not yield to external approaches.

Alzheimer's Disease. Much attention has focused on the *abnormal* deterioration of the brain called senile dementia of the Alzheimer's type (SDAT), which affects 5 percent of U.S. adults by age 65 and 20 percent of those over 80.[60] Diagnosis of SDAT depends on its characteristic symptoms: The victim gradually loses memory and reasoning, the ability to communicate, physical capabilities, and eventually life itself. Nerve cells in the brain die, and communication between the cells breaks down.

The causes of SDAT continue to elude researchers, although genetic factors are apparently involved. A newly established connection to a human gene that makes part of a lipoprotein has sparked new hope for preventing SDAT.*[61] Researchers hope to use these genetic findings to develop early detection tests and a cure for this devastating disease.

Treatment involves providing care to clients and support to their families. One drug (trade-named Tacrine) seems to slow the disease's advance in about 20 percent of those who use it, but it does not reverse the damage already done. Meanwhile, some drugs seem to improve the ability to remember and so hold promise for improving the lives of those with SDAT. Other drugs may be used to control depression or behavior problems.

One abnormality of interest involves the extremely low concentrations of the enzyme that catalyzes production of the neurotransmitter acetylcholine from choline and acetyl CoA. Acetylcholine is essential to memory. To date, supplements of choline (or of lecithin, which contains choline) have had no effect on memory or on the progression of the disease.

Most people have heard of an association between aluminum and the development of SDAT, although a causal connection seems unlikely.

neuron: a nerve cell; the structural and functional unit of the nervous system. Neurons initiate and conduct nerve transmissions.

cerebral cortex: the outer surface of the cerebrum.

senile dementia: the loss of brain function beyond the normal loss of physical adeptness and memory that occurs with aging.

senile dementia of the Alzheimer's type (SDAT): a degenerative disease of the brain involving memory loss and major structural changes in neuron networks; often simply called **Alzheimer's disease.**

---

*The gene associated with Alzheimer's disease codes for apo E4, one of three apolipoprotein E varieties. A report on the genetic and other aspects of Alzheimer's disease is available from Alzheimer's Disease Education and Referral Center, P.O. Box 8250, Silver Spring, MD 20907-8250.

Brain concentrations of aluminum in SDAT people exceed normal brain concentrations by some 10 to 30 times, but blood and hair aluminum remains normal, indicating that the accumulation is caused by something in the brain itself, not by an overload of aluminum in the body. Thus the high brain aluminum must be at least partly a result, rather than a cause, of SDAT.

Maintaining appropriate body weight may be the most important nutrition concern for the person with SDAT. Depression and forgetfulness can lead to poor food intake, and restlessness may increase energy needs. Perhaps the best that a caretaker can do nutritionally for an SDAT client is to supervise food planning and mealtimes. Providing well-liked and well-balanced meals and snacks in a cheerful atmosphere encourages food consumption. To minimize confusion, offer a few ready-to-eat foods, in bite-size pieces, with seasonings and sauces. To avoid mealtime disruptions, control distractions such as television, children, and the telephone.

SDAT is an identifiable disease, the course of which is probably not influenced by nutrition. But poor nutrition in general affects the brain in other ways.

Nutrient Deficiencies and Brain Function. Moderate, long-term nutrient deficiencies may contribute to the loss of memory and cognition that some older adults experience.[62] For example, the ability of neurons to synthesize specific neurotransmitters depends in part on the availability of precursor nutrients that are obtained from the diet. The neurotransmitter serotonin derives from the amino acid tryptophan. To function properly, the enzymes involved in neurotransmitter synthesis require vitamins and minerals. Severe dietary deficiencies of thiamin, vitamin B_6, vitamin B_{12}, folate, and vitamin C impair mental ability, including memory. Minerals such as iron and zinc also support normal brain function. Table 10-5 summarizes some of the better known connections between impaired brain function and severe nutrient deficiencies. If long-term, moderate nutrient deficiencies influence the loss of cognitive function that accompanies aging, then the loss may be preventable or at least diminished or delayed through diet.

Table 10-5 Summary of Nutrient–Brain Relationships

Brain Function	Nutrient Deficiency
Short-term memory loss	Vitamin B_{12}, vitamin C
Poor performance in problem-solving tests	Riboflavin, folate, vitamin B_{12}, vitamin C
Dementia	Thiamin, niacin, zinc
Cognition	Folate, vitamin B_6, vitamin B_{12}, iron
Degeneration of brain tissue	Vitamin B_6

These brief discussions of cataracts, arthritis, osteoporosis, and the aging brain show that nutrition may provide at least some protection against certain conditions associated with aging. In fact, nutrition through the prime years may play a greater role than formerly realized in preventing many changes once thought to be inevitable consequences of growing older.

In addition, there is much people can do, besides obtaining adequate nutrition, to support a high quality of life into old age. By practicing stress management skills, maintaining physical fitness, participating in activities of interest, and cultivating spiritual health, a person can grow old gracefully (see Table 10-6 for some strategies).

Table 10-6 Strategies for Growing Old Gracefully

- Choose nutrient-dense foods.
- Maintain appropriate body weight.
- Reduce stress.
- For women, see a physician about estrogen replacement.
- For people who smoke, quit.
- Expect to enjoy sex, and learn new ways of enhancing it.
- Use alcohol only moderately, if at all; use drugs only as prescribed.
- Take care to prevent accidents.
- Expect good vision and hearing throughout life; obtain glasses and hearing aids if necessary.
- Be alert to confusion as a disease symptom, and seek diagnosis.
- Control depression through activities and friendships.
- Drink 6 to 8 glasses of water every day.
- Practice mental skills. Keep on solving math problems and crossword puzzles, playing cards or other games, reading, writing, imagining, and creating.
- Make financial plans early to ensure security.
- Accept change. Work at recovering from losses; make new friends.
- Cultivate spiritual health. Cherish personal values. Make life meaningful.
- Go outside for sunshine and fresh air as often as possible.
- Be physically active. Walk, run, dance, swim, bike, row, or climb for aerobic activity. Lift weights, do calisthenics, or pursue some other activity to tone, firm, and strengthen muscles. Change activities to suit changing abilities and tastes.
- Be socially active—play bridge, join an exercise group, take a class, teach a class, eat with friends, volunteer time to help others.
- Stay interested in life—pursue a hobby, spend time with grandchildren, take a trip, grow a garden, or go to the movies.
- Enjoy life.

Food Choices and Fitness during Late Adulthood

Taking time to nourish your body well is a gift you give yourself.

Older people are an incredibly diverse group, and for the most part they are independent, socially sophisticated, mentally lucid, fully participating members of society who report themselves to be happy and healthy. In fact, chronic disabilities among the elderly have declined dramatically over the past decade.[63] Older people spend more money per person on foods to eat at home than other age groups and less money on foods away from home. Manufacturers would be wise to cater to the preference of older adults by providing good-tasting, nutritious foods in easy-to-open, single-serving packages with labels that are easy to read. Such services enable older adults to maintain their independence and to feel a sense of control and involvement in their own lives. Another way older adults can take care of themselves is by remaining or becoming physically active. Physical activity helps preserve one's ability to perform daily tasks and so promotes independence.[64]

Diets Tailored to Individual Preferences

Familiarity, taste, and health beliefs are most influential on older people's food choices. Eating foods that are familiar, especially those that recall family meals and pleasant times, can be comforting. People 65 and over are less likely to diet to lose weight than younger people are, but are more likely to diet in pursuit of medical goals such as controlling blood glucose, cholesterol, and sodium. The importance of diet and health beliefs in food selection is evidenced by surveys indicating that older adults are choosing low-fat poultry and fish, low-fat milk and milk products, and high-fiber breads and grains.[65] Few older adults, however, consume the recommended amounts of milk products.[66]

Food Assistance Programs

Nutrition services are an integral part of health care, and different subgroups of the aging population need different programs designed to meet their specific needs.[67] People living alone can benefit from congregate meal programs; people confined to their homes need meals delivered.

The U.S. government funds programs to provide nutritious meals to older adults at congregate meal sites. These meals are a valuable source

---

🏃 **Healthy People 2000**

Increase to at least 80% the receipt of home foodservices by people aged 65 and older who have difficulty in preparing their own meals or are otherwise in need of home-delivered meals.

of nutrients for many older adults. Like school lunches, though, such congregate meals typically do not meet current dietary recommendations to limit sodium, fat, and cholesterol.[68] Table 10-7 describes food assistance programs for older adults.

Unfortunately, federal programs reach only about one-third of those needy older adults who do not have access at all times to nutritious foods. Some of those in need may be unaware that food assistance programs exist; others may live in areas where such programs are unavailable or difficult to get to. The American Dietetic Association stresses the importance of food and nutrition as sustenance and in disease prevention and therapy for the growing population of older adults.[69]

Fitness during Late Adulthood

The many and remarkable benefits of regular physical activity are not limited to the young: Older adults who are active weigh less and have greater flexibility, more endurance, and better balance than those who are inactive.[70] They reap additional benefits as well; for example, moderate endurance activities improve the quality of sleep, and strength training significantly improves mobility and resistance to injury.[71] In fact, regular physical activity is the most powerful predictor of a person's mobility in the later years.[72]

Interestingly, research shows that in addition to improving muscle strength, strength training can significantly improve walking endurance in older adults.[73] Men and women 65 years of age or older who participated in strength-training activities, but not aerobic activities, improved their "walking to exhaustion time" by 38 percent, and blood pressure, heart rate, and perceived exertion were all lower than in their sedentary peers.

Activities of all kinds are recommended to maintain and promote health: Strength training can build muscles, and aerobic activity can improve cardiorespiratory endurance and lower blood lipid concentrations.[74] Although aging affects both speed and endurance to some degree, older adults can still train and achieve exceptional performances.

Ideally, physical activity should be part of each day's schedule and should be intense enough to prevent muscle atrophy and to speed up the heartbeat and respiration rate. Healthy older adults who have not been active can ease into a suitable routine. They can start by walking short distances until they can walk at least a mile three times a week, and then they can gradually increase their pace to achieve a 20- to 25-minute mile.[75]

Muscle mass and muscle strength tend to decline with aging, making older people vulnerable to falls and immobility. Falls are a major cause of fear, injury, disability, dependence, and even death among older adults. Regular physical activity tones, firms, and strengthens muscles, helping to improve confidence, reduce the risk of falling, and minimize the risk of injury should a fall occur. Strength training, even

Social interactions at a congregate meal site can be as nourishing as the foods served.

Table 10-7 Food Assistance Programs for Older Adults

Title IIIc Nutrition Program for Older Americans

- *Funding*: U.S. Department of Health and Human Services.
- *Services*: Congregate and home-delivered meals, therapeutic diets. Supportive services include transportation to congregate meal sites; shopping assistance; information and referral; and, to some extent, nutrition counseling and education.
- *Impact*: The Title IIIc program improves the nutrient content of high-risk older adults' diets and offers socialization and recreation. Many of the nutrition programs around the country go above and beyond federal requirements of congregate and home meals by offering lunch clubs, ethnic meals, acceptance of food stamps for meal payment, and meals for older homeless people.

Food Stamps

- *Funding*: U.S. Department of Agriculture.
- *Services*: Income supplement to low-income households in the form of coupons to purchase food.
- *Impact*: Some research results suggest that food stamps serve more as an income supplement to elderly participants than as a device to improve nutrition status.[a] Other research indicates that food stamp participants' nutrient intakes are higher than those of nonparticipants with similar incomes.[b]

Meals on Wheels

- *Funding*: Private funding to supplement the Title IIIc program; an example of a private and public sector partnership responding to the needs of the growing numbers of older adults.
- *Services*: Direct meal delivery to the homebound elderly, integrated into the meal delivery services provided by the Title IIIc program.
- *Impact*: Meals on Wheels focuses on filling the need for weekend and holiday meals for homebound elderly people, a service that is limited in the Title IIIc program.

---

[a]J. S. Butler, J. C. Ohls, and B. M Posner, The effect of the food stamp program on the nutrient intake of the eligible elderly, *Journal of Nutrition for the Elderly* 4 (1985): 25–51, as cited in M. B. Kohrs, Effectiveness of nutrition intervention programs for the elderly, in *Nutrition and Aging*, eds. M. L. Hutchinson and H. N. Munro (New York: Academic Press, 1986), pp. 139–167.

[b]J. S. Akin and coauthors, The impact of federal transfer programs on the nutrient intake of elderly individuals, *Journal of Human Resources* 20 (1985): 382–404, as cited in B. M. Posner and E. Levine, Nutrition services for older Americans, in *Geriatric Nutrition: The Health Professional's Handbook*, ed. R. Chernoff (Gaithersburg, Md.: Aspen, 1991), pp. 415–447.

Source: Adapted from B. M. Posner and E. Levine, Nutrition services for older Americans, in *Geriatric Nutrition: The Health Professional's Handbook*, ed. R. Chernoff (Gaithersburg, Md.: Aspen, 1991), pp. 415–447.

in frail, elderly people over 85 years of age, has been shown not only to improve muscle strength and mobility but also to increase energy expenditure and energy intake, thereby enhancing nutrient intakes.[76] This finding highlights another reason to be physically active: A person spending energy on physical activity can afford to eat more food and with it, receive more nutrients. People who are committed to an ongoing fitness program have higher energy and nutrient intakes than more sedentary people.[77]

One expert suggests the following physical activity program to maintain good health and function in older adults:[78]

■ *Every day.* 60 minutes of some physical activity: gardening, walking, climbing stairs, or simply moving about. This can be for 5 minutes at a time, 12 times a day; 12 minutes at a time, 5 times a day; or any combination of activity to total 60 minutes.

■ *Three days a week.* 30 to 45 minutes of vigorous and continuous physical activity, such as swimming, dancing, rowing, or brisk walking.

Strength training promotes strong muscles and bones and healthy appetites.

With persistence, people can achieve great improvements at any age. Training not only tones, firms, and strengthens muscles but also increases the blood flow to the brain, thereby improving mental ability.

Researchers studying nutrition and aging are challenged by the physiological and psychosocial diversity of older adults. Life factors such as nutrition, genetics, physical activity, and stress contribute to the diversity and make the study of the aging process complex. Many of the health problems older people experience are presently attributed to normal, age-related processes, perhaps to a greater extent than is valid. Research that focuses on how life factors affect aging and disease processes is vital to ensuring that more and more people can look forward to long, healthy lives.

CHAPTER TEN NOTES

1. R. Chernoff, Demographics of aging, in *Geriatric Nutrition: The Health Professional's Handbook*, ed. R. Chernoff (Gaithersburg, Md.: Aspen, 1991), pp. 1–9.

2. Centers for Disease Control and Prevention, National Center for Health Statistics, *Monthly Vital Statistics Report*, October 1996, p. 4.

3. K. G. Kinsella, Changes in life expectancy 1900–1990, *American Journal of Clinical Nutrition* 55 (1992): 1196S–1202S.

4. W. J. Evans, Effects of aging and exercise on nutrition needs of the elderly, *Nutrition Reviews* (II) 54 (1996): 35–39; C. M. Morganti and coauthors, Strength improvements with

1 year of progressive resistance training in older women, *Medicine and Science in Sports and Exercise* 27 (1995): 906–912.

5. G. Paolisso and coauthors, Body composition, body fat distribution, and resting metabolic rate in healthy centenarians, *American Journal of Clinical Nutrition* 62 (1995): 746–750.

6. A. L. de Weck, Immune response and aging: Constitutive and environmental aspects, in *Nutrition of the Elderly*, eds. H. Munro and G. Schlierf (New York: Raven Press, 1992), pp. 89–97.

7. R. K. Chandra, Nutrition and immunity in

the elderly, *Nutrition Reviews* 50 (1992): 367–371.

8. K. Hirokawa, Understanding the mechanism of the age-related decline in immune function, *Nutrition Reviews* 50 (1992): 361–366.

9. S. Hosoda and coauthors, Age-related changes in the gastrointestinal tract, *Nutrition Reviews* 50 (1992): 374–377.

10. C. Murphy, Age-associated changes in taste and odor sensation, perception, and preference, in *Nutrition of the Elderly*, eds. H. Munro and G. Schlierf (New York: Raven Press, 1992), pp. 79–87; V. B. Duffy, J. R. Backstrand, and A. M. Ferris, Olfactory dysfunction and related nutritional risk in free-living, elderly women, *Journal of the American Dietetic Association* 95 (1995): 879–884.

11. C. O. Mitchell and R. Chernoff, Nutritional assessment of the elderly, in *Geriatric Nutrition: The Health Professional's Handbook*, ed. R. Chernoff (Gaithersburg, Md.: Aspen, 1991), pp. 363–395.

12. J. V. White and coauthors, Consensus of the Nutrition Screening Initiative: Risk factors and indicators of poor nutritional status in older Americans, *Journal of the American Dietetic Association* 91 (1991): 783–787.

13. A. K. Kant and G. Block, Dietary vitamin B-6 intake and food sources in the US population: NHANES II, 1976–1980, *American Journal of Clinical Nutrition* 52 (1990): 707–716.

14. M. A. Davis and coauthors, Living arrangements and dietary quality of older U.S. adults, *Journal of the American Dietetic Association* 90 (1990): 1667–1672.

15. E. Foder-Levitt and coauthors, Utilization of home-delivered meals by recipients 75 years of age or older, *Journal of the American Dietetic Association* 95 (1995): 552–557.

16. D. Walker and R. E. Beauchene, The relationship of loneliness, social isolation, and physical health to dietary adequacy of independently living elderly, *Journal of the American Dietetic Association* 91 (1991): 300–304.

17. Committee on Dietary Reference Intakes, *Dietary Reference Intakes for Calcium, Phosphorus, Magnesium, Vitamin D, and Fluoride* (Washington, D.C.: National Academy Press, 1977); R. M. Russell, New views on the RDAs for older adults, *Journal of the American Dietetic Association* 97 (1997): 515–518.

18. A. Bendich, Criteria for determining recommended dietary allowances for healthy older adults, *Nutrition Reviews* 53 (1995): S105–S110.

19. J. V. G. A. Durnin, Energy metabolism in the elderly, in *Nutrition of the Elderly*, eds. H. Munro and G. Schlierf (New York: Raven Press, 1992), pp. 51–63.

20. Durnin, 1992; L. DiPietro, The epidemiology of physical activity and physical function in older people, *Medicine and Science in Sports and Exercise* 28 (1996): 596–660.

21. E. T. Poehlman and E. S. Horton, Regulation of energy expenditure in aging humans, *Annual Review of Nutrition* 10 (1990): 255–275.

22. D. Kritchevsky, Protein requirements of the elderly, in *Nutrition of the Elderly*, eds. H. Munro and G. Schlierf (New York: Raven Press, 1992), pp. 109–118; W. D. Campbell, Dietary protein requirements of older people: Is the RDA adequate? *Nutrition Today* 31 (1996): 192–197.

23. A. A. Abbase and D. Rudman, Undernutrition in the nursing home: Prevalence, consequences, causes and prevention, *Nutrition Reviews* 52 (1994): 113–122.

24. P. J. Nestel, Dietary fat for the elderly: What are the issues? in *Nutrition of the Elderly*, eds. H. Munro and G. Schlierf (New York: Raven Press, 1992), pp. 119–127.

25. P. A. Phillips and coauthors, Reduced thirst after water deprivation in healthy elderly men, *New England Journal of Medicine* 311 (1984): 753–759; B. J. Rolls and P. A. Phillips, Aging and disturbances of thirst and fluid balance, *Nutrition Reviews* 48 (1990): 137–144.

26. R. Chernoff, Physiologic aging and nutritional status, *Nutrition in Clinical Practice*, February 1990, pp. 5–8.

27. J. C. Chidester and A. A. Spangler, Fluid intake in the institutionalized elderly, *Journal of the American Dietetic Association* 97 (1997): 23–28; S. A. Gilmore and coauthors, Clinical indicators associated with unintentional weight loss and pressure ulcers in elderly residents of nursing facilities, *Journal of the American Dietetic Association* 95 (1995): 984–992.

28. S. D. Krasinski and coauthors, Vitamin A and E intake: Relationship to fasting plasma retinol; retinol binding protein, retinyl ester, carotene, and alpha tocopherol levels in the elderly and young adults, *American Journal of Clinical Nutrition* 49 (1989): 112–120.

29. R. M. Russell and P. M. Suter, Vitamin requirements of elderly people: An update, *American Journal of Clinical Nutrition* 58 (1993): 4–14.

30. A. R. Webb and coauthors, An evaluation of the relative contributions of exposure to sunlight and of diet to the circulating concentrations of 25-hydroxyvitamin D in an elderly nursing home population in Boston, *American Journal of Clinical Nutrition* 51 (1990): 1075–1081.

31. Committee on Dietary Reference Intakes, 1997.

32. K. Tucker, Micronutrient status and aging, *Nutrition Reviews* 53 (1995): 9–15; M. M. Manore and coauthors, Plasma pyridoxal 5'-phosphate concentration and dietary vitamin B_6 intake in free-living, low income elderly people, *American Journal of Clinical Nutrition* 50 (1989): 339–345.

33. de Weck, 1992.

34. Russell and Suter, 1993.

35. J. Selhub and coauthors, Association between plasma homocysteine concentrations and extracranial carotid-artery stenosis, *New England Journal of Medicine* 332 (1995): 286–291.

36. Tucker, 1995.

37. Russell and Suter, 1993.

38. Tucker, 1995.

39. C. A. Swanson and coauthors, Zinc status of elderly adults: Response to supplement, *American Journal of Clinical Nutrition* 48 (1988): 343–349; R. J. Wood, P. M. Suter, and R. M. Russell, Mineral requirements of elderly people, *American Journal of Clinical Nutrition* 62 (1995): 493–505.

40. G. J. Fosmire, Trace mineral requirements, in *Geriatric Nutrition: The Health Professional's Handbook*, ed. R. Chernoff (Gaithersburg, Md.: Aspen, 1991), pp. 77–105.

41. D. V. Porter, Washington update: NIH consensus development conference statement optimal calcium intake, *Nutrition Today*, September–October 1994, pp. 37–40.

42. Committee on Dietary Reference Intakes, 1997.

43. W. A. McIntosh and coauthors, The relationship between beliefs about nutrition and dietary practices of the elderly, *Journal of the American Dietetic Association* 90 (1990): 671–675; H. Payette and K. Gray-Donald, Do vitamin and mineral supplements improve the dietary intake of elderly Canadians? *Canadian Journal of Public Health* 82 (1993): 58–60.

44. J. Dwyer and coauthors, Screening older Americans' nutritional health: Future possibilities, *Nutrition Today*, September–October 1991, pp. 21–24.

45. G. E. Bunce, J. Kinoshita, and J. Horwitz, Nutritional factors in cataract, *Annual Review of Nutrition* 10 (1990): 233–254.

46. Bunce, Kinoshita, and Horwitz, 1990; S. D. Varma, Scientific basis for medical therapy of cataracts by antioxidants, *American Journal of Clinical Nutrition* 53 (1991): 335S–345S.

47. P. F. Jacques and L. T. Chylack, Epidemiologic evidence of a role for the antioxidant vitamins and carotenoids in cataract prevention, *American Journal of Clinical Nutrition* 53 (1991): 352S–355S; G. E. Bunce, Antioxidant nutrition and cataract in women: A prospective study, *Nutrition Reviews* 51 (1993): 84–86.

48. J. M. Robertson, A. P. Donner, and J. R. Trevithick, A possible role for vitamins C and E in cataract prevention, *American Journal of Clinical Nutrition* 53 (1991): 346S–351S.

49. E. D. Harris, Rheumatoid arthritis: Pathophysiology and implications for therapy, *New England Journal of Medicine* 322 (1990): 1277–1289.

50. R. S. Panush, Nutritional therapy for rheumatic diseases, *Annals of Internal Medicine* 106 (1987): 619–621.

51. J. M. Kremer and coauthors, Fish-oil fatty acid supplementation in active rheumatoid arthritis: A double-blind, controlled crossover study, *Annals of Internal Medicine* 106 (1987): 497–503.

52. P. Merry and coauthors, Oxidative damage to lipids within the inflamed human joint provides evidence of radical-mediated hypoxic-reperfusion injury, *American Journal of Clinical Nutrition* 53 (1991): 362S–369S.

53. R. Roubenoff and coauthors, Catabolic effects of high-dose corticosteroids persist despite therapeutic benefit in rheumatoid arthritis, *American Journal of Clinical Nutrition* 52 (1990): 1113–1117.

54. D. T. Felson and coauthors, The effect of postmenopausal estrogen therapy on bone density in elderly women, *New England Journal of Medicine* 329 (1993): 1141–1146.

55. J. L. Hopper and E. Seeman, The bone density of female twins discordant for tobacco use,

New England Journal of Medicine 330 (1994): 387–392.

56. C. W. Slemenda, Cigarettes and the skeleton, *New England Journal of Medicine* 330 (1994): 430–431.

57. B. Dawson-Hughes and coauthors, A controlled trial of the effect of calcium supplementation on bone density in postmenopausal women, *New England Journal of Medicine* 323 (1990): 878–883.

58. B. Dawson-Hughes, Calcium supplementation and bone loss: A review of controlled clinical trials, *American Journal of Clinical Nutrition* 54 (1991): 2745–2805; I. R. Reid and coauthors, Effect of calcium supplementation on bone loss in postmenopausal women, *New England Journal of Medicine* 328 (1993): 460–464.

59. C. D. Arnaud and S. D. Sanchez, The role of calcium in osteoporosis, *Annual Review of Nutrition* 10 (1990): 397–414; B. Ettinger, H. K. Genant, and C. E. Cann, Long-term estrogen replacement therapy prevents bone loss and fractures, *Annals of Internal Medicine* 102 (1985): 319–324.

60. R. N. Butler, Senile dementia of the Alzheimer type (SDAT), in *The Merck Manual of Geriatrics* (Rahway, N.J.: Merck, 1990), pp. 933–938.

61. National Institute of Aging, *Progress Report on Alzheimer's Disease, 1994*, NIH Publication number 94-3885 (Washington, D.C.: U.S. Government Printing Office, 1994); E. M. Reiman and coauthors, Preclinical evidence of Alzheimer's disease in persons homozygous for the ϵ4 allele for apolipoprotein E, *New England Journal of Medicine* 334 (1996): 752–758.

62. A. La Rue and coauthors, Nutritional status and cognitive functioning in a normally aging sample: A 6-year reassessment, *American Journal of Clinical Nutrition* 65 (1997): 20–29.

63. K. G. Manton, L. Corder, and E. Stallard, Chronic disability trends in elderly United States population: 1982–1994, *Proceedings of the National Academy of Sciences of the USA* 94 (1997): 2593–2598.

64. DiPietro, 1996.

65. Are older Americans making better food choices to meet diet and health recommendations? *Nutrition Reviews* 51 (1993): 20–22.

66. J. G. Fischer and coauthors, Dairy product in-take of the oldest old, *Journal of the American Dietetic Association* 95 (1995): 918–921.

67. Position of The American Dietetic Association: Nutrition, aging, and the continuum of care, *Journal of the American Dietetic Association* 96 (1996): 1048–1052.

68. M. B. Moran and E. Reed, Are congregate meals meeting clients' needs for "heart healthy" menus? *Journal of Nutrition for the Elderly* 13 (1993): 3–10.

69. Position of The American Dietetic Association, 1996.

70. L. E. Voorrips and coauthors, The physical condition of elderly women differing in habitual physical activity, *Medicine and Science in Sports and Exercise* 25 (1993): 1152–1157.

71. M. A. Fiatarone and coauthors, High-intensity strength training in nonagenarians: Effects on skeletal muscle, *Journal of the American Medical Association* 263 (1990): 3029–3034.

72. A. Z. LaCroix and coauthors, Maintaining mobility in late life: Smoking, alcohol consumption, physical activity, and body mass index, *American Journal of Epidemiology* 137 (1993): 858–869; L. Breslow and N. Breslow, Health practices and disability: Some evidence from Alameda County, *Preventive Medicine* 22 (1993): 86–95.

73. P. A. Ades and coauthors, Weight training improves walking endurance in healthy elderly persons, *Annals of Internal Medicine* 124 (1996): 568–571.

74. Fiatarone and coauthors, 1990; D. E. Danforth and coauthors, Report on the fourth conference for federally supported human nutrition research units and centers, *American Journal of Clinical Nutrition* 54 (1991): 164–168; M. Whitehurst and E. Menendez, Endurance training in older women, *The Physician and Sportsmedicine* 19 (1991): 95–102.

75. J. Posner, M.D., Professor of Medicine and Chief of the Divisions of Geriatric Medicine at the Medical College of Pennsylvania in Philadelphia, as cited in C. L. Pollock, Breaking the risk of falls, *The Physician and Sportsmedicine* 20 (1992): 146–156.

76. A. C. King and coauthors, Moderately intense exercise and self-rated quality of sleep in older adults: A randomized controlled trial, *Journal of the American Medical Association*

277 (1997): 32–37; M. A. Fiatarone and coauthors, Exercise training and nutritional supplementation for physical fraility in very elderly people, *New England Journal of Medicine* 330 (1994): 1769–1775; W. W. Campbell and coauthors, Increased energy requirements and changes in body composition with resistance training in older adults, *American Journal of Clinical Nutrition* 60 (1994): 167–175.

77. D. E. Butterworth and coauthors, Exercise training and nutrient intake in elderly women, *Journal of the American Dietetic Association* 93 (1993): 653–657.

78. P. Astrand, Physical activity and fitness, *American Journal of Clinical Nutrition* (supplement) 55 (1992): 1231–1236.

Nutrient–Drug Interactions

People over the age of 65 take about 25 percent of all the over-the-counter and prescription drugs sold in the United States. They often go to different doctors for different conditions and receive different prescriptions from each. The drugs prescribed enable many older adults to enjoy long, healthy lives, but they also bring side effects and risks.

The Actions of Drugs

Most people think of drugs either as medicines that help them recover from illnesses or as illegal substances that lead to bodily harm and addiction. Actually, both uses of the term *drug* are correct, because any substance that modifies one or more of the body's functions is, technically, a drug. Even when people use medical drugs, the drugs set in motion not only desirable, but also undesirable, events within the body.

Consider aspirin. One action of aspirin is to retard the production of certain prostaglandins. Some prostaglandins help to produce fevers, some sensitize pain receptors, some cause contractions of the uterus, some stimulate digestive tract motility, some control nerve impulses, some regulate blood pressure, some promote blood clotting, and some cause inflammation. By interfering with prostaglandin actions, aspirin reduces fever and inflammation, relieves pain, and slows blood clotting, among other things.

A person cannot use aspirin to produce one of its effects without being subject to all its other effects. Someone who is prone to strokes and heart attacks might take aspirin to prevent blood clotting, but it would also dull that person's sense of pain. In another person, who took aspirin only for pain, it would also slow blood clotting. For the second person, the anticlotting effect might be dangerous because it might cause abnormal bleeding. A single two-tablet dose of aspirin doubles the bleeding time of wounds, an effect that lasts from four to seven hours. For this

Taking several drugs over long periods intensifies the risk of nutrient–drug interactions.

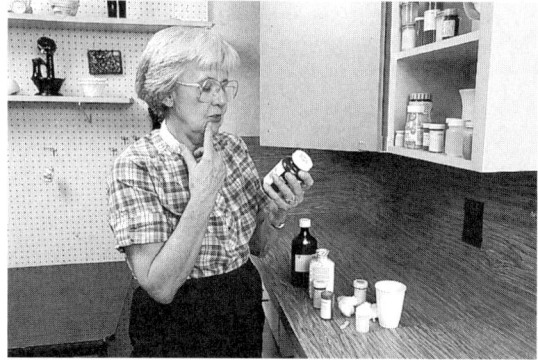

reason, physicians caution clients to refrain from taking aspirin before surgery.

This discussion focuses on some of the nutrition-related consequences of medical drugs, both prescription drugs and nonprescription, or over-the-counter, drugs. (Focal Point 9 described the relationships between nutrition and the drug alcohol.)

The Interactions between Drugs and Nutrients

Hundreds of drugs and nutrients interact, which can lead to nutrient imbalances or interfere with drug effectiveness.[1] Adverse drug–nutrient interactions are most likely if drugs are taken over long periods, if several drugs are taken, or if nutrition status is poor or deteriorating. Understandably, then, elderly people with chronic diseases and people with illnesses that dramatically raise nutrient needs are most at risk. Studies of institutionalized elderly people suggest that multiple drug use may significantly affect nutrition status in this population.[2]

Nutrients and drugs may interact in many ways:

■ Drugs can alter food intake and the absorption, metabolism, and excretion of nutrients.

■ Foods and nutrients can alter the absorption, metabolism, and excretion of drugs.

The following sections describe these interactions, and Table FP10-1 summarizes this information and provides specific examples.

Drugs and Food Intake

Many drugs can lead to malnutrition by interfering with food intake. Drugs can influence appetite, alter taste or smell, cause sores or irritation in the mouth, reduce the flow of saliva, or induce nausea or vomiting. Drugs used to treat hypertension provide an example of how drug side effects can interfere with food intake. Antihypertensive drugs sometimes cause nausea, diarrhea, and vomiting.

Conversely, some drugs stimulate appetite and lead to undesirable weight gain. An example is astemizole (Hismanal), an antihistamine used by some people to relieve allergy symptoms.

Absorption and Drug–Nutrient Interactions

Foods can enhance, delay, or reduce drug absorption, and drugs can have the same effects on nutrient absorption. Laxatives provide an example of how drugs can interfere with nutrient absorption. Laxatives cause foods to move rapidly through the intestine, so that many vitamins have too little time to be absorbed. The use of mineral oil as a laxative robs the person of the fat-soluble vitamins, notably vitamin D. The vitamins dissolve in the indigestible oil and are excreted; calcium, too, is excreted. A person who uses laxatives daily for a long time may find that the intestines can no longer function without them. The dependence can lead to malnutrition.

A classic example of how foods can interfere with drug absorption is the relationship between the antibiotic tetracycline and the minerals calcium and iron. People are advised not to take tetracycline with milk, milk products, or calcium-containing antacids such as Tums. When calcium and tetracycline are taken at the same time, they bind to each other, thus reducing the

absorption of both. Iron has a similar effect. Therefore, iron supplements should be taken two hours apart from tetracycline doses.

Another example of how foods can interfere with the absorption of a drug is that of acidic foods and nicotine gum. Nicotine gum is used to help people quit smoking cigarettes. Certain acidic foods and beverages interfere with the absorption of nicotine through the lining of the mouth into the blood.[3] For maximum effectiveness, people should refrain from ingesting foods and beverages for 15 minutes before, and while, chewing nicotine gum. When a food or beverage blocks nicotine's absorption from the mouth, the person swallows the nicotine, and this may cause nausea and hiccups as well as interference with the drug's effectiveness.

Conversely, foods can enhance the absorption of some drugs. For this reason, the antifungal drug griseofulvin is always given with meals. Similarly, some drugs can enhance the absorption of nutrients. Enzyme replacements, designed specifically to improve the absorption of proteins, carbohydrates, and fats, are an example.

The ways foods and drugs influence absorption determine how drugs are administered. Drugs are absorbed rapidly when they are taken on an empty stomach; foods delay the rate at which the stomach empties. An aspirin works faster when given on an empty stomach, for example, than when it is given with food. But aspirin can irritate the stomach to such an extent that repeated over time it can cause iron deficiency anemia by inducing bleeding in the GI tract. The presence of food can minimize this irritation. As a guideline, then, if rapid action is not essential, to help reduce nausea and protect the GI lining, take food with aspirin or other drugs that irritate the GI tract.

Metabolism and Drug–Nutrient Interactions

Foods can affect drug metabolism in two basic ways: first, nutrients can change the way the body uses a drug, and second, substances in foods can alter the drug's action. To appreciate how drug–nutrient interactions can affect me-

Table FP10-1 Mechanisms and Examples of Food–Drug Interactions

---

Drugs Can Alter Food Intake by

- Altering the appetite (amphetamines suppress the appetite).
- Interfering with taste or smell (methotrexate changes taste sensations).
- Inducing nausea or vomiting (digitalis can do both).
- Changing the oral environment (phenobarbital can cause dry mouth).
- Irritating the GI tract (cyclophosphamide induces mucosal ulcers).
- Causing sores or inflammation of the mouth (methotrexate can cause painful mouth ulcers).

Drugs Can Alter Nutrient Absorption by	*Foods Can Alter Drug Absorption by*
• Changing the acidity of the digestive tract (antacids can interfere with iron absorption).	• Changing the acidity of the digestive tract (candy can change the acidity, thereby causing slow-acting asthma medication to dissolve too quickly).
• Altering digestive juices (cimetidine can improve fat absorption).	• Stimulating secretion of digestive juices (griseofulvin is absorbed better when taken with foods that stimulate the release of digestive enzymes).
• Altering motility of the digestive tract (laxatives speed motility, causing the malabsorption of many nutrients).	• Altering rate of absorption (aspirin is absorbed more slowly when taken with food).
• Inactivating enzyme systems (neomycin may reduce lipase activity).	• Binding to drugs (calcium binds to tetracycline, limiting drug absorption).
• Damaging mucosal cells (chemotherapy can damage mucosal cells).	• Competing for absorption sites in the intestines (dietary amino acids interfere with levodopa absorption this way).
• Binding to nutrients (some antacids bind phosphorus).	

Drugs and Nutrients Can Interact and Alter Metabolism by

- Acting as structural analogs (as anticoagulants and vitamin K do).
- Competing with each other for metabolic enzyme systems (as phenobarbital and folate do).
- Altering enzyme activity and contributing pharmacologically active substances (as monoamine oxidase inhibitors and tyramine do).

Drugs Can Alter Nutrient Excretion by	*Foods Can Alter Drug Excretion by*
• Altering reabsorption in the kidneys (some diuretics increase the excretion of sodium and potassium).	• Changing the acidity of the urine (vitamin C can alter urinary pH and limit the excretion of aspirin).
• Displacing nutrients from their plasma protein carriers (aspirin displaces folate).	

---

tabolism, consider drugs that resemble vitamins in structure. Vitamin K and the anticlotting medication warfarin (Coumadin) provide an example. Warfarin opposes clotting by interfering with vitamin K. To be effective, the warfarin dose must be large enough to counteract the vitamin K in the diet. If a person's vitamin K intake increases, then the physician must increase the drug dose. Other examples are methotrexate, used to treat cancer and psoriasis, and pyri-

Table FP10-2 Foods Restricted in a Tyramine-Controlled Diet

Beverages:	Red wines including chianti, sherry[a]
Cheeses:	Aged cheese, American, camembert, cheddar, gouda, gruyère, mozzarella, parmesan, provolone, romano, roquefort, stilton[b]
Meats:	Liver; dried, salted, smoked, or pickled fish; sausage; pepperoni; salami; dried meats
Vegetables:	Fava beans; Italian broad beans; sauerkraut; snow peas; fermented pickles and olives
Other:	Brewer's yeast;[c] all aged and fermented products; soy sauce in large amounts; cheese-filled breads, crackers, and desserts; salad dressings containing cheese

Note: The tyramine contents of foods vary from product to product depending on the methods used to prepare, process, and store the food. In some cases, as little as 1 ounce of cheese can cause a severe hypertensive reaction in people taking monoamine oxidase inhibitors. In general, the following foods contain small enough amounts of tyramine that they can be consumed in small quantities: ripe avocado, banana, yogurt, sour cream, acidophilus milk, buttermilk, raspberries, and peanuts.

[a]Most wine and domestic beer can be consumed in small quantities.

[b]Unfermented cheeses, such as ricotta, cottage cheese, and cream cheese, are allowed.

[c]Products made with baker's yeast are allowed.

methamine, used to prevent malaria. Both are structurally similar to the B vitamin folate and can cause severe folate deficiency.

Aspirin can also alter folate metabolism, but in a different way. Aspirin competes with folate for its protein carrier, thus interfering with the body's use of the vitamin.

An example of a substance in foods that alters a drug's action is tyramine, a substance in some foods that interacts with monoamine oxidase (MAO) inhibitors, drugs that physicians sometimes prescribe to treat some forms of severe depression. Normally, the enzyme monoamine oxidase inactivates tyramine, and MAO inhibitors block the action of this enzyme. Thus tyramine remains active and stimulates the release of the neurotransmitter norepinephrine. This action can lead to severe hypertension and headaches. If blood pressure rises high enough, it can be fatal. For this reason, people taking MAO inhibitors must restrict their intakes of foods rich in tyramine (see Table FP10-2).

Excretion and Drug–Nutrient Interactions

The acidity of the urine affects drug reabsorption from the kidneys back into the body. For example, acidic urine limits the excretion of acidic drugs such as aspirin. Some nutrients, such as vitamin C in large doses, can lower the pH of urine, making it more acidic. Therefore, when large doses of vitamin C are taken with aspirin, aspirin remains in the blood longer.

Drugs can also alter the excretion of nutrients. For example, some diuretics and laxatives accelerate the excretion of the minerals calcium, potassium, magnesium, and zinc.

The Health Professional and Drug–Nutrient Interactions

Hundreds of drug–nutrient interactions have been identified, and information continues to accumulate. It would be difficult, if not impossible, to remember all the potential effects of drug–nutrient interactions on nutrition status. Instead, health care professionals serve their clients best if they:[4]

■ Keep in mind that drug–nutrient interactions can and do occur, especially when drug use is long term.

■ Record the complete drug and diet histories of clients; review these histories with potential interactions in mind.

■ Be aware of groups of people who are likely to develop nutrient deficiencies, and be prepared to look up the nutrition-related effects of drugs these clients are taking.

■ Reassess nutrition status frequently for high-risk clients.

■ Become familiar with the nutrient interactions of drugs commonly used to treat the disorders of their clients.

Nutrient interactions and risks are not unique to prescription drugs. People who buy over-the-counter drugs also need to protect themselves. About 300,000 nonprescription drug products are marketed in the United States; more than 400 of these were available only by prescription 15 years ago.[5] The increasing availability of over-the-counter drugs allows people to treat themselves for many ailments from arthritis to yeast infections. Many older adults take over-the-counter laxatives and antacids regularly. Excessive use of either of these drugs may impair nutrition status by interfering with nutrient absorption, increasing nutrient excretion, or both. Consumers need to ask their physicians about potential interactions and check with their pharmacists for instructions on taking drugs with foods. Should problems arise, they should seek professional care without delay.

FOCAL POINT 10 NOTES

1. J. A. Thomas, Drug–nutrient interactions, *Nutrition Reviews* 53 (1995): 271–282.
2. C. W. Lewis, E. A. Frongillo, and D. A. Roe, Drug–nutrient interactions in long-term care facilities, *Journal of the American Dietetic Association* 95 (1995): 309–315; R. N. Varma, Risk for drug-induced malnutrition is unchecked in elderly patients in nursing homes, *Journal of the American Dietetic Association* 94 (1994): 192–194.
3. J. E. Henningfield and coauthors, Drinking coffee and carbonated beverages blocks absorption of nicotine from nicotine polacrilex gum, *Journal of the American Medical Association* 264 (1990): 1560–1564.
4. Varma, 1994.
5. Nonprescription Drug Manufacturers Association, *Facts and Figures* (Washington, D.C.: Nonprescription Drug Manufacturers Association, 1993).

Nutrition Assessment: Procedures, Standards, and Forms

This appendix provides a sample of the procedures, standards, and forms commonly used in nutrition assessment. The appropriate uses of each of these are discussed in the opening chapter of this text.

Chapter 1 described the interviews used in collecting *historical data*. To go along with that discussion, Forms A-1 through A-4 ascertain pertinent information.

Chapter 1 also described the standard *anthropometric measurements* used in nutrition assessment. To go with that discussion, Tables A-1 through A-4 and Figures A-1 and A-2 present standards and procedures.

Chapter 1 described the *physical examinations* used in nutrition assessment. To go with that discussion, Table A-5 lists symptoms of vitamin and mineral imbalances.

Chapter 1 also described *biochemical analyses* as part of nutrition assessment. Table A-6 lists biochemical tests useful in nutrition assessment of vitamin and mineral status, and Tables A-7 through A-12 provide standards for assessing iron status.

Form A-1 Historical Data

Name _____ Date _____

Address _____ Date of last medical checkup _____

_____ Age _____ Sex _____

_____ Height _____ Weight _____

Phone _____ Usual weight _____

Reason for admission _____ Desirable weight range _____

_____

Health History

1. Have you been told that you have (check any that apply):

   ❏ Diabetes ❏ Hardening of arteries ❏ Kidney disease ❏ Cancer

   ❏ GI disorders ❏ Heart disease ❏ Liver disease ❏ Other _____

   ❏ High blood pressure ❏ Lung disease ❏ Ulcers

2. Do you have complaints about any of the following:

   ❏ Lack of appetite ❏ Diarrhea ❏ Nausea

   ❏ Difficulty chewing or swallowing ❏ Indigestion ❏ Vomiting

   ❏ Constipation ❏ Fever ❏ Other

3. Do you use tobacco in any way? _____ How much? _____

4. For females:

   Are you pregnant? _____ How many months? _____

   How many pregnancies have you carried to term? _____

   When was your last child born? _____

   Are your menstrual periods normal? _____ If not, please explain: _____

   _____

Drug History

1. Do you take medication, either prescribed by a doctor or over-the-counter?

Name of Drug	Reason for Taking	Dose	Frequency	Duration of Intake
_____	_____	_____	_____	_____
_____	_____	_____	_____	_____
_____	_____	_____	_____	_____

2. Have you noticed any side effects from taking these medications? _____

   If so, please explain: _____

(continued)

Form A-1 Historical Data (*continued*)

3. Do you take vitamins or any kind of supplements? _____

Which ones? _____ How often? _____

For what reason? _____

Socioeconomic History

1. Last grade of school completed _____ Still in school? _____

2. Are you employed? _____ Occupation _____

3. Does someone else live with you? _____ Who? _____

4. Do you regularly eat alone or with others? _____

5. Do you have a refrigerator? _____ Stove? _____

6. How often do you shop for food? _____

Where? _____

Diet History

1. Have you recently lost or gained more than 10 lb? _____ If yes, explain the surrounding circumstances (including associated illness, dietary changes, and time frame): _____

_____

2. Do you eat at regular times each day? _____ How many times per day? _____

3. Where do you eat most of your meals? _____

4. Do you usually eat snacks? _____ When? _____

5. What foods do you particularly like? _____

6. Are there foods you don't eat for other reasons? _____

7. Do you have difficulty eating? _____

8. How would you describe your feelings about food? _____

9. How do your eating habits change when you are emotionally upset? _____

10. Are you, or any member of your family, on a special diet? _____

If yes, who and what kind? _____

11. Do you drink alcohol? _____ How much? _____ How often? _____

12. How would you describe your exercise habits? _____ Type of exercise _____

Intensity _____ Duration _____ Frequency _____

13. Are there any other facts about your lifestyle that you think might be related to your nutritional health? _____

Explain _____

Note: Use the appropriate form to record food intake data (Form A-2, A-3, and A-4).

Form A-2   Food Intake for a 24-Hour Recall or Usual Intake Pattern

Name and address _____   Date _____

_____

_____

Did or do you take a vitamin-mineral supplement? _____

If yes, what kind? _____   Dose _____

Please record the amount and type of foods and beverages consumed today. [Or: Please record the amounts and types of foods and beverages you typically consume each day.]

Time of Day	*Food*	*Amount* (c, tbs, or piece)	*Description* (how cooked, how served)

Form A-3 Food Frequency Checklist

The assessor helps the client estimate portion sizes and frequency of use.

	Number of of Servings	Frequency[a]
1. How often do you eat the following foods?		
Bread, toast, rolls, muffins	_____	_____
Cereal (type?)	_____	_____
Rice or other cooked grains	_____	_____
Noodles (macaroni, spaghetti)	_____	_____
Pancakes or waffles	_____	_____
Crackers or pretzels	_____	_____
Fruits or fruit juices	_____	_____
Vegetables other than potatoes	_____	_____
Vegetable juice	_____	_____
Potatoes	_____	_____
Dried beans and peas	_____	_____
Beef	_____	_____
Pork or ham	_____	_____
Veal	_____	_____
Poultry	_____	_____
Fish	_____	_____
Organ meats (such as liver)	_____	_____
Bacon	_____	_____
Sausage	_____	_____
Lunch meat	_____	_____
Hot dogs	_____	_____
Other meats (type?)	_____	_____
Eggs	_____	_____
Peanut butter or nuts	_____	_____
Milk (including on cereal)	_____	_____
Cheese or cheese dishes	_____	_____
Yogurt or tofu	_____	_____
Other milk products (type?)	_____	_____

[a]Number of servings per day, week, month, or year.

(continued)

Form A-3   Food Frequency Checklist (*continued*)

Butter or margarine (type?)   _____   _____

Salt pork   _____   _____

Mayonnaise or salad dressing
(type?)   _____   _____

Oil (type?)   _____   _____

Cream   _____   _____

Sugar, jam, jelly, syrup, honey   _____   _____

Bakery goods (type?)   _____   _____

Candy   _____   _____

Soft drinks (type?)   _____   _____

Potato or snack chips (type?)   _____   _____

Coffee or tea (type?)   _____   _____

Alcoholic beverages (type?)   _____   _____

Fast foods eaten out (type?)   _____   _____

TV dinners, pot pies, other prepared
meals (type?)   _____   _____

Instant meals such as breakfast bars or
diet meal beverages (type?)   _____   _____

2. What specific kinds of the following foods do you eat? Include the
   name of the food; whether it is fresh, canned, or frozen; and how
   it is prepared.

   Fruits and fruit juices _____

   Vegetables _____

   Milk and milk products _____

   Meats and meat alternates _____

   Breads and cereals _____

   Desserts _____

   Snack foods _____

3. Please list the names of any liquid, powder, or pill forms of
   vitamin or mineral products you take, and state how often you
   take them. Please also list any diet supplement you use (such as
   protein milkshakes or brewer's yeast), how much you use, and
   how often you use it.

   _____

4. Is there anything else you can relate about your food/nutrient
   intake?

   _____

Form A-4 Food Record

Time	Place	With Whom	Emotional State	Hungry or Not Hungry	Food Eaten (amount)
				Name _____	
				Date _____	

Table A-1   1983 Metropolitan Height and Weight Tables

Men					Women				
Height		Frame			Height		Frame		
Feet	Inches	Small	Medium	Large	Feet	Inches	Small	Medium	Large
5	2	128–134	131–141	138–150	4	10	102–111	109–121	118–131
5	3	130–136	133–143	140–153	4	11	103–113	111–123	120–134
5	4	132–138	135–145	142–156	5	0	104–115	113–126	122–137
5	5	134–140	137–148	144–160	5	1	106–118	115–129	125–140
5	6	136–142	139–151	146–164	5	2	108–121	118–132	128–143
5	7	138–145	142–154	149–168	5	3	111–124	121–135	131–147
5	8	140–148	145–157	152–172	5	4	114–127	124–138	134–151
5	9	142–151	148–160	155–176	5	5	117–130	127–141	137–155
5	10	144–154	151–163	158–180	5	6	120–133	130–144	140–159
5	11	146–157	154–166	161–184	5	7	123–136	133–147	143–163
6	0	149–160	157–170	164–188	5	8	126–139	136–150	146–167
6	1	152–164	160–174	168–192	5	9	129–142	139–153	149–170
6	2	155–168	164–178	172–197	5	10	132–145	142–156	152–173
6	3	158–172	167–182	176–202	5	11	135–148	145–159	155–176
6	4	162–176	171–187	181–207	6	0	138–151	148–162	158–179

Note: To use the table, add an inch to your barefoot height (you are assumed to be wearing shoes with 1-inch heels), and adjust for clothing (the tables assume 5 pounds of clothes for men and 3 pounds for women). Weights are at age 25 to 29 based on lowest mortality, in pounds according to frame size.

Source: Reproduced courtesy of Metropolitan Life Insurance Company. Source of basic data: Society of Actuaries and Association of Life Insurance Medical Directors of America, *1979 Build Study*, 1980.

Table A-2 How to Determine Body Frame by Elbow Breadth

To make a simple approximation of frame size, do the following: Extend the arm, and bend the forearm upward at a 90-degree angle. Keep the fingers straight, and turn the inside of the wrist away from the body. Place the thumb and index finger on the two prominent bones on *either side* of the elbow. Measure the space between the fingers against a ruler or a tape measure.[a] Compare the measurements with the following standards.

These standards represent the elbow measurements for medium-framed men and women of various heights. Measurements smaller than those listed indicate a small frame, and larger measurements indicate a large frame.

Men		Women	
Height in 1-Inch Heels	*Elbow Breadth*	*Height in 1-Inch Heels*	*Elbow Breadth*
5 ft 2 in to 5 ft 3 in	2 1/2 to 2 7/8 in	4 ft 10 in to 4 ft 11 in	2 1/4 to 2 1/2 in
5 ft 4 in to 5 ft 7 in	2 5/8 to 2 7/8 in	5 ft 0 in to 5 ft 3 in	2 1/4 to 2 1/2 in
5 ft 8 in to 5 ft 11 in	2 3/4 to 3 in	5 ft 4 in to 5 ft 7 in	2 3/8 to 2 5/8 in
6 ft 0 in to 6 ft 3 in	2 3/4 to 3 1/8 in	5 ft 8 in to 5 ft 11 in	2 3/8 to 2 5/8 in
6 ft 4 in and over	2 7/8 to 3 1/4 in	6 ft 0 in and over	2 1/2 to 2 3/4 in

[a]For the most accurate measurement, measure elbow breadth with calipers.

Source: Metropolitan Life Insurance Company.

Figure A-1 Wrist Circumference

The wrist circumference is measured just above the wrist bone.

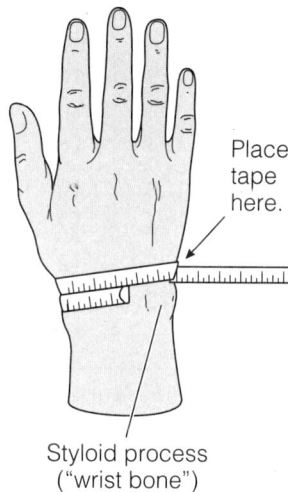

Place tape here.

Styloid process ("wrist bone")

Table A-3 Frame Size from Height–Wrist Circumference Ratios (*r*)[a]

Frame Size	Male r Values	Female r Values
Small	>10.4	>11.0
Medium	9.6–10.4	10.1–11.0
Large	<9.6	<10.1

$$^a r = \frac{\text{height (cm)}}{\text{wrist circumference (cm)}}.$$

The wrist is measured where it bends (distal to the styloid process), on the right arm (see Figure A-1).

Table A-4 Triceps Fatfold Percentiles (Millimeters)

Age	Male 5th	25th	50th	75th	95th	Female 5th	25th	50th	75th	95th
1–1.9	6	8	10	12	16	6	8	10	12	16
2–2.9	6	8	10	12	15	6	9	10	12	16
3–3.9	6	8	10	11	15	7	9	11	12	15
4–4.9	6	8	9	11	14	7	8	10	12	16
5–5.9	6	8	9	11	15	6	8	10	12	18
6–6.9	5	7	8	10	16	6	8	10	12	16
7–7.9	5	7	9	12	17	6	9	11	13	18
8–8.9	5	7	8	10	16	6	9	12	15	24
9–9.9	6	7	10	13	18	8	10	13	16	22
10–10.9	6	8	10	14	21	7	10	12	17	27
11–11.9	6	8	11	16	24	7	10	13	18	28
12–12.9	6	8	11	14	28	8	11	14	18	27
13–13.9	5	7	10	14	26	8	12	15	21	30
14–14.9	4	7	9	14	24	9	13	16	21	28
15–15.9	4	6	8	11	24	8	12	17	21	32
16–16.9	4	6	8	12	22	10	15	18	22	31
17–17.9	5	6	8	12	19	10	13	19	24	37
18–18.9	4	6	9	13	24	10	15	18	22	30
19–24.9	4	7	10	15	22	10	14	18	24	34
25–34.9	5	8	12	16	24	10	16	21	27	37
35–44.9	5	8	12	16	23	12	18	23	29	38
45–54.9	6	8	12	15	25	12	20	25	30	40
55–64.9	5	8	11	14	22	12	20	25	31	38
65–74.9	4	8	11	15	22	12	18	24	29	36

Note: If measurements fall between the percentiles shown here, the percentile can be estimated from the information in this table. For example, a measurement of 7 millimeters for a 27-year-old male would be about the 20th percentile.

Source: Adapted from A. R. Frisancho, New norms of upper limb fat and muscle areas for assessment of nutritional status, *American Journal of Clinical Nutrition* 34 (1981): 2540–2545.

Figure A-2 How to Measure the Triceps Fatfold

A. Find the midpoint of the arm:

1. Ask the subject to bend his or her arm at the elbow and lay the hand across the stomach. (If he or she is right-handed, measure the left arm, and vice versa.)

2. Feel the shoulder to locate the acromial process. It helps to slide your fingers along the clavicle to find the acromial process. The olecranon process is the tip of the elbow.

3. Place a measuring tape from the acromial process to the tip of the elbow. Divide this measurement by 2, and mark the midpoint of the arm with a pen.

B. Measure the fatfold:

1. Ask the subject to let his or her arm hang loosely to the side.

2. Grasp a fold of skin and subcutaneous fat between the thumb and forefinger slightly above the midpoint mark. Gently pull the skin away from the underlying muscle. (This step takes a lot of practice. If you want to be sure you don't have muscle as well as fat, ask the subject to contract and relax the muscle. You should be able to feel if you are pinching muscle.)

3. Place the calipers over the fatfold at the midpoint mark, and read the measurement to the nearest 1.0 millimeter in two to three seconds.

4. Repeat steps 2 and 3 twice more. Add the three readings, and then divide by 3 to find the average.

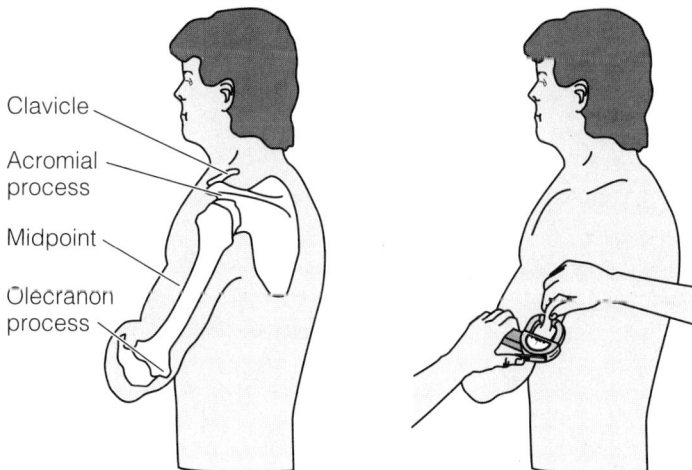

Table A-5   Vitamins and Minerals—Deficiency and Toxicity Symptoms

Fat-Soluble Vitamins		
Nutrient	*Deficiency Symptoms*	*Toxicity Symptoms*
Vitamin A	***Blood/Circulatory System*** Anemia (small-cell types)[a]	Red blood cell breakage, nosebleeds; increased blood lipids, increased blood calcium
	Bones/Teeth Cessation of bone growth, painful joints; impaired enamel formation, cracks in teeth, tendency to decay	Joint pain, thickening of long bones
	Digestive System Diarrhea, changes in lining	Abdominal cramps and pain, nausea, vomiting, diarrhea, weight loss
	Immune System Suppression of immune reactions; frequent respiratory, digestive, bladder, vaginal, and other infections	Overstimulation of immune reactions
	Nervous/Muscular Systems Loss of taste, smell, and balance	Increased fluid pressure in brain and spine, causing headaches, nausea, poor coordination, and loss of appetite
	Eyes Night blindness (retinal); corneal degeneration leading to blindness[b]	
	Skin Keratinization, rashes	Dry skin, rashes, loss of hair
	Other Kidney stones, impaired growth	Cessation of menstruation, liver and spleen enlargement, birth defects, calcification of soft tissues
Vitamin D[c]	***Blood/Circulatory System***	Raised blood calcium
	Bones/Teeth Abnormal growth, misshapen bones (bowing of legs), soft bones, joint pain, malformed teeth	

[a] Small-cell anemia is termed *microcytic anemia*; large-cell type is *macrocytic* or *megaloblastic anemia*.

[b] Corneal degeneration progresses from *keratinization* (hardening) to *xerosis* (drying) to *xerophthalomia* (thickening, opacity, and irreversible blindness).

[c] The vitamin D deficiency diseases are rickets and osteomalacia.

(continued)

Table A-5 Vitamins and Minerals—Deficiency and Toxicity Symptoms (*continued*)

Nutrient	Deficiency Symptoms	Toxicity Symptoms
Vitamin D (*continued*)	**Nervous/Muscular Systems** Muscle spasms	Excessive thirst, headaches, irritability, loss of appetite, weakness, nausea
	Other	Calcification of soft tissues
Vitamin E	**Blood/Circulatory System** Red blood cell breakage, anemia	Augments the effects of anticlotting medication
	Nervous/Muscular Systems Degeneration, weakness, difficulty walking, leg cramps	
Vitamin K	**Blood/Circulatory System** Hemorrhaging	Interference with anticlotting medication; vitamin K analogues may cause jaundice

Water-Soluble Vitamins

Nutrient	Deficiency Symptoms	Toxicity Symptoms
Thiamin[d]	**Blood/Circulatory System** Edema, enlarged heart, abnormal heart rhythms, heart failure	(No symptoms reported)
	Nervous/Muscular Systems Degeneration, wasting, weakness, pain, low morale, difficulty walking, loss of reflexes, mental confusion, paralysis	
Riboflavin	**Mouth, Gums, Tongue** Cracks at corners of mouth,[e] magenta tongue	(No symptoms reported)
	Nervous System and Eyes Hypersensitivity to light,[f] reddening of cornea	
	Other Skin rash	

[d] The thiamin deficiency disease is beriberi.

[e] Cracks at the corners of the mouth are termed *cheilosis* (kee-LOH-sis).

[f] Hypersensitivity to light is *photophobia*.

(*continued*)

Table A-5 Vitamins and Minerals—Deficiency and Toxicity Symptoms (*continued*)

Nutrient	Deficiency Symptoms	Toxicity Symptoms
Niacin[g]	**Digestive System** Diarrhea	Diarrhea, heartburn, nausea, ulcer, irritation, vomiting
	Mouth, Gums, Tongue Inflamed, swollen, smooth tongue[h]	
	Nervous System Irritability, loss of appetite, weakness, dizziness, mental confusion progressing to psychosis or delirium	Fainting, dizziness
	Skin Flaky skin rash on areas exposed to sun	Painful flush and rash ("niacin flush"), sweating
	Other	Abnormal liver function, low blood pressure
Vitamin B_6	**Blood/Circulatory System** Anemia (small-cell type)[a]	Bloating
	Mouth, Gums, Tongue Smooth tongue[h]	
	Nervous/Muscular Systems Abnormal brain wave pattern, irritability, muscle twitching, convulsions	Depression, fatigue, impaired memory, irritability, headaches, numbness, damage to nerves, difficulty walking, loss of reflexes, weakness, restlessness
	Skin Irritation of sweat glands, rashes, greasy dermatitis	
	Other Kidney stones	
Folate	**Blood/Circulatory System** Anemia (large-cell type)[a]	
	Digestive System Heartburn, diarrhea, constipation	
	Immune System Suppression, frequent infections	

[a] Small-cell anemia is termed *microcytic anemia*; large-cell type is *macrocytic* or *megaloblastic anemia*.

[g] The niacin deficiency disease is pellagra.

[h] Smoothness of the tongue is caused by loss of its surface structures and is termed *glossitis* (gloss-EYE-tis).

(*continued*)

Table A-5 Vitamins and Minerals—Deficiency and Toxicity Symptoms (*continued*)

Nutrient	Deficiency Symptoms	Toxicity Symptoms
Folate (*continued*)	***Mouth, Gums, Tongue*** Smooth red tongue[h]	
	Nervous System Depression, mental confusion, fainting, insomnia, irritability, forgetfulness	
	Other	Masks vitamin B_{12} deficiency
Vitamin B_{12}[i]	***Blood/Circulatory System*** Anemia (large-cell type)[a]	(No symptoms reported)
	Mouth, Gums, Tongue Smooth tongue[h]	
	Nervous System Fatigue, degeneration progressing to paralysis, insomnia, irritability, forgetfulness	
	Skin Hypersensitivity	
Pantothenic Acid	***Digestive System*** Vomiting, intestinal distress	Occasional diarrhea
	Nervous System Insomnia, fatigue	
	Other	Water retention (infrequent)
Biotin	***Blood/Circulatory System*** Abnormal heart action	(No symptoms reported)
	Digestive System Loss of appetite, nausea	
	Nervous/Muscular Systems Depression, muscle pain, weakness, fatigue	
	Skin Drying, rash, loss of hair	

[a] Small-cell anemia is termed *microcytic anemia*; large-cell type is *macrocytic* or *megaloblastic anemia*.

[h] Smoothness of the tongue is caused by loss of its surface structures and is termed *glossitis* (gloss-EYE-tis).

[i] The name *pernicious anemia* refers to the vitamin B_{12} deficiency caused by lack of intrinsic factor, but not to that caused by inadequate dietary intake.

(*continued*)

Table A-5 Vitamins and Minerals—Deficiency and Toxicity Symptoms (*continued*)

Nutrient	Deficiency Symptoms	Toxicity Symptoms
Vitamin C[i]	**Blood/Circulatory System** Anemia (small-cell type),[a] atherosclerotic plaques, pinpoint hemorrhages	
	Digestive System	Nausea, abdominal cramps, diarrhea, excessive urination
	Immune System Suppression, frequent infections	
	Mouth, Gums, Tongue Bleeding gums, loosened teeth	
	Nervous/Muscular Systems Muscle degeneration and pain, hysteria, listlessness, malaise, personality changes	Headache, fatigue, insomnia
	Bones/Teeth Bone fragility, joint pain	
	Skin Rough skin, blotchy bruises	Rashes
	Other Failure of wounds to heal	Interference with medical tests; aggravation of gout symptoms; deficiency symptoms may appear at first on withdrawal of high doses

Major Minerals

Nutrient	Deficiency Symptoms	Toxicity Symptoms
Calcium	Stunted growth in children; adult bone loss (osteoporosis)	Urinary stone formation and kidney dysfunction
Phosphorus	Weakness, bone pain	Calcium excretion
Magnesium	Weakness; confusion; depressed pancreatic hormone secretion; if extreme, convulsions, bizarre movements (especially of eyes and face), hallucinations, and difficulty in swallowing; in children, growth failure[k]	Confusion, muscle weakness, low blood pressure, irregular heartbeat, vomiting, respiratory failure

[a] Small-cell anemia is termed *microcytic anemia*; large-cell type is *macrocytic* or *megaloblastic anemia.*

[i] The vitamin C deficiency disease is scurvy.

[k] A still more severe deficiency causes tetany, an extreme, prolonged contraction of the muscles similar to that caused by low blood calcium.

(*continued*)

Table A-5 Vitamins and Minerals—Deficiency and Toxicity Symptoms (*continued*)

Nutrient	Deficiency Symptoms	Toxicity Symptoms
Sodium	Muscle cramps, mental apathy, loss of appetite	Hypertension, edema
Potassium	Muscular weakness, paralysis, and confusion	Muscular weakness; vomiting; if given into a vein, can stop the heart

Trace Minerals

Nutrient	Deficiency Symptoms	Toxicity Symptoms
Iodine	Goiter, cretinism	Depressed thyroid activity; goiterlike thyroid enlargement
Iron	Anemia: weakness, pallor, headaches, reduced resistance to infection, inability to concentrate, lowered cold tolerance	Iron overload: infections, liver injury, bloody stools, shock
Zinc	Growth failure in children, sexual retardation, loss of taste, poor wound healing, loss of appetite	Fever, nausea, vomiting, diarrhea, muscle incoordination, dizziness, anemia, kidney failure
Selenium	Muscle discomfort, weakness, pancreas damage, heart disease (cardiomyopathy)	Nausea, abdominal pain, nail and hair changes, nerve damage
Fluoride	Susceptibility to tooth decay	Fluorosis (discoloration) of teeth, nausea, vomiting, diarrhea, chest pain, itching
Chromium	Abnormal glucose metabolism	Unknown
Copper	Anemia	Vomiting, diarrhea

Table A-6 Biochemical Tests Useful for Assessing Nutrition Status

Nutrient	Assessment Tests
Protein	Urinary creatinine excretion, serum albumin, serum prealbumin, serum transferrin, retinol-binding protein, total lymphocyte count, nitrogen balance
Vitamins	
Vitamin A	Retinol-binding protein, serum carotene
Thiamin	Erythrocyte (red blood cell) transketolase activity, urinary thiamin
Riboflavin	Erythrocyte glutathione reductase activity, urinary riboflavin
Vitamin B_6	Urinary xanthurenic acid excretion after tryptophan load test, urinary vitamin B_6, erythrocyte transaminase activity
Niacin	Urinary metabolites NMN (N-methyl nicotinamide) or 2-pyridone, or preferably both expressed as a ratio
Folate	Free folate in the blood, erythrocyte folate (reflects liver stores), urinary formiminoglutamic acid (FIGLU), vitamin B_{12} status (because folate assessment tests alone do not distinguish between the two deficiencies)
Vitamin B_{12}	Serum vitamin B_{12}, erythrocyte vitamin B_{12}, urinary methylmalonic acid synthesis or DUMP test (from the abbreviation for the chemical name of DNA's raw material, deoxyuridine monophosphate), Schilling test
Biotin	Serum biotin, urinary biotin
Vitamin C	Serum or plasma vitamin C,[a] leukocyte vitamin C, urinary vitamin C
Vitamin D	Serum alkaline phosphatase
Vitamin E	Serum tocopherol, erythrocyte hemolysis
Vitamin K	Blood-clotting time (prothrombin time)
Minerals	
Potassium	Serum potassium
Magnesium	Serum magnesium
Iron	Hemoglobin, hematocrit, serum ferritin, total iron-binding capacity (TIBC), transferrin saturation, erythrocyte protoporphyrin, mean corpuscular volume (MCV), serum iron
Iodine	Serum protein-bound iodine, radioiodine uptake
Zinc	Plasma zinc, hair zinc

[a]Vitamin C shifts unpredictably between the plasma and the white blood cells known as leukocytes; thus a plasma or serum determination may not accurately reflect the body's pool. The appropriate clinical test may be a measurement of leukocyte vitamin C. A combination of both tests may be more reliable than either one alone.

Source: Adapted from A. Grant and S. DeHoog, *Nutritional Assessment and Support*, 3rd ed., 1985 (available from Anne Grant and Susan DeHoog, Box 25057, Northgate Station, Seattle, WA 98125).

Table A-7   Standards for Hemoglobin Test Results

Age (yr)	Sex	Deficient (g/dL)	Acceptable (g/dL)
0.5–2	M–F	<11	≥11
2–6	M–F	<11	≥11
6–12	M–F	<11.5	≥11.5
12–18	M	<13	≥13
	F	<12	≥12
Adult	M	<13.5	≥13.5
	F	<12	≥12
Pregnancy			
Trimester 1		<11	≥11
Trimester 2		<10.5	≥10.5
Trimester 3		<11	≥11

Table A-8   Standards for Hematocrit Test Results

Age (yr)	Sex	Deficient %	Acceptable %
0.5–2	M–F	<33	≥33
2–6	M–F	<34	≥34
6–12	M–F	<35	≥35
12–18	M	<38	≥38
	F	<36	≥36
Adult	M	<41	≥41
	F	<36	≥36

Table A-9   Standards for Serum Ferritin

Group	Deficient (ng/mL)
Children (3–14 years of age)	<10
Adolescents and adults	<12
Pregnant women	<10

Table A-10   Standards for Serum Iron

Age (yr)	Sex	Deficient		Acceptable	
		($\mu g/$ 100 mL)	($\mu mol/L$)	($\mu g/100$ mL)	($\mu mol/L$)
<2	M–F	<30	<5.3	30 or >	5.3 or >
2–5	M–F	<40	<7.1	40 or >	7.1 or >
6–12	M–F	<50	<8.9	50 or >	8.9 or >
>12	M	<60	<10.7	60 or >	10.7 or >
	F	<40	<7.1	40 or >	7.1 or >

Table A-11   Standards for Percent Transferrin Saturation

Age (yr)	Sex	Deficient	Acceptable
<2	M–F	<15%	15% or >
2–12	M–F	<20%	20% or >
≥13	M	<20%	20% or >
	F	<15%	15% or >

Table A-12   Standards for Erythrocyte Protoporphyrin and Mean Corpuscular Volume

Age (yr)	Erythrocyte Protoporphyrin ($\mu g/dL$ RBC)	MCV (fL)
1–2	>80	<73
3–4	>75	<75
5–10	>70	<76
11–14	>70	<78
15–74	>70	<80

United States: Recommendations World Health Organization: Recommendations

Chapter 1 introduced Recommended Dietary Allowances (RDA), Daily Values, and Healthy People 2000. This appendix provides additional details. (See Appendix C for Canada's nutrition recommendations.) The last page of this appendix presents nutrition recommendations from the World Health Organization (WHO).

RDA

Some of the recommendations for nutrient intakes appear in the table on the inside front cover, left. The remaining RDA are here, in Tables B-1, B-2, and B-3.

Daily Values

Food labels use another set of standards that derive from the RDA. From the late 1960s to the early 1990s, the set of standards used on food labels was called the U.S. RDA. The U.S. RDA were derived from the 1968 RDA and were established by the Food and Drug Administration (FDA) so that labels could express the nutrient contents of foods as percentages of those standards (see Table B-4). The intent was to help consumers evaluate the nutrient contents of foods for themselves and at the same time to spare them the burden of learning the different units in which nutrient amounts are expressed. Thus all nutrient amounts in a food, whether originally measured in micrograms, milligrams, or grams, could be expressed as "percent of U.S. RDA."

With the 1990 labeling regulations came a name change—the Daily Values. With the exception of protein, the current Daily Values continue to use the same values as the old U.S. RDA. The FDA continues to consider their revision and has added several nutrients to the table (see inside front cover, right).

Table B-1 Estimated Safe and Adequate Daily Dietary Intakes of Additional Selected Vitamins and Minerals (United States)[a]

Age (yr)	Vitamins		Trace Elements[b]				
	Biotin (μg)	Panto-thenic Acid (mg)	Chromium (μg)	Molybdenum (μg)	Copper (mg)	Manganese (mg)	Fluoride (mg)
Infants							
0.0–0.5	10	2	10–40	15–30	0.4–0.6	0.3–0.6	0.1–0.5
0.5–1.0	15	3	20–60	20–40	0.6–0.7	0.6–1.0	0.2–1.0
Children							
1–3	20	3	20–80	25–50	0.7–1.0	1.0–1.5	0.5–1.5
4–6	25	3–4	30–120	30–75	1.0–1.5	1.5–2.0	1.0–2.5
7–10	30	4–5	50–200	50–150	1.0–2.0	2.0–3.0	1.5–2.5
11+	30–100	4–7	50–200	75–250	1.5–2.5	2.0–5.0	1.5–2.5
Adults	30–100	4–7	50–200	75–250	1.5–3.0	2.0–5.0	1.5–4.0

[a]Less information is available on which to base allowances for these nutrients. Therefore, they are not included in the main table of the RDA, and the figures provided here are in the form of ranges of recommended intakes.

[b]The toxic levels for many trace elements may be only several times usual intakes, so the upper levels for the trace elements given in this table should not be habitually exceeded.

Source: Recommended Dietary Allowances. © 1989 by the National Academy of Sciences, National Academy Press, Washington, D.C. Used by permission.

Table B-2 Estimated Minimum Requirments of Sodium, Chloride, and Potassium

Age (yr)	Weight (kg)	Sodium[a] (mg)	Chloride (mg)	Potassium[b] (mg)
Infants				
0.0–0.5	4.5	120	180	500
0.5–1.0	8.9	200	300	700
Children				
1	11.0	225	350	1000
2–5	16.0	300	500	1400
6–9	25.0	400	600	1600
Adolescents	50.0	500	750	2000
Adults	70.0	500	750	2000

[a]Sodium requirements are based on estimates of needs for growth and for replacement of obligatory losses. They cover a wide variation of physical activity patterns and climatic exposure but do not provide for large, prolonged losses from the skin through sweat.

[b]Dietary potassium may benefit the prevention and treatment of hypertension, and recommendations to include many servings of fruits and vegetables would raise potassium intakes to about 3500 milligrams per day.

Source: Recommended Dietary Allowances. © 1989 by the National Academy of Sciences, National Academy Press, Washington, D.C. Used by permission.

Table B-3 Median Heights and Weights and Recommended Energy Intakes (United States)

Age (yr)	Weight		Height		Average Energy Allowance			
	kg	lb	cm	in	REE[a] (kcal/day)	Multiples of REE[b]	kcal/kg	kcal/day[c]
Infants								
0.0–0.5	6	13	60	24	320		108	650
0.5–1.0	9	20	71	28	500		98	850
Children								
1–3	13	29	90	35	740		102	1300
4–6	20	44	112	44	950		90	1800
7–10	28	62	132	52	1130		70	2000
Males								
11–14	45	99	157	62	1440	1.70	55	2500
15–18	66	145	176	69	1760	1.67	45	3000
19–24	72	160	177	70	1780	1.67	40	2900
25–50	79	174	176	70	1800	1.60	37	2900
51+	77	170	173	68	1530	1.50	30	2300
Females								
11–14	46	101	157	62	1310	1.67	47	2200
15–18	55	120	163	64	1370	1.60	40	2200
19–24	58	128	164	65	1350	1.60	38	2200
25–50	63	138	163	64	1380	1.55	36	2200
51+	65	143	160	63	1280	1.50	30	1900
Pregnant (2nd and 3rd trimesters)								+300
Lactating								+500

[a]REE (resting energy expenditure) represents the energy expended by a person at rest under normal conditions.

[b]Recommended energy allowances assume light-to-moderate activity and were calculated by multiplying the REE by an activity factor.

[c]Average energy allowances have been rounded.

Source: Recommended Dietary Allowances. © 1989 by the National Academy of Sciences, National Academy Press, Washington, D.C. Used by permission.

Table B-4 U.S. Recommended Daily Allowances (U.S. RDA)

Nutrient	Adults and Children over 4 Years	Infants	Children under 4 Years	Pregnant or Lactating Women
Protein (g)	45[a]	18[a]	20[a]	
Vitamin A (RE)	1000	300	500	1600
Vitamin D[b] (IU)	400	400	400	400
Vitamin E[b] (IU)	30	5	10	30
Vitamin C (mg)	60	35	40	60
Folate (mg)	0.4	0.1	0.2	0.8
Thiamin (mg)	1.5	0.5	0.7	1.7
Riboflavin (mg)	1.7	0.6	0.8	2
Niacin (mg)	20	8	9	20
Vitamin B_6[b] (mg)	2	0.4	0.7	2.5
Vitamin B_{12}[b] (μg)	6	2	3	8
Biotin[b] (mg)	0.3	0.5	0.15	0.3
Pantothenic acid[b] (mg)	10	3	5	10
Calcium (g)	1	0.6	0.8	1.3
Phosphorus[b] (g)	1	0.5	0.8	1.3
Iodine[b] (μg)	150	45	70	150
Iron (mg)	18	15	10	18
Magnesium[b] (mg)	400	70	200	450
Copper[b] (mg)	2	0.6	1	2
Zinc[b] (mg)	15	5	8	15

Note: Four sets of U.S. RDA were developed for different groups of people—infants, children, adults, and pregnant and lactating women. The most commonly used set was the U.S. RDA for adults. The one for infants was used for formulas. Supplements designed for children and for pregnant and lactating women used the U.S. RDA for these groups on their labels.

[a]If protein efficiency ratio of protein is equal to or better than that of casein.

[b]Optional for adults and children 4 years or over in vitamin and mineral supplements.

Source: U.S. Department of Health and Human Services, Public Health Service, Food and Drug Administration, Office of Public Affairs, 5600 Fishers Lane, Rockville, Maryland 20857, HHS publication no. (FDA) 81-2146, revised March 1981.

Healthy People 2000

In 1990, the U.S. Department of Health and Human Services established a set of almost 300 health objectives for the nation called Healthy People 2000.[1] The 21 objectives that have a nutrition component were presented throughout this text wherever their topic was discussed. Table B-5 presents them in full.

Table B-5 Healthy People 2000 Nutrition Objectives

Health-Related Objectives

- Reduce coronary heart disease deaths to no more than 100 per 100,000 people.
- Reverse the rise in cancer deaths to achieve a rate of no more than 130 per 100,000 people.
- Reduce overweight to a prevalence of no more than 20% among people aged 20 years and older and maintain prevalence at no more than 15% among adolescents aged 12 through 19 years.
- Reduce growth retardation amoung low-income children aged five years and younger to less than 10%.

Nutrient Intake Objectives

- Reduce dietary fat intake to an average of 30% of energy or less and average saturated fat intake to less than 10% of energy among people aged two years and older.
- Increase complex carbohydrate and fiber-containing foods in the diets of adults to five or more daily servings for vegetables (including legumes) and fruits and to six or more daily servings for grain products.
- Increase to at least 50% the proportion of overweight people aged 12 years and older who have adopted sound dietary practices combined with regular physical activity to attain an appropriate body weight.
- Increase calcium intake, so that at least 50% of youth aged 12 through 24 years and 50% of pregnant and lactating women consume three or more servings of calcium-rich foods daily and at least 50% of people aged 25 years and older consume two or more servings of calcium-rich foods daily.
- Decrease salt and sodium intake so at least 65% of home meal preparers prepare foods without adding salt, at least 80% of people avoid using salt at the table, and at least 40% of adults regularly purchase foods modified or lower in sodium.
- Reduce iron deficiency to less than 3% among children aged 1 to 4 and women of childbearing age.
- Increase to at least 75% the proportion of mothers who breastfeed their babies in the early weeks and to at least 50% the proportion who continue breastfeeding until their babies are five to six months old.
- Increase to at least 75% the proportion of parents and caregivers who use feeding practices that prevent nursing bottle tooth decay.
- Increase to at least 85% the proportion of people aged 18 and older who use food labels to make nutritious food selections.

Services and Information Objectives

- Achieve useful and informative nutrition labeling for virtually all processed foods and at least 40% of fresh meats, poultry, fish, fruits, vegetables, baked goods, and ready-to-eat carry-away foods.
- Increase to at least 5000 brand items the number of processed food products that are reduced in fat and saturated fat.
- Increase to at least 90% the proportion of restaurants and institutional foodservice operations that offer identifiable low-fat, low-kcalorie food choices, consistent with the *Dietary Guidelines for Americans*.
- Increase to at least 90% the proportion of school lunch and breakfast services and increase to at least 50% the proportion of child care foodservices with menus that are consistent with the nutrition principles in the *Dietary Guidelines for Americans*.
- Increase to at least 80% the receipt of home foodservices by people aged 65 and older who have difficulty in preparing their own meals or are otherwise in need of home-delivered meals.
- Increase to at least 75% the proportion of the nation's schools that provide nutrition education from preschool through grade 12, preferably as part of quality school health education.
- Increase to at least 50% the proportion of worksites with 50 or more employees that offer nutrition education and/or weight management programs for employees.
- Increase to at least 75% the proportion of primary care providers who provide nutrition assessment and counseling and/or referral to qualified nutritionists or dietitians.

Nutrition Recommendations from WHO

Like the Committee on Diet and Health in the United States, the World Health Organization (WHO) has also assessed the relationships between diet and the development of chronic diseases.[2] Its recommendations are expressed in average daily ranges that represent the lower and upper limits:

■ Total energy: sufficient to support normal growth, physical activity, and body weight (body mass index = 20 to 22).

■ Total fat: 15 to 30 percent of total energy.
   — Saturated fatty acids: 0 to 10 percent total energy.
   — Polyunsaturated fatty acids: 3 to 7 percent total energy.
   — Dietary cholesterol: 0 to 300 milligrams per day.

■ Total carbohydrate: 55 to 75 percent total energy.
   — Complex carbohydrates: 50 to 75 percent total energy.
   — Dietary fiber: 27 to 40 grams per day.
   — Refined sugars: 0 to 10 percent total energy.

■ Protein: 10 to 15 percent total energy.

■ Salt: upper limit of 6 grams/day (no lower limit set).

Notes

1. *Healthy People 2000: National Health Promotion and Disease Prevention Objectives* (Washington, D.C.: U.S. Department of Health and Human Services, 1990).
2. Diet, nutrition and the prevention of chronic diseases: A report of the WHO Study Group on Diet, Nutrition and Prevention of Noncommunicable Diseases, *Nutrition Reviews* 49 (1991): 291–301.

Canada: Recommendations and Food Labels

Chapter 1 introduced Recommended Nutrient Intakes (RNI) and food labels. This appendix presents additional details for Canadians. Appendix F includes addresses of Canadian governmental agencies and professional organizations that may provide additional information.

RNI

The Canadian equivalent of the RDA is the Recommended Nutrient Intakes (RNI). The Canadian RNI are presented in Tables C-1 and C-2. The inside front cover (left) includes selected DRI values, which were developed for both the United States and Canada.

Table C-1 Recommended Nutrient Intakes for Canadians, 1990

Age	Sex	Weight (kg)	Protein (g/day)[a]	Fat-Soluble Vitamins Vitamin A (RE/day)[b]	Vitamin D (μg/day)[c]	Vitamin E (mg/day)[d]
Infants (months)						
0–4	Both	6	12[e]	400	10	3
5–12	Both	9	12	400	10	3
Children and Adults (years)						
1	Both	11	13	400	10	3
2–3	Both	14	16	400	5	4
4–6	Both	18	19	500	5	5
7–9	M	25	26	700	2.5	7
	F	25	26	700	2.5	6
10–12	M	34	34	800	2.5	8
	F	36	36	800	5	7
13–15	M	50	49	900	5	9
	F	48	46	800	5	7
16–18	M	62	58	1000	5	10
	F	53	47	800	2.5	7
19–24	M	71	61	1000	2.5	10
	F	58	50	800	2.5	7
25–49	M	74	64	1000	2.5	9
	F	59	51	800	2.5	6
50–74	M	73	63	1000	5	7
	F	63	54	800	5	6
75+	M	69	59	1000	5	6
	F	64	55	800	5	5
Pregnancy (additional amount needed)						
1st trimester			5	0	2.5	2
2nd trimester			20	0	2.5	2
3rd trimester			24	0	2.5	2
Lactation (additional amount needed)			20	400	2.5	3

[a]The primary units are expressed per kilogram of body weight. The figures shown here are examples.

[b]One retinol equivalent (RE) corresponds to the biological activity of 1 microgram of retinol, 6 micrograms of beta-carotene, or 12 micrograms of other carotenes.

[c]Expressed as cholecalciferol or ergocalciferol; see inside front cover (left) for 1997 DRI values.

[d]Expressed as δ-α-tocopherol equivalents, relative to which β- and γ-tocopherol and α-tocotrienol have activities of 0.5, 0.1, and 0.3, respectively.

[e]The assumption is made that the protein is from breast milk or has the same biological value as breast milk and that, between 3 and 9 months, adjustment for the quality of the protein is made.

Note: Recommended intakes of energy and of certain nutrients are not listed in this table because of the nature of the variables on which they are based. The figures for energy are estimates of average requirements for expected patterns of activity (see Table C-2). For nutrients not shown, the following amounts are recommended based on at least 2000 kcalories per day and body weights as given: thiamin, 0.4 milligrams per 1000 kcalories (0.48 milligrams/5000 kilojoules); riboflavin, 0.5 milligrams per 1000 kcalories (0.6 milligrams/ 5000 kilojoules); niacin, 7.2 niacin equivalents per 1000 kcalories (8.6 niacin equivalents/5000 kilojoules); vitamin B_6, 15 micrograms, as pyridoxine, per gram of protein. Recommended intakes during periods of growth are taken as appropriate for individuals representative of the midpoint in each age group. All recommended intakes are designed to cover individual variations in essentially all of a healthy population subsisting on a variety of common foods available in Canada.

Source: Health and Welfare Canada, *Nutrition Recommendations: The Report of the Scientific Review Committee* (Ottawa: Canadian Government Publishing Centre, 1990), Table 20, p. 204.

Table C-1 Recommended Nutrient Intakes for Canadians, 1990 (*continued*)

Water-Soluble Vitamins			Minerals					
Vitamin C (mg/day)[f]	Folate (μg/day)	Vitamin B$_{12}$ (μg/day)	Calcium[j] (mg/day)	Phosphorus[j] (mg/day)	Magnesium[j] (mg/day)	Iron (mg/day)	Iodine (μg/day)	Zinc (mg/day)
20	25	0.3	250	150	20	0.3[g]	30	2[h]
20	40	0.4	400	200	32	7	40	3
20	40	0.5	500	300	40	6	55	4
20	50	0.6	550	350	50	6	65	4
25	70	0.8	600	400	65	8	85	5
25	90	1.0	700	500	100	8	110	7
25	90	1.0	700	500	100	8	95	7
25	120	1.0	900	700	130	8	125	9
25	130	1.0	1100	800	135	8	110	9
30	175	1.0	1100	900	185	10	160	12
30	170	1.0	1000	850	180	13	160	9
40	220	1.0	900	1000	230	10	160	12
30	190	1.0	700	850	200	12	160	9
40	220	1.0	800	1000	240	9	160	12
30	180	1.0	700	850	200	13	160	9
40	230	1.0	800	1000	250	9	160	12
30	185	1.0	700	850	200	13[i]	160	9
40	230	1.0	800	1000	250	9	160	12
30	195	1.0	800	850	210	8	160	9
40	215	1.0	800	1000	230	9	160	12
30	200	1.0	800	850	210	8	160	9
0	200	0.2	500	200	15	0	25	6
10	200	0.2	500	200	45	5	25	6
10	200	0.2	500	200	45	10	25	6
25	100	0.2	500	200	65	0	50	6

[f]Cigarette smokers should increase intake by 50 percent.

[g]Based on the assumption that breast milk is the source of iron.

[h]Based on the assumption that breast milk is the source of zinc.

[i]After menopause, the recommended intake is 8 milligrams per day.

[j]See inside front cover (left) for 1997 DRI values.

Table C-2 Average Energy Requirements for Canadians

Age	Sex	Average Height (cm)	Average Weight (kg)	Requirements[a] (kcal/kg)[b]	(MJ/kg)[b]	(kcal/day)	(MJ/day)	(kcal/cm)	(MJ/cm)
Infants (months)									
0–2	Both	55	4.5	120–100	0.50–0.42	500	2.0	9	0.04
3–5	Both	63	7.0	100–95	0.42–0.40	700	2.8	11	0.05
6–8	Both	69	8.5	95–97	0.40–0.41	800	3.4	11.5	0.05
9–11	Both	73	9.5	97–99	0.41	950	3.8	12.5	0.05
Children and Adults (years)									
1	Both	82	11	101	0.42	1100	4.8	13.5	0.06
2–3	Both	95	14	94	0.39	1300	5.6	13.5	0.06
4–6	Both	107	18	100	0.42	1800	7.6	17	0.07
7–9	M	126	25	88	0.37	2200	9.2	17.5	0.07
	F	125	25	76	0.32	1900	8.0	15	0.06
10–12	M	141	34	73	0.30	2500	10.4	17.5	0.07
	F	143	36	61	0.25	2200	9.2	15.5	0.06
13–15	M	159	50	57	0.24	2800	12.0	17.5	0.07
	F	157	48	46	0.19	2200	9.2	14	0.06
16–18	M	172	62	51	0.21	3200	13.2	18.5	0.08
	F	160	53	40	0.17	2100	8.8	13	0.05
19–24	M	175	71	42	0.18	3000	12.6		
	F	160	58	36	0.15	2100	8.8		
25–49	M	172	74	36	0.15	2700	11.3		
	F	160	59	32	0.13	1900	8.0		
50–74	M	170	73	31	0.13	2300	9.7		
	F	158	63	29	0.12	1800	7.6		
75+	M	168	69	29	0.12	2000	8.4		
	F	155	64	23	0.10	1500	6.3		

[a]Requirements can be expected to vary within a range of ±36 percent.

[b]First and last figures are averages at the beginning and end of the three-month period.

Source: Health and Welfare Canada, *Nutrition Recommendations: The Report of the Scientific Review Committee* (Ottawa: Canadian Government Publishing Centre, 1990), Tables 5 and 6, pp. 25, 27.

Food Labels

Consumers can gather a lot of information from a nutrition label. Table C-3 defines terms, and Figure C-1 demonstrates the reading of a food label.

Table C-3   Terms on Food Labels

Energy

---

kCalorie reduced: 50% or fewer kcalories than the regular version.

Light: term may be used to describe anything (for example, light in colour, texture, flavour, taste, or kcalories); read the label to find out what is "light" about the product.

Low kcalorie: kcalorie-reduced and no more than 15 kcalories per serving.

Fat and Cholesterol

Low cholesterol: no more than 3 mg of cholesterol per 100 g of the food and low in saturated fat; *does not* always mean low in total fat.

Low fat: no more than 3 g of fat per serving; *does not* always mean low in kcalories.

Lower fat: at least 25% less fat than the comparison food; be aware that "80% fat free" still means the food is 20% fat.

Carbohydrates: Fibre and Sugar

Carbohydrate reduced: not more than 50% of the carbohydrate found in the regular version; *does not* always mean the product is lower in kcalories, because other ingredients such as fat may have increased.

Source of dietary fibre: a product that provides 2–4 g of fibre.

High source of dietary fibre: a product that provides 4–6 g of fibre.

Very high source of fibre: a product that provides 6 g (or more) of fibre.

Sugar free: low in carbohydrates and kcalories; can be used as an extra food in the exchange system.

Unsweetened or no sugar added: no sugar was added to the product; sugar may be found naturally in the food (for example, fruit canned in its own juice).

Figure C-1   How to Read a Canadian Food Label

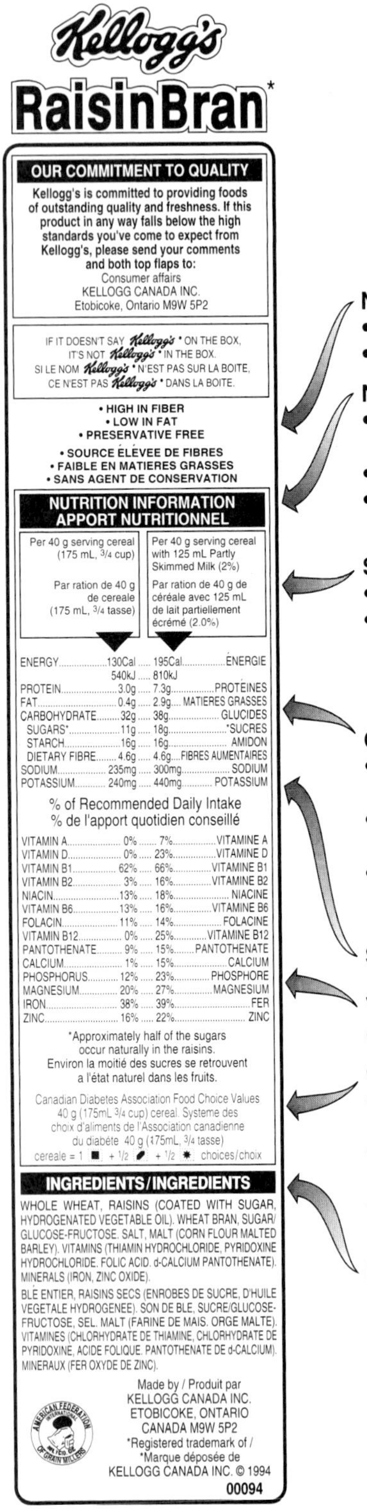

Kellogg's
RaisinBran*

OUR COMMITMENT TO QUALITY

Kellogg's is committed to providing foods of outstanding quality and freshness. If this product in any way falls below the high standards you've come to expect from Kellogg's, please send your comments and both top flaps to:
Consumer affairs
KELLOGG CANADA INC.
Etobicoke, Ontario M9W 5P2

IF IT DOESN'T SAY *Kellogg's* ON THE BOX,
IT'S NOT *Kellogg's* IN THE BOX.
SI LE NOM *Kellogg's* N'EST PAS SUR LA BOITE,
CE N'EST PAS *Kellogg's* DANS LA BOITE.

- HIGH IN FIBER
- LOW IN FAT
- PRESERVATIVE FREE
- SOURCE ÉLEVÉE DE FIBRES
- FAIBLE EN MATIERES GRASSES
- SANS AGENT DE CONSERVATION

NUTRITION INFORMATION
APPORT NUTRITIONNEL

	Per 40 g serving cereal (175 mL, 3/4 cup)	Per 40 g serving cereal with 125 mL Partly Skimmed Milk (2%)
	Par ration de 40 g de cereale (175 mL, 3/4 tasse)	Par ration de 40 g de céréale avec 125 mL de lait partiellement écrémé (2.0%)

ENERGY	130Cal	195Cal	ENERGIE
	540kJ	810kJ	
PROTEIN	3.0g	7.3g	PROTEINES
FAT	0.4g	2.9g	MATIERES GRASSES
CARBOHYDRATE	32g	38g	GLUCIDES
SUGARS*	11g	18g	SUCRES
STARCH	16g	16g	AMIDON
DIETARY FIBRE	4.6g	4.6g	FIBRES AUMENTAIRES
SODIUM	235mg	300mg	SODIUM
POTASSIUM	240mg	440mg	POTASSIUM

% of Recommended Daily Intake
% de l'apport quotidien conseillé

VITAMIN A	0%	7%	VITAMINE A
VITAMIN D	0%	23%	VITAMINE D
VITAMIN B1	62%	66%	VITAMINE B1
VITAMIN B2	3%	16%	VITAMINE B2
NIACIN	13%	18%	NIACINE
VITAMIN B6	13%	16%	VITAMINE B6
FOLACIN	11%	14%	FOLACINE
VITAMIN B12	0%	25%	VITAMINE B12
PANTOTHENATE	9%	15%	PANTOTHENATE
CALCIUM	1%	15%	CALCIUM
PHOSPHORUS	12%	23%	PHOSPHORE
MAGNESIUM	20%	27%	MAGNESIUM
IRON	38%	39%	FER
ZINC	16%	22%	ZINC

*Approximately half of the sugars occur naturally in the raisins.
Environ la moitié des sucres se retrouve à l'état naturel dans les fruits.

Canadian Diabetes Association Food Choice Values
40 g (175mL 3/4 cup) cereal. Système des choix d'aliments de l'Association canadienne du diabète 40 g (475mL, 3/4 tasse)
cereale = 1 ■ + 1/2 ✦ + 1/2 ● choices/choix

INGREDIENTS/INGREDIENTS

WHOLE WHEAT, RAISINS (COATED WITH SUGAR, HYDROGENATED VEGETABLE OIL). WHEAT BRAN, SUGAR/GLUCOSE-FRUCTOSE. SALT, MALT (CORN FLOUR MALTED BARLEY). VITAMINS (THIAMIN HYDROCHLORIDE, PYRIDOXINE HYDROCHLORIDE. FOLIC ACID. d-CALCIUM PANTOTHENATE). MINERALS (IRON, ZINC OXIDE).

BLE ENTIER, RAISINS SECS (ENROBES DE SUCRE, D'HUILE VEGETALE HYDROGENEE). SON DE BLE, SUCRE/GLUCOSE-FRUCTOSE, SEL. MALT (FARINE DE MAIS. ORGE MALTE). VITAMINES (CHLORHYDRATE DE THIAMINE, CHLORHYDRATE DE PYRIDOXINE, ACIDE FOLIQUE. PANTOTHENATE DE d-CALCIUM). MINERAUX (FER OXYDE DE ZINC).

Made by / Produit par
KELLOGG CANADA INC.
ETOBICOKE, ONTARIO
CANADA M9W 5P2
*Registered trademark of /
*Marque déposée de
KELLOGG CANADA INC. © 1994
00094

WHAT YOU WILL FIND ON A LABEL:

Nutrition claims
- In Canada, it is optional for a company to decide to use claims.
- When claims appear on a label, they must follow government laws.

Nutrition Information
- Gives detailed nutrition facts about the product, including serving size and core list.
- Does not have to appear by law on food products in Canada.
- Refers to the food as packaged, so if you add milk, eggs, or other food, the nutritional content of the food you eat can be very different.

Serving Size
- The amount of food for which the information is given.
- Check the serving size: The serving size on the label may not be the same as the serving size you would actually eat (for example, the serving size of cereal may be 3/4 cup, much smaller than your regular serving).

Core List
- The energy (in calories and kilojoules), grams of protein, fat, and carbohydrate for each serving.
- Some products break down fat into monounsaturates, polyunsaturates, saturates, and cholesterol.
- Carbohydrates may include the amount of sugars, starch, and fibre, or may list these items separately.

Sodium and Potassium (in milligrams)

Vitamins and Minerals (as percentage of your recommended daily intake)

Canadian Diabetes Association Food Choice Values and Symbols
- The Values and Symbols are tools to help you fit the food into your meal plan; they are not an endorsement by CDA.
- It is up to the food company to decide if it wants its foods analyzed and assigned symbols.
- When they are on a label, they have been assigned by a dietitian working for CDA, so you can be sure the information is correct.

Ingredients
- Must be found on all food labels by law.
- Ingredients are listed in decreasing order by weight, so what you see first is what you get the most of.

Infant Formulas

Table D-1 compares the nutrient composition of three cow's milk–based infant formulas and two soy-based infant formulas. All infant formulas are designed to resemble breast milk and must meet an American Academy of Pediatrics standard for nutrient composition. Milk-based formulas are intended for full-term, healthy infants. Soy-based formulas are designed for infants with milk sensitivity or lactose intolerance. Special formulas are available for premature infants or infants with medical conditions requiring special nutrition treatment.

Table D-1 Comparison of Nutrients in Infant Formulas

Nutrient	Cow's Milk–Based Infant Formulas			Soy-Based Infant Formulas	
	Mead Johnson Enfamil with Iron	Ross Similac with Iron	Nestlé Carnation	Ross Isomil	Mead Johnson Prosobee
Water (g)	134	133	134	134	134
Carbohydrate (g)	10.9 (lactose)	10.7 (lactose)	13.2 (lactose)	10.3 (corn syrup solids, sucrose)	10 (corn syrup solids)
Protein (g)	2.1	2.14	2.6	2.45	3
Fat (g)	5.3	5.40	4.1	5.46	5.3
Linoleic acid (mg)	860	1300	680	1300	860
Minerals					
Calcium (mg)	78	73	135	105	105
Chloride (mg)	63	64	90	62	80
Iron (mg)[a]	1.8	1.8	1.9	1.8	1.8
Phosphorus (mg)	53	56	90	108	83
Potassium (mg)	108	105	135	108	120
Sodium (mg)	27	27	39	44	36
Magnesium (mg)	8	6	8.4	7.5	11
Zinc (mg)	1	0.75	0.63	0.75	1.2
Manganese (μg)	15	5	8.4	30	25
Copper (μg)	75	90	76	75	75
Iodine (μg)	10	9	5.7	15	15
Vitamins					
Vitamin A (IU)	300	300	250	300	300
Thiamin (μg)	80	100	80	60	80
Riboflavin (μg)	140	150	96	90	90
Niacin (μg)	1000	1050	1280	1350	1000
Vitamin B_6 (μg)	60	60	66	60	60
Vitamin B_{12} (μg)	0.3	0.25	0.32	0.45	0.3
Folate (μg)	16	15	16	15	16
Vitamin C (mg)	12	9	8	9	12
Vitamin D (IU)	60	60	65	60	60
Vitamin E (IU)	2	3	2	3	2
Vitamin K (μg)	8	8	8.1	15	8
Pantothenic acid (μg)	500	450	480	750	500
Biotin (μg)	3	4.4	2	4.5	3
Choline (mg)	12	16	12	8	12
Inositol (mg)	6	4.7	18	5	17

[a]Low iron formulas are also available, but not recommended. They provide less than half the iron of iron-fortified formulas.

Vitamin-Mineral Supplements

The following tables are useful for comparing the essential vitamin and mineral contents of supplements commonly available in the United States for pregnant women, infants, and children (Tables E-1 and E-2). Notice that an "Other" column has been provided for the addition of locally available products you may wish to compare with those shown here.

Not all ingredients in vitamin-mineral preparations are of proven benefit. To facilitate meaningful comparison, the tables list only the nutrients known to be essential in human nutrition. Other nutrients and compounds found on the labels of these supplements are listed in the table notes.

When a supplement that supplies certain nutrients is needed, these tables will ease the task of selecting an appropriate one. Notice, for example, that the iron and calcium contents of the supplements listed here for children (Table E-2) vary considerably. Some contain no iron at all, while others provide more than 100 percent of a child's RDA. Many of the supplements contain no calcium, either.

Table E-1 Prenatal Supplements for Pregnant Women

	Ethex Prenatal Maternal Tablets[a]	Ross Pramilet FA Tablets	Roberts Sigtab-M Tablets[b]	ME Pharm Mynatal P.N. Forte Caplets	Lederle Materna Tablets[c]
Availability	R_x	R_x	R_x	R_x	R_x
Vitamins					
Vitamin A (IU)	5000	4000	6000	5000	5000
Vitamin D (IU)	400	400	400	400	400
Vitamin E (IU)	30[d]	—	45[d]	30[d]	30[e]
Vitamin C (mg)	100	60	100	80	100
Thiamin (mg)	2.9	3	5	3	3
Riboflavin (mg)	3.4	2	5	3.4	3.4
Vitamin B_6 (mg)	12.2	3	3	4	10
Vitamin B_{12} (μg)	12	3	—	12	12
Niacin (mg)	20	10	25	20	20
Folate (mg)	1	1	0.4	1	1
Biotin (μg)	30	—	45	—	30
Minerals					
Calcium (mg)	250	250	200	250	250
Iron (mg)	60	40	18	60	60
Magnesium (mg)	f	f	f	f	25
Zinc (mg)	25	f	15	25	25
Copper (mg)	f	f	f	f	2
Iodine (μg)	f	f	f	f	150
Manganese (mg)	f	—	f	—	5
Cost per day[g]					

[a]This product also contains chromium and molybdenum, but values were not available.

[b]This product also contains vitamin K, phosphorus, potassium, chlorine, molybdenum, selenium, chromium, nickel, tin, vanadium, silicon, and boron, but values were not available.

[c]This product also contains chromium, molybdenum, lactose, parabens, and sucrose.

[d]Form of vitamin E unknown.

[e]As alpha-tocopherol acetate.

[f]Supplement contains this mineral, but values were not available.

[g]Divide the total retail price for the container by the number of doses per container. For example, XYZ Vitamins are sold in bottles of 100 tablets, and the recommended dose is 2 tablets per day; there are 50 doses in the bottle. At $5.00 per bottle, XYZ Vitamins cost $.10 per day.

Table E-1   Prenatal Supplements for Pregnant Women (*continued*)

	Wyeth-Ayerst Stuartnatal Plus Tablets	Mead-Johnson Natalins R_x Tablets	Geneva Prenatal with Folic Acid Tablets	Fielding Nestabs FA Tablets	Pasadena Nutricon Tablets	Mission Prenatal R_x Tablets	Other
Availability	R_x	R_x	OTC	R_x	OTC	R_x	
Vitamins							
Vitamin A (IU)	4000	4000	4000	5000	2500	8000	
Vitamin D (IU)	400	400	400	400	200	400	
Vitamin E (IU)	11 [d]	15 [d]	11 [d]	30 [d]	15 [e]	0	
Vitamin C (mg)	120	80	100	120	50	240	
Thiamin (mg)	1.5	1.5	1.5	3	1.5	4	
Riboflavin (mg)	3	1.6	1.7	3	1.5	2	
Vitamin B_6 (mg)	10	4	2.6	3	2	20	
Vitamin B_{12} (μg)	12	2.5	4	8	5	8	
Niacin (mg)	20	17	18	20	10	20	
Folate (mg)	1	1	0.8	1	0.4	1	
Biotin (μg)	—	30	—	—	150	—	
Minerals							
Calcium (mg)	200	200	200	200	200	—	
Iron (mg)	65	60	60	36	20	—	
Magnesium (mg)	—	f	—	—	f	—	
Zinc (mg)	25	25	25	15	3.75	15	
Copper (mg)	f	f	—	—	f	f	
Iodine (μg)	—	—	—	f	f	f	
Manganese (mg)	—	—	—	—	—	—	
Cost per day [g]							

[d]Form of vitamin E unknown.

[e]As alpha-tocopherol acetate.

[f]Supplement contains this mineral, but values were not available.

[g]Divide the total retail price for the container by the number of doses per container. For example, XYZ Vitamins are sold in bottles of 100 tablets, and the recommended dose is 2 tablets per day; there are 50 doses in the bottle. At $5.00 per bottle, XYZ Vitamins cost $.10 per day.

Source: Drug Facts and Comparisons (St. Louis: Facts and Comparisons, 1997), pp. 31a–31c.

Table E-2 Supplements for Infants and Children

	Company/Product					
	Lederle[a] Centrum Jr. with Iron	Miles Laboratories Flintstones with Iron	Mead-Johnson Poly-Vi-Sol with Iron	Ciba Self-Medication, Inc. Sunkist Multivitamins 1 Extra C	Chocks Bugs Bunny plus Iron	Other
Age of Intended Users	Children over 4	Children over 2	Infants	Adults and Children over 2	Children over 2	
Recommended Daily Dose	1 Chewable[b]	1 Chewable	1-mL Dropper	1 Chewable	1 Chewable	
Vitamins						
Vitamin A (IU)	5000	2500	1500	2500	2500	
Vitamin D (IU)	400	400	400	400	400	
Vitamin E (IU)	30	15	5	15	15	
Vitamin K (mg)	0.010	—	—	0.005	—	
Vitamin C (mg)	60	60	35	250	60	
Thiamin (B_1) (mg)	1.5	1.05	0.5	1.1	1.05	
Riboflavin (B_2) (mg)	1.7	1.2	0.6	1.2	1.2	
Vitamin B_6 (mg)	2	1.05	0.4	1	1.05	
Vitamin B_{12} (μg)	6	4.5	2	5	4.5	
Niacin (mg)	20	13.5	8	14	13.5	
Folate (mg)	0.4	0.3	—	0.3	0.3	
Minerals						
Calcium (mg)	108	—	—	—	—	
Phosphorus (mg)	50	—	—	—	—	
Iron (mg)	18	15	10	—	15	
Potassium (mg)	—	—	—	—	—	
Magnesium (mg)	40	—	—	—	—	
Zinc (mg)	15	—	—	—	—	
Copper (mg)	2	—	—	—	—	
Iodine (μg)	0.15	—	—	—	—	
Manganese (mg)	1	—	—	—	—	
Cost per day[c]						

[a]This product also contains 1.4 milligrams chlorine, 0. 045 milligrams biotin, 10 milligrams pantothenic acid, 0.020 milligrams chromium, and 0.020 milligrams molybdenum.

[b]Recommended daily dose for children ages 2 to 4 is 1/2 tablet.

[c]Divide the total retail price for the container by the number of doses per container. For example, XYZ Vitamins are sold in bottles of 100 tablets, and the recommended dose is 2 tablets per day; there are 50 doses in the bottle. At $5.00 per bottle, XYZ Vitamins cost $.10 per day.

Addresses of Nutrition-Related Resources

Reliable nutrition information and support are not always easy to come by, especially if you do not know where to look. The agencies, groups, and organizations listed here can provide reliable information or offer support for those with problems related to specific topics.

U.S. GOVERNMENT

- Federal Trade Commission (FTC)
  Public Reference Branch
  (202) 326-2222
  Internet address: http://www.ftc.gov

- Food and Drug Administration (FDA)
  Office of Consumer Affairs
  HFE 881 Room 16-63
  5600 Fishers Lane
  Rockville, MD 20857
  Internet address:
  http://www.fda.gov/fdahomepage.html

- FDA Consumer Information Line
  (301) 443-3170

- FDA Office of Food Labeling (HFS-150)
  200 C Street SW
  Washington, DC 20204
  (202) 205-4561; fax: (202) 205-4564

- FDA Office of Nutrition and Food Sciences
  200 C Street SW
  Washington, DC 20204
  (202) 205-4561

- FDA Office of Plant and Dairy Foods and
  Beverages (HFS-300)
  200 C Street SW
  Washington, DC 20204
  (202) 205-4064; fax: (202) 205-4422

- FDA Office of Special Nutritionals (HFS-450)
  200 C Street SW
  Washington, DC 20204
  (202) 205-4168; fax: (202) 205-5295

- Food and Nutrition Information Center
  National Agricultural Library, Room 304
  10301 Baltimore Blvd.
  Beltsville, MD 20705-2351
  (301) 504-5719; fax: (301) 504-6409
  Internet address:
  http://www.nal.usda.gov/fnic

- Food Research and Action Center
  1875 Connecticut Avenue, NW, Suite 540
  Washington, DC 20009
  (202) 986-2200

- Superintendent of Documents
  U.S. Government Printing Office
  Washington, DC 20402

- U.S. Department of Agriculture (USDA)
  14th Street SW and Independence Avenue
  Washington, DC 20250
  Internet address:
  http://www.usda.gov/fcs/cnpp
  (202) 720-2791

- USDA Center for Nutrition Policy
  and Promotion
  1120 20th Street, NW, Suite 200,
  North Lobby
  Washington, DC 20036
  (202) 418-2321

- USDA Food Safety and Inspection Service
  Food Safety Education Office
  Room 1180-S
  Washington, DC 20250
  (202) 690-0351

- USDA Health Inspection Service
  4700 River Road
  Riverdale, MD 20737-1228

- USDA Meat and Poultry Hotline
  (800) 535-4555
- U.S. Department of Education (DOE)
  Accreditation Agency Evaluation Branch
  7th and D Street SW
  Building 3, Room 336
  Washington, DC 20202
  (202) 708-7417
- U.S. Environmental Protection Agency (EPA)
  401 M Street NW
  Washington, DC 20460
  (202) 382-3535
- U.S. EPA Safe Drinking Water Hotline
  (800) 426-4791
- U.S. Public Health Service Public
  Affairs Office
  Hubert H. Humphrey Building
  Room 725-H
  200 Independence Avenue SW
  Washington, DC 20201
  (202) 245-6867

CANADIAN GOVERNMENT
Federal

- Bureau of Nutritional Sciences,
  Food Directorate
  Health Protection Branch
  Department Health Canada
  Banting Building
  Tunney's Pasture, Ottawa, Ontario K1A 0L2
- Nutrition Programs Unit
  Health Promotion Directorate
  Health Canada
  4th Floor, Jeanne Mance Building
  Tunney's Pasture, Ottawa, Ontario K1A 1B4
  Internet address: http://www.hwc.ca
- Food Production and Inspection Branch
  Agriculture and Agri-Food Canada
  59 Camelot Drive
  Nepean, Ontario KIA OY9
  Internet address:
  http://www.cfia.acia.agr.com
- Nutrition Specialist, Health Support Services
  Indian and Northern Health Services
  Directorate
  Health Canada
  11th Floor, Jeanne Mance Building
  Tunney's Pasture, Ottawa, Ontario K1A 0L3

Provincial & Territorial

- Special Services Coordinator
  Health & Community Services Agency
  P.O. Box 2000, 4 Sydney Street
  Charlottetown, Prince Edward Island,
  Canada C1A 7N8
- Director
  Health Promotion Division
  Department of Health
  P.O. Box 8700
  Confederation Bldg., West Block
  St. John's, Newfoundland, AIB 4J6
- Senior Nutrition Consultant
  Public Health Services
  Department of Health and Community
  Services
  P.O. Box 5100
  Fredericton, New Brunswick E3B 5G8
- Nutrition Coordinator, Nutrition
  Program Planning, Public Health and
  Health Promotion
  Department of Health
  P.O. Box 488
  Halifax, Nova Scotia B3J 2R8
- Responsable des programmes de nutrition
  Direction de la Santé
  Ministère de la Santé et des Services sociaux
  3ᵉ étage, 1075 chemin Saint-Foy
  Québec, Québec G1S 2M1
- Senior Nutrition Consultant
  Public Health Branch
  Ministry of Health, 8th Floor
  5700 Yonge Street
  New York, Ontario M2M4K2
- Program Specialist Nutrition
  Program Development Branch
  Healthy Public Policy Program Division
  Manitoba Health
  599 Empress Street
  Box 925
  Winnipeg, Manitoba R3C2T6
- Provincial Nutritionist, Health
  Promotion Unit
  Population Health
  Saskatchewan Health Branch
  3475 Albert Street
  Regina, Saskatchewan S4S 6X6

- Population Health and Program Development, Prevention and Promotion Branch, Alberta Health
Jasper Avenue Building, 24th Floor
10025 Jasper Avenue
Edmonton, Alberta T5J2P4

- Ministry of Health
Nutrition Section
1520 Blanshard Street, 1st Floor
Victoria, British Columbia V8W 3C8

- Director Nutrition Services, Yukon Hospital Corporation
Department of Health and Social Services
5 Hospital Road
Whitehorse, Yukon Y1A 2C6

- Consultant, Infant/Child Nutrition, Community Health Program
Department of Health and Social Services
6th Floor, P.O. Box 1320
Yellowknife, Northwest Territories X1A 2L9

INTERNATIONAL AGENCIES

- Food and Agriculture Organization of the United Nations (FAO)
Liaison Office for North America
1001 22nd Street NW
Washington, DC 20437
(202) 653-2400

- World Health Organization (WHO)
Regional Office
525 23rd Street NW
Washington, DC 20037
(202) 861-3200
Internet address: http://www.who.ch

- Internet Health Resources
Internet address: http://www.ihr.com

- International Food Information Council Foundation
Internet address:
http://www.ificinfo://health.org

PROFESSIONAL NUTRITION ORGANIZATIONS

- American Dietetic Association (ADA)
216 West Jackson Boulevard, Suite 800
Chicago, IL 60606-6995
(312) 899-0040
Internet address: http://www.eatright.org

- ADA, The National Center for Nutrition and Dietetics' Food and Nutrition Information Hotline
(800) 366-1655

- American Academy of Nutritional Sciences
Internet address:
http://www.nutrition.org

- American Society for Clinical Nutrition
9650 Rockville Pike
Bethesda, MD 20814-3998
Internet address:
http://www.faseb.org/ascn

- Dietitians of Canada
480 University Avenue, Suite 604
Toronto, Ontario M5G 1V2, Canada
(416) 596-0857

- National Academy of Sciences/National Research Council (NAS/NRC)
2101 Constitution Avenue NW
Washington, DC 20418

- National Institute of Nutrition
302-265 Carling Avenue
Ottawa, Ontario K1S 2E1

- Nutrition Foundation, Inc. (INACG)
1126 Sixteenth Street NW, Suite 111
Washington, DC 20036

- Nutrition Information Service
University of Alabama at Birmingham
Room 447 Webb Building
UAB Station
Birmingham, AL 35294-3360

- Society for Nutrition Education
2850 Metro Drive, Suite 416
Minneapolis, MN 55425-1412
(612) 854-0035

ALCOHOL AND DRUG ABUSE

- Al-Anon Family Group Headquarters
P.O. Box 862
Midtown Station
New York, NY 10018-0862
(800) 356-9996

- Alateen
1600 Corporate Landing Parkway
Virginia Beach, VA 23454
(800) 356-9996

■ Alcohol & Drug Abuse Information Line
(800) 252-6465

■ Alcoholics Anonymous (AA)
General Service Office
475 Riverside Drive
New York, NY 10115
(212) 870-3400

■ Narcotics Anonymous (NA)
P.O. Box 9999
Van Nuys, CA 91409
(818) 780-3951

■ National Clearinghouse for Alcohol
and Drug Information (NCADI)
P.O. Box 2345
Rockville, MD 20847-2345
(800) 729-6686

■ National Council on Alcoholism and
Drug Dependence
12 West 21st Street
New York, NY 10010
(800) NCA-CALL

■ OSAP's National Clearinghouse for Alcohol
and Drug Information (ONCADI)
P.O. Box 2345
Rockville, MD 20847-2345
(800) 729-6686

FITNESS

■ American College of Sports Medicine
P.O. Box 1440
Indianapolis, IN 46204
(317) 637-9200

■ President's Council on Physical Fitness
and Sports
701 Pennsylvania Avenue NW
Suite 250
Washington, DC 20004
(202) 272-3421

■ Sport Medicine and Science Council
of Canada
1600 James Naismith Drive
Gloucester, Ontario K1B 5N4

HEALTH AND DISEASE

■ Alzheimer's Disease Education and
Referral Center
P.O. Box 8250
Silver Spring, MD 20907-8250
(800) 438-4380

■ National Institutes of Health (NIH)
9000 Rockville Pike
Bethesda, MD 20892
(301) 496-2433
Internet address: http://www.nih.gov

■ Alzheimer's Disease Information and
Referral Service
919 North Michigan Avenue
Chicago, IL 60611
(800) 272-3900

■ American Academy of Allergy, Asthma,
and Immunology
611 East Wells Street
Milwaukee, WI 53202
(414) 272-6071; fax: (414) 276-3349

■ American Cancer Society
Cancer Information Center
1701 Rickenbacker Drive, Suite 5B
Sun City Center, FL 33573-5361
(800) ACS-2345
Internet address: http://www.cancer.org

■ American Council on Science and Health
1995 Broadway, 16th Floor
New York, NY 10023-5860

■ American Dental Association
Division of Communications
211 East Chicago Avenue
Chicago, IL 60611-2678

■ American Diabetes Association
1660 Duke Street
Alexandria, VA 22314
(703)549-1500
(800) 232-3472
Internet address: http://www.diabetes.org

■ American Heart Association
Box BHG, National Center
7320 Greenville Avenue
Dallas, TX 75231
(800) 242-8721
Internet address: http://www.amhrt.org

■ American Institute for Cancer Research
1759 R Street NW
Washington, DC 20009
Internet address: http://www.aicr.org

■ American Medical Association
515 North State Street
Chicago, IL 60610
(312) 464-5000
Internet address:
http://www.ama-assn.org

■ American Public Health Association
1015 Fifteenth Street NW
Washington, DC 20005
Internet address: http://www.apha.org

■ American Red Cross AIDS Education Office
1730 D Street NW
Washington, DC 20006
(202) 737-8300

■ Canadian Diabetes Association
15 Toronto St., Suite 1001
Toronto, Ontario M5C 2E3
(416) 362-4440

■ Canadian Public Health Association
Publications, Suite 400
1565 Carling Ave.
Ottawa, Ontario K1Z 8R1

■ Centers for Disease Control and
Prevention (CDC)
Information Hotline
(404) 332-4555
Internet address: http://www.cdc.gov

■ Disease Prevention and Health Promotion's
National Health Information Center,
Office of
(800) 336-4797

■ The Food Allergy Network
10400 Eaton Place, Suite 107
Fairfax, VA 22030-5647
(703) 691-3179
(800) 929-4040

■ National AIDS Hotline (CDC)
(800) 342-AIDS (English)
(800) 344-SIDA (Spanish)
(800) 2437-TTY (Deaf)
(900) 820-2437

■ National Cancer Institute
Office of Cancer Communications
Building 31, Room 10824
Bethesda, MD 20892
(800) 4-CANCER
Internet address: http://www.nci.nih.gov

■ National Digestive Disease Information
Clearinghouse (NDDIC)
2 Information Way
Bethesda, MD 20892-3570
(301) 654-3810

■ National Heart, Lung, and Blood Institute
National High Blood Pressure Education
Program Information Center
P.O. Box 30105
Bethesda, MD 20824-0105
(301) 951-3260

■ National Institute of Allergy
and Infectious Diseases
Office of Communications, Building 31,
Room 7A50
31 Center Drive, MSC 2520
Bethesda, MD 20892-2520
(301) 496-5717
Internet address:
http://www.niaid.nih.gov

■ National Institute of Dental Research (NIDR)
Building 31, Room 2C35
31 Center Drive, MSC 2290
Bethesda, MD 20892
(301) 496-4261

■ National Osteoporosis Foundation
2100 M Street NW, Suite 602
Washington, DC 20037
(202) 223-2226

■ New England Journal of Medicine
Internet address: http://www.nejm.org

■ ODPHP National Health Information
Center (ONHIC)
(800) 336-4797
Internet address:
http://NHIC-NT.Health.org

■ Smoking and Health Office (CDC)
Technical Information Center
Mail Stop K-12
1600 Clifton Road NE
Atlanta, GA 30333

INFANCY AND CHILDHOOD

■ American Academy of Pediatrics
P.O. Box 927
141 Northwest Point Boulevard
Elk Grove Village, IL 60009-0927

■ Association of Birth Defect Children, Inc.
827 Irma Street
Orlando, FL 32803
(407) 245-7035

■ Canadian Paediatric Society
410 Smyth Rd.
Ottawa, Ontario K1H 8L1

■ National Center for Education in
Maternal & Child Health
2000 15th Street North, Suite 701
Arlington, VA 22201-2617
(703) 524-7802

■ National Maternal and Child Health
Clearinghouse
8201 Greensboro Drive
Suite 600
McLean, VA 22102
(703) 821-8955, Ext. 254 or 255

■ Nurture/Center to Prevent Childhood
Malnutrition
1840 18th Street, NW
Washington, DC 20009
(202) 797-9244; fax: (202) 797-9257

PREGNANCY AND LACTATION

■ American College of Obstetricians and
Gynecologists Resource Center
409 12th Street SW
Washington, DC 20024-2188

■ La Leche League International, Inc.
1400 N. Meacham Rd.
P.O. Box 4079
Schaumburg, IL 60168-4079
(847) 519-7730

■ March of Dimes Birth Defects Foundation
(National Headquarters)
1275 Mamaroneck Avenue
White Plains, NY 10605

WEIGHT CONTROL AND EATING DISORDERS

■ American Anorexia & Bulimia
Association, Inc.
418 East 76th Street
New York, NY 10021
(212) 734-1114

■ Anorexia Nervosa and Related Eating
Disorders (ANRED)
P.O. Box 5102
Eugene, OR 97405
(503) 344-1144

■ National Association of Anorexia Nervosa
and Associated Disorders, Inc. (ANAD)
P.O. Box 7
Highland Park, IL 60035
(708) 831-3438

■ National Eating Disorder Information Centre
200 Elizabeth St., College Wing 1-328
Toronto, Ontario M5G 2C4

■ Overeaters Anonymous (OA)
383 Van Ness Avenue, Suite 1601
Torrance, CA 90501

■ T.O.P.S. (Take Off Pounds Sensibly)
P.O. Box 07360
Milwaukee, WI 53207

Index

Note: Boldface numbers indicate pages on which definitions appear.

PHOTO CREDITS

Acceptable Weight for Height Based on Body Mass Index (BMI)

To determine your acceptable weight range, find your height in the top line. Look down the column below it and find the range represented by the white strip. Look to the left column to see what weights are acceptable for you.

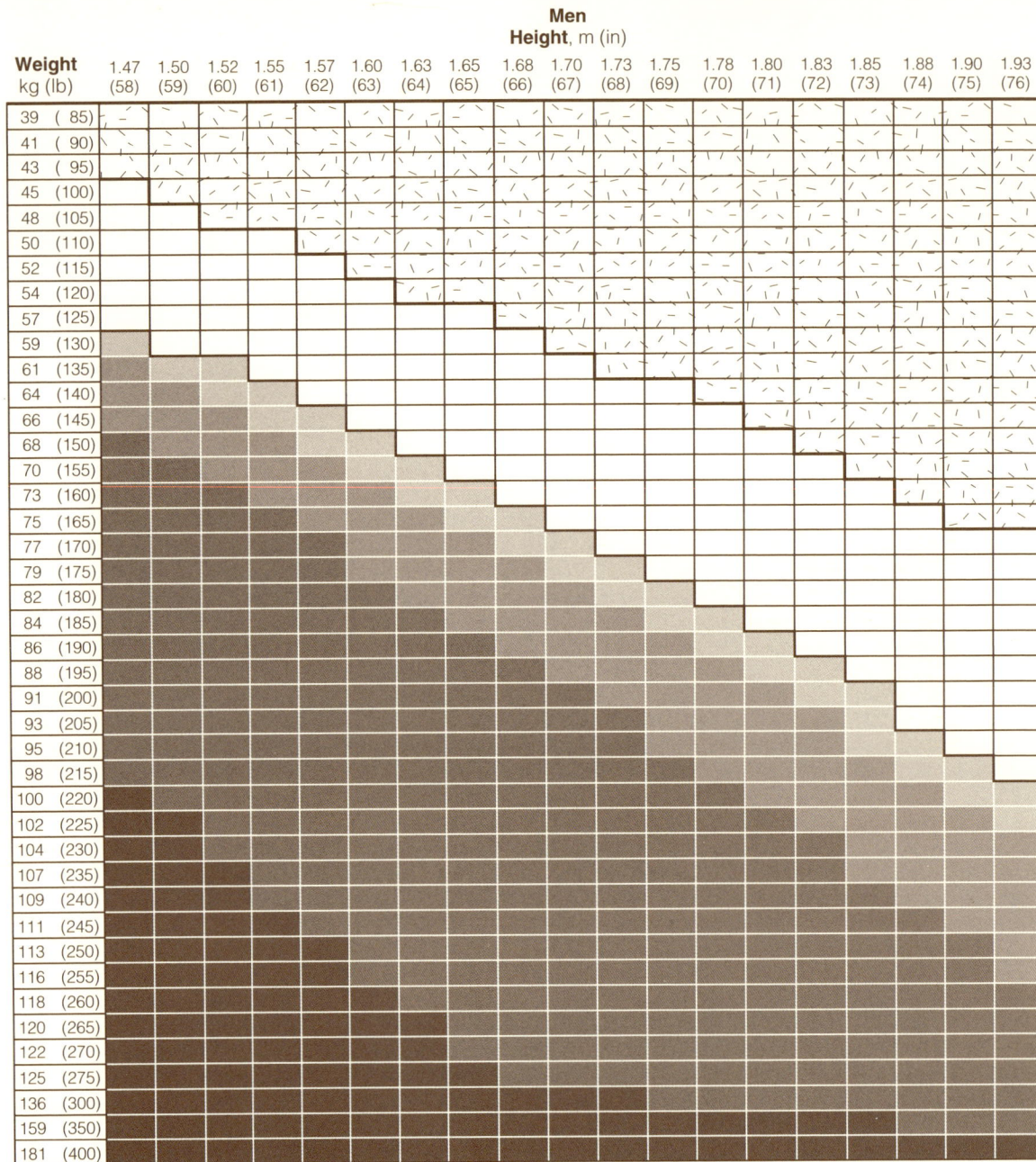

Men
Height, m (in)

Weight kg (lb)	1.47 (58)	1.50 (59)	1.52 (60)	1.55 (61)	1.57 (62)	1.60 (63)	1.63 (64)	1.65 (65)	1.68 (66)	1.70 (67)	1.73 (68)	1.75 (69)	1.78 (70)	1.80 (71)	1.83 (72)	1.85 (73)	1.88 (74)	1.90 (75)	1.93 (76)
39 (85)																			
41 (90)																			
43 (95)																			
45 (100)																			
48 (105)																			
50 (110)																			
52 (115)																			
54 (120)																			
57 (125)																			
59 (130)																			
61 (135)																			
64 (140)																			
66 (145)																			
68 (150)																			
70 (155)																			
73 (160)																			
75 (165)																			
77 (170)																			
79 (175)																			
82 (180)																			
84 (185)																			
86 (190)																			
88 (195)																			
91 (200)																			
93 (205)																			
95 (210)																			
98 (215)																			
100 (220)																			
102 (225)																			
104 (230)																			
107 (235)																			
109 (240)																			
111 (245)																			
113 (250)																			
116 (255)																			
118 (260)																			
120 (265)																			
122 (270)																			
125 (275)																			
136 (300)																			
159 (350)																			
181 (400)																			

Underweight (BMI = < 20.7)
Acceptable weight (BMI = 20.7 to 26.4)
Marginal overweight (BMI = 26.4 to 27.8)
Overweight (BMI = 27.8 to 31.1)
Severe overweight (BMI = 31.1 to 45.4)
Morbid obesity (BMI = > 45.4)

Source:
Adapted from M. I. Rowland, A nomogram for computing body index, *Dietetic Currents* 16 (1989): 8–9, used with permission from Ross Laboratories, Columbus, OH 43216.
Copyright 1989 Ross Laboratories.